● : WELL DISCUSSED   ✗ : LESS DIS

CHAPTER NUMBER

10	11	12	13	14	15	16	17	18	19	20	21	22	23				27	28
–	–	–	–	–	–	–	–	–	–	–	–	–	–	–	–	–	–	–
–	–	–	✗	–	✗	–	–	–	–	–	●	–	–	–	–	–	–	–
●	●	●	–	●	–	–	✗	✗	–	–	●	✗	–	–	●	–	●	–
●	–	–	–	●	–	–	–	–	–	–	–	–	–	–	–	–	–	–
–	–	–	✗	✗	●	–	–	–	–	–	–	–	–	–	–	–	–	–
✗	–	–	–	●	–	✗	–	–	–	–	–	–	–	–	–	–	–	–
–	–	–	–	–	–	–	–	–	–	–	–	–	–	–	–	–	–	–
–	–	–	–	–	–	–	–	–	–	–	–	–	–	–	–	–	–	–
–	–	–	✗	–	–	–	–	–	✗	–	–	–	–	–	–	–	✗	–
–	–	–	●	–	–	–	–	–	●	–	–	–	–	–	–	–	–	–
●	–	–	✗	–	–	–	✗	✗	–	–	–	–	–	–	–	–	●	–
✗					●	●								–	–	–	–	–
✗	–	–	●	–	●	–	–	●	●	✗	–	●	–	–	–	–	–	–
●	–	–	–	–	–	–	–	–	–	–	–	–	–	–	–	–	–	–
–	●	●	–	–	–	–	–	–	–	–	–	–	–	–	–	–	–	–
–	●	–	–	–	–	–	–	–	–	–	–	–	–	–	–	–	●	–
–	●	–	✗	●	–	–	–	–	●	–	–	–	–	–	–	–	–	–
●	●	●	●	●	✗	●	–	●	●	●	●	●	✗	–	–	–	–	–
●	–	–	–	●	✗	–	–	–	–	●	●	●	✗	●	●	–	–	–
–	–	✗	✗	●	✗	–	–	–	–	✗	–	●	–	✗	–	–	–	–
–	–	✗	●	●	–	–	–	–	–	✗	✗	✗	●	–	–	–	–	–
–	–	●	✗	✗	–	●	–	–	–	–	–	✗	–	–	–	–	–	–
–	–	–	●	●	–	✗	–	–	✗	●	–	✗	–	●	–	–	–	–
–	–	–	–	●	–	✗	–	–	●	–	–	–	–	–	–	–	–	–
–	–	–	–	✗	●	✗	–	–	●	–	–	–	–	●	–	–	–	–
–	–	–	–	–	–	–	–	–	✗	–	●	–	–	–	–	–	–	–
–	–	–	–	–	–	–	–	–	–	–	–	–	●	–	–	–	–	–
✗	–	–	–	–	–	–	–	–	–	–	–	–	–	●	●	–	–	–
–	●	●	–	●	✗	●	✗	–	●	✗	–	–	–	✗	–	–	–	–
–	●	●	●	–	✗	–	●	●	–	●	✗	–	–	●	–	–	●	●
–	●	–	–	●	–	✗	–	–	●	●	●	✗	–	●	–	–	–	–
–	–	●	–	–	–	✗	●	–	–	–	–	✗	–	–	–	●	●	●
–	–	–	–	–	–	–	–	–	–	–	–	–	–	–	–	●	–	–

FINE-GRAINED TURBIDITE SYSTEMS

Edited by

Arnold H. Bouma

and

Charles G. Stone

AAPG Memoir 72

SEPM Special Publication No. 68

Published jointly by
The American Association of Petroleum Geologists and
SEPM (Society for Sedimentary Geology)
Tulsa, Oklahoma, U.S.A.
Printed in the U.S.A.

Printed in the U.S.A.
Published March 2000

ISBN: 0-89181-353-5

This book and other AAPG and SEPM publications are available from:

The AAPG Bookstore
P.O. Box 979
Tulsa, OK 74101-0979
Telephone: 1-918-584-2555 or 1-800-364-AAPG (USA)
Fax: 1-918-560-2652 or 1-800-898-2274 (USA)
www.aapg.org

SEPM
1731 E. 71st St.
Tulsa, OK 74136-5108
Telephone: 1-918-493-3361 or 1-800-865-9765 (USA)
Fax: 1-918-493-2093 (USA)
www.sepm.tulsa.net

Geological Society Publishing House
Unit 7, Brassmill Enterprise Centre
Brassmill Lane, Bath, U.K.
BA1 3JN
Tel +44-1225-445046
Fax +44-1225-442836
www.geolsoc.org.uk

AAPG publications are also available from:

Australian Mineral Foundation
AMF Bookshop
63 Conyngham Street
Glenside, South Australia 5065
Australia
Tel. +61-8-8379-0444
Fax +61-8-8379-4634

Affiliated East-West Press Private Ltd.
G-1/16 Ansari Road Darya Ganj
New Delhi 110 002
India
Tel +91 11 3279113
Fax +91 11 3260538
e-mail: affiliat@nda.vsnl.n

Preface

The lack of a comprehensive text on fine-grained turbidite systems, combined with confusion on the exact nomenclature of several terms, were reasons to suggest a meeting with a field trip for the AAPG/SEPM Annual Convention in 1997 in Dallas, Texas. No space was available on the schedule for a half-day session, but we were able to have an SEPM-sponsored field trip. More than 50 people participated in the field trip to the Pennsylvanian Jackfork Formation's fine-grained deepwater turbidite systems in Arkansas. Lee Krystinik, then president-elect of SEPM, requested that an edited book on fine-grained turbidite systems should result, with several of the field trip participants becoming contributors.

Several organizational steps resulted in finding persons to share the burden. Chuck Stelting (Chevron) provided valuable advice and Charlie Stone (Arkansas Geological Commission) accepted the challenge to be co-editor. Rather than continue the proven arrangement of ancient, modern, and subsurface groupings, we established a theme concept. The following persons were willing to share one of the seven themes: Paul Weimer on 3-D seismic, John Armentrout on sequence stratigraphy, Roger Slatt on outcrops, Randy Miller on lithologies, Tom Fett and Steve Hansen on logging, Mike DeVries on reservoir characterization, Gary Parker on experiments, and the editors on any remaining topics. This program did not work out completely as planned, but we retained the theme concept. We are very thankful to the authors and reviewers for accepting those ideas and thus they contributed to the success of this volume.

No publication can obtain a good scientific standard without the tremendous help provided by reviewers. Authors are too close to their material to be completely critical about the mode of presentation and clarity of the text. The reviewers realized those difficulties and were able to inform the authors about improvements that should be made in order to produce a properly revised manuscript. Table 1 shows the list of all the persons who helped with this important task. Some of them were even willing to review several manuscripts within a few weeks or less, in order to stay within the extension of the contract period. All these reviewers deserve a very strong "thank you very, very much" for their constructive help.

This is the first joint publication between the SEPM and the AAPG, and we hope that it will be the start of a continued cooperation. The societies agreed that AAPG would oversee the publication process. Working with Ken Wolgemuth, Anne Thomas, and Neil Hurley from the AAPG; and Bob Dalrymple, Theresa Scott and Dave Pettyjohn from the SEPM, was great. They deserve special thanks for all their help. Back in the office I encountered much help from many of my students. Special thanks should go to Kelly LaGrange, Erik Scott, and Don Rehmer. Last, but not least, I want to thank my wife Leineke for all her support and administrative assistance.

The long list of thank-you's will not be complete if the large number of authors and coauthors are left out. Several of them will long-remember the phone calls and e-mails showing no mercy from my side regarding their busy schedules. The authors are the persons bringing their knowledge to the readers.

Many of the topics require color figures or large format illustrations to adequately present the data. In order to keep the publication price under control, it was decided to go partially digital. Therefore, some of these expensive illustrations and some additional material have been placed on a CD-ROM, which is part of the volume. The book and the CD complement one another, but each can be read/viewed independently.

The readers are the important group that should benefit from this publication. The wealth of topics may seem overwhelming, initially. Therefore, the editors have constructed a table (endsheets) to assist the reader in finding the papers that are of prime interest to him or her personally. The table has two axes: one with topics and the other with the chapters of this volume. By the use of symbols, it should be easy to find the chapters that deal with a certain topic, and at the same time see which chapters are the most important on a certain topic.

January 2000
Arnold H. Bouma

TABLE 1. Reviewers of the manuscripts.

Daniel J. Acquaviva	David Ormerod
Wayne M. Ahr	Frank J. Peel
Philip J. Bart	Ronald L. Phair
Roger B. Bloch	Andrew J. Pulham
John F. Bratton	John B. Southard
Richard T. Buffler	John R. Suter
George R. Clemenceau	John B. Thomas*
James L. Coleman, Jr.	Alan Thomson
James M. Coleman	David C. Twichell
Michael B. DeVries	Grant D. Wach
Senira S. Kattah	Paul Weimer
Gerald J. Kuecher*	Lesli J. Wood
Cooper B. Land	Don Ying
D. Bradford Macurda*	Jianjun Zeng

*Colleagues who reviewed more than five manuscripts.

AAPG and SEPM
wish to thank the following
for their generous contributions
to

Fine-Grained Turbidite Systems

◆

Arnold H. Bouma

◆

BP Amoco Upstream Technology

◆

Texaco Upstream Technology

◆

U.S. Office of Naval Research

◆

Contributions are applied against the production costs of publication, thus directly reducing the book's purchase price and making the volume available to a greater audience.

Table of Contents

About the Editors

Arnold H. Bouma

Arnold H. Bouma has the McCord Endowed Professorship at Louisiana State University. He received his B.S. degree from the University of Groningen (the Netherlands) in 1956, and his M.S. and Ph.D degrees in 1959 and 1961, respectively, from the University of Utrecht, also in the Netherlands. A Fulbright post-doctoral fellowship at the Scripps Institute of Oceanography from 1962–1963 finished his formal training. In 1966, the Bouma family emigrated to the United States where he held various positions to make it possible to alternate studies on consolidated and unconsolidated sediments: Texas A&M University (Geological Oceanography), U.S. Geological Survey (Marine Branch), Gulf Oil and Chevron, and Louisiana State University.

His interest in marine sediments always moved him back to the turbidite systems and later specifically to the fine-grained ones. This important group in deepwater sands has received little attention, which is one of the reasons that the publication was suggested.

He is a member of many professional societies, holds a number of positions (presently President of SEPM), loves teaching and working with students, and publishes extensively.

Charles G. Stone

Charlie Stone, a geologist (Arkansas Geological Commission, 1957–present), has dealt with the structure, stratigraphy, sedimentology, economic, and environmental geology in the Ouachita Mountains and Arkoma basin of Arkansas. Recent activities include geological mapping of 178 7 1/2-minute quadrangles for the COGEOMAP program and geochemistry of rocks in west-central Arkansas. Charlie earned his B.S. (1956) and M.S. (1957) degrees in geology from the University of Arkansas.

Stelting, C. E., A. H. Bouma, and C. G. Stone, 2000, Fine-grained turbidite systems: overview, *in* A. H. Bouma and C. G. Stone, eds., Fine-grained turbidite systems, AAPG Memoir 72/SEPM Special Publication No. 68, p. 1–8.

Chapter 1

Fine-Grained Turbidite Systems: Overview

Charles E. Stelting
Chevron Petroleum Technology Company
New Orleans, Louisiana, U.S.A.

Arnold H. Bouma
Department of Geology and Geophysics, Louisiana State University
Baton Rouge, Louisiana, U.S.A.

Charles G. Stone
Arkansas Geological Commission
Little Rock, Arkansas, U.S.A.

ABSTRACT

Fine-grained, mud-rich turbidite systems primarily occur in basins with a large fluvial input. Depositional models derived from sand-rich turbidite systems are not appropriate because the large volume of mud in fine-grained turbidite systems produces different sediment distribution patterns, geomorphic features, and internal architecture at bed-to-sequence scales. Many of the chapters in this volume demonstrate that understanding fine-grained turbidite systems requires a number of steps and degrees of resolution, very similar to the range of data utilized in the oil industry. Industrial examples include 2-D and 3-D seismic, cores, and well logs. To refine the understanding of a turbidite field, the earth scientist must integrate the most applicable models with subsurface data, outcrop analyses, modern analogs, and experimental results.

INTRODUCTION

The explosion of research efforts focused on submarine fan, or turbidite system, deposits during the 1990s are principally the result of exploration for and the discovery of large hydrocarbon accumulations in fine-grained, mud-rich turbidite systems deposited in passive margin settings (e. g., Brazil, Gulf of Mexico, West Africa, North Sea).

In an effort to maximize exploitation of these deepwater reserves, industry and academia have joined to advance the understanding of sediment transport and depositional processes, sediment distribution patterns, processes of fan construction, and reservoir architecture in fine-grained turbidite systems. This has been accomplished by (1) reexamining numerous outcrop successions with emphasis on sequence stratigraphic concepts and reservoir architectural elements, (2) conducting new experimental studies (e.g., flume) and numerical modeling to investigate sediment gravity flow transport and depositional processes, and (3) cooperative geological-geophysical-engineering studies to achieve integrated definition and characterization of subsurface sand bodies. These studies, which ultimately questioned many paradigms (Shanmugan and Moiola, 1988; Shanmugan, 1999), have demonstrated that fundamental differences exist in where and how deepwater sediments, particularly the sand-grade fractions, are deposited in fine-grained, mud-rich turbidite systems as compared with the more thoroughly studied coarse-grained, sand-rich turbidite systems (Normark et al., 1986).

The results of some of these studies have been included in various scientific publications over the past decade and a half (Bouma et al., 1985a; Weimer and Link, 1991; Weimer et al., 1994; Pickering et al., 1995; Prather et al., 1998). However, no publication has specifically focused on the characteristics of fine-grained deepwater turbidite systems. The collection of experimental results and case examples presented in this volume constitutes the first effort to provide information exclusively on fine-grained systems in a single publication. These data and results reported are intended to provide a framework for improved understanding of the many aspects of these depositional systems. These results do not represent an end; they are a beginning, a springboard from which to direct continuing research studies and to refine our understanding of the many factors that control system growth and sediment distribution. The oil industry is concentrating large efforts on the advancement of technical expertise necessary to meet economically viable exploitation needs. As a result, technology in the deepwater settings around the world is expanding at a much more rapid rate than the associated scientific knowledge. Publication of this volume and others similar in scope are, therefore, critical components in any attempt to bridge the gap between technology and our understanding of fine-grained turbidite depositional systems.

DEEPWATER DEPOSITIONAL SYSTEMS: TERMINOLOGY

Deepwater depositional systems represent deposition primarily by sediment gravity flows, which transport clastic sediment down a slope and onto a basin floor. Although sedimentation can occur in these systems at any time, the largest volume of sediment is transported into the basin during lowstands and initial rise of relative sea level (Shanmugan and Moiola, 1982; Bouma et al., 1989; Posamentier et al., 1991). These systems are called submarine fans when referring to a "modern" deepwater accumulation exposed on the present-day sea floor (Menard, 1955), or even in some cases to any unconsolidated sedimentary succession. They are called turbidite systems when referring to subsurface and/or outcrop occurrences (Mutti and Normark, 1987, 1991) and commonly to consolidated deposits.

Mutti and Normark (1991) define a turbidite (fan) system as "a body of genetically related mass-flow and turbidity-current facies and facies associations that were deposited in virtual stratigraphic continuity." Bouma et al. (1985c) called individual unconformity-bounded turbidite systems "fanlobes." When stacked, turbidite (fan) systems and their bounding basinal shales are defined as a turbidite (submarine fan) complex. If the shales or silty mudstones are about the same thickness as the sand-rich beds of the individual turbidite systems or if they comprise at least 70% of the total succession, the system is referred to as being "mud-rich"(Reading and Richards, 1994). "Sand-rich" systems, in contrast, are characterized by a high net-to-gross ratio and by shale accumulations that are much thinner than the sand-dominated successions (Bouma, this volume).

Because nonrigorous usage of these terms can result in confusion, each investigator needs to define how a term is used for a particular case. Accordingly, we use *turbidite system* (complex) to refer to a composite succession of sand and mud gravity flow deposits that form depositional units as a series of second-, third-, and/or fourth-order depositional sequences as defined by Posamentier et al. (1988; also Goldhammer et al., this volume, and Gardner and Borer, this volume).

"Deep-water" or "deepwater" is another term that is used inconsistently throughout the scientific community. Dingler (1999) discussed the root derivation of deepwater and associated terms and suggested that usage of the term as a single word is most consistent with definition in several dictionaries. We are in agreement with Dingler's arguments and recommend the use of *deepwater* in this volume and for subsequent use elsewhere.

CHARACTERISTICS OF FINE-GRAINED TURBIDITE SYSTEMS

Fundamental Differences between Fine- and Coarse-Grained Turbidite Systems

The earliest submarine fan models were based on studies of coarse-grained, sand-rich fans in the California Borderland (Normark, 1970) and in small Piedmont basins in Europe (Mutti and Ricci Lucchi, 1972). The first models for the larger, fine-grained, mud-rich submarine fans were based on oceanographic studies of the Amazon Fan (Damuth and Kumar, 1975) and the Mississippi Fan (Moore et al., 1978). It was not until the drilling results of the Deep Sea Drilling Project Leg 96 on the Mississippi Fan in 1983 were analyzed that the differences between the two end-member systems could be fully appreciated (Bouma et al., 1985b, 1995a, b).

The principal characteristics of coarse-grained and fine-grained systems are summarized by Bouma (this volume) as follows. *Coarse-grained, sand-rich* turbidite systems typically occur in small basins on continental crust, have short terrestrial transport distance, narrow shelf, canyon-sourced, nonefficient basin transport, progradational depositional style, and a net-to-gross percentage that decreases lateral from sediment pathway. *Fine-grained, mud-rich* turbidite systems, by contrast, are found in large basins on passive margins, have long terrestrial transport distance and a broad shelf, are delta-sourced, have efficient basin transport resulting in bypass of much of the sand to the outer fan, and areally variable net-to-gross ratio patterns. Another group can be classified as fine-grained, sand-rich turbidite systems as found in the west Texas Permian Basin (Gardner and Borer, this volume; Carr and Gardner, this volume). See Bouma (this volume) for a more detailed treatise.

Reading and Richards (1994) devised a classification scheme for deepwater turbidite systems based on grain size and type of feeder system that further contrasts fine-grained and coarse-grained turbidite systems. The classification of various modern and ancient turbidite systems is shown in Table 1. All the case examples presented in this volume occur within the mud-rich or mud/sand-rich turbidite systems fed by a point source.

Table 1. Classification of modern and ancient turbidite systems (modified from Reading and Richards, 1994).

		Mud-Rich Systems	Mud/Sand-Rich Systems	Sand-Rich Systems	Gravel-Rich Systems
Submarine Fan, Point Source	Modern Systems	Amazon, Astoria, Bengal, Indus, Laurentian, Magdalena, Mississippi, Monterey, Mozambique, Nile, N.W. Africa	Delgada, La Jolla, Navy, Rhone	Avon, Calabar, Redondo	Bear Bay, Yallahs, Gulf of Corinth
	Ancient Systems	Jackfork Group	Ferrelo, Hecho Type I, Kongsfjord, Laga, Lainsburg, Macigno, Marnosa-Arenacea, Peira-Cava, Stevens, Tanqua	Annot, Balder, Bullfrog, Cengio, Chatsworth, Hercho Type II, Magnus, Miller	North Brae, Cap Enrage
Multiple Source, Ramps	Modern Systems	Cape Ferret, Nitinat, Wilmington	Ebro, Natal Coast, C. America Trench, San Lucas		
	Ancient Systems	Catskill, Hareelv, Forbes	Butano, Everest, Forties, Gottero	Campos, Matilija, Montrose, Tyee	Blanca, Brae
Slope Apron, Linear Source	Modern Systems	Nova Scotia and Grand Banks highstand, N.W. Africa, S.W. Africa		Sardinia-Tyrrhenian	
	Ancient Systems	Gull Island	Alba, Hareelv (Trans), Nova Scotia lowstand	Rocky Gulch, Tonkawa	Helmsdale, Marambio

Fan Divisions

Although many aspects of the early fan models have fallen out of favor with most geoscientists over the years, the use of physiographic regions in defining turbidite systems continues to the present day. These terms—inner, middle, and outer—for modern submarine fans, and upper, middle, and lower for ancient turbidite systems reflect the differences in methodology used to study these very different data types. On modern fans, these subdivisions are based on a characterization of sea-floor bathymetry and geomorphologic features at progressively greater distances from the shelf edge. Definition of subdivisions in ancient systems is based on facies associations observed in outcrop exposures and reflects the progressive changes in a vertical or stratigraphic sense of these facies associations.

The distinct geomorphologic divisions and/or facies associations that have been defined in fine-grained, mud-rich turbidite systems differ from those described in coarse-grained, sand-rich turbidite systems (see discussion by Normark et al., 1986). Figure 3 in Bouma (this volume) notes the distinguishing depositional features that occur in fine-grained, mud-rich systems constructed in an unconfined basin setting. In this case, the upper/inner fan region is dominated by an erosive canyon that changes to an erosional-constructional channel complex ("fan valley") approaching the middle fan region. The middle fan starts at or near the base-of-slope, is a constructional or aggradational part of the fan, and is dominated by a channel-levee complex that is typically sinuous and decreases in size upward and in a downchannel direction (Peakall et al., this volume). The outer/lower fan consists of small, ephemeral channels (distributaries) that grade into sheet-sand complexes that dominate the depositional style in the distal region of the system.

Although fan divisions are useful distinguishing characteristics of turbidite systems in unconfined basins, where downfan variations in the channel system may be more readily apparent, they are less appropriate—if at all—in constricted basins such as those that dominate the Gulf of Mexico and west Africa continental slopes. In these basins, the gradation from sheet sand to channel-levee systems does not occur longitudinally. Instead it occurs vertically in response to changes in sea-floor gradient and rate of sediment supply as the basin fills and sediment spills into the next basin downdip. Steffens (1993) proposed the use of "deepwater depositional elements" as an alternative to fan divisions to enhance reservoir predictability in confined basins. The four elements are: leveed channel sands; amalgamated channel sands; amalgamated and layered sheet sands; and slumps, debris flows, and marine shales. The widespread acceptance and usage of this descriptive concept is demonstrated within this volume; these depositional elements have been applied to the sandstone bodies in all of the outcrop case examples.

Sediments

Sedimentation and the distribution of facies in turbidite systems is not simply a product of the dominant depositional processes. The processes active on any deepwater depositional system respond to a variety of external forcing agents: sediment type and supply, climate, tectonic setting and activity both in the hinterland and in the receiving basin, and sea-level fluctuations (Stow et al., 1985).

Although deepwater depositional systems are commonly referred to as turbidite systems, these systems are actually composed of a continuum of sediment gravity flow deposits. Depositional processes range from mass movement (creep) at one end of the spectrum through cohesive flows (slumps, debris flows) and turbulent flow (turbidites) to laminar flow and settling from suspension (contourites, hemipelagites) on the other end of the spectrum (Middleton and Hampton, 1976).

In the past few years, an additional type of turbidite bed has been recognized. Slurry beds are muddy turbidites that are intermediate between clean well-graded sandy turbidites and muddy debris flow beds, sometimes referred to as debrites. Slurry beds are deposited from slurry flows in which the mud and organic components have not been fully winnowed because the flow has not evolved into a fully turbulent state. Extensive studies by Hickson (1999) show that slurry beds have a combination of features that indicate both turbulent and cohesive behavior of the flows, and some unique characteristics such as sheared dewatering structures.

Combined, the various types of sediment gravity flows move large volumes of clastic detritus from shelfal areas onto the slope and basin floor to form submarine fans or turbidite complexes (Mutti and Normark, 1987). Turbidite facies schemes (Mutti and Ricci Lucchi, 1972; Walker, 1978; Pickering et al., 1986) have been devised that relate deepwater facies to transport and depositional processes. These classification schemes and associated fan models can be useful tools to aid in the interpretation of depositional environments. However, extreme caution in their application must be used because facies and facies associations are not indicative of a single depositional environment (Shanmugan and Moiola, 1985, Figure 4; Yielding and Apps, 1994, Figure 10).

The bulk of sand deposition results from low- and high-concentration turbidity currents (graded beds and massive beds, respectively) and takes place mainly within sinuous, leveed channels or as sheet sands. Secondary depositional sites include levees, overbank areas, and the lower reaches of submarine canyons. In addition to turbidity currents, sands may be transported to their depositional site by slurry flows, debris flows, or slumps (less common). Muds or shales comprise the largest volume of sediments (>70%) in fine-grained turbidite systems and occur in all depositional environments and/or elements. Dilute, muddy turbidity currents, spilling from channel confinement, carry most of the muds that are deposited in overbank and basin-plain areas. Debris flows and slumps are common, but their deposits tend to be areally restricted.

Recent debate over transport and depositional mechanisms of massive turbidite sand beds has polarized geoscientists working on deepwater depositional systems (Bouma et al., 1997; Lowe, 1997; Slatt et al., 1997). Shanmugan and Moiola (1994) proposed that these beds are the product of cohesive, sandy debris flows rather than high-concentration turbidity currents (Lowe, 1982). They cited the lack of graded bedding, indicative of waning turbulent flow, in the Jackfork Group of Arkansas as evidence that these "turbidite" beds are actually the product of cohesive sandy debris flows. Lack of visually observable graded bedding, however, is not a sufficient criterion to distinguish between sandy debrites and turbidites. In a recent reevaluation of these Jackfork Group beds, Hickson (1999) used detailed point-count analysis to document that the beds are indeed normally graded and were deposited by turbidity currents.

Although Hickson's research and other studies cast doubts on the Shanmugan and Moiola (1994) model, they do not preclude the presence and importance of sandy debris flows in turbidite systems. The controversy over sandy debris flows demonstrates that proper characterization of outcrop successions where weathering and postdepositional alteration can greatly subdue sedimentary structures requires detailed laboratory analysis of sedimentary features to properly diagnose the depositional origin of "turbidite" beds. It also points out the need for more experimental work on turbulent flows of various types to be able to accurately interpret the origin of deepwater sands in the subsurface. To this end, Pratson et al. (this volume) used numerical models to specifically evaluate the differences between sandy debris flows and turbidity currents. Their studies show significant differences in depositional geometries and stratigraphic stacking patterns between the two flow types.

Misunderstanding or misuse of geologic terminology (e.g., debrites or turbidites) can result in critical errors with scientific and economic implications. The majority of earth scientists adhere to the original meaning of debrite: a two-component deposit of coarser grains floating in a finer-grained matrix. Unfortunately, this definition suggests a "finer-grained" deposit that often is understood to be nearly impermeable. In this volume, we prefer the term turbidite rather than sandy debris flow deposit in order to avoid serious misunderstanding.

Because of the wide range of sedimentary processes and sediment types, predicting depositional environments and facies distribution in fine-grained turbidite systems can be a challenging endeavor. Successful exploitation of hydrocarbons requires knowledge of the stratigraphic framework, coupled with an integrated, multidisciplinary analysis (e.g., seismic, well logs, electrical images, cores, engineering and production data) of sediments and their geometric forms.

COVERAGE OF TOPICS

This publication, which is the first volume dedicated exclusively to fine-grained turbidite systems, consists of experimental, outcrop, and subsurface case studies that document the principal characteristics of these systems and their depositional elements.

The chapters are presented in a topical rather than geographic or other customary format because the objective is to illustrate the depositional and stratigraphic characteristics of fine-grained turbidite systems at a variety of scales rather than to focus on the study approach or location. The topics are arranged so that the larger-scale aspects are discussed first, followed by progressively more detailed aspects. Major topic groups and corresponding chapters include: General (Chapters 1–7), Seismic (Chapters 8–12), Stratigraphy (Chapters 13–16), Reservoir Characterization (Chapters 17 and 18), Sedimentology (Chapters 19–26), and Logging (Chapters 27 and 28).

Chapters in the General topic section cover a variety of subjects, beginning with a depositional model for fine-grained turbidite systems (Chapter 2) followed by discussions of basin evolution in three of the basins from which the case examples originated (Chapters 3–5) and concludes with the results of two experimental studies—one comparing sandy debris flows with turbidite deposits (Chapter 6) and the other a flume study addressing the evolution of sinuous, meandering submarine channels (Chapter 7).

The Seismic topic section begins with an overview article that documents state-of-the-art techniques and outlines future directions (Chapter 8). Subsurface case examples are presented from West Africa and the Gulf of Mexico (Chapters 9 and 10), with outcrop seismic models from the Jackfork Group (Arkansas) and Brushy Canyon Formation (West Texas) in Chapters 11 and 12.

Stratigraphy case examples detail sequence stratigraphic components and basinal framework in the Mount Messenger Formation of New Zealand (Chapter 13) and for the Tanqua Karoo Basin in South Africa (Chapters 14–16).

In the Reservoir Characterization topic section, the importance of outcrop characterization in geologic and subsurface reservoir modeling is examined (Chapter 17) and is illustrated with an outcrop case example (Chapter 18).

Sedimentologic-based outcrop case examples (from Arkansas, South Africa, and West Texas) define and characterize the sedimentary and architectural elements of the primary depositional environments. Subjects include: systematic downfan changes (Chapters 19 and 20), channel-levee complex (Chapter 21), channel-lobe transition deposits on the middle fan (Chapter 22), thin-bedded turbidite occurrences (Chapter 23), and sheet sand deposits in the outer fan and fan fringe areas (Chapters 24 and 25). Chapter 26 defines and discusses the significance of modern turbidity current sediment waves off the Canary Islands.

Two very different approaches are presented in the Logging section. In Chapter 27, a method is outlined for the construction of "pseudo-well" logs from outcrop lithologies used for synthetic seismograms. The value of borehole images in the stratigraphic and structural analysis of deepwater sediments, especially thin-bedded sands, is illustrated in Chapter 28.

Outcrop studies comprise more than half (~60%) of the case examples. Emphasis on outcrop datasets collected from fine-grained turbidite systems is critical to increase our understanding of turbidite reservoirs because subsurface data are either discontinuous or at too large a scale to resolve reservoir architecture (Slatt, this volume). Outcrop analogs are valuable tools to resolve internal architectural elements and sand body geometries. The ability to use outcrop data to aid resolution of depositional and reservoir features, not readily interpreted from seismic and borehole data, greatly outweigh the fact that outcrop sediments are lithified and porosity and permeability measurements are not directly related to subsurface conditions.

REFERENCES CITED

Bouma, A. H., W. R. Normark, and N. E. Barnes, eds., 1985a, Submarine fans and related turbidite systems: New York, Springer-Verlag, 351 p.

Bouma, A. H., C. E. Stelting, and J. M. Coleman, 1985c, Mississippi Fan, Gulf of Mexico, *in* A. H. Bouma, W. R. Normark, and N. E. Barnes, eds., 1985a, Submarine fans and related turbidite systems: New York, Springer-Verlag, p. 143–150.

Bouma, A. H., et al., 1985b, Mississippi Fan: Leg 96 program and principal results, *in* A. H. Bouma, W. R. Normark, and N. E. Barnes, eds., 1985a, Submarine fans and related turbidite systems: New York, Springer-Verlag, p. 247–252.

Bouma, A. H., J. M. Coleman, C. E. Stelting, and B. Kohl, 1989, Influence of relative sea level changes on the construction of the Mississippi Fan: Geo-Marine Letters, v. 9, p. 161–170.

Bouma, A. H., G. H. Lee, O. van Antwerpen, and T. C. Cook, 1995a, Channel complex architecture of fine-grained submarine fans at the base-of-slope: Gulf Coast Association of Geological Societies Transactions, v. 65, p. 65–70.

Bouma, A. H., H. D. Wickens, and J. M. Coleman, 1995b, Architectural characteristics of fine-grained submarine fans: a model applicable to the Gulf of Mexico: Gulf Coast Association of Geological Societies Transactions, v. 65, p. 71–75.

Bouma, A. H., M. B. DeVries, and C. G. Stone, 1997, Reinterpretation of depositional processes in a classic flysch sequence (Pennsylvanian Jackfork Group), Ouachita Mountains, Arkansas and Oklahoma: Discussion: AAPG Bulletin, v. 81, p. 470–472.

Damuth, J. E., and N. Kumar, 1975, Amazon Cone: morphology, sediments, growth pattern: Geological Society of America Bulletin, v. 86, p. 863–878.

Dingler, C., 1999, Deep thinking: Editor's letter in Houston Geological Society Bulletin (March 1999), p. 7.

Hickson, T. A., 1999, A study of deep-water deposition: constraints on the sedimentation mechanics of slurry flows and high concentration turbidity currents, and the facies architecture of a conglomeratic channel-overbank system: Ph.D. dissertation, Stanford University, 470 p.

Lowe, D. R., 1982, Sediment gravity flows: II. Depositional models with special reference to the deposits of high-density turbidity currents: Journal Sedimentary Petrology, v. 52, p. 279–297.

Lowe, D. R., 1997, Reinterpretation of depositional processes in a classic flysch sequence (Pennsylvanian Jackfork Group), Ouachita Mountains, Arkansas and Oklahoma: Discussion: AAPG Bulletin, v. 81, p. 460–465.

Menard, H. W., 1955, Deep-sea channels, topography, and sedimentation: AAPG Bulletin, v. 39, p. 236–255.

Middleton, G. V., and M. A. Hampton, 1976, Subaqueous sediment transport and deposition by sediment gravity flows, *in* D. J. Stanley and D. J. P. Swift, eds., Marine sediment transport and environmental management: New York, Wiley Interscience, p. 197–218.

Moore, G. T., G. W. Starke, L. C. Bonham, and H. O. Woodbury, 1978, Mississippi Fan, Gulf of Mexico—physiography, stratigraphy, and sedimentation patterns, *in* A. H. Bouma, G. T. Moore, and J. M. Coleman, eds., Framework, facies, and oil-trapping characteristics of the upper continental margin: AAPG Studies in Geology 7, p. 155–191.

Mutti, E., and F. Ricci Lucchi, 1972, Le torbiditi dell'Appennino settentrionale: introduzione all'analisi de facies: Memorie della Societa Geologia Italiana, v. 11, p. 161–199.

Mutti, E., and W. R. Normark, 1987, Comparing examples of modern and ancient turbidite systems: problems and concepts, *in* J. K. Leggett and G. G. Zuffa, eds., Marine clastic sedimentology: concepts and case studies: London, Graham and Troutman, p. 1–38.

Mutti, E., and W. R. Normark, 1991, An integrated approach to the study of turbidite systems, *in* P. Weimer, and M. H. Link, eds., Seismic facies and sedimentary processes of submarine fans and turbidite systems: New York, Springer-Verlag, p. 75–106.

Normark, W. R., 1970, Growth patterns of deep-sea fans: AAPG Bulletin, v. 54, p. 2170–2195.

Normark W. R., et al., 1986, Summary of drilling results for the Mississippi Fan and considerations for application to other turbidite systems, *in* A. H. Bouma, J. M. Coleman, and A. W. Meyer, eds., Initial reports of the Deep Sea Drilling Project, v. 96, Washington, D.C., U.S. Government Printing Office, p. 425–436.

Pickering, K. T., D. A. V. Stow, M. P. Watson, and R.N. Hiscott, 1986, Deep-water facies, processes and models: a review and classification scheme for modern and ancient sediments: Earth Science Reviews, v. 23, p. 75–174.

Pickering, K. T., R. N. Hiscott, N. H. Kenyon, F. Ricci Lucchi, and R. D. A. Smith, eds., 1995, Atlas of deep water environments: architectural style in turbidite systems: London, Chapman and Hall, 333 p.

Posamentier, H. W., M. T. Jervey, and P.R. Vail, 1988, Eustatic controls on clastic deposition 1—conceptual framework, *in* C. K. Wilgus, B. S. Hasting, C. G. S. C. Kendall, H. W. Posamentier, C. A. Ross, and J. C. Van Wagoner, eds., Sea level changes: an integrated approach: SEPM Special Publication 42, p. 109–124.

Posamentier, H. W., R. D. Erskine, and R. M. Mitchum, Jr., 1991, Models for submarine-fan deposition within a sequence-stratigraphic framework, *in* P. Weimer and M. H. Link, eds., Seismic facies and sedimentary processes of submarine fans and turbidite systems: New York, Springer-Verlag, p. 127–136.

Prather, B. E., J. R. Booth, G. S. Steffens, and P. A. Craig, 1998, Classification, lithologic calibration, and stratigraphic succession of seismic facies of intraslope basins, deep-water Gulf of Mexico: AAPG Bulletin, v. 82, p. 701–728.

Reading, H. G., and M. Richards, 1994, Turbidite systems in deep-water basin margins classified by grain size and feeder system: AAPG Bulletin, v. 78, p. 792–822.

Shanmugan, G., 1999, 50 years of the turbidite paradigm (1950s–1990s): deep-water processes and facies models—a critical perspective: Marine and Petroleum Geology, in press.

Shanmugan, G., and R. J. Moiola, 1982, Eustatic control of turbidites and winnowed turbidites: Geology, v. 10, p. 213–235.

Shanmugan, G., and R. J. Moiola, 1985, Submarine fan models: problems and solutions, *in* A. H. Bouma, W. R. Normark, and N. E. Barnes, eds., Submarine fans and related turbidite systems: New York, Springer-Verlag, p. 29–34.

Shanmugan, G., and R. J. Moiola, 1988, Submarine fans: characteristics, models, classification, and reservoir potential: Earth Sciences Review, v. 24, p. 383–428.

Shanmugan, G., and R. J. Moiola, 1994, An unconventional model for the deep-water sandstones of the Jackfork Group (Pennsylvanian), Ouachita Mountains, Arkansas and Oklahoma, *in* P. Weimer, A. H. Bouma, and B. F. Perkins, eds., Submarine fans and turbidite systems, sequence stratigraphy, reservoir architecture and production characteristics, Gulf of Mexico and international: Gulf Coast Section of the SEPM Foundation 15th Annual Research Conference Proceedings, p. 311–326.

Slatt, R. M., P. Weimer, and C. G. Stone, 1997, Reinterpretation of depositional processes in a classic flysch sequence (Pennsylvanian Jackfork Group), Ouachita Mountains, Arkansas and Oklahoma: Discussion: AAPG Bulletin, v. 81, p. 449–459.

Steffens, G. S., 1993, Gulf of Mexico deepwater seismic stratigraphy: AAPG Annual Convention Official Program, p. 186.

Stow, D. A. V., D. G. Howell, and C. H. Nelson, 1985, Sedimentary, tectonic, and sea-level controls, *in* A. H. Bouma, W. R. Normark, and N. E. Barnes, eds.,

Submarine fans and related turbidite systems: New York, Springer-Verlag, p. 15–22.
Walker, R. G., 1978, Deep-water sandstone facies and ancient submarine fans: Models for exploration for stratigraphic traps: AAPG Bulletin, v. 62, p. 932–966.
Weimer, P., and M. H. Link, eds., 1991, Seismic facies and sedimentary processes of submarine fans and turbidite systems: New York, Springer-Verlag, 447 p.
Weimer, P., A. H. Bouma, and B. F. Perkins, eds., 1994, Submarine fans and turbidite systems, sequence stratigraphy, reservoir architecture and production characteristics, Gulf of Mexico and international: Gulf Coast Section of the SEPM Foundation 15th Annual Research Conference Proceedings, 440 p.
Yielding, C. A., and G. M. Apps, 1994, Spatial and temporal variations in the facies associations of depositional sequences on the slope: Examples from the Miocene–Pleistocene of the Gulf of Mexico, *in* P. Weimer, A. H. Bouma, and B. F. Perkins, eds., Submarine fans and turbidite systems, Proceedings of Gulf Coast Section of SEPM 15th Annual Research Conference, p. 425–437.

Bouma, Arnold H., 2000, Fine-grained, mud-rich turbidite systems: model and comparison with coarse-grained, sand-rich systems, *in* A. H. Bouma and C. G. Stone, eds., Fine-grained turbidite systems, AAPG Memoir 72/SEPM Special Publication No. 68. p. 9–20.

Chapter 2

Fine-Grained, Mud-Rich Turbidite Systems: Model and Comparison with Coarse-Grained, Sand-Rich Systems

Arnold H. Bouma
Department of Geology and Geophysics, Louisiana State University
Baton Rouge, Louisiana, U.S.A.

ABSTRACT

Several models of submarine fans/turbidite systems have been published, based on tectonic setting, basin characteristics, grain size, types of gravity flows, relative sea-level fluctuations, and so on. Among the various general and specific models are two siliciclastic end members that are important guides for many turbidite studies: fine-grained, mud-rich, and coarse-grained, sand-rich turbidite systems.

Fine-grained, mud-rich complexes are typical for passive margin settings, with long fluvial transport, fed by deltas, wide shelf, efficient basin transport, resulting in a bypassing system. A high sand-to-shale ratio occurs at the base-of-slope, changes to an overall low ratio on the middle fan and to a high ratio on the outer fan. Coarse-grained, sand-rich complexes are typical for regions in active margin setting, characterized by a short continental transport distance, narrow shelf, and a canyon-sourced, nonefficient basin transport system that results in a prograding type of fan. The high sand-to-shale ratio slowly decreases away from channels and in the fan fringe region.

INTRODUCTION

A submarine fan, called turbidite system by Mutti and Normark (1987, 1991), fan sequence by Feeley et al. (1985) and Weimer and Buffler (1985), and fanlobe by Bouma et al. (1985), represents the deposition by gravity flows into a basin. A stratigraphic succession of a number of turbidite systems is known as a submarine fan complex or turbidite complex (Mutti and Normark, 1991). The term submarine fan typically relates to a modern accumulation exposed at the present sea floor, and the term turbidite system is more used for a subsurface occurrence or outcrop. Many earth scientists, however, use the terms submarine fan and turbidite system interchangeably.

Submarine fans/turbidite systems have received considerable interest both for research purposes and economic reasons (Shanmugam, 1999, and references therein). As a result numerous models and descriptions have been published, sometimes very general and other times site-specific. However, no one has been able to organize this varied assemblage of deposits into one neat model. In fact, the COMFAN conference held in 1982 discouraged attempts to develop a unifying model (Normark and Barnes, 1983/1984). Although a unified classification is impossible, some guidelines on major trends and characteristics are in order. A large number of factors influence the transport and deposition of the sediments. The influencing factors are variable within themselves and between them with regard to the relative significance for any fan. That means that each fan occurrence is unique and differs from other occurrences. However, some general guidelines for studies can be provided.

The general classification by Mutti (1985, 1992) focuses on turbidite hierarchy and classification, physical scale comparisons, channel types, transport efficiency, and so forth, and today serves as a great guideline for a first approach to an area (Figure 1). A significant next step was provided by Reading and Richards (1994), who divided submarine fans/turbidite systems into twelve classes, based on grain size and feeder system. Their grain size components are mud-rich, mud/sand-rich, sand-rich, and gravel-rich. The feeder systems are point-source, multiple source submarine ramp, and linear-source slope apron. From a possible polyaxial system two grain size systems are selected as end members: mud-rich and sand-rich, both with point-source feeding with each one being variable with respect to texture.

INFLUENCES ON TRANSPORT OF SEDIMENT

Stow et al. (1985) presented a block diagram that demonstrates the major factors that influence sediment source area, transport, and deposition. The major factors are influenced by tectonics, climate, sediments, and relative sea-level fluctuations. They can be subdivided into several subgroups.

Tectonics has its influence everywhere, normally with changes in intensity over time and from location to location (Scott et al., this volume). Tectonic activity will influence the orogeny, which in turn influences climate and fluvial gradients. It also dictates whether a basin is located on oceanic or continental crust, whether the shelf is narrow or wide, whether the basin is small or large, and the gradient of the basin slope (often the continental slope).

World climate may be more important than heretofore recognized. Major climatic changes can dictate an

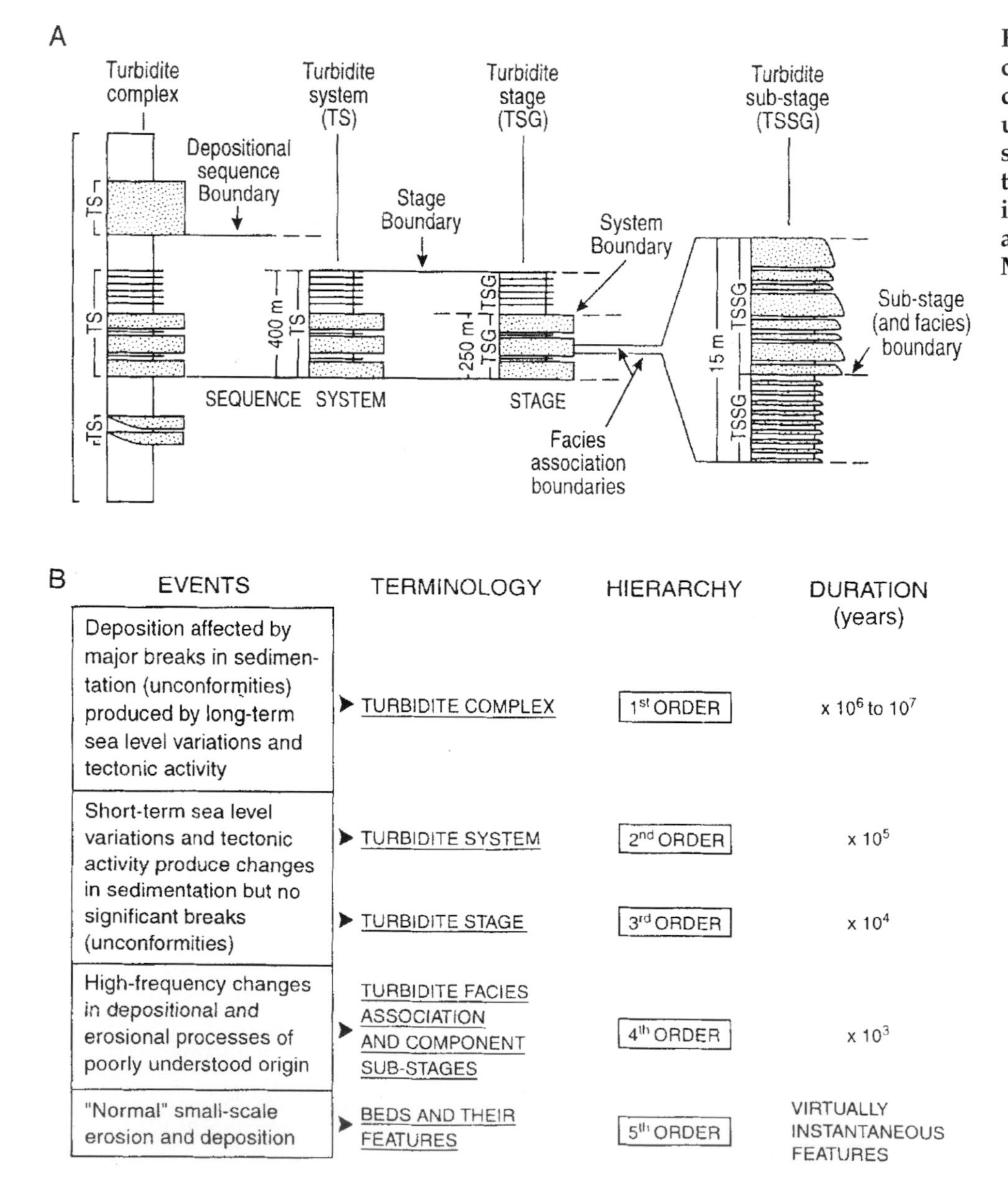

Figure 1—Conceptional classification for turbidite deposits. (A) Hierarchy of units based on physical scale: decreasing in size to the right. (B) Classification in a time frame (after Mutti and Normark, 1987, 1991; Mutti, 1992).

ice age with water becoming tied up as glacial ice, a long period of melting ice, or more or less a balance between freezing and thawing. Regional climate is strongly influenced by the elevation of the mountains where the new sediment will be generated. Physical and chemical forces break down rock into smaller particles. The amount of rainfall, and its distribution through time, determines the time required for subsequent fluvial transport.

Sedimentary characteristics are determined by a number of criteria, such as (1) the composition of source material, (2) the physical and chemical forces during transport, and (3) the time required to get particles from the mountains to the basin and thus the amount of weathering that can take place. The percentage and types of unstable minerals released from the host rock dictate the sensitivity to diagenesis and the degree of maturity of the final sediment. The length of transport and the gradient and current force of the fluvial system determine the final grain size distribution.

Sea-level fluctuations can be eustatic or more local, and the influence of these fluctuations varies significantly with the type and width of the coastal plain and the coast itself, the shelf width, and the slope characteristics toward the basin.

Figure 2 presents two block diagrams, based on Stow et al. (1985) and Reading and Richards (1994), that show a coarse-grained, sand-rich and fine-grained, mud-rich turbidite system. It should be kept in mind that each of the above-mentioned major influencing factors operates within its own spatial and temporal domain. As a result not only a significant number of variations can be encountered, but also seemingly contradictory combinations can result, such as fine-grained, mud-rich turbidite systems mimicking passive margin conditions while being deposited in an

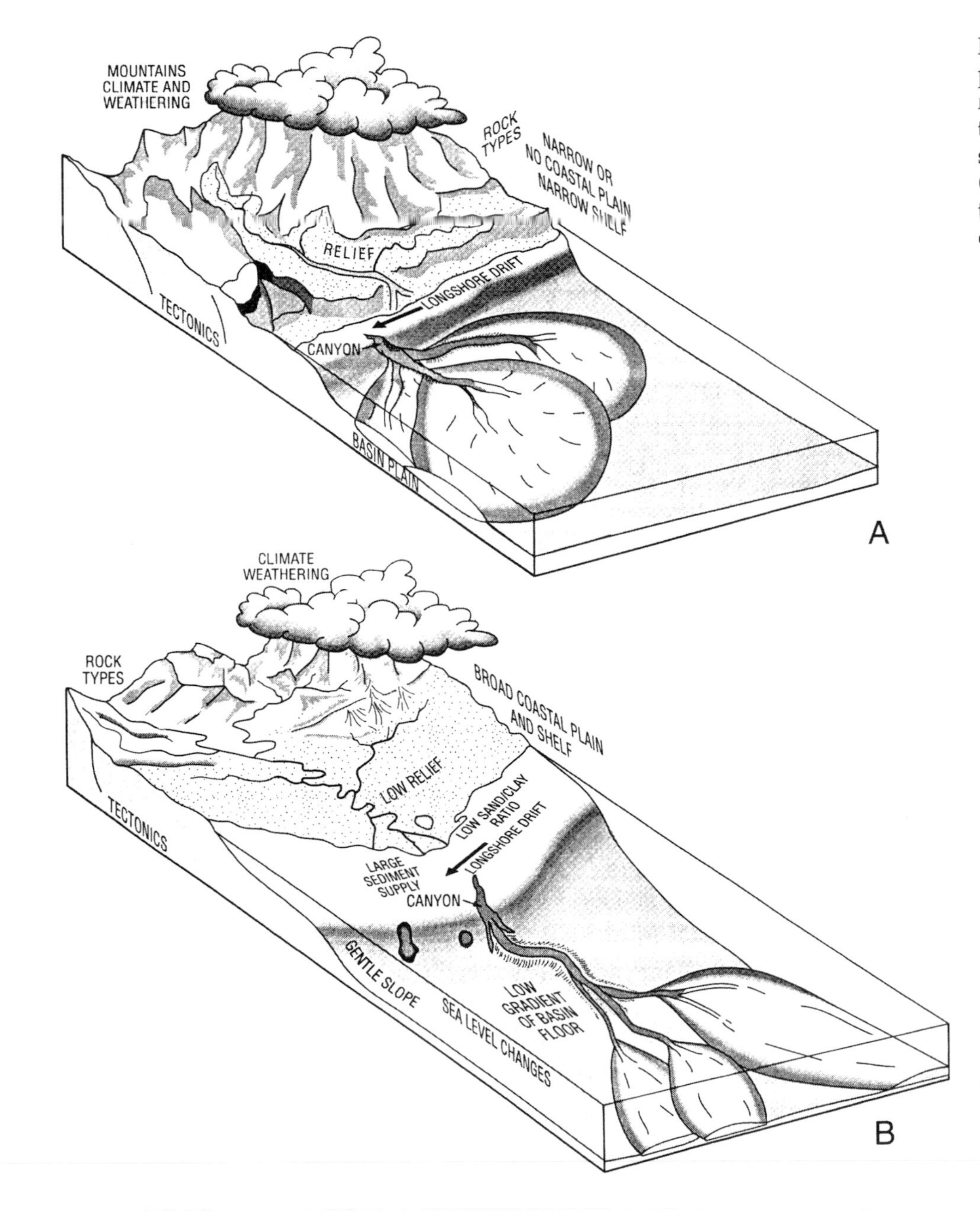

Figure 2—Block diagram presenting the two end members discussed in the text. (A) Coarse-grained, sand-rich turbidite system. (B) Fine-grained, mud-rich turbidite system. For explanation, see text.

active margin setting (Scott et al., this volume). It is therefore advisable to have an understanding of the various factors that act on the transport and on deposition of the sediment (Pratson et al., this volume).

COARSE-GRAINED, SAND-RICH SUBMARINE FAN SYSTEM

The characteristic of coarse-grained, sand-rich submarine fan/turbidite systems is the high net-to-gross percentage of any turbidite system itself, as well as of the entire complex, because of the relative thin sandy/silty shales dividing the successive systems. The bulk of the sediment normally falls within the sand size, often medium and coarser. The sediment source is rather close to the shoreline, commonly with several smaller streams rather than one major river bringing the sediment to the shore. Distance and time of terrestrial transport are relatively short, reducing the chance for active breakdown of particles and the unstable minerals (Figure 2A).

Once the sediment reaches the coastline, there is little possibility of forming a significant deltaic complex because a wide coastal plain is missing. Most of the sediment will be transported along the coast by longshore drift. Where it becomes piled up in front of a beach head or encounters an offshore-directed swale, the sediment will creep seaward, gradually deepening that swale into a canyon head. The upper part of the canyon will fill until instability is reached and sediment moves downcanyon into the basin. Because of the scarcity of very fine-grained sediment, the transport in the basin itself should be classified as nonefficient. As a result the outbuilding of the fan is a prograding one (Figure 2A). The overbank deposits will decrease in size away from the channels but in many cases will have sufficient porosity, permeability, and connectivity to constitute a reservoir.

Sea-level fluctuations, in this case, have a minor effect because the shelf is often narrow to start with. Climatic changes can have a major effect when rain-induced discharge increases, bringing more sediment to the coast (Gorsline and Emery, 1959). The list of excellent publications on coarse-grained, sand-rich submarine fan/turbidite systems is extensive. The interested reader should start with edited publications and the references provided therein (Bouma et al., 1985; Weimer and Link, 1991; Weimer et al., 1994; Pickering et al., 1995).

FINE-GRAINED, MUD-RICH SUBMARINE FAN SYSTEM

This end member is characterized by a low sand-to-shale ratio when dealing with a total complex that is comprised of two major components: high net-to-gross individual submarine fans/turbidite systems and very low net-to-gross basinal shales separating the fan systems. The basinal shales commonly are comprised of very thin bedded (2–10 mm), fine and coarse silt-rich, or very fine-grained sand-rich deposits resulting from low-density turbidity currents (Bouma, 1962), separated by laminae commonly rich in clay minerals and organics. These shale series can be as thick as the individual turbidite systems. The individual turbidite systems basically consist of fine- to medium-grained sand, either forming massive sandstones or being bedded with thin silt-rich shales in between. The fine-grain size makes it difficult to visually observe graded bedding and/or many amalgamated contacts. The lack of visible graded bedding does not automatically place the deposit in the debris flow category (Shanmugam, 1999), because in most cases the sediment shows a grain-supported texture in thin sections.

Because of the long transport route from the sediment source to the coastline, and the low-gradient fluvial system, coarser sediment is left behind. Fluvial transport is often characterized by seasonal variations in flow strength, causing short-distance transport alternating with longer periods of burial. During burial periods, diagenesis becomes an active process.

A wide, low-gradient, coastal plain supports frequent delta switching, which results in a major delta complex (Figure 2B). Sea level fluctuations are very important in the transport of deltaic sediment to the deeper basin. When sea level starts to fall, the active distributary channel of the delta will follow the basinward movement of the coastline, and the delta will prograde to the shelf-edge area. When fine-grained sediment rapidly accumulates at the shelf-edge, high pore pressures will result. High sediment pore pressures cause instability and commonly initiate shelf-edge failure. Sliding and slumping will transport sediment across the remaining shelf and down the slope, carving out a major valley (lower canyon, fan valley, or upper fan). The basin slope (often being the continental slope) is the steepest part of the marine transport route and causes an increase in velocity of the gravity flows. However, once a flow enters the zone of the base-of-slope, the reduction in gradient can result in initial deposition. A rather "dirty" turbidite-type deposit may result. It is not uncommon that parts of the fan deposits are from debris flows. The bulk of the gravity flow will continue onto the basin floor, where it also becomes better organized. Partial fallout of sediment from the head of the flow will result in channel deposits, flanked by somewhat finer-grained material from the body of the flow that overbanked the leveed sides of the channel. Levee overbanking changes at least the bottom of the flow dynamics from turbulence to traction, resulting in levee deposits that include current ripple lamination and climbing ripples (Bouma and Wickens, 1994). The active channel gradually becomes shallower, bifurcation may result and the gravity flow changes slowly from confined to unconfined flow. The unconfined flow will deposit its load as sheet sands or depositional lobes.

The sand-to-shale ratio at the base-of-slope is rather high, and low at the channelized midfan. The latter is comprised of a very fine sand-rich channel fill and mud- (shale-) rich overbank deposits. A high sand-to-shale ratio results again on the outer fan, where the sheet sands were deposited. This highly efficient transport system results in a sand-bypassing constructional mode. The very low basin floor gradients (0.1–0.3°)

force gravity currents to find steeper routes, and avulsion becomes normal. Major changes in the orientation of sand bodies in a turbidite system may result, a phenomenon commonly observed in 3-D seismic.

Each fan system likely progrades very rapidly and relocates somewhat laterally with regard to the underlying turbidite system as a result of bottom compensation. However, progradation likely slows down in favor of avulsion (Bouma and Wickens, 1994; Wickens and Bouma, this volume). During the next highstand, the shelf may prograde and the active deltaic distributary normally has moved to a different position. Basin shape and size have a tremendous influence on the effectiveness of the lateral switching.

Combining an understanding of the transport-depositional processes and the concepts portrayed in Figure 2, one may be able to place the deposits under study along a bar that connects the two end members. The results of experiments reveal that adding small amounts of fine-grained sediment to a sand-rich gravity current has a pronounced effect on enhancing the flow velocities (Gladstone et al., 1998). Therefore, an investigator should have an idea of how to use the two end members (Table 1).

Table 1. Generalized, often relative, differences between coarse-grained, sand-rich and fine-grained, mud-rich submarine fans/turbidite systems.

Type of Controls and Characteristics	Coarse-Grained, Sand-Rich Systems	Fine-Grained, Mud-Rich Systems
Tectonic influence on basin	High	Low
Location of receiving basin	Typically on continental crust	Typically on oceanic crust
Size of basin	Often medium to small	Often medium to very large
Length of continental transport	Rather short	Long
Width and type of coastal plain	Narrow: mountainous	Wide and flat
Width of shelf	Normally narrow	Normally wide
Volume of sediment input	Medium to small	Large
Sand/clay ratio of input sediment	High	Relatively low
Main grain size of input	Medium sand & coarser	Fine sand & finer
Types of basin transport	Nonefficient	Efficient
Type of depositional system	Prograding system	Bypassing system
Size of gravity flows	Normally small	Normally medium to large
Thickness of shale between turbidite systems	Thin	Medium to thick
Sand/shale ratio of turbidite complex	High	Low to medium
Sand/shale ratio of individual turbidite systems	High	High
Tendency for major slumps	Small to good	Excellent
Persistence of major channels	Low	Good
Size of main channels	Medium to small	Large
Number of canyon tributaries	Few to several	One to few
Development of distributaries	Good	Reasonable
Type of feeding system	Canyon-fed	Delta-fed
Development of levees	Moderate to good	Good
Type sediment on levees	Bedded sands, some silts	Thin-bedded sand-mud
Sedimentary structures in levees	Parallel & current ripple lamination, climbing ripples	Parallel & current ripple lamination, climbing ripples
Sediment type of overbank	Sand, muddy or silty sand	Silty mud
Main sediment of lower fan	Sand	Sand
Chance of fan to become exposed	Rather good	Poor
Sand/shale ratio at base-of-slope	High	High
Sand/shale ratio on middle fan	High	Low
Sand/shale ratio on outer fan	Becoming lower	High
Direction of gravity flows	Turning parallel to shore	Often directed offshore
Influence of sea level fluctuations	Low to moderate	High
Type of canyon fill	Sand or mud-rich	Normally mud
Reservoir potential: presence of reservoir sand	Excellent	Very good
Conductivity between sands	Often high	Medium to high
Stability of basin sides	Slumping occurs	Slumping normal

MODEL OF FINE-GRAINED TURBIDITE SYSTEM

The Deep Sea Drilling Project results on the Mississippi Fan (Bouma et al., 1985, 1986) were combined with studies on the Pennsylvania Jackfork Formation (Cook et al., 1994; Bouma et al.,1994) and the Permian Tanqua Karoo in South Africa (Bouma and Wickens, 1991, 1994), and then compared with the extensive literature. That resulted in the development of a preliminary model of a fine-grained submarine fan/turbidite system (Bouma, 1995, 1997; Bouma et al., 1995a, b). The tripartite division of a submarine fan, as used in Bouma et al. (1985), into upper (inner), middle (mid-), and lower (outer) was used because it fits so well. Three major end members can be identified with transition zones in between. The base-of-slope, forming the transition from upper to middle fan (basin slope to basin floor) is characterized by a wide channel complex. The middle fan is comprised of a leveed channel with extensive overbank deposits. The lower part of the midfan area contains a transition made up by a distributary channel system. The lower fan is characterized by sheet sands or depositional lobes (Figure 3).

It should be kept in mind that no two succeeding gravity flows have the same volume, density, or velocity. Therefore, the transition from one end member to the next one for succeeding deposits cannot be found stratigraphically at the same locality. As a result, vertical cores and well logs can encounter a wide variety of sedimentary characteristics. Lateral switching due to compensation will complicate these vertical changes. However, a general seismic reconnaissance, covering an entire fan, should help to reduce confusion. Figure 3 illustrates a few stacking arrangements that may be encountered when drilling overlapping sedimentary deposits. Schematic cross sections and cores present characteristics of each end member.

Canyon and Upper Fan Area

Figure 3 shows that the coastline had moved landward, revealing similar characteristics. In this example, the canyon and the outer part of the active distributary are not connected. The basin slope is a typical mud province with some local slumps. A major channel feature was formed by the sliding and slumping of large volumes of sediment that accumulated in the upper canyon. This upper fan channel or fan valley serves as the conduit for the transport of sediment toward the basin. It is the last feature to be filled, normally with fine-grained sediment (fine silty clays), although some local sandy channel fills and slumps of shelf sands may be present.

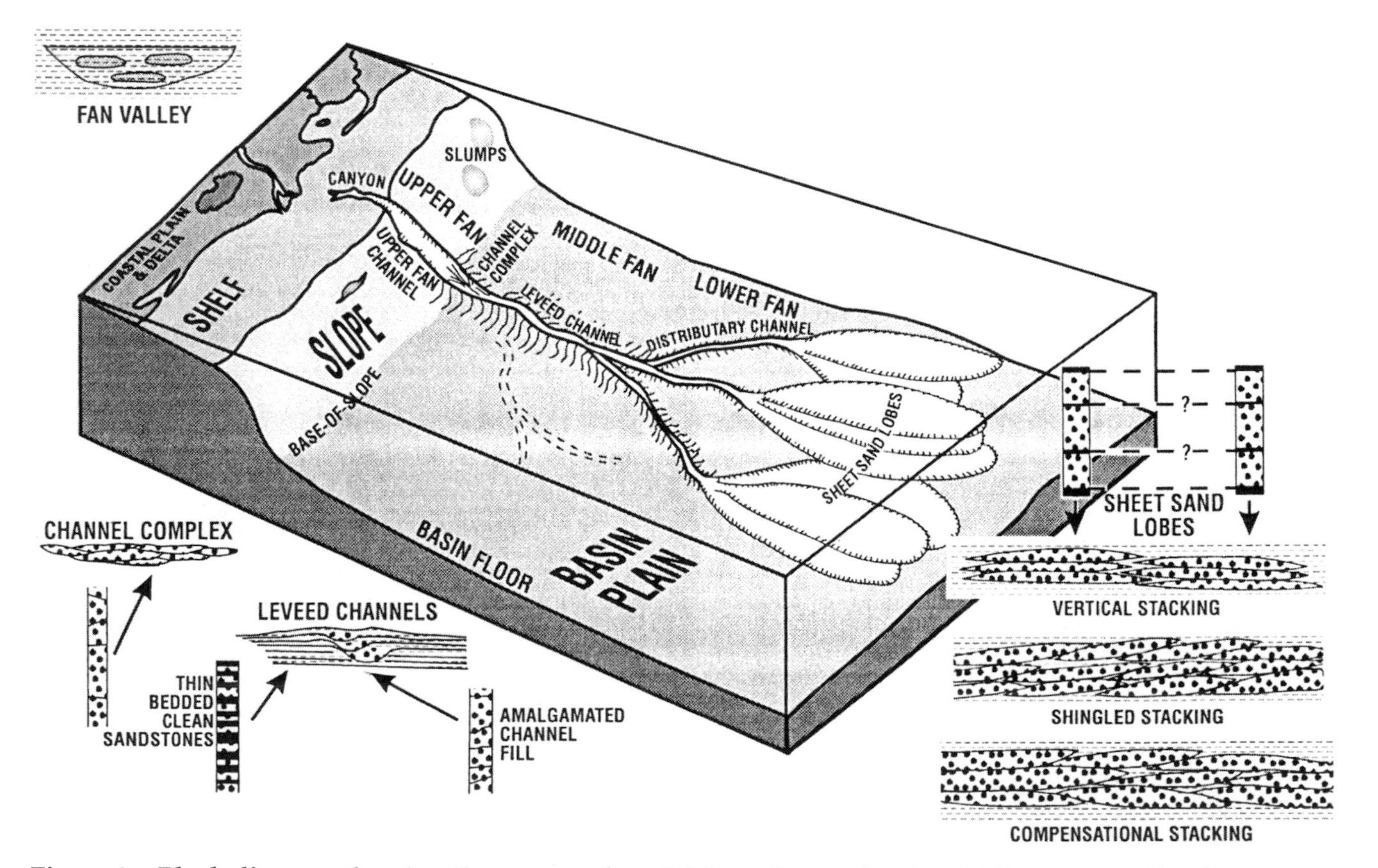

Figure 3—Block diagram showing the proposed model for a fine-grained turbidite system. The three end components of the model (channel complex at the base-of-slope, leveed channel with overbank areas on the middle fan, and sheet sands on the lower fan) are shown in cross section and as an idealized core. Not to scale (modified after Bouma et al., 1995b).

The base-of-slope position will move updip as the basin-floor fan aggradates, giving rise to the idea that the fan valley can be sand filled.

Channel Complex at the Base-of-Slope

The base-of-slope is the zone where the gradient of the slope decreases to that of the upper basin floor. That change in gradient can force several of the gravity flows to start deposition in that zone. It is assumed in this model (Figure 3) that the initial slides and slumps were larger than the later gravity flows. As a consequence the conduit is too wide for many of the flows that move onto the basin floor (Cook et al., 1994).

The fill of the conduit in the base-of-slope area is comprised of a stack of wide, thin channel fills with some remnants of sand-rich levees and sometimes minor overbank deposits (Figure 4). Width-to-thickness ratios can range from 40–80:1. Locally the channels may erode part of the older ones. The width of the base-of-slope part of the conduit can be larger than 10 km (Van Antwerpen, 1992; Cook et al., 1994; Lee et al., 1996; Bouma et al., 1995b). Very silty shales may separate the channel sandstones. They normally show very thin beds of very fine sand, separated by thin silt layers and laminae of very fine silt, organics, and some clays. The channel sandstones are dirty compared with those in the basin because the initial gravity flows had not been able to become well organized. The net-to-gross ratio of the base-of-slope fill is high.

Leveed Channels

Once the "constricted" base-of-slope conduit changes to the more unrestricted basin floor, the type of deposition changes into a well-developed channel with levee and overbank sedimentation. Levees focus the direction of the head of a gravity current; the upper part of the main body of the current often spills over the levee, resulting in accumulation on the levee and on the extensive overbank area. The part of the gravity flow that crosses the levee will, at least partially, change from a turbulent flow to a traction current. The forward momentum of the overall current and the outside gradient of the levee result in two vectors that cause the direction of the flow axis to differ from that of the main current (Kirschner and Bouma, this volume). In the case of Fan 5 in the Blaukop area of the Tanqua Karoo, an angle of at least 32° was measured (D. Basu, personal communication, 1995; Kirschner and Bouma, this volume). The traction current often carries a high volume of sediment, as evidenced from the abundant presence of climbing ripples. Traction currents also seem to winnow the sediment, resulting in levee

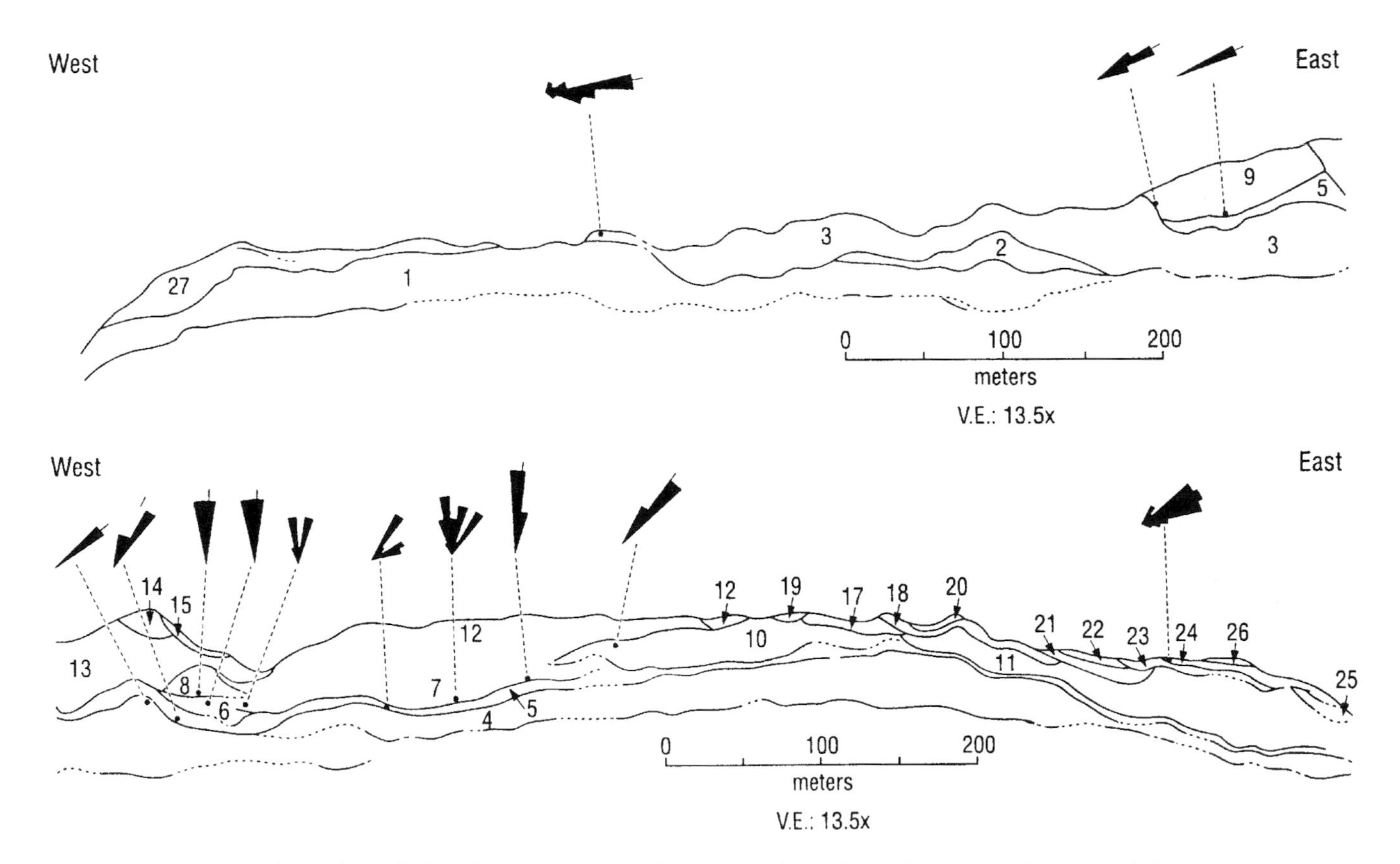

Figure 4—Drawing, based on field observations and measured section of Fan 3 at the Ongeluks River section, Tanqua Karoo, South Africa. Upper section is the western part of this 2.7-km-long section; it continues in the lower section. Notice the high width-to-thickness ratio of individual channel fills in spite of high vertical exaggeration. Paleocurrents show changes in direction during fill (redrawn from Van Antwerpen, 1992).

sands that commonly exhibit the highest porosity and permeability values within a fan system.

The channel fill commonly appears to be massive, being comprised of amalgamated sandstones. Because of the fine-grained nature of the sandstones, such amalgamated contacts are hard to recognize and commonly are overlooked. Occasionally one may detect fluid-escape structures. The levees consist of alternating sandstones and shales. The sandstones commonly contain foreset bedding, climbing ripples, and parallel lamination. The shales are normally silt-rich and may contain very thin laminae of very fine sand. The sand-to-shale ratio of the sandstones can vary from 30 to 60%. The overbank deposits also have sandstone-shale couplets that gradually become finer and thinner away from the channel. At the same time the net-to-gross percentage slowly decreases. Within a geologically short time, the overall sand-to-shale ratio of the channel-levee-overbank system is low because of the extensive lateral coverage of the overbank deposits compared with the coverage of the actual channel fill.

The levee deposits are also known as low-resistivity, low-contrast, thin-bedded sandstones (Darling and Sneider, 1992). The individual layers are too thin to detect on traditional electrical logs. For that reason they are misclassified as sandy/silty shales. However, it has been shown that they can be excellent reservoirs because of the high porosities and permeabilities (Bouma and Wickens, 1994).

Many outcrops may show well-bedded deposits, which under closeup examination contain climbing ripples, current ripple lamination, and parallel lamination. Using interpretations from the more sand-rich systems, such bedded deposits are commonly placed in the outer fan in the fan-fringe area. The filled channel in an outcrop then suggests progradation of an erosional channel (Figure 5). However, a number of examples were found in the Tanqua Karoo, where the contact between channel and levee sandstones could be documented (Basu and Bouma, this volume; Kirschner and Bouma, this volume). Because of the extensive occurrence of overbank areas, it is expected that bedded levee–overbank deposits are more common than channel sands in a stratigraphic sequence. The bedded deposits often show thin-to-thick lenses of massive sandstones, noticeable because of a slightly different color. Such bedded sandstones could have been deposited in distributary channels or crevasse channels (Kirschner and Bouma, this volume).

The channel gradually becomes shallower and a little narrower. Bifurcation takes place, sometimes once or twice, other times very frequently (Weimer, 1989). Commonly, one distributary will be active at a given time, and its life span can be relatively short. The distributaries continue to become shallower and finally are no longer capable of containing the head of the gravity currents. Gradual overflowing will take place, and sheet-sand deposition commences.

Sheet Sands or Depositional Lobes

A major part of the input sediment bypasses the middle fan and ends up in the outer fan as sheet sands or depositional lobes. The gravity current that starts to

Figure 5—Photograph of the turbidite complex with three turbidite systems (#1,2,3) at Kanaalkop, Tanqua Karoo, South Africa. The channel (Fan 3) is 508 m wide and cuts into levee deposits of other channels (not visible in this section). The three fan systems are separated by bedded basin shales—primarily fine-grained turbidites. At the contact between basin shale and overlying turbidite system one may find a condensed section (C.S.: Fan 3), often comprised of a number of closely spaced levels of concretions.

overflow the low terminal levees will fan out, forming an elongate sheet sand. The length-to-width ratio of any sheet will likely depend on flow velocity and density, in addition to bottom roughness and shape of the basin floor.

The overall architecture of the sheet sand package is characterized by sheet-shaped sandstone layers. Each layer looks more or less plane parallel, except at the sides and distal terminus, where a rapid thinning takes place together with a fining of the grain size, followed by merging with the basin shales (Bouma and Rozman, this volume). A minor convexity of the upper surface of a sheet exists but is so slight that it is not observable in the field, even when dealing with long outcrops. A number of layers may stack on top of one another before lateral switching takes place and another stack will be constructed. Figure 3 shows three theoretical stacking patterns, none of which have been completely proven so far (Bouma et al., 1995b). Similarly there are few hard data at this time on lateral relationships between individual sheets.

Individual sheet sands may be deposited by separate gravity currents or by a number of slightly eroding currents, as witnessed by the presence of amalgamated contacts. Individual sheets are separated stratigraphically by thin silty shales that are often discontinuous (Figure 6). Individual sheet sands are commonly deposited by turbidity currents, and the fine grain size prevents observation of graded bedding in the field. The major portion of these sandstones is massive in character, although incomplete Bouma Sequences (T_b and T_c intervals) can be found occasionally. The T_d interval is difficult to identify and the shaly top may be a T_d and/or T_e. In the field we observed that two successive sandstone beds can be in contact with each other, although on both sides a shale comes in that thickens away from that amalgamated contact (Figure 6). A possible interpretation is that the next gravity flow, coming from the same distributary,

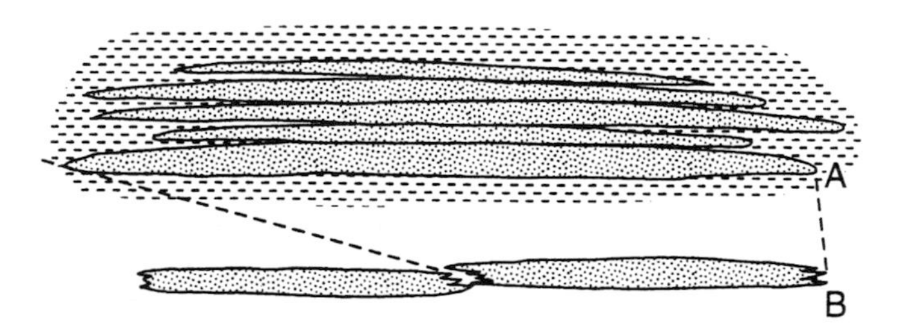

Figure 6—Schematic explanation of the deposition of sheet sands. (A) A succession of individual sandstone layers. Note the amalgamated (or very thin shale) contact between the centers of the slightly convexed-shaped sandstones. (B) Two packages of sandstones (as shown in A) onlap onto each other. These packages are called sheet sands. No scale provided: individual sandstones may be 10–80 cm thick, a package can consist of six to ten layers and can range in width from 500 to >1000 m. The internal stacking pattern is not completely understood (see the three possibilities suggested in Figure 3).

eroded the top silty mud from the underlying deposit. Because the flow strength decreases laterally, the erosion will become less and less in that direction, and shales can be seen that separate succeeding sandstones (Bouma and Wickens, 1994; Bouma et al., 1995b).

Occasionally one can observe a channel-shaped depression at the base of a stack of sheet sands, and sometimes channel-shaped cross sections can be found inside a stack of sheet sands. Whether these represent real channels or local depressions is unknown at this time.

The model presented in Figure 3 highlights the major end components. The transition zones between those end components often cover more surface area than the end components themselves. In addition, the ratios between the influencing factors sometimes change continuously, and succeeding deposits therefore may not look alike. Finally it must be understood that most turbidite sequences are comprised of deposits that fall between the two end members presented in Figure 2.

TIMING OF SEQUENCE STRATIGRAPHIC EVENTS/SYSTEMS TRACTS

Sequence stratigraphy had a major influence on understanding the stacking patterns of successive depositional units. The concepts are applicable in spite of the fact that artificially forced applications and simplifications may cause serious disagreements regarding their validity. The spatial and temporal variations in the interaction of influential factors (e.g., tectonics, climate, sediment, and sea level fluctuations) are too often neglected. Those large variations make it impossible to contain the resulting deposits in one detailed framework such as was initially presented by Bouma et al. (1989). The literature is extensive and will not be reviewed here. Some of the latest publications also have extensive references lists (Weimer and Posamentier, 1994; Shanley and McCabe, 1998).

One aspect will be discussed here: time and transport variations in fine-grained turbidite systems. Normally it is accepted that major transport from the continent to a basin is initiated in the lowstand systems tract, and turbidite deposition proceeds into the transgressive systems tract (Perlmutter, 1985). If the sea level change is glacio-eustatically controlled, this may not be the case, making climate then the most important factor. The eustatic lowering is due to the stacking of glacial ice on continents, meaning that fluvial flow becomes strongly reduced. Once melting becomes stronger than freezing, the fluvial transport can increase to major proportions. The runoff results in a marine transgression. The coastline will move landward and depressions, such as incised valleys, will be filled. However, rapid fill of fine-grained sediment prevents equal rapid escape of pore fluid. As a consequence, high pore pressure in rapidly deposited sediments is commonly followed by failure, initiating a mass movement continuum downslope (Shanmugam and Moiola, 1991). The above scenario assumes normal hypopycnal (fluvial flow less dense than that of the basin water) fluvial flow conditions, which are more or less typical for the present highstand conditions. However, major melting can result in very large fluvial systems with hyperpycnal (fluvial flow with dispersed sediment denser than basin water) flow conditions, at least for the bottom part of the fluid column. If the density of those turbulent flows is higher than that of the shelfal-basin waters, those turbidity currents will not deposit much sediment at the shoreline but keep on going to the basin floor. Bypassing may even become a more important issue, resulting in ever more deposition on the outer fan.

Climatic conditions will determine the intensity of melting of ice and thus fluvial flow conditions (Blum, 1998). Hypopycnal and hyperpycnal flow conditions likely alternate and/or change from one mode to the next through time. Although such considerations may have an insignificant effect on the concept of sequence stratigraphy, it may influence deposition, such as on the continental slope of the northern Gulf of Mexico with its salt-withdrawal basins (Perlmutter, 1985; Kolla and Perlmutter, 1993). Transport across sills that separate those basins may have been enhanced because of higher turbulent flow velocities.

CONCLUSIONS

The purpose of this article is to emphasize that there is no one model that can represent all turbidite systems. Likewise, no two submarine fans/turbidite systems are completely alike, and squeezing the deposits into one model results in more misinterpretations than are acceptable. Combining an understanding of the various influencing factors and sediment transport processes results in finding the proper spot for the formation of interest. In this paper only two end members have been discussed: fine-grained, mud-rich turbidite complex and coarse-grained, sand-rich turbidite complex. Those two end members represent delta-fed vs. canyon-fed, efficient vs. nonefficient sediment transport, or bypassing vs. progradational.

When dealing with fine-grained fans constructed under glacial-eustatic conditions, it may be correct to assume that of the four major influential factors (tectonics, climate, sediment, sea level fluctuation), climate is much more important than is often realized. It affects the breakdown of rocks as well as the fluvial transport conditions (hypopycnal vs. hyperpycnal). It can determine local and/or global sea level. It is likely that under glacial conditions rather little sediment is available for the construction of a submarine fan during the lowstand systems tract because much is covered by land ice. However, when melting far exceeds freezing, enormous fluvial transport can result; this will result in a transgression. Often the assumed transport can be hypopycnal, resulting in distributary mouth bars and incised valley fill. The deposits will fail due to high pore pressure, setting sediment in motion toward the basin floor. At other times the fluvial flow may be hyperpycnal and turbulent. Many of these turbidity currents will cross the shelf and slope and deposit their sediment load onto the basin floor.

An understanding of the strength and interactions of the different factors that influence transport to and deposition in the deepwater basin is essential before a selection should be made to construct a model that best represents the formation under study. Only two end member models have been discussed because these presently represent most of the fields of interest.

ACKNOWLEDGMENTS

I thank the many colleagues and students for the discussions that gradually transformed into the model on fine-grained turbidites. Brad Macurda and Jerry Kuecher did a fantastic job on the review of the manuscript and any errors remaining are mine. Thanks to Kelly LaGrange and Lieneke Bouma for the typing of the various drafts.

REFERENCES CITED

Blum, M. D., 1994, Genesis and architecture of incised valley fill sequences: A late Quaternary example from the Colorado River, Gulf coastal plain of Texas, *in* P. Weimer and H. W. Posamentier, eds., Siliciclastic sequence stratigraphy: Recent developments and applications: AAPG Memoir 58, p. 259–283.

Bouma, A. H., 1962, Sedimentology of some flysch deposits: a graphic approach to facies interpretation: Amsterdam, Elsevier, 168 p.

Bouma, A. H., J.M. Coleman, A. M. Meyer, eds., 1986, Initial reports of the Deep Sea Drilling Project Leg 96: Washington, D.C., U.S. Government Printing Office, 824 p.

Bouma, A. H., 1995, Geological architecture of fine-grained submarine fans: Corpus Christi Geological Society and Coastal Bend Geophysical Society Bulletin, March 1995, p. 11–19.

Bouma, A. H., 1997, Comparison of fine-grained, mud-rich and coarse-grained, sand-rich submarine fans for exploration-development purposes: Gulf Coast Association of Geological Societies Transactions, v. 47, p. 59–64.

Bouma, A. H., W. R. Normark, and N. E. Barnes, 1985, Submarine fans and related turbidite systems: New York, Springer-Verlag, 351 p.

Bouma, A. H., J. M. Coleman, C. E. Stelting, and B. Kohl, 1989, Influence of relative sea level changes on the construction of the Mississippi Fan: Geo-Marine Letters, v. 9, p. 161–170.

Bouma, A. H., and H. deV. Wickens, 1991, Permian passive margin submarine fan complex, Karoo Basin, South Africa: possible model to Gulf of Mexico: Gulf Coast Association of Geological Societies Transactions, v. 41, p. 30–42.

Bouma, A. H., and H. deV. Wickens, 1994, Tanqua Karoo, ancient analog for fine-grained submarine fans, *in* P. Weimer, A. H. Bouma, and B. F. Perkins, eds., Submarine fans and turbidite systems: sequence stratigraphy, reservoir architecture and production characteristics; Gulf of Mexico and international: Gulf Coast Section SEPM 15th

Annual Research Conference Proceedings, p. 23–34.
Bouma, A. H., T. W. Cook, and M. A. Chapin, 1994, Architecture of a marine channel complex, Jackfork Formation, Arkansas: Basin Research Institute Bulletin, Louisiana State University, v. 4, p. 10–21.
Bouma, A. H., G. H. Lee, O. Van Antwerpen, and T. W. Cook, 1995a, Channel complex architecture of fine-grained submarine fans at the base-of-slope: Gulf Coast Association of Geological Societies Transactions, v. 45, p. 65–70.
Bouma, A. H., H. deV. Wickens, and J. M. Coleman, 1995b, Architectural characteristics of fine-grained submarine fans: a model applicable to the Gulf of Mexico: Gulf Coast Association of Geological Societies Transactions, v. 45, p. 71–75.
Cook, T. W., A. H. Bouma, M. A. Chapin, and H. Zhu, 1994, Facies architecture and reservoir characterization of a submarine fan channel complex, Jackfork Formation, Arkansas, *in* P. Weimer, A. H. Bouma, and B. F. Perkins, eds., Submarine fans and turbidite systems: sequence stratigraphy, reservoir architecture and production characteristics; Gulf of Mexico and international: Gulf Coast Section SEPM 15th Annual Research Conference Proceedings, p. 69–81.
Darling, H. L., and R. M. Sneider, 1992, Production of low resistivity, low-contrast reservoirs, offshore Gulf of Mexico basin: Gulf Coast Association of Geological Societies Transactions, v. 42, p. 73–88.
Feeley, M. H., R. T. Buffler, and W. R. Bryant, 1985, Depositional units and growth pattern of the Mississippi Fan, *in* A. H. Bouma, W. R. Normark, and N. E. Barnes, eds., Submarine fans and related turbidite systems: New York, Springer-Verlag, p. 253–257.
Gladstone, C., J. C. Phillips, and R. S. J. Sparks, 1998, Experiments on bidisperse, constant-volume gravity currents: propagation and sediment deposition: Sedimentology, v. 45, p. 833–843.
Gorsline, D. S., and K. O. Emery, 1959, Turbidity current deposits in San Pedro and Santa Monica basins off southern California: Geological Society of America Bulletin, v. 70, p. 279–290.
Kolla, V., and M. A. Perlmutter, 1993, Timing of turbidite sedimentation on the Mississippi Fan: AAPG Bulletin, v. 77, p. 1129–1141.
Lee, G. H., J. S. Watkins, and W. R. Bryant, 1996, Bryant Canyon Fan System: an unconfined, large-river turbidite system in the northwestern Gulf of Mexico: AAPG Bulletin, v. 80, p. 340–358.
Mutti, E., 1985, Turbidite systems and their relations to depositional sequences, *in* G. G. Zuffa, ed., Provenance of arenites: NATO-AST Series, Dordrecht, Riedel, p. 65–93.
Mutti, E., and W. R. Normark, 1987, Comparing examples of modern and ancient turbidite systems: problems and concepts, *in* J. K. Legget, and G. G. Zuffa eds., Marine clastic sedimentology: London, Graham and Trotman, p. 1–38.
Mutti, E., and W. R. Normark, 1991, An integrated approach to the study of turbidite systems, *in* P. Weimer and M.H. Link, eds., Seismic facies and sedimentary processes of submarine fans and turbidite systems: New York, Springer-Verlag, p. 75–106.
Mutti, E., 1992, Turbidite sandstones: Milan, Agip S.A., 275 p.
Normark, W. R., and N. E. Barnes, 1983/1984, Aftermath of COMFAN—Comments, not solutions: Geo-Marine Letters, v. 3, p. 223–224.
Perlmutter, M. A., 1985, Deep water clastic reservoirs in the Gulf of Mexico: a depositional model: Geo-Marine Letters, v. 5, p. 105–112.
Pickering K. T., R. N. Hiscott, N. H. Kenyon, F. Ricci Lucchi, and R. D. A. Smith, eds., 1995, Atlas of deep water environments: architectural style in turbidite systems: London, Chapman & Hall, 333 p.
Reading, H. G., and M. Richards, 1995, Turbidite systems in deepwater basin margins classified by grain size and feeder system: AAPG Bulletin, v. 78, p. 792–822.
Shanley, K. W., and P. J. McCabe, eds., 1998, Relative role of eustasy, climate, and tectonism in continental rocks: SEPM (Society for Sedimentary Petrology) Special Publication 59, 234 p.
Shanmugam, G., and R. J. Moiola, 1991, Types of submarine fan lobes: Models and implications: AAPG Bulletin, v. 75, p. 156–179.
Shanmugam, G., 1999, 50 years of the turbidite paradigm (1950s–1990s): Deep-water processes and facies models—a critical perspective: Marine and Petroleum Geology, in press.
Stow, D. A. V., D. G. Howell, and C. H. Nelson, 1985, Sedimentary, tectonic, and sea-level controls, *in* A. H. Bouma, W. R. Normark, and N. E. Barnes, eds., Submarine fans and related turbidite systems: New York, Springer-Verlag, p. 15–22.
Van Antwerpen, O., 1992, Ongeluks River channel fills, Skoorsteenberg Formation: Unpublished Honors thesis, University of Port Elisabeth, South Africa, 69 p.
Weimer, P., 1989, Sequence stratigraphy of the Mississippi Fan (Plio-Pleistocene), Gulf of Mexico: Geo-Marine Letters, v. 9, p. 185–272.
Weimer, P., and R. T. Buffler, 1985, Distribution and seismic facies of Mississippi Fan Channels: Geology, v. 16, p. 900–903.
Weimer, P., and M. H. Link, eds., 1991, Seismic facies and sedimentary processes of submarine fans and turbidite systems: New York, Springer-Verlag, 447 p.
Weimer, P., A. H. Bouma, and B. F. Perkins, eds., 1994, Submarine fans and turbidite systems: sequence stratigraphy, reservoir architecture and production characteristics, Gulf of Mexico and international: Gulf Coast Section SEPM 15th Annual Research Conference Proceedings, 440 p.
Weimer, P., and H. W. Posamentier, 1994, Recent developments and applications in siliciclastic sequence stratigraphy, *in* P. Weimer, and H. W. Posamentier, eds., Siliciclastic sequence stratigraphy: recent developments and applications: AAPG Memoir 58, p. 3–12.

Coleman Jr., J. L., 2000, Carboniferous Submarine Basin development of the Ouachita mountains of Arkansas and Oklahoma, *in* A. H. Bouma and C. G. Stone, eds., Fine-grained turbidite systems, AAPG Memoir 72/SEPM Special Publication No. 68, p. 21–32.

Chapter 3

Carboniferous Submarine Basin Development of the Ouachita Mountains of Arkansas and Oklahoma

J. L. Coleman, Jr.
BP Amoco
Houston, Texas, U.S.A.

ABSTRACT

The Paleozoic stratigraphic succession of the Ouachita Basin is dominated by deepwater siliciclastics, carbonates, and chert. Within the Carboniferous, the Stanley fan complex is a thick shale interval, with upper and lower sandstone sections, that was deposited during an overall sea level highstand. The overlying Jackfork Formation is predominantly a sandstone section, with no shelf equivalent. The Johns Valley Formation, a unit of turbidite sandstone, shales, and unusual boulder beds, overlies the Jackfork. The 6100-m-thick Atoka Formation succeeds the Johns Valley. This thick sandstone and shale interval is divisible into a central basin (or axial) fan complex, a series of slope (or intraslope basin) fans, and thick shelf margin deltaic complex.

INTRODUCTION

The Carboniferous stratigraphic succession of the Ouachita Mountains of Arkansas and Oklahoma is a 10,000+-m interval of deepwater siliciclastic and cherty sedimentary rocks deposited during foreland basin subsidence immediately prior to late Carboniferous (late Pennsylvanian) thrust fault and fold deformation of the Ouachita Orogeny. This thick sequence is highly representative of foreland basin deepwater detrital fill. Because sedimentation took place in an evolving structural basin and as a result of continent-continent collision in the vicinity of the paleoequator, examination of this succession enables a review of controlling factors in deepwater deposition.

THE OUACHITA BASIN

The Ouachita Basin lies immediately marginal to and south of the generally undeformed Paleozoic continental shelf of mid-continent United States. It extends from the Black Warrior Basin of Mississippi on the east to the Ardmore-Anadarko Basin of Oklahoma on the southwest (Figure 1). Its true southern extent is unknown; however, post-Ouachita orogenic, Paleozoic-age sedimentary rocks are known from the subsurface of northwestern Louisiana and eastern Texas (Nicholas and Waddell, 1989).

Late Cambrian to Carboniferous deepwater sedimentary rocks dominate the Ouachita Basin. Volumetrically, siliciclastics of the Carboniferous Stanley, Jackfork, John Valley, and Atoka Formations make up more than 80% of this interval (Roberts, 1982). Well-documented shelf biostratigraphic and paleoenvironmental indicators help constrain age and depositional environment interpretations of the basinal sedimentary rocks (e.g., Gordon, 1970; Lane and Straka, 1974; Sutherland and Manger, 1977; Gordon and Stone, 1977; Sutherland and Manger, 1979; Groves, 1983; Miller et al., 1989). Prior to thrusting, uplift, and erosion, cumulative stratigraphic thicknesses of deepwater sedimentary rocks in the Ouachita Basin probably exceeded 17,000 m (Roberts, 1982). Of this, as much as 14,250 m have been attributed to the Carboniferous section (Arbenz, 1989).

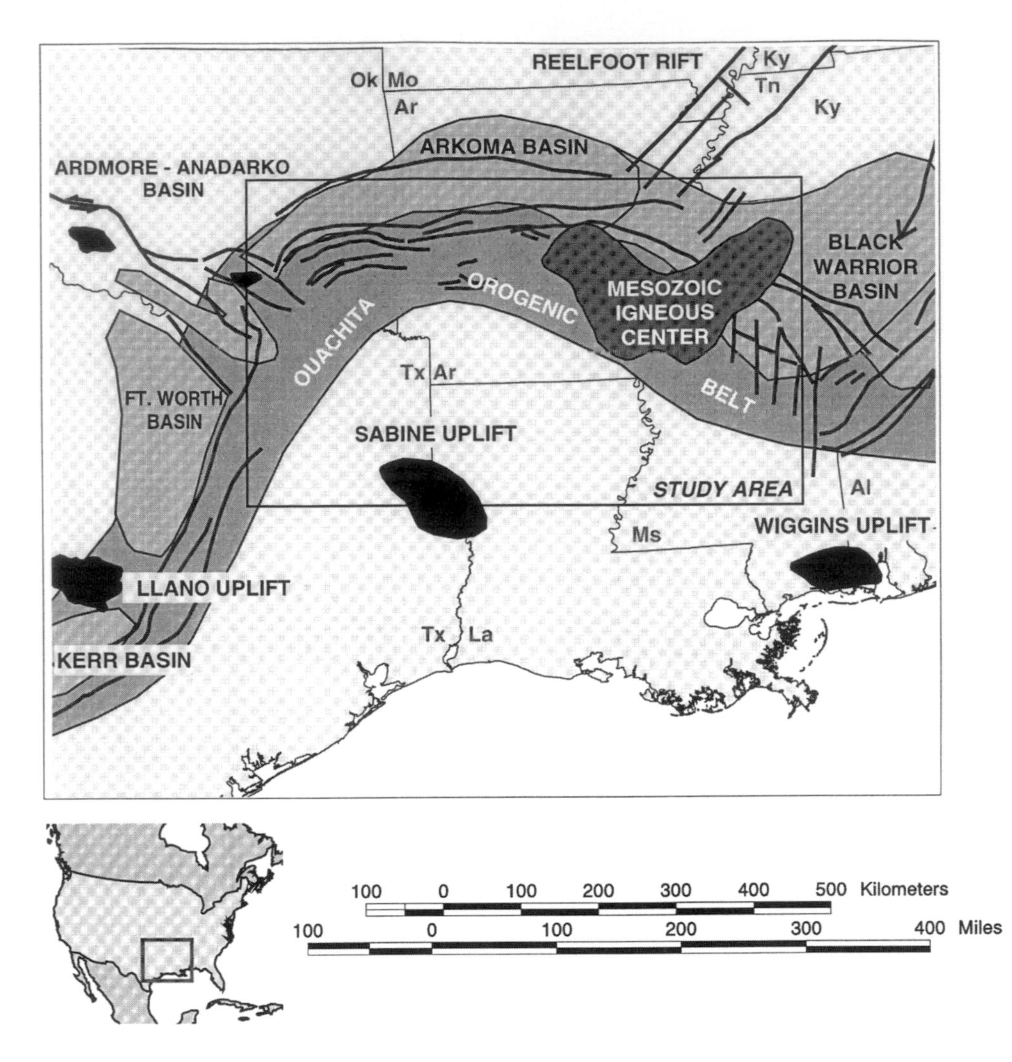

Figure 1—Location map of study area, showing Ouachita orogenic trend, adjacent foreland basins, and associated basement uplifts.

The deepwater Ouachita Basin formed during the middle(?) Cambrian as southern North America and an as-yet unidentified continental mass rifted apart. This block retreated only partially from North America, leaving an intervening deepwater basin, but was still capable of providing a southern sediment source for the Ouachita Basin deepwater Paleozoic strata. During the Paleozoic an extensive carbonate bank developed on the southern margin of North America, with minor punctuations of siliciclastic sedimentation from the interior of North America.

For most of its duration, the Ouachita Basin was essentially a starved basin adjacent to the carbonate shelf margin to the north. However, as thrust-sheet loading proceeded northward during the late Carboniferous, the locus of basin sedimentation shifted northward as large-displacement normal faults downdropped the shelf margin, forming new, deepwater basins on top of the Cambrian to upper Carboniferous (Pennsylvanian) shallow water sequences. The prethrusting size of the Ouachita Basin is difficult to determine accurately. However regional seismic profiles, gravity and aeromagnetic data, and deep oil and gas exploration wells, in conjunction with outcrop data, suggest that, at the onset of the Carboniferous, the Ouachita Basin extended approximately 450 km north to south and 550 km east to west (Coleman, 1990; Coleman et al., 1994). The northern paleoshelf extended from eastern Oklahoma across central Arkansas and into Mississippi. The northern, preorogenic shelf margin is presently buried beneath slope and shelf sedimentary rocks of the synorogenic Atoka Formation, which makes up the bulk of the Arkoma Basin. This margin is imaged on seismic lines across the northern margin of the Arkoma Basin and lies approximately beneath the surface trace of the Ross Creek–Choctaw Fault Zone. Structural restoration of faults and folds indicates that Ouachita frontal compression is about 50% (Arbenz, 1989; Vanarsdale and Schweig, 1990; Coleman, 1993a, 1994; Roberts, 1994).

The eastern paleoshelf continued to the southeast from northern Mississippi into western Alabama. Today the eastern margin lies beneath Mesozoic and Cenozoic sediments in Mississippi, and appears to be highly deformed, consisting of a complex zone of relict, right-lateral, strike-slip faulting, with a general up-to-the-northeast sense of structural displacement, primarily of late Carboniferous age. Structural restoration of these faults and folds is significantly more problematic, as both dip-slip reverse faulting as well as strike-slip faulting are present (Coleman, 1990; Hale-Erlich and Coleman, 1993; Coleman and Hale-Erlich, 1994).

The western paleoshelf ran from southwestern Oklahoma southward into east Texas, where it is interrupted by the large displacement faults bounding the Anadarko Basin beneath the present Oklahoma–Texas border. The western paleoshelf margin is present in southeastern Oklahoma and northeastern Texas, both on the surface and in the subsurface. Where exposed, this margin is dominated by a narrow fault zone separating a large area of open synclines from generally undeformed platform margin sedimentary rocks. In the subsurface of northeastern Texas, this margin appears to be more highly faulted, with substantial structural offset at the Cambro-Ordovician level. This general type of high-relief, structural style suggests that the western margin in Texas may have a strong strike-slip component.

The southern paleoshelf has not been clearly identified, since its postulated presence is at great depths beneath a thick post-Ouachitian Paleozoic, Mesozoic, and Cenozoic sedimentary and volcanic section in southern Arkansas and/or northern Louisiana. A structural reconstruction of the Ouachitas based on the 50% compression of the frontal zone places the predeformation location of the southernmost occurrences of Ouachita pre- and synorogenic sedimentary rocks near the Arkansas–Louisiana border (Coleman, 1990; Coleman et al., 1994). Although this does not clearly define the location of the prethrusted southern shelf, it does suggest a possible minimum limit, and as such gives a minimum areal extent for the Ouachita Basin. Regional aeromagnetic and Bouguer gravity maps of this area (Keller, 1989) are not especially helpful, because large, postorogenic, Mesozoic salt and rift basins, and extrusive and intrusive igneous areas control both the gravity and the magnetic signature of the area. Given these observations, interpretations, and assumptions, the areal extent of the predeformation Ouachita Basin was at least 250,000 km^2.

BASIN DRAINAGE SYSTEMS

Major drainages were probably controlled by the major structural trends peripheral to the basin: the Reelfoot Rift draining the mid-continent area and the Ouachita–Appalachian strike-slip juncture zone draining the southern Appalachians fold belt and adjacent Black Warrior foreland basin. The Ardmore–Anadarko aulocogen appears to have been a significant sediment sink during the time of deposition of the Carboniferous of the Ouachita Basin and was not a major conduit for sediment delivery (McKee and Crosby, 1975). Additional sediment delivery systems were probably associated with lesser fault and fracture zones, potentially accentuated by the increasing normal fault displacement of the northern shelf during foreland basin subsidence, and by uplift and cannibalization of older rocks in the encroaching orogenic terrane to the south.

PALEOWATER DEPTH OF DEPOSITION

Details of the water depths in the Carboniferous sedimentary basin is not known; however, most of the Carboniferous of the Ouachita Basin have been considered deepwater deposits for more than 60 years. Likewise, almost all pre-Carboniferous strata in the Ouachita Basin have sedimentary characteristics indicative of deepwater deposition.

Paleobathymetric faunal indicators are generally thought to be unreliable in the Paleozoic. Chamberlain (1978) suggested Ouachita Basin paleowater depths of 2000 m based on trace fossils, but Ekdale and Mason (1988) and others have shown that trace fossils more accurately measure oxygen content and substrate faunal viability than water depth. Coleman (1990) suggested that the Jackfork was deposited in water depths of 1500 to 2000 m, based on an interpreted sequence stratigraphic geometry. Facies successions, stacking patterns, and sedimentary structures in the Stanley and the bulk of the Atoka suggest that these units were also deposited in generally similar water depths.

The determination that the Stanley, Jackfork, Johns Valley, and Atoka formations are of deepwater origin is based on a relative preponderance of sedimentary structures, bedding characteristics, and lithofacies successions more indicative, overall, of deepwater than shallow, the substantially thicker stratigraphic intervals of these four as compared with the thickness of their more obvious shallow water counterparts, and the more structurally basinward (when restored) position of the four when compared with their shelf equivalents.

EUSTATIC SEA LEVEL

Global sea level was generally very high during the lower Carboniferous (Mississippian), with only minor lowerings. Ross and Ross (1987) indicate that during this time interval, sea level was as high as anytime during the late Paleozoic (Figure 2). Third-order fluctuations throughout this highstand may have helped create and maintain a basinal clastic sediment delivery system throughout Stanley time. During and following a major lowstand at about 324 m.y., the style of Ouachita sedimentation completely changed from carbonate shelf with relatively minor shale and basinal siliciclastic sedimentation to thick basinal sandstone and shale deposition, with a gradually increasing shelf deltaic complex component. Sea level gradually rose throughout Atokan deposition, flooding the adjacent shelves. However, tectonic uplift and sediment supply overwhelmed the rising sea level, and more than 6100 m of basinal Atoka was deposited. By the time global sea level returned to its pre-324 m.y. low stand level, Ouachita basin sedimentary rocks were exposed and being eroded.

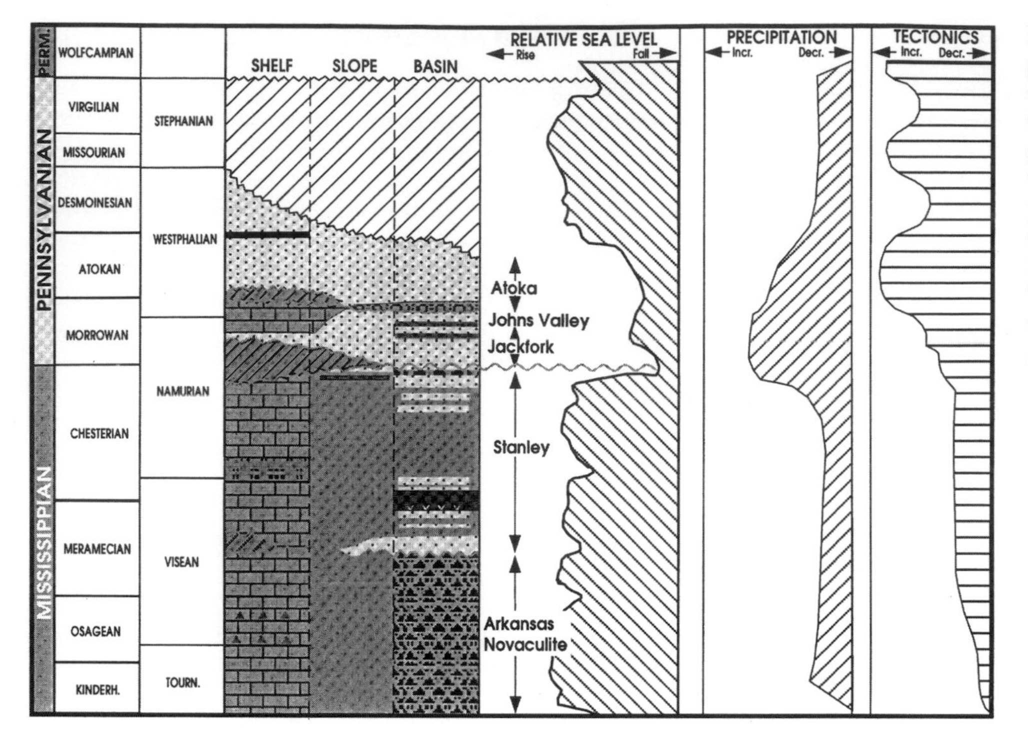

Figure 2—Chronostratigraphic column of the Carboniferous of the Ouachita Basin, showing general lithostratigraphy with eustatic sea level, precipitation, tectonics curves (developed from Sutherland and Manger, 1979; Ross and Ross, 1987; Cecil and Eble, 1989; Arbentz, 1989).

PALEOCLIMATE CONDITIONS

The Ouachita Basin during the Carboniferous was located at between approximately 15°S and the paleoequator, with a general east–west trend (Figure 3). Climatic conditions were generally dictated by the relative position within the global climatic system, drifting from dominantly arid subequatorial zone between 15° and 30°S into the humid equatorial zone, which typically extends from 15°S to 15°N (Trewartha et al., 1967; Parrish, 1982). Paleoclimatic interpretations from exposures within the Appalachians and the United States's mid-continent support these general continental climate conditions (Cecil and Eble, 1989). The upper Devonian and lower Carboniferous (Mississippian) strata were deposited during overall arid climatic conditions. These conditions are typified by evaporites within the Illinois and Appalachian basins and widespread carbonate bank development (Figure 4A). Development of limestone-capped mesas and buttes during periods of

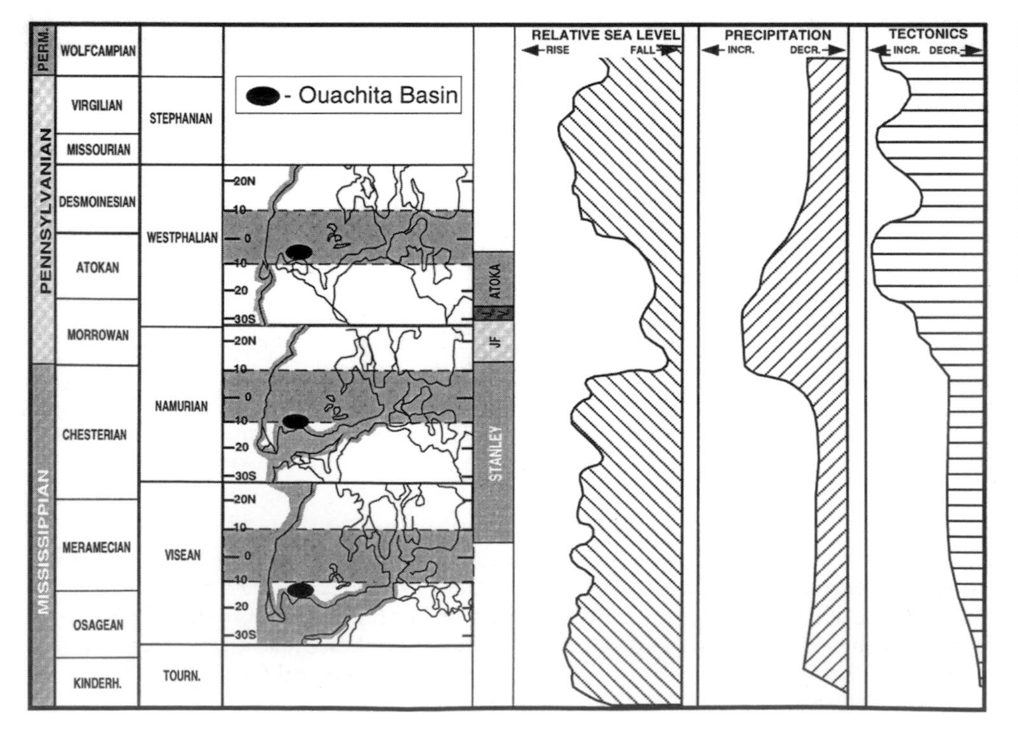

Figure 3—Plate tectonic and paleogeographic position of the Ouachita Basin during the Carboniferous (developed from Sutherland and Manger, 1979; Ross and Ross, 1987; Cecil and Eble, 1989; Arbentz, 1989; Scotese and McKerrow, 1990).

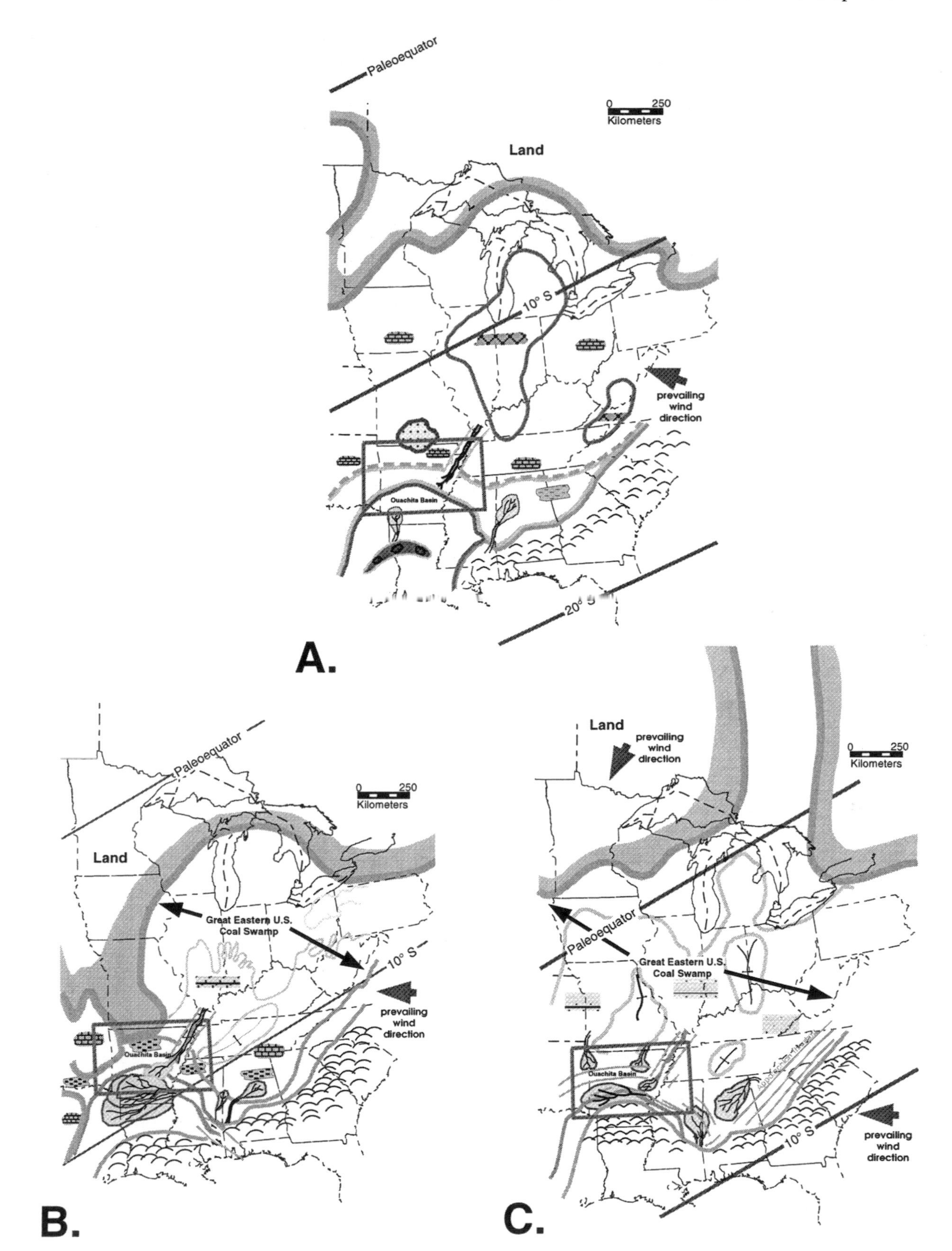

Figure 4—Paleogeographic setting of the Ouachita Basin and eastern North America during the early Carboniferous. (A) Visean (developed, in part, from Houseknecht and Kacena, 1983; Thomas, 1988; Witzke, 1990; Scotese and McKerrow, 1990; Coleman and Hale-Erlich, 1994). (B) Namurian (developed, in part, from McKee and Crosby, 1975; Houseknecht and Kacena, 1983; Thomas, 1989; Johnson et al., 1988; Witzke, 1990; Scotese and McKerrow, 1990; Coleman and Hale-Erlich, 1994). (C) Westphalian (developed, in part, from McKee and Crosby, 1975; Houseknecht and Kacena, 1983; Witzke, 1990; Scotese and McKerrow, 1990).

subaerial erosion in the Illinois Basin indicate arid climate conditions persisted even during periods of erosion and nondeposition (Bristol and Howard, 1974). About or slightly prior to the 324 m.y. sea level fall, the Ouachita Basin drifted out of the arid paleolatitudes of the southern hemisphere into the humid equatorial zone (Figure 4B). The accompanying increase in precipitation increased subaerial weathering and stream carrying capacity, producing a more quartz-rich, less-lithic siliciclastic sediment. This new sediment character is present in the first, and most subsequent, Jackfork sands, which were deposited in the basin. High sediment runoff and relatively lower sea level delivered substantial amounts of sand size and coarser clastics into the basin, while at the same time creating a substantial regional unconformity across the central and eastern United States. As the continent continued to move across the major climate belts, it reached a more temperate zone, with increasing seasonal variations. This development, in conjunction with foreland basin loading and subsidence help produce the expansive coal swamps across the eastern United States (Figure 4C). As the Appalachians and Ouachita Mountains rose, they began to dictate regional climate patterns more than global climate belts. Precipitation levels in the area declined substantially from the beginning of deposition of the late synorogenic Hartshorne Formation, with its thin coals, sandstones, and shales.

TECTONIC CONDITIONS

Tectonic conditions were relatively quiet from the time of Arkansas Novaculite deposition through most of Stanley sedimentation. Volcanic detritus within the Stanley indicate island arc activity on the southern flank of the basin; however, evidence of direct structural control of sedimentation patterns is lacking. This relative quiescence continued though Jackfork time, with a substantial change indicated at the end of Jackfork sedimentation by an apparently low amplitude, but large-scale, regional basinal unconformity between the Jackfork and the overlying Johns Valley shale. The presence of exotic boulders of Cambrian to Pennsylvanian age shelf, slope, and basinal(?) sedimentary rocks within the Johns Valley indicate major uplift along one or more of the shelf margins of the basin. Seismic profiles do not indicate sufficient uplift to erode this stratigraphic sequence entirely from the northern, eastern, or western margins, so some component of southern margin uplift, erosion, and sediment transport is postulated. The presence and distribution of these exotic boulders, coupled with their stratigraphic range and paleodepositional environment character, suggest that the basin was beginning to structurally invert. Foreland basin loading stresses increased throughout the Atoka deposition, with the formation of major down-to-the-south normal faults, near and north of the pre-Atokan shelf margin zone. These normal faults compartmentalized the northern slope and its attendant deposition. Foreland basin compression apparently compartmentalized the basinal Atoka deepwater deposits as well, ultimately lifting the basin to storm wave base and above, resulting in the deposition of upper Atoka thin coals and sandstones in deltaic margin paleoenvironments. Orogenic forces culminated in the late Carboniferous (late Pennsylvanian)–early Permian with the uplift of the main thrust and fold belt.

AGE OF CARBONIFEROUS FORMATIONS

In the deep basin, thick, rapid deposition during the Carboniferous produced a severe dilution of significant biostratigraphic indicator fossils, resulting in a very poorly preserved biostratigraphy within the submarine fan complex. Sequence stratigraphic concepts have been used to infer chronostratigraphic ages where biostratigraphic zones are not definitive (Coleman, 1993a, 1994) (Figure 2).

The Stanley unconformably overlies the Arkansas Novaculite and extends upward from the Visean to the middle Namurian (late Chester), where it is topped by another unconformity and its correlative basinal conformity. The Jackfork–Stanley contact is conventionally chosen as the Pennsylvanian–Mississippian boundary (Namurian A–B boundary) and is present a few meters above the Chickasaw Creek siliceous shale. The Jackfork extends upward into the late Namurian (late Morrowan Pennsylvanian), where it is overlain by the Johns Valley shale. The Atoka Formation overlies the Johns Valley in the Ouachita Basin and ranges in age from latest Namurian(?)/earliest Westphalian to middle Westphalian (late Morrowan to late Atokan Pennsylvanian). The Atoka Formation is overlain by upper Westphalian (Desmoinesian Pennsylvanian) shallow water clastics of the Hartshorne Formation (Sutherland and Manger, 1977, 1979; Ross and Ross, 1987; Coleman, 1993a, 1994).

CARBONIFEROUS SEDIMENTATION

Carboniferous sedimentation began with continued chert and novaculite deposition across the Upper Devonian boundary as the Arkansas Novaculite (Figure 2). Within the Novaculite are intervals of terrigenous debris flow deposition (Stone and Bush, 1982). However, the bulk of late Paleozoic siliciclastic deepwater sedimentation began in the Visean Carboniferous (Meramecian Mississippian) with the onset of the Stanley shale deposition. Although the majority of the Stanley is a lithic mudstone shale, thick intervals of thin- and medium-bedded turbidite sandstones and associated thin-bedded siltstones exist throughout the entire formation, with the thickest sandy intervals near the base and the top of the unit (Morris, 1989). Much of the detrital material that makes up the Stanley was apparently derived from volcanic, metamorphic, granitic, and sedimentary terranes to the south (Niem, 1977; Morris, 1989). The thick shales within the Stanley typically host some of the major thrust faults of the Ouachitas. As such, stratigraphically extensive or continuous outcrops of the Stanley sandstones are rare, although some undeformed intervals are locally present.

The lowermost Stanley submarine fan complex consists of sandstones, siltstones, and shales of the Hot

Springs Sandstone, a northerly derived deepwater complex, and the southerly derived lower Tenmile Creek Formation. The Stanley is capped by the shale-prone upper Tenmile Creek and sandy Moyers formations. Sufficient detail is lacking to make definitive statements, but the Stanley sandstones appear to be more channel-like in Arkansas and more sheetlike in Oklahoma, with a general zone of transition near the state boundary (Coleman, 1993b). Original stratigraphic thicknesses are difficult to assess due to significant structural thickening with the Stanley shales; however, the original thickness of the Stanley may have been as much as 3350 m (Figure 5A). As such, the Stanley accumulated at an average rate of about 200 m/m.y.

At (or near) the top of the Stanley is a key basinal marker bed: the Chickasaw Creek siliceous shale. It is composed of gray shale with some intervals of siliceous shale (or "chert"). Small, white siliceous globules occur in the siliceous shale and are reported by Cline (1968) to have nuclei of sponge spicules, diatoms, and radiolarians. Because of its lithologic content, relative thinness, and regional extent, it is the only reliable stratigraphic marker between the lower Stanley and the Hartshorne sandstones above the top of the Atoka. Coleman (1990) and Roberts (1994) concluded the Chickasaw Creek represented a major condensed section interval near the top of the Chesterian Mississippian (upper Namurian "A" Carboniferous).

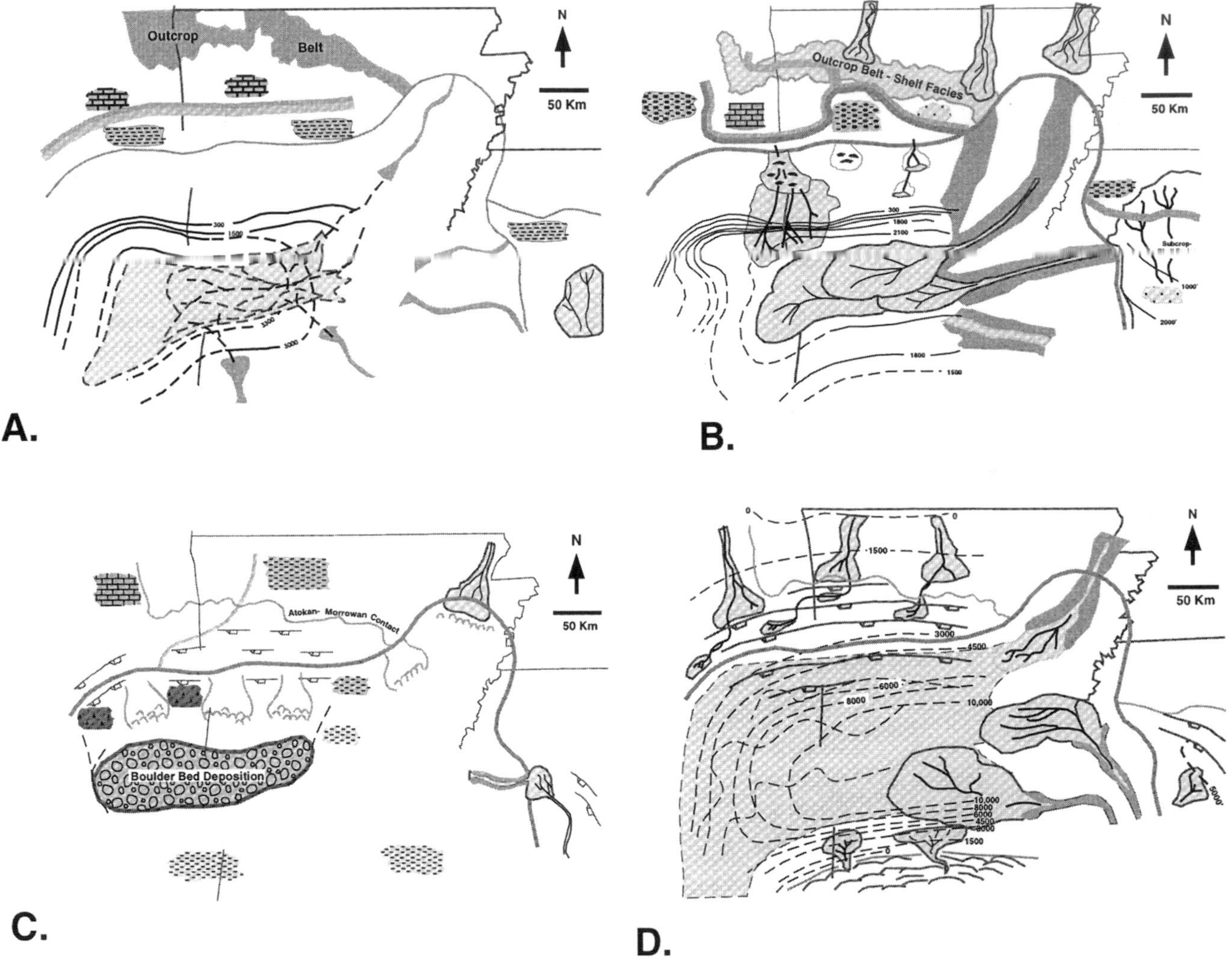

Figure 5—Paleogeography of the Ouachita Basin. (A) Stanley Formation (Chesterian Mississippian), with general isopachous map, contours in meters (developed in part from Morris et al., 1979; Houseknecht and Kacena, 1983; Johnson et al., 1988; Coleman, 1990). (B) Jackfork Formation (Morrowan Pennsylvanian), with general isopachous map, contours in meters (developed in part from McKee and Crosby, 1975; Morris et al., 1979; Houseknecht and Kacena, 1983; Roberts, 1995; Johnson et al. in part, 1988; Coleman, 1990; Pauli, 1994). (C) Johns Valley Formation (end of the Morrowan Pennsylvanian)(developed from Shideler, 1970; English, 1984; Johnson et al., 1988; Coleman, 1990). (D) Atoka Formation (Atokan Pennsylvanian), with general isopachous map, contours in meters; not all local sediment compartments shown (developed in part from McKee and Crosby, 1975; Joslin, 1980; Houseknecht and Kacena, 1983; Houseknecht, 1986; Johnson et al., 1988; Coleman, 1990).

Commonly overlying the Chickasaw Creek are a few thin sandstone beds similar to beds immediately beneath it. They are relatively dirty, containing a relatively high percentage of lithic, volcanic, and metamorphic rock fragments. Approximately 3 to 4 m above the Chickasaw Creek, slightly thicker, substantially cleaner sandstones (95% quartz grains) appear. The first of these beds is the lowest unit of the Jackfork Formation. For simplicity and uniformity, the top of the Chickasaw Creek is used as the base of the Jackfork Formation. Where the Chickasaw Creek is absent or covered in outcrop, the base of the Jackfork Formation is picked as the base of the lowest "clean" sandstone.

The Jackfork Formation is a complex basinal unit, with sedimentation primarily from the northern and eastern (or southeastern) shelves (Figure 5B). It is distinctly different from the underlying Stanley in that most sandstones are nearly 95% quartz (Figure 6) and substantially thicker on a bed-for-bed basis. Biostratigraphically, the Jackfork Formation is younger than the youngest Mississippian rocks and older than the oldest Pennsylvanian rocks exposed on northern shelf, and, therefore, does not have a stratigraphic equivalent along the northern basin margin. It best represents the basin equivalent of the hiatus represented by the Mississippian-Pennsylvanian unconformity, which extends throughout most of the interior eastern United States. The interpreted change to humid conditions, with resulting increased chemical and mechanical weathering, provides a possible explanation for the more mature nature of the Jackfork Formation (Figure 6). This change in paleoclimatic conditions roughly coincided with the global eustatic lowstand at the Namurian A-B (Mississippian–Pennsylvanian) boundary.

The Jackfork Formation shows minimal thickness changes across the basin, indicating that the basin floor had not been affected substantially by foreland basin thrusting inversion (Figure 5B). The majority of the Jackfork was deposited by muddy and sandy turbidity currents and debris flows that emanated from systems to the northeast and southeast of the present-day outcrop belt. Additional sediments were contributed from the southern and northern shelves (Danielson et al., 1988; Link and Roberts, 1986; Arbenz, 1989; Pauli, 1994). The Jackfork approaches 2100 m in original stratigraphic thickness and was accumulated at an average rate of 400 m/m.y.

Succeeding the Jackfork Formation is the lower Pennsylvanian Johns Valley Formation, a problematic unit of shale, thin turbidites, and boulder bed debrites, which contain clasts of Cambrian- to Pennsylvanian-age shelf, slope, and basinal(?) strata. The stratigraphic content of the Johns Valley boulders indicates a major structural reorganization of the Ouachita Basin occurred at the end of Jackfork deposition (Figure 5C). Preliminary evidence suggests that at this time the Ouachita Basin thrust and fold belt system took on a more strike-slip component, resulting in basement involvement and uplift and erosion of early syn- and preorogenic strata from the northern margin as well as the center of the basin. This basin inversion (or at least reorganization) set up the compartmentalization that controlled the Atoka deposition. Overall, the Johns Valley accumulated at an average rate of about 115 m/m.y.

Succeeding the Johns Valley is the Atoka Formation, a thick succession of lower and middle Pennsylvanian sandstone and shale that may have originally exceeded 9140 m in depositional thickness (Figure 5D). Atoka sedimentation shallowed upward from deepwater turbidites to fluvial deltaic strata before being capped by the shallow-water Desmoinesian Hartshorne Formation (Figures 2, 7A, 7B). The basinal Atoka consists of

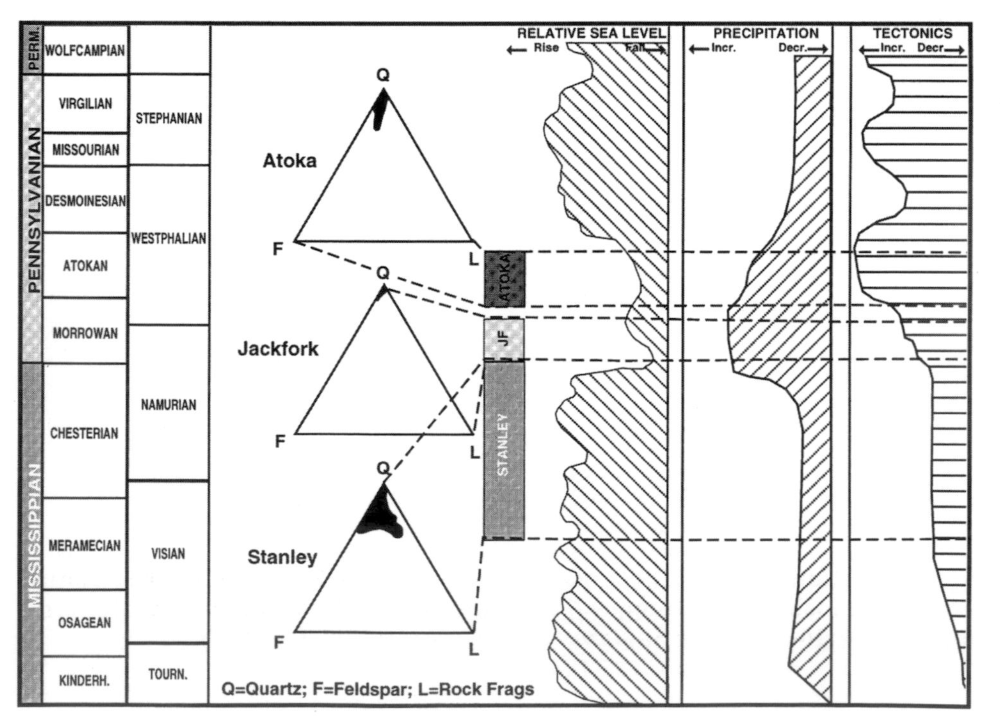

Figure 6—Sandstone compositional diagrams of Stanley, Jackfork, and Atoka sandstones and their relationship to eustasy, climate, and tectonics (developed from Sutherland and Manger, 1979; Ross and Ross, 1987; Cecil and Eble, 1989; Arbentz, 1989; Morris, 1989).

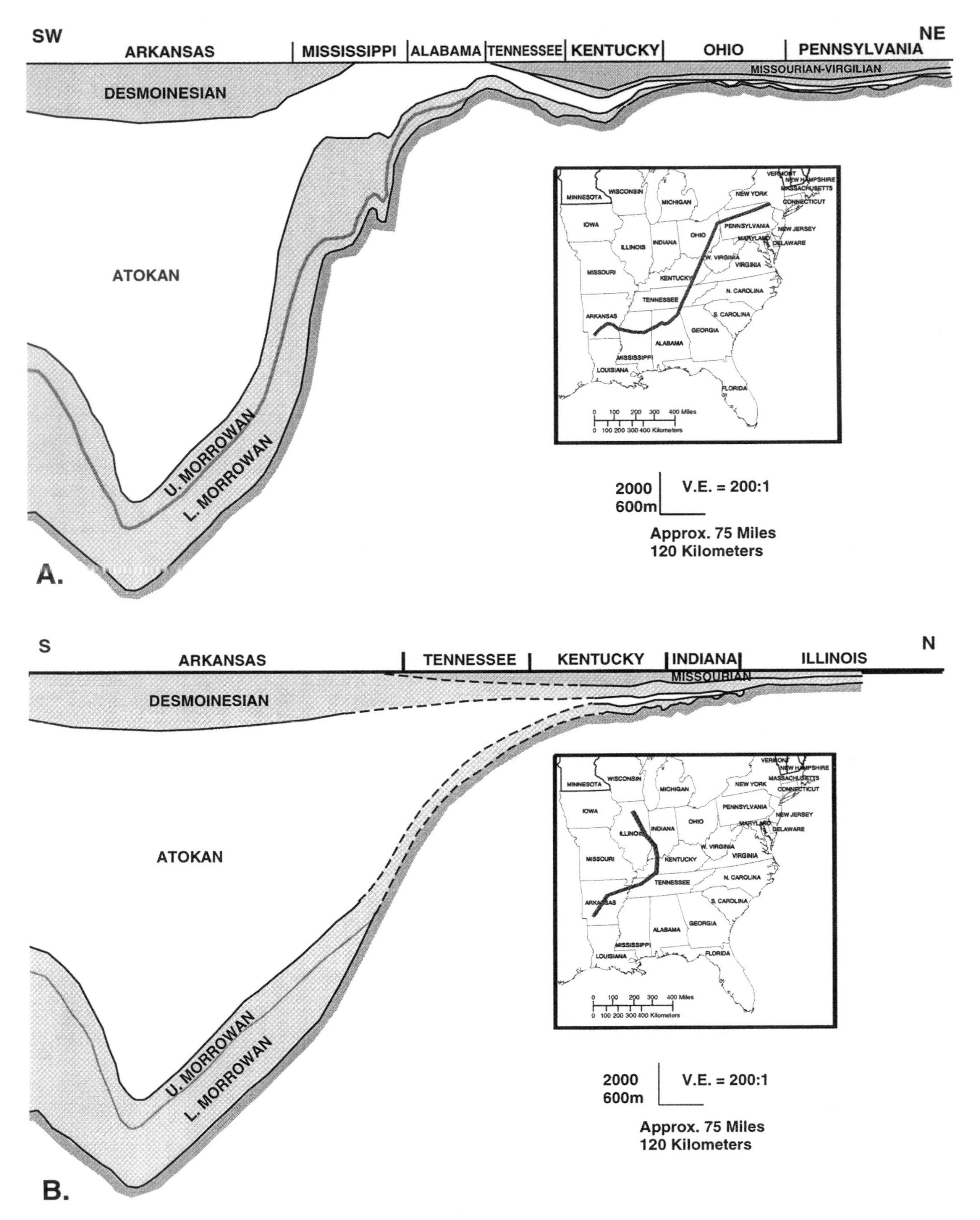

Figure 7—Restored stratigraphic cross sections of the Pennsylvanian System. A: Arkansas to Pennsylvania (developed from McKee and Crosby, 1975; Joslin, 1980; Sprague, 1985; Coleman, 1990); B: Arkansas to Illinois (developed from McKee and Crosby, 1975; Joslin, 1980; Sprague, 1985; Coleman, 1990).

tectonically compartmentalized, slope channel turbidite complexes and a main axial channel complex. Overall the Atoka accumulated at an average rate of approximately 1350 m/m.y, three to ten times the rate of the three lower Carboniferous deepwater units.

The Atoka is insufficiently understood from a regional perspective, although numerous wells produce from it and numerous reports have been written on it. It is generally divided into three parts: the upper, middle, and lower divisions, with a further stratigraphic subdivision of informal subsurface reservoir names. The Atoka is essentially Westphalian, ranging in age from late Morrowan at the base to Atokan Pennsylvanian throughout the majority of the formation. Within the turbidite units, most of the formation is devoid of biostratigraphically useful fossils. A clear correlation between Atokan rocks of the Ouachita Basin and similarly aged strata of the adjacent Arkoma Basin has not been clearly established. It appears, however, that the middle Atoka of the Arkoma Basin may equate to the lower Atoka of the Ouachita Mountains (Ouachita Basin) (Sprague, 1985).

CONCLUSIONS

The primary factors controlling deposition of the Ouachita basinal sediments are tectonics, eustasy, and climate. Coleman (1993a) (Table 1) compared the Ouachita Carboniferous deepwater sediment complexes and found that each of these factors varied in their degree of significance in controlling deposition.

The Stanley was deposited during a long period of high sea level, under arid conditions and quiet intrabasinal tectonics. Contemporaneous carbonate banks and minor siliciclastic marine shelf sands were deposited along the adjacent continental shelves. Occasional major storms and runoff, possibly reinforced by short-term, third-order sea level lowstands contributed to basinal sedimentation transport. The Jackfork was deposited during a major eustatic lowstand following and during a period of maximum precipitation and accompanying mechanical and chemical weathering. This lowstand shifted the accommodation space basinward off the Black Warrior and Ozark shelves into the Ouachita Basin (Figure 4). Third-order cycles of sea level fluctuations may have produced the four sequences of deepwater clastics within the Jackfork, which occur as mappable units in Oklahoma. Regional tectonic activity within the basin was probably minor, although it gradually increased throughout the Jackfork deposition. The Johns Valley shale was deposited as a major flooding event over the Jackfork during rising sea level, decreasing precipitation, and increasing tectonic activity in the form of foreland basin faulting. The Atoka Formation was deposited on the shelf, slope, and in the basin during rising sea level, decreasing (and seasonal) precipitation, and substantially increasing levels of tectonic activity, which resulted in rapid foreland basin subsidence and cratonic uplift.

Table 1. *Influence of primary factors controlling deposition.

	Eustatic Sea Level	Climate	Tectonics
Atoka	Rising	Transitional	High
Johns Valley	Medium	Wet	Increasing
Jackfork	Low	Wet	Increasing
Stanley	High	Arid	Low

*from Coleman (1993a).

ACKNOWLEDGMENTS

The author thanks BP Amoco for permission to publish this paper and for financial support for publication. Thanks are also extended to Arnold Bouma, Charles Stone, Alan Thomson, and Cooper Land for their helpful review.

REFERENCES CITED

Arbenz, J. K., 1989, The Ouachita system, *in* A. W. Bally, and A. R. Palmer, eds., The geology of North America—an overview: Geological Society of America, The Geology of North America, v. A, p. 372–396.

Bristol, H. M., and R. H. Howard, 1974, Sub-Pennsylvanian valleys in the Chesterian surface of the Illinois Basin and related Chesterian slump blocks, *in* G. Briggs, ed., Carboniferous of the southeastern United States: Geological Society of America Special Paper 148, p. 315–335.

Cecil, C. B., and C. Eble, 1989, Carboniferous geology of the eastern United States: American Geophysical Union Field Trip Guidebook T143, 154 p.

Chamberlain, C. K., 1978, Trace fossil ichnofacies of an American flysch, *in* C. K. Chamberlain and P. B. Basan, eds., Trace fossils and paleoecology of the Ouachita Geosyncline: SEPM Field Trip Guidebook, p. 23–37.

Cline, L. M., 1968, Comparison of main geologic features of Arkoma Basin and Ouachita Mountains, southeastern Oklahoma, *in* L. M. Cline, ed., A guidebook to the geology of the western Arkoma Basin and Ouachita Mountains, Oklahoma: Oklahoma City Geological Society Guidebook for AAPG-SEPM annual meeting, p. 63–74.

Coleman, J. L., Jr., 1990, Comparison of depositional elements of an ancient and a "modern" submarine fan complex: Early Pennsylvanian Jackfork and Late Pleistocene Mississippi fans (abs.): AAPG Bulletin, v. 74, p. 631.

Coleman, J. L., Jr., 1993a, Controls on variability of depositional style in Carboniferous submarine fan complexes of the Ouachita Basin of Oklahoma and Arkansas (abs.): AAPG 1993 Annual Convention Program with Abstracts, p. 86–87.

Coleman, J. L., Jr., 1993b, Geology of deep water sandstones in the Mississippian Stanley shale at Cossatot Falls, Arkansas: Transactions of Gulf Coast Association of Geological Societies, v. 43, p. 71–75.

Coleman, J. L., Jr., 1994, Factors controlling the deposition and character of submarine fan complexes: an illustration from the Carboniferous Ouachita Basin (abs.): Geological Society of America abstracts with programs, v. 26, no. 1. p. 4.

Coleman, J. L., Jr., and W. S. Hale-Erlich, 1994, Annotated bibliography of selected references for the Ouachita–Appalachian structural belt, as pertains to the Black Warrior Basin of Mississippi: Mississippi Geology, v. 15, p. 21–34.

Coleman, J. L., Jr., G. Van Swearingen, C. E., and Breckon, 1994, The Jackfork Formation of Arkansas: a test of the Walker-Mutti-Vail models for deep-sea fan deposition: Arkansas Geological Commission Guidebook 94-2, 56 p.

Danielson, S. E., P. K. Hankinson, K. D. Kitchings, and A. Thomson, 1988, Provenance of the Jackfork sandstone, Ouachita Mountains, Arkansas and eastern Oklahoma, *in* J. D. McFarland, III, ed., Contributions to the geology of Arkansas: Arkansas Geological Commission Miscellaneous Publication 18-C, p. 95–112.

Ekdale, A. A., and T. R. Mason, 1988, Characteristic trace-fossil associations in oxygen-poor sedimentary environments, Geology, v. 16, p. 720–723.

English, W. J., 1984, Coarse clastics of the Johns Valley Formation, south-central Arkansas: Unpublished Masters' thesis, Memphis State University, 108 p.

Gordon, M., Jr., 1970, Carboniferous ammonoid zones of the south-central and western United States: Compte rendu, 6e Congrès International Stratigraphie Géologie Carbonifère, v. 2, p. 817–826.

Gordon, M., Jr., and C. G. Stone, 1977, Correlation of the Carboniferous rocks of the Ouachita Trough with those of the adjacent foreland, *in* C. G. Stone, ed., Symposium on the geology of the Ouachita Mountains, v. 1, p. 70–91.

Groves, J. R., 1983, Calcareous foraminifers and algae from the type Morrow (Lower Pennsylvanian) region of northeastern Oklahoma and northwestern Arkansas: Oklahoma Geological Survey Bulletin 133, p. 65 p.

Hale-Erlich, W. S., and J. L. Coleman, Jr., 1993, Ouachita–Appalachian juncture: a Paleozoic transpressional zone in the southeastern USA: AAPG Bulletin, v. 77, p. 552–568.

Houseknecht, D. W., and J. A. Kacena, 1983, Tectonic and sedimentary evolution of the Arkoma foreland basin, *in* D. W. Houseknecht, ed., Tectonic-sedimentary evolution of the Arkoma Basin and guidebook to deltaic facies, Hartshorne sandstone: SEPM Midcontinent Section, v. 1, p. 3–52.

Houseknecht, D. W., 1986, Evolution from passive margin to foreland basin: the Atoka Formation of the Arkoma Basin, south-central USA: International Association of Sedimentologists Special Publication 8, p. 327–345.

Johnson, K. S., T. W. Amdsen, R. E. Denison, S. P. Dutton, A. G. Goldstein, B. Rascoe, Jr., P. K. Sutherland, and D. M. Thompson, 1988, Southern midcontinent region, *in* L. L. Sloss, ed., The geology of North America, volume D-2, Sedimentology cover—North American craton: U.S., Geological Society of America, Boulder, Colorado, p. 307–359.

Joslin, P. S., 1980, The stratigraphy and petrology of the Atoka Formation, west central Arkansas: Unpublished MS thesis, Northern Illinois University, 123 p.

Keller, G. R., 1989, Geophysical maps of the Ouachita Region, *in* R. D. Hatcher, Jr., W. A. Thomas, and G. W. Viele, eds., The geology of North America, volume F-2, The Appalachian–Ouachita Orogen in the United States: Geological Society of America, Boulder, Colorado, plate 10.

Lane, H. R., and J. J. Straka, II, 1974, Late Mississippian and Early Pennsylvanian conodonts, Arkansas and Oklahoma: Geological Society of America Special Paper 152, 144 p.

Link, M. H., and M. T. Roberts, 1986, Pennsylvanian paleogeography for the Ozarks, Arkoma, and Ouachita Basins in east-central Arkansas, *in* C. G. Stone and B. R. Haley, eds., Sedimentary and igneous rocks of the Ouachita Mountains of Arkansas: a guidebook with contributed papers part 2: Arkansas Geological Commission Guidebook 86-2, p. 37–60.

McKee, E. D., and E. J. Crosby, coordinators, 1975, Paleotectonic investigations of the Pennsylvanian System in the United States: U.S. Geological Survey Professional Paper 853, part 1, 349 p.; part 2, 192 p.; part 3, 17 plates.

Miller, M. A., L. E. Eames, and D. R. Prezbindowski, 1989, Upper Mississippian and lower Pennsylvanian lithofacies and palynology from northeastern Oklahoma: a field excursion: American Association of Stratigraphic Palynologists Foundation Guidebook, 64 p.

Morris, R. C., K. E. Proctor, and M. R. Koch, 1979, Petrology and diagenesis of deep-water sandstones, Ouachita Mountains, Arkansas and Oklahoma: SEPM Special Publication 26, p. 263–279.

Morris, R. C., 1989, Stratigraphy and sedimentary history of post-Arkansas Novaculite Carboniferous rocks of the Ouachita Mountains, *in* R. D. Hatcher, Jr., W. A. Thomas, and G. W. Viele, eds., The geology of North America, volume F-2, The Appalachian–Ouachita Orogen in the United States: Geological Society of America, Boulder, Colorado, p. 591–602.

Nicholas, R. L., and D. W. Waddell, 1989, The Ouachita system in the subsurface of Texas, Arkansas, and Louisiana, *in* R. D. Hatcher, Jr., W. A. Thomas, and G. W. Viele, eds., The Geology of North America, volume F-2, The Appalachian–Ouachita Orogen in the United States: Geological Society of America, Boulder, Colorado, p. 661–672.

Niem, A. R., 1977, Mississippian pyroclastic flow and ash-fall deposits in the deep-marine Ouachita flysch basin, Oklahoma and Arkansas: Geological Society of America Bulletin, v. 88, p. 49–61.

Parrish, J. T., 1982, Upwelling and petroleum source beds, with reference to Paleozoic: AAPG Bulletin, v. 66, p. 750–774.

Pauli, D., 1994, Friable submarine channel sandstones in the Jackfork Group, Lynn Mountain Syncline, Pushmataha and Le Flore Counties, Oklahoma, *in* N. H. Suneson and L. A. Hemish, eds., Geology and resources of the eastern Ouachita Mountains frontal belt and southeastern Arkoma Basin, Oklahoma: Oklahoma Geological Survey Guidebook 29, p. 179–202.

Roberts, M. T., 1982, Chart showing sedimentation rates of the Paleozoic rocks in the Ouachita Mountains, Arkansas, *in* C. G. Stone and J. D. McFarland, III, eds., Field guide to the Paleozoic rocks of the Ouachita Mountain and Arkansas Valley Provinces, Arkansas: Arkansas Geological Commission Guidebook 81-1, p. 4.

Roberts, M. T., 1994, Geologic relations along a regional cross section from Spavinaw to Broken Bow, eastern Oklahoma, *in* N. H. Suneson and L. A. Hemish, eds., Geology and resources of the eastern Ouachita Mountains frontal belt and southeastern Arkoma Basin, Oklahoma: Oklahoma Geological Survey Guidebook 29, p. 137–160.

Ross, C. A., and J. R. P. Ross, 1987, Late Paleozoic sea levels and depositional sequences: Cushman Foundation for Foraminiferal Research Special Publication 24, p. 137–149.

Scotese, C. R., and W. S. McKerrow, 1990, Revised world maps and introduction, *in* W. S. McKerrow and C. R. Scotese, eds., Paleozoic palaeogeography and biogeography: Geological Society Memoir 12, p. 1–21.

Shideler, G. L., 1970, Provenance of Johns Valley boulders in late Paleozoic Ouachita facies, southeastern Oklahoma and southwestern Arkansas: AAPG Bulletin, v. 54, p. 789–806.

Sprague, A. R. G., 1985, Depositional environment and petrology of the lower member of the Pennsylvanian Atoka Formation, Ouachita Mountains, Arkansas and Oklahoma: Unpublished Ph.D. dissertation, University of Texas (Dallas), 323 p.

Stone, C. G., and W. V. Bush, 1982, Guidebook to the geology of the eastern Ouachita Mountains Arkansas: Arkansas Geological Commission Guidebook 82-2, 24 p.

Sutherland, P. K., and W. L. Manger, eds., 1977, Upper Chesterian–Morrowan stratigraphy and the Mississippian–Pennsylvanian boundary in northeastern Oklahoma and northwestern Arkansas, Oklahoma Geological Survey Guidebook 18, 183 p.

Sutherland, P. K., and W. L. Manger, eds., 1979, Mississippian–Pennsylvanian shelf-to-basin transition Ozark and Ouachita regions, Oklahoma and Arkansas: Oklahoma Geological Survey Guidebook 19, 81 p.

Thomas, W. A., 1989, The Appalachian–Ouachitian orogen beneath the Gulf Coastal Plain between the outcrops in the Appalachian and Ouachita Mountains, *in* R. D. Hatcher, Jr., W. A. Thomas, and G. W. Viele, eds., The geology of North America, volume F-2, The Appalachian–Ouachita Orogen in the United States: Geological Society of America, Boulder, Colorado, p. 537–553.

Trewartha, G. T., A. H. Robinson, and E. H. Hammond, 1967, Physical elements of geography (5th edition): New York, McGraw-Hill, 525 p.

Vanarsdale, R. B., and E. S. Schweig, III, 1990, Subsurface structure of the eastern Arkoma Basin: AAPG Bulletin, v. 74, p. 1030–1037.

Witzke, B. J., 1990, Palaeoclimatic constraints for Palaeozoic palaeolatitudes of Laurentia and Euramerica, *in* W. S. McKerrow and C. R. Scotese, eds., Palaeozoic palaeogeography and biogeography: Geological Society Memoir 12, p. 57–73.

Liu, J. Y., and W. R. Bryant, 2000, Sea floor morphology and scientific paths of the northern Gulf of Mexico deepwater, *in* A. H. Bouma and C. G. Stone, eds., Fine-grained turbidite systems, AAPG Memoir 72/ SEPM Special Publication No. 68, p. 33–46.

Chapter 4

Sea Floor Morphology and Sediment Paths of the Northern Gulf of Mexico Deepwater[1]

Jia Y. Liu
William R. Bryant
Department of Oceanography, Texas A&M University
College Station, Texas, U.S.A.

ABSTRACT

The recently compiled multibeam and digitized seismic data detail the complex bathymetry of the continental slope of the northern Gulf of Mexico slope and deepwater areas. Detailed bathymetry data, together with a watershed basin analysis model, generate landlike drainage paths in the submarine environment. Four drainage systems were identified and coincide with the major sediment sources in the west (Rio Grande River system), the northwest (Brazos and Colorado river systems), and the north (Mississippi River system). The carbonate-dominated platforms in the eastern and southern Gulf of Mexico show few drainage paths.

On a regional scale, these drainage paths were the primary conduits for density currents during periods of low sea stand, which may help our understanding of the distribution of turbidite-derived sediments in the Gulf of Mexico.

INTRODUCTION

Although the Gulf of Mexico is an Atlantic type passive continental margin, it displays one of the most complicated morphologies on Earth (Martin, 1978). This is especially true in the northern Gulf of Mexico, where folded, uplifted, depressed, fractured, and faulted features are common. The rugged topography of the slope is the interplay of shelf-edge progradation, sediment-loading–induced subsidence, shale and salt diapirism with withdrawal basins, and slope instability induced mass movement (Bryant et al., 1995).

Since the 1950s, studies on the Gulf of Mexico continental slope mainly emphasized large-scale sediment variations in the northern Gulf, salt tectonics, and seismic stratigraphy (Fisk et al., 1954; Moore and Scruton, 1957; Ewing and John, 1966; Lehner, 1969; Davis, 1972; Woodbury et al., 1973; Bouma, 1981; Salvador, 1991; Galloway et al., 1991; Simmons, 1992; Peel et al., 1995). Prior discussions of sediment paths and subaerial erosion on the slope are rare (Martin and Bouma, 1982; Roberts et al., 1986; van den Bold et al., 1987). This is mainly due to our lack of understanding the morphology of the seafloor, and because the available bathymetric charts, published by Bryant et al. (1990) and Bouma and Bryant (1994), were based on coarse-spaced (8–14 km) seismic data.

In 1997, the National Geophysical Data Center (NGDC) released full-resolution multibeam data, and Liu was able to map the bathymetry of the northwestern of the Gulf of Mexico deep water area in more detail than was previously possible (Liu, 1997). Instead of showing the bathymetry in contours, Liu's map used simulated sun-shading and solid surface to

[1]Selected color illustrations are provided on CD-ROM.

display the complexity of the offshore Texas–Louisiana continental slope. In 1999, Liu and Bryant expanded the 1997 bathymetry coverage by adding more multibeam data and digitized seismic information that covers the slope and deep water of northern Gulf of Mexico (Figure 1; Liu and Bryant, 1999). The water depths range from 150 m at the shelf-edge to about 4,000 m in the abyssal plain. In the areas with multibeam data coverage, some areas as shallow as 24 m are also mapped. This bathymetry covers a region of more than 675,000 km^2 and extends from the West Florida Terrace in the east, East Mexico Slope in the west, Texas–Louisiana Slope in the north, to the Campeche Terrace in the south. The objective of this paper is to detail the morphology of the northern Gulf of Mexico deepwater areas and to run a watershed model (GRASS, 1993) to predict the potential sediment paths. The results show a complexity that demonstrates that widely distributed source points and a large number of sediment paths can be active during a relative short geological time period. It also demonstrates that detailed dating may be needed to separate the individual transport/deposition periods.

GEOLOGICAL SETTING AND SEDIMENT DISTRIBUTION

The Gulf of Mexico rift basin was formed in the Triassic due to crustal extension (Martin, 1978). During the middle Jurassic, red beds and more than 3 km of salt were deposited (Coleman et al., 1986). A massive carbonate reef was formed along the shelf margin during the middle Cretaceous (Martin, 1978). In the early Tertiary, the Laramide orogeny uplifted the continental interior and a large amount of terrigenous clastics began to flux into the basin from the north and the west. The sediments deposited in the northern Gulf of Mexico are estimated to exceed 15 km (Murray, 1952; Hardin, 1962; Martin and Bouma, 1978; Bouma et al., 1978). In the southern and eastern part of the Gulf, carbonate platforms have dominated deposition since the Cretaceous (Coleman et al., 1986).

The terrigenous sediments in the northern and western part of the Gulf of Mexico were mainly deposited during sea level lowstands. During those periods, coarse-grained sediments were transported to the outer shelf and upper slope, causing shelf-edge progradation, subaerial erosion, slope instability, and sediment gravity flows (Beard et al., 1982). During periods of sea level highstand, fine-grained sediments formed a thin hemipelagic and pelagic layer in the deepwater areas (Bryant et al., 1995). According to Hardin (1962), Shinn (1971), Woodbury et al. (1973), and Martin (1978), the main sediment depocenter shifted from the Rio Grande embayment in the Paleocene, to south Texas in the Eocene, to south Louisiana in the Miocene, to the mouth of the present-day Mississippi River in the Pliocene, to 160 km south of the Texas–Louisiana border in the Pleistocene. Although the location of a depocenter is generally in response to the sediment supply, and represents where the shelf edge progrades the fastest, small depocenters may also occur within local salt-withdrawal basins (Simmons, 1992).

In spite of the changes of sediment sources, shifting of depocenters, transgression/regression cycles, and basin subsidence in the northern Gulf, the continental shelf has moved nearly 400 km seaward since the early Tertiary.

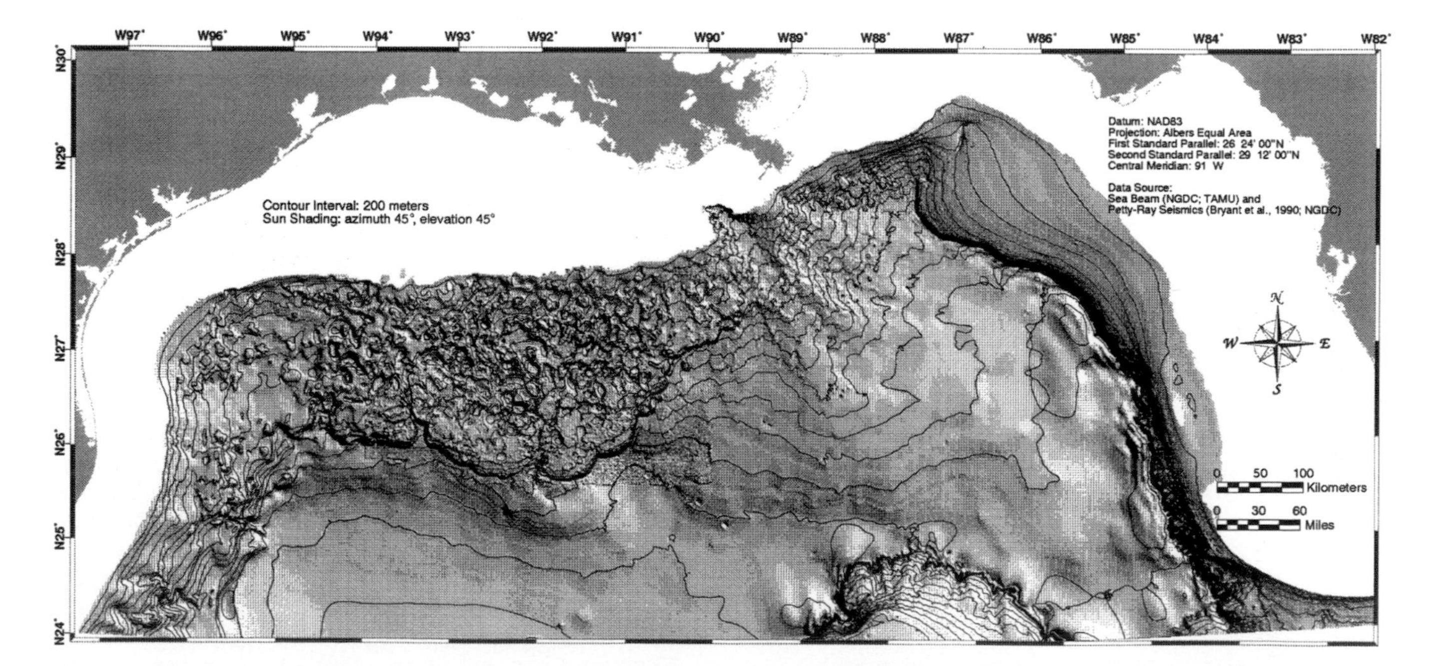

Figure 1—Sea-floor relief of northern Gulf of Mexico deepwater (modified from Liu and Bryant, 1999). The data were based on multibeam and digitized seismic data. The water depths range from 150 m at the shelf edge to about 4,000 m on the abyssal plain. The sun-shaded relief reveals the complexity of the sea floor, especially offshore Texas–Louisiana.

This occurred at an average rate of about 5–6 km/m.y., however, the progradation rate was less than 1 km/m.y. in the eastern Gulf (Hardin, 1962; Martin, 1978; Coleman et al., 1986). During the Pliocene and Pleistocene, the shelf edge prograded 80 km south (Woodbury et al., 1973). During the Pleistocene, up to 3,600 m of sediments were deposited on the upper slope offshore Texas–Louisiana, and up to 3,000 m of sediments were deposited in the Mississippi Fan (Lehner, 1969; Coleman et al., 1986). The Holocene sea level rise occurred about 15,000 years ago, and the average sediment thickness deposited since that time is approximately 7 m in the Louisiana upper continental slope (van den Bold et al., 1987). The sedimentation rate on the continental slope during the Holocene was estimated to be about 20–30 cm/1000 yr (Roberts et al., 1986; Beard, 1973).

To find the pathways of Pleistocene turbidity currents and the final depositional sites of coarse-grained sediments (sands in particular), one can examine present-day bathymetry of the continental slope and gain a considerable amount of insight into where sands may be present in the deepwater portions of the Gulf of Mexico. In some areas such as the Mississippi Fan, the turbidity current pathways are obvious, the major pathway being the Mississippi Canyon. In other areas of the Gulf of Mexico, turbidity current pathways and thus the likely location of coarse grained sediments may not be so obvious. The continental slope off Texas and Louisiana is the most complex passive continental slope in the world. The halokenesis of allochthaneous salt, salt tectonics, has created geological structures in the time and size frame of centimeters per year. Submarine canyons with relief of 800 m have been infilled with salt over a period of 8000 to 15,000 years, during the early and middle Holocene. The rates of emergence of salt in certain areas of the Gulf of Mexico equal those of sea-floor spreading and arc subduction along the active margins of the world. This rapid movement of salt has partially filled some of the major Pleistocene turbidity current pathways, the best example being Bryant Canyon (Lee, 1990). Major portions of Bryant Canyon have been totally filled with salt, present-day salt being within meters of the sea floor. Intraslope interlobal and intraslope supralobal basins, both of which have been formed by salt migration, are also being deformed by present-day salt migration are not only local depocenters but were major portions of former turbidity current pathways. The ability to trace Pleistocene turbidity current pathways on the northwestern continental slope of the northwestern Gulf of Mexico has been greatly reduced by the effects of halokenesis. Nevertheless a study of the bathymetry of the northern Gulf of Mexico is the most simple and useful means of determining such pathways.

SEA FLOOR MORPHOLOGY

The early Cretaceous carbonates that comprise the Florida Terrace and the Campeche Terrace, with dips less than 2° and 4°, respectively, are relatively flat. On the seaward side they are bounded by the Florida Escarpment and the Campeche Escarpment (Figure 2). These two escarpments represent a depth difference of up to 2300 m, covering a 40-km-wide area and display the steepest slope gradients (up to 40°) of the region. In the southern part of the Florida Escarpment and along the Campeche Escarpment one finds a series of gullies and small canyons, spaced about 5 km apart, cut through the slope. At 27.3°N and 25.7°N, two 4000-km^2-sized, dual-folded zones lie parallel to the Florida Escarpment. The Vernon Basin is located between these two folded zones.

The northern end of the Florida Escarpment is the NNE–SSW-aligned De Soto Canyon, which separates the carbonate-dominated Florida platform from the terrigeneous-dominated environment in the west (Figure 2). West of the De Soto Canyon are many NNW–SSE canyons that may have formed during the late Wisconsin sea level lowstand. These canyons include the Dorsey and Sounder canyons. Except for a canyon west of the De Soto Canyon, which cuts through the shelf break, all other canyons seem to originate at a water depth of 400 m or deeper and extend more than 60 km basinward. To the west of these canyons are a series of pancake-shaped, 5- to 15-km-wide salt domes, which become more scattered and smaller in size in a basinward direction.

The U-shaped Mississippi Canyon is a late Pleistocene erosional feature that cuts through the shelf break (Coleman et al., 1986; Figure 2). The canyon has a width of about 30 km. The Holocene Mississippi Fan appears approximately 120 km down the canyon mouth. The canyon/fan extends for more than 500 km past the shelf edge into the abyssal plain. The Mississippi Fan occupies an area of about 145,000 km^2. It extents to the Florida Escarpment and the Vernon Basin in the east, terminates at about 50 km north of the Campeche Escarpment in the south, and is just west of the Farnella Canyon (Figure 3).

The Texas–Louisiana Slope is located west of the Mississippi Canyon (Figure 3). It occupies an area of about 120,000 km^2 and contains the widest slope (≤230 km) and the most rugged morphology in the northern Gulf of Mexico. There are more than 100 domes and basins that have been named and approved by the U.S. Board of Geographic Names (BGN) (Bouma and Bryant, 1994). Although the average slope gradient is less than 1°, a local slope gradient can exceed 20°. These domes and basins range from 5 to 30 km in diameter. The domes are prominent on the upper slope, they increase in size to become salt massifs that surround basins on the middle slope and lie under basins on the lower slope (Simmons, 1992). The basinward termination of the slope is the Sigsbee Escarpment, which is a surficial expression of the basinward salt front (Moore et al., 1978). This salt front displays an extruded tongue feature with an elevation of about 600 m above the continental rise and comprises gradients ranging from 10 to 20°. Several canyons break the semicontinuous Sigsbee Escarpment: Green Canyon, Farnella Canyon, Cortez Canyon, Bryant Canyon, Keathley Canyon, and Alaminos Canyon (Figures 3, 4). There are fan deposits between the Green and Farnella canyons that merge with the Mississippi Fan (Figures 2, 3). Some channels are visible on top of these coalesced fans that radiate outward in a basinward direction. Southward of

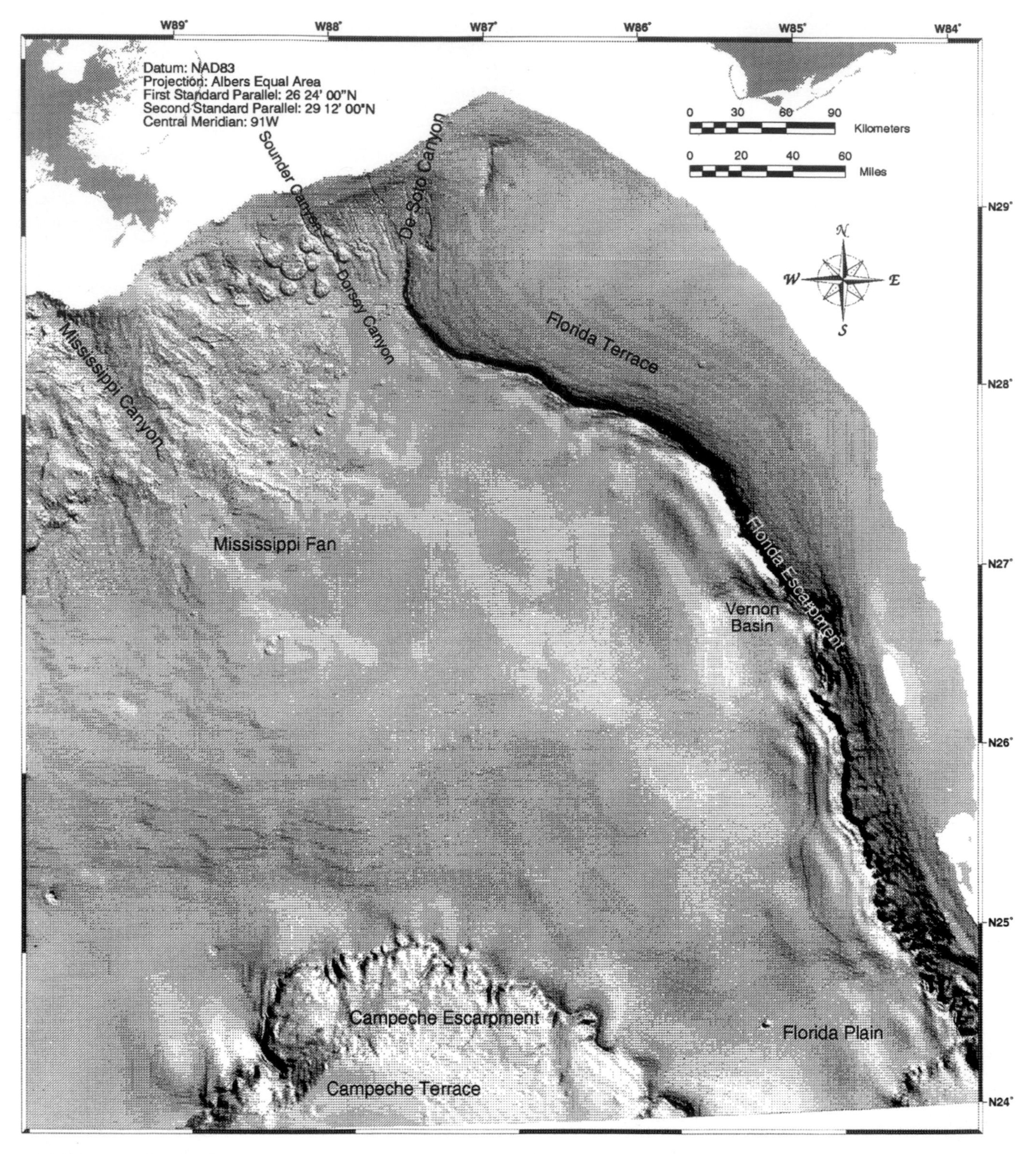

Figure 2—Sun-shaded sea-floor relief of the northeastern Gulf of Mexico deepwater. The simulated sun has an azimuth of 45° and an elevation of 45°. The Florida Terrace/Escarpment and the Campeche Terrace/Escarpment are mainly composed of Cretaceous carbonates. The area west of the De Soto Canyon is dominated by the terrigenous sediment. The late Pleistocene Mississippi Canyon and the Holocene Mississippi Fan extend more than 500 km from the shelfedge to about 50 km north of the Campeche Escarpment.

Bryant Canyon lies the Bryant Fan, which is about 25,000 km^2 in size (Figure 3). This fan reaches to the Cortez Canyon in the east, Keathley Canyon in the west, and extends about 170 km south from the canyon mouth. The lack of fan deposits in the mouth of Keathley Canyon suggests that this canyon is structurally controlled (Lee, 1990).

In the northwest, the Rio Grande Slope separates

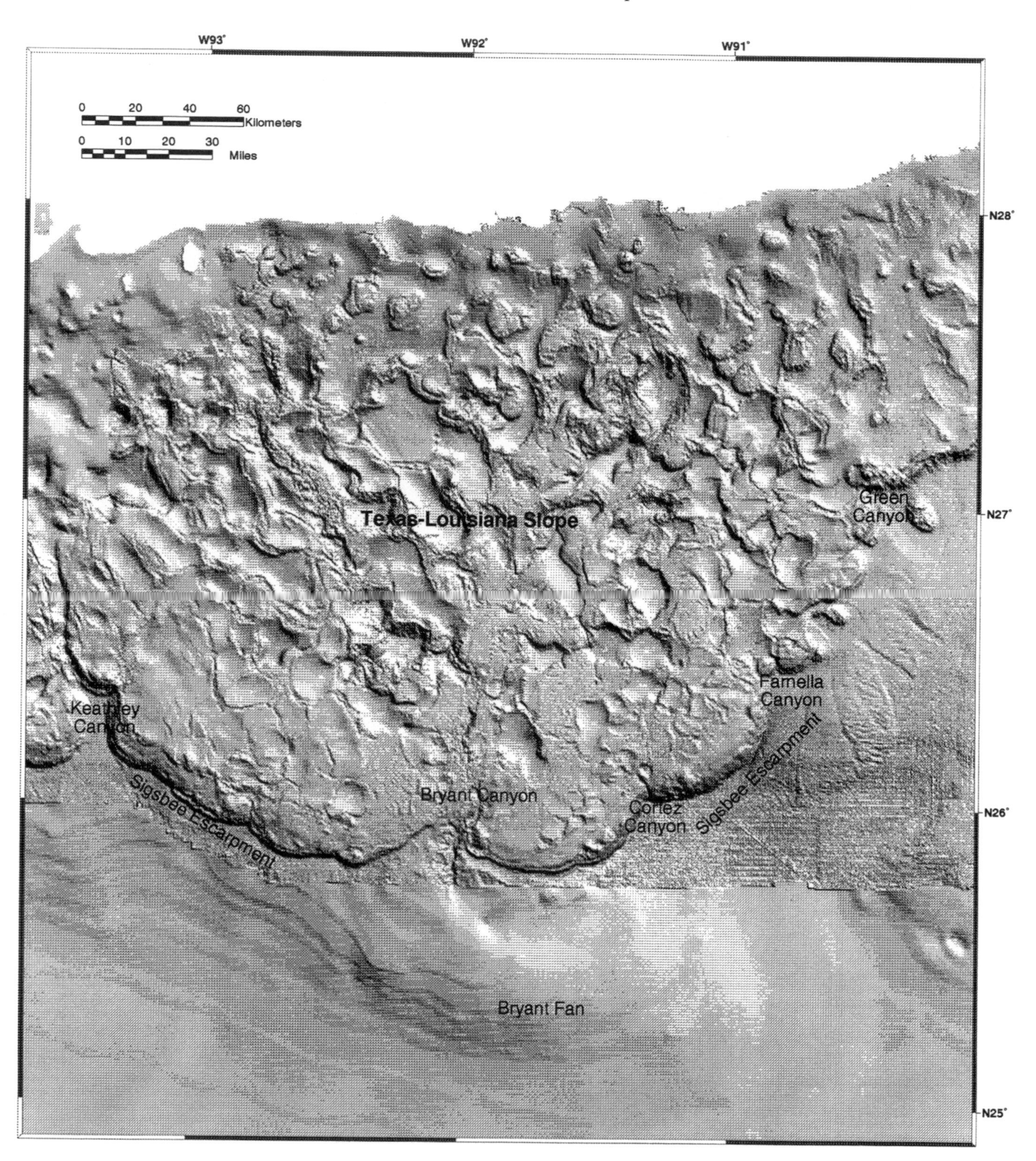

Figure 3—Sun-shaded sea-floor relief of the north-central Gulf of Mexico deepwater. The Texas–Louisiana slope has the most rugged sea floor in the northern Gulf of Mexico. More than 90 features have been officially named in this area. The bathymetry becomes smoother south of N26° and is due to lack of multibeam data. The projection and the direction of the simulated sun are the same as Figure 2.

the Texas–Louisiana Slope from the East Mexico Slope (Figure 4). Similar to offshore Mississippi and Alabama, the Rio Grande Slope displays more than 15 pancake-shaped domes. The canyons cut through the shelf break and are much broader and tend to be located between the domes or banks. In between Price Spur and Calhoun Dome, there are many tightly spaced canyons that form a 22-km-wide

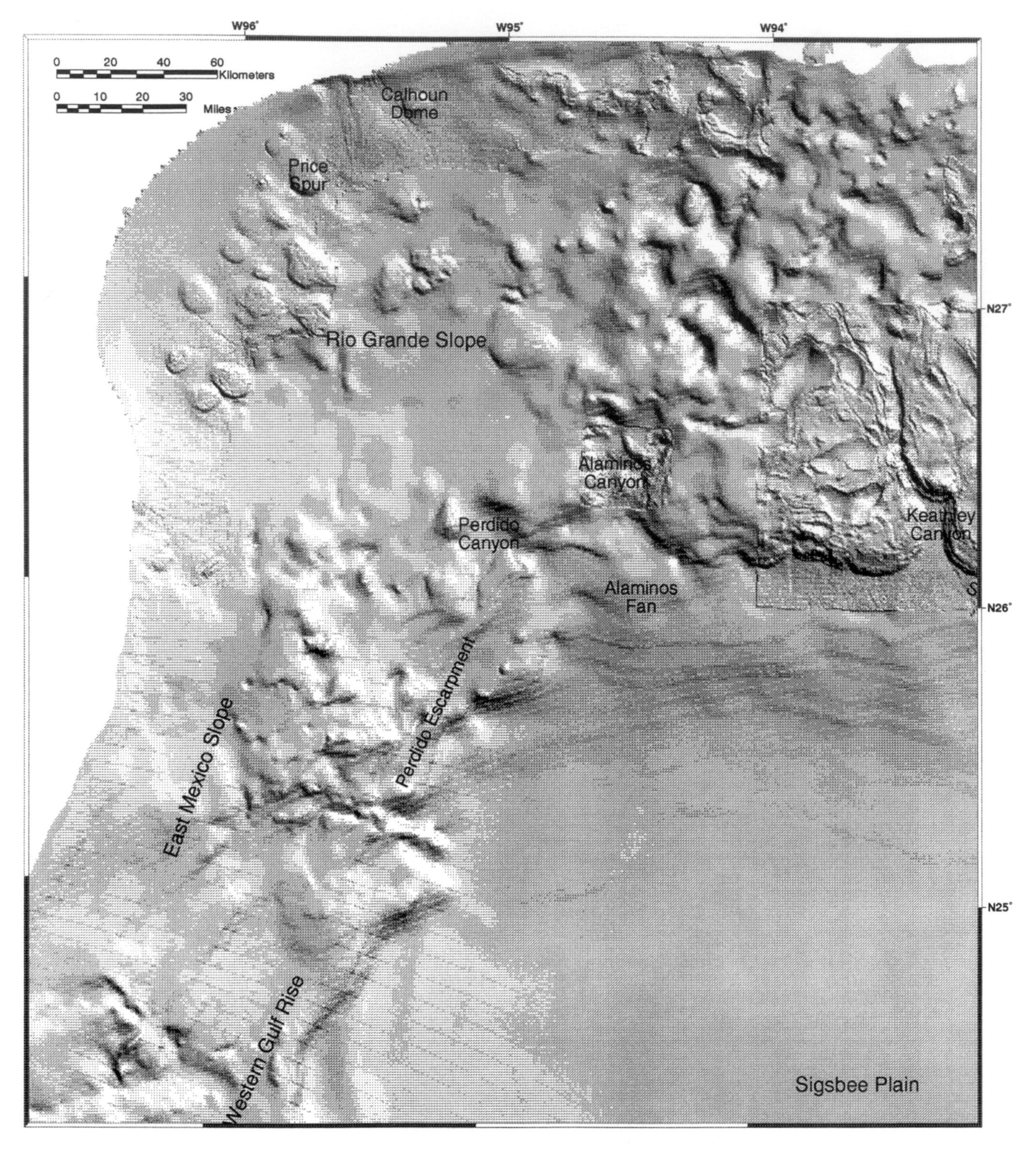

Figure 4—Sun-shaded sea-floor relief of the northwestern Gulf of Mexico deepwater. The multibeam data cover Keathley Canyon, northwest of the Rio Grande Slope, and the Alaminos Canyon; the rest of the areas are based on digitized seismic data. Note the pancake-shaped domes in the area northwest of the Rio Grande Slope and the much smoother bathymetry on the East Mexico Slope. The projection and the direction of the simulated sun are the same as Figure 2.

valley and a 7-km-wide bank. There are only a few canyons, hundreds of meters wide, cut through the Rio Grande Slope. The number of canyons may be an artifact due to the lack of detailed multibeam data in the west. At the seaward end of the slope, the Perdido and Alaminos canyons appear. The fans downdip from these two canyons merge basinward and are much smaller in size than the Bryant Fan (about

5,000 km²; Figures 3, 4).

To the south of the Rio Grande Slope is the East Mexico Slope (Figure 4). The northern portion of this slope is mainly controlled by diapiric activity. The southern side of the East Mexico Slope represents the northern end of the Mexican Ridges (Bryant, 1986). At the basinward limit of the Texas–Louisiana, Rio Grande, and East Mexico Slopes is the Western Gulf Rise, which is about 40 km wide. In between the Western Gulf Rise, Mississippi Fan, and the Campeche Escarpment is the Sigsbee Plain, which has a water depth of about 3700 m (Figure 4). In between the Campeche Escarpment and the Florida Escarpment is the Florida Plain, with a water depth of about 3400 m (Figure 2). Except for the Sigsbee Knolls, which are up to 250 m above the Sigsbee Plain, these two abyssal plains have an average slope gradient of less than 0.5° and are the flattest mapped regions in the deepest part of the Gulf of Mexico.

WATERSHED MODEL

This review of the complicated bathymetry of the northern Gulf of Mexico is needed to understand the reason to use a watershed model to detect bathymetric details required to map out sediment pathways.

Two of the frequently asked questions related to the deepwater environment are where is all the sand, and how did it get there. Ground-truth samples (Davies, 1972) and seismic data (Weimer, 1989) have been used to answer these questions in the Gulf of Mexico. However, they either had limited coverage or were widely spaced seismic lines and have not being helpful to delineate the drainage paths for the entire Gulf. In this study, we used one of the common hydrologic models in the land elevation data, called the watershed model, to estimate stream flow runoffs (GRASS, 1993). Instead of representing rivers on land, these drainage paths represent potential sediment transport conduits that were active during sea level lowstands or acted as trails constructed by local mass movement. Unlike the land data, our controlling parameters are limited to the bathymetry data and the size of the drainage basins. Two drainage basin sizes were tested: 2000 m and 300 m across. Thc 2000-m basin gave higher-order drainage and overall sediment paths. The 300-m basin gave lower order drainage with shorter streams, which represent local sediment paths.

DRAINAGE PATTERNS AND SEDIMENT PATHS

For the higher-order drainage basin model of the Gulf of Mexico, there are four drainage systems identified: western, central, northeastern, and southeastern continental slope areas (Figure 5). These drainages merge basinward and can be up to 500 km long. The western system includes E–W-oriented patterns on the East Mexico Slope and sinuous NW–SE-oriented patterns on the Perdido Slope (Figure 6). In the East Mexico Slope area, the paths coincide with the gullies and canyons. On the Perdido Slope, the Perdido and Alaminos canyons are the major tributaries at the base of the slope. The paths tend to move around the salt domes on the upper slope of the Perdido Canyon. On the upper slope of the Alaminos Canyon, the paths tend to go through

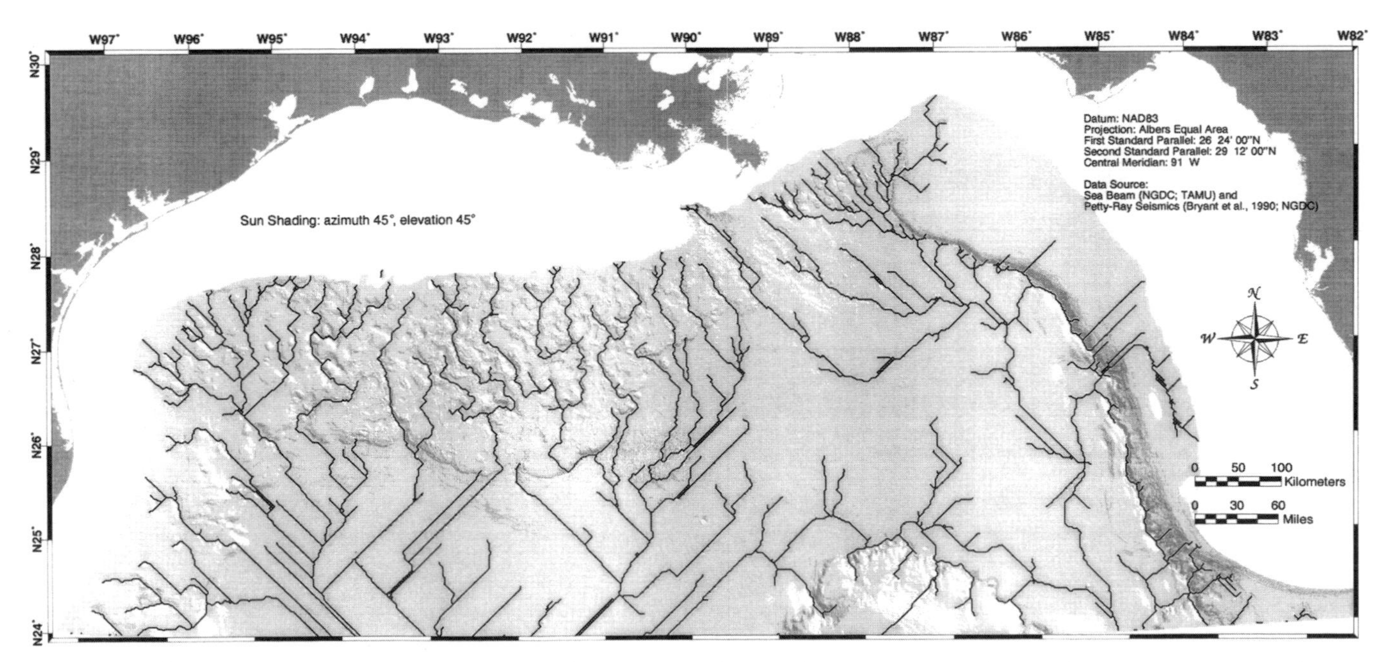

Figure 5—High-order drainage paths (black lines) draped over the sea-floor relief of the northern Gulf of Mexico deepwater. Four drainage systems were identified: western, central, northeastern, and southeastern. The drainage paths were generated from GRASS (1993).

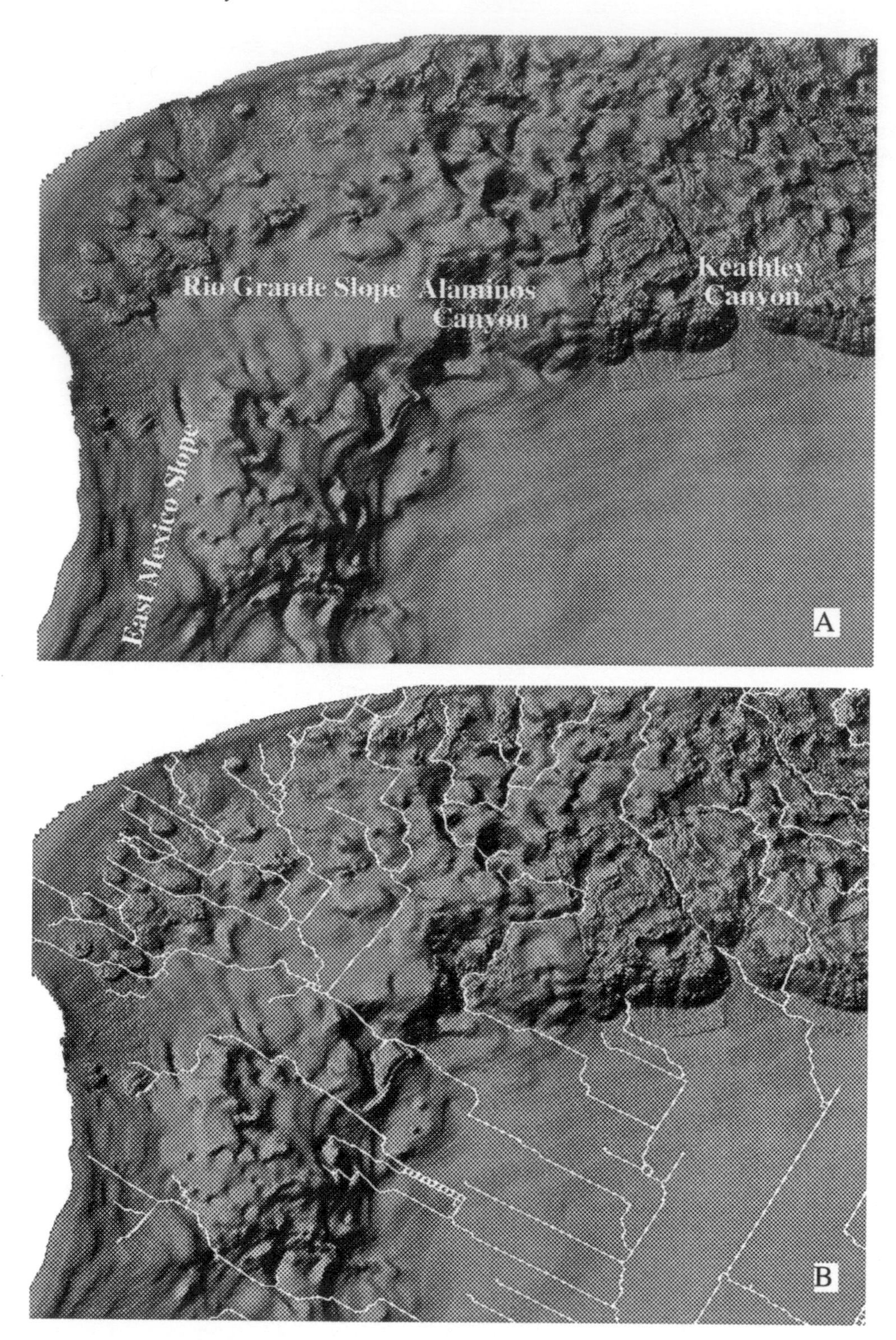

Figure 6—Perspective view of sea-floor relief and drainage paths in the northwestern Gulf of Mexico. (A) Shaded sea-floor relief, and (B) high-order drainage paths (white lines) draped over the sea-floor. The vertical exaggeration is 10X.

the intraslope basins.

The central area is located offshore Texas and Louisiana (Figures 6, 7). The sinuous patterns are mainly in N–S directions. The path that crosses the Keathley Canyon is straighter and follows a major fault, which formed the canyon (Lee, 1990). The drainage paths to the east of the Keathley Canyon are highly irregular and are located between salt domes on the upper slope, and traverse the intraslope basins on the lower slope. Except for the reentrant north of the Farnella Canyon, where the path cuts through the slope and the canyon, other canyons, such as Bryant, Farnella, and Green, the paths originate at the head of the canyons.

In the northeastern area, many paths align in a NW–SE direction on the Louisiana–Mississippi Slope

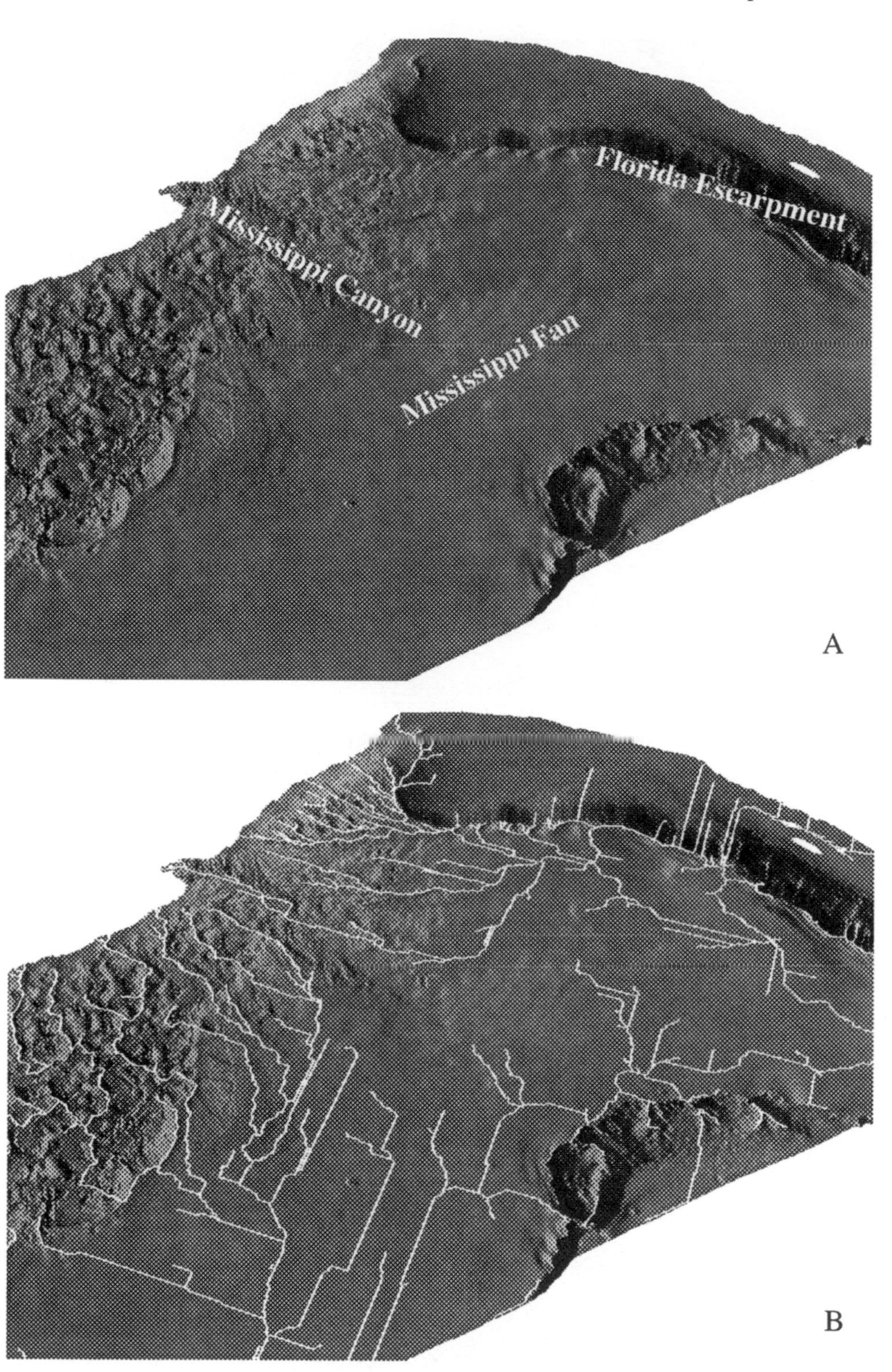

Figure 7—Perspective view of sea-floor relief and drainage paths in the Mississippi Canyon/Fan area. (A) Shaded sea-floor relief, and (B) high-order drainage paths (white lines) draped over the sea floor. The vertical exaggeration is 10X.

and few paths align in an E–W direction on the West Florida Slope (Figures 7, 8). In the Mississippi Canyon, the path goes down the canyon and swerves to the east at a water depth of about 2800 m. To the east of the Mississippi Canyon, paths tend to go around the salt domes and the base of the Florida Escarpment, and merge at a water depth of about 3100 m. This merged path then goes in between the Mississippi Fan and topographic highs west of the Florida Escarpment.

In the southeastern part, either due to the carbonate platforms or to the less detailed topographical data available, the paths are less sinuous (Figure 7). Generally, they come down from the Florida Escarpment, the Campeche Escarpment, and the Mississippi Fan. The paths go around the escarpments, fan, and then move south. There are also a few paths that come down the

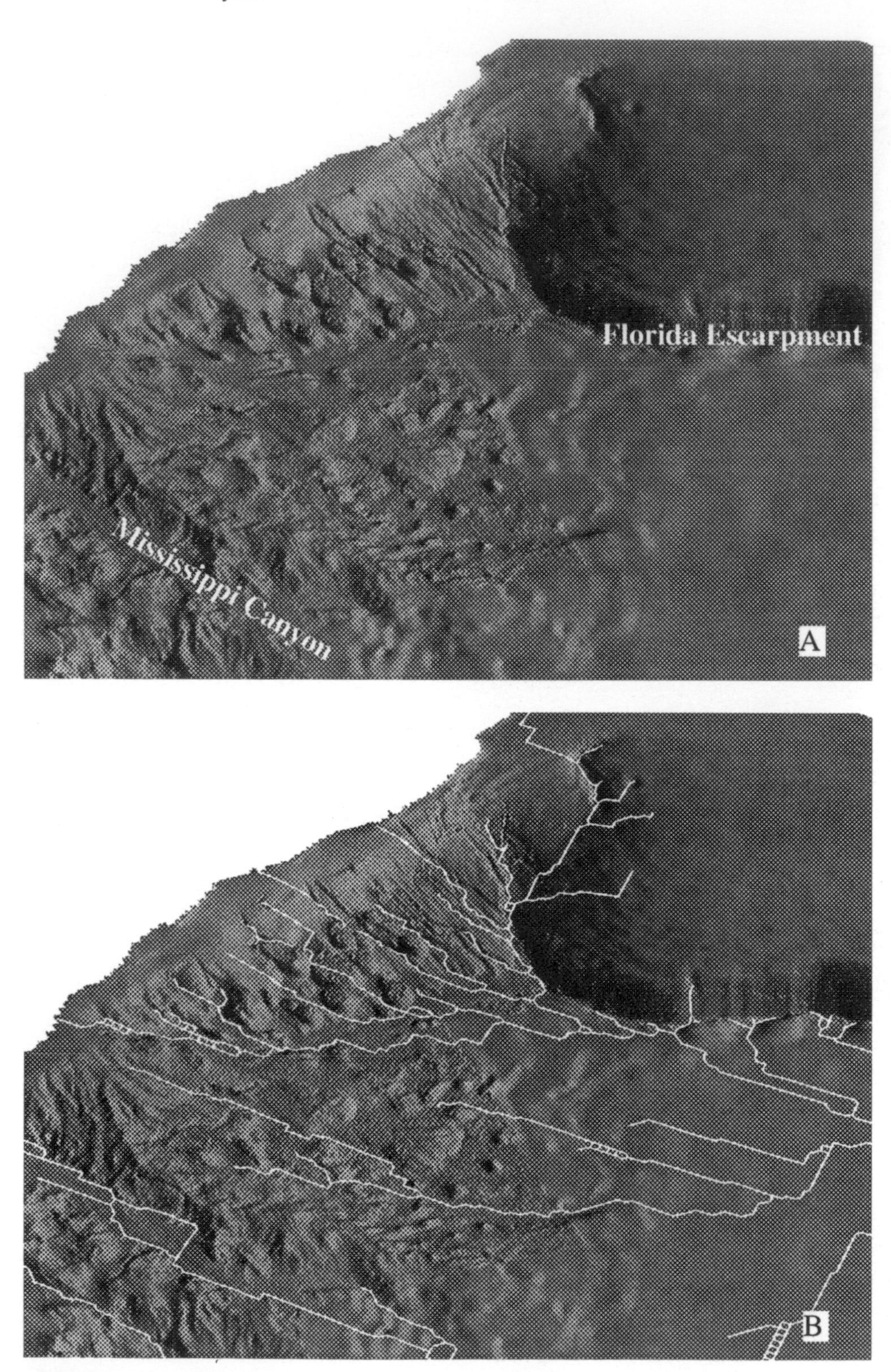

Figure 8—Perspective view of sea-floor relief and drainage paths in the northeastern Gulf of Mexico. (A) Shaded sea-floor relief, and (B) high-order drainage paths (white lines) draped over the sea floor. The vertical exaggeration is 10X.

Pourtales Escarpment, as well as the Tortugas Terrace.

For the lower-order streams, in the northwestern area, the computer had difficulty tracing the paths on a relatively smoothed area. As a result, it shows parallel 1-km-spaced paths on the upper slope (150–500 m water depth). In the Alaminos Canyon, the drainage is not limited to the only path that runs across from NE to the SW, but also comes from two reentrants in the northwestern part of the canyon. Another separate drainage system occurs in the southeastern part of the canyon that follows the fault planes. In Bryant Canyon, the lower-order streams could be traced through the canyon mouth and into the Iberia Basin (Figure 9). On the Louisiana–Alabama upper slope, the smoothed topography shows

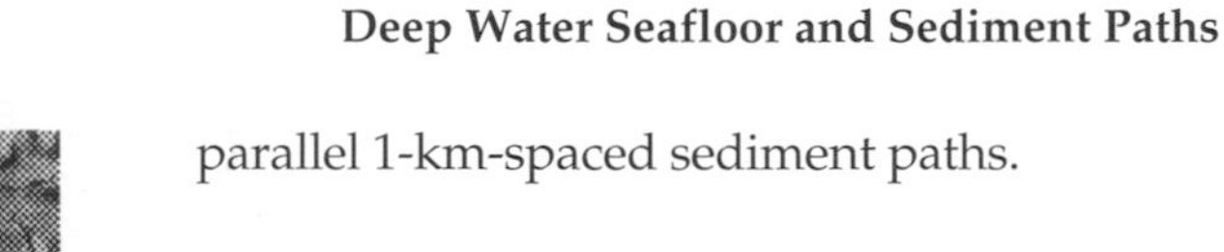

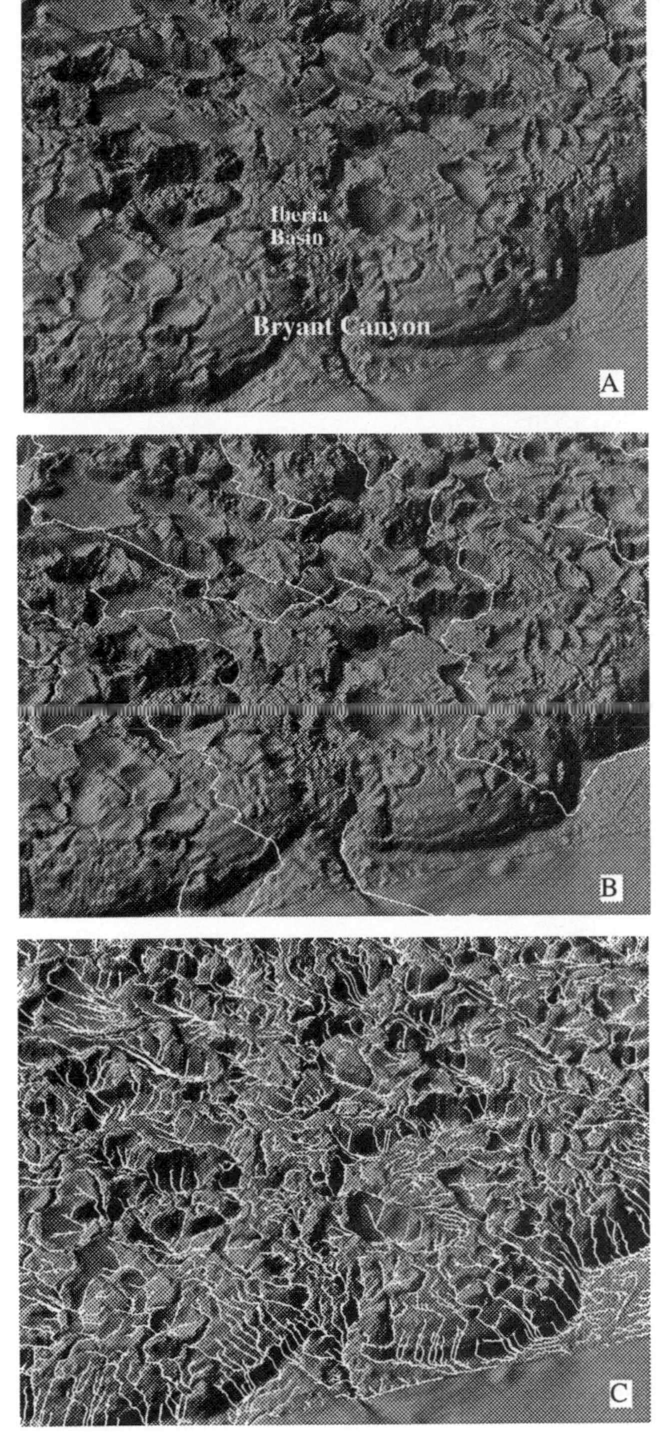

Figure 9—Perspective view of sea-floor relief and drainage paths in the Bryant Canyon area. (A) Shaded sea-floor relief (B) high-order drainage paths (white lines) draped over the sea floor, and (C) low-order drainage paths (white lines) draped over sea floor. The vertical exaggeration is 10X.

parallel 1-km-spaced sediment paths.

SEDIMENT SOURCES AND TRANSPORTING MECHANISMS

All drainage patterns align well to today's major river systems in the northern Gulf: sediment sources from the west (Rio Grande system), the northwest (Brazos and Colorado river systems), and the north (Mississippi River system). The carbonate-dominated platforms in the eastern and southern Gulf of Mexico show fewer drainage paths and suggest less contribution from the southern rim.

Except for the carbonate platforms offshore Florida and Campeche, the detrital sediments in the deepwater environment are mainly transported subaerially by river systems during sea level lowstands (Beard et al., 1982). During lowstands, the canyons cut into the outer shelf and upper slope and became conduits that carried coarser-grained sediments to the deep water. The transporting mechanisms are gravity-induced mass movements (Middleton and Hampton, 1976). Depending on the interaction of the grains with the density flow, they may be turbulence-supported turbidity currents or matrix-strength-supported debris flows. Movement can also result from less deformed whole body movement along a surface or slides, or from slump deposits. During sea level highstand, the majority of the coarse-grained sediments were trapped within the shelf province and hemipelagic sediment settling became dominant in the deepwater (Bryant et al., 1995). Sea level highstand deposits tend to be parallel laminated but are highly bioturbated and comparatively thin compared with sea level lowstand deposits. Other mechanisms that might influence the deposition in the deepwater of Gulf of Mexico are salt movements, faulting, and bottom currents. The interplay between sediments, salt, and faults can be very complicated, and in each area it may only be effective at a local scale. The influence of bottom currents has been controversial in the deepwater. Roberts et al. (1982) suggested that bedform erosions were the results of oceanic currents. Martin and Bouma (1982) suggested slumping was the cause of bed truncations, and van den Bold et al. (1987) believed bottom currents were insignificant in a water depth deeper than 200 m in the Gulf of Mexico. Recent Texas A&M University Deep Tow surveys in the Gulf of Mexico lower continental slope confirmed that there were deepwater processes that produced 20-m-spaced, 5-m-wide megafurrows that were sub-parallel to the bathymetric contour lines southward of the Sigsbee Escarpment. These mega bedforms indicate swift bottom currents in water depths of more than 3,000 m.

CONCLUSIONS

Using up-to-date detailed multibeam and digitized data from the northern Gulf of Mexico deep-water bathymetry the presence of broad and diversified physiographic provinces was determined. That data, along with simulated sun shading, reveal vivid and more recognizable sea floor topography than is possible to obtain from our traditional contour maps. Although a differentiation between different sediment-transporting mechanisms has to be decided on with the help of high-resolution seismic and ground-truth sediment samples, it is possible to predict present-day sediment paths based on the bathymetry data using the drainage-basin model. By using palinspastic reconstructions it should be possible to obtain sediment pathways for older stratigraphic periods.

Generally speaking, the computer-generated paths agree with the routes that can be manually picked from bottom topography. The drainage patterns show higher sinuosity on the northern upper and middle continental slopes, where the paths cut through domes, especially in the area that has multibeam data. On the lower continental slope, the paths mainly follow fault planes and traverse intraslope basins. The paths that climb basin walls may seem unreasonable, but the fast salt movement (6 cm/yr) in the Gulf of Mexico (Lee, 1990) suggests these are either older drainage paths, now plugged by salt, or potential paths of slump deposits. On the western upper continental slope and Mississippi Canyon, the sinuosity is low. The lower sinuosity on the western slope may be due to lack of detailed bathymetric data. The drainage displays mainly tributary patterns that feed into the main channel and merge on the lower slope and abyssal plain. There are very few distributaries that branch out from the main channel. These patterns may be explained by the smaller than 1° gradient from the slope to the abyssal plains. The scattered domes on the upper continental slope divert sediment input, while the instraslope basins on the middle and lower slope trap denser sediments.

The complexity described above shows the importance the various tectonic and sedimentary processes can exert on the bathymetry and consequently on the locations of the various sediment pathways over a short geologic time, and even over longer time. To better understand the sandstone distribution in an area, it is recommended that a larger region is studied before being able to evaluate the significance of the sand distributions in the area of interest.

ACKNOWLEDGMENTS

The authors thank Texas Sea Grant College Program and National Science Foundation for the support of this project. We also thank Arnold H. Bouma, James M. Coleman, and Philip J. Bart for their comments and review of the manuscript.

REFERENCES CITED

Beard, J. H., 1973, Pleistocene-Holocene boundary and Wisconsinan substages, Gulf of Mexico: Geological Society of America Memoir 136, p. 277–316.

Beard, J. H., J. B. Sangree, and L. A. Smith, 1982, Quaternary chronology, paleoclimate, depositional sequences, and eustatic cycles: AAPG Bulletin, v. 66, p. 158–169.

Bouma, A. H., G. T. Moore, and J. M. Coleman, eds., 1978, Framework, facies, and oil-trapping characteristics of the upper continental margin: AAPG Studies in Geology 7, 326 p.

Bouma, A. H., 1981, Depositional sequences in clastic continental slope deposits, Gulf of Mexico: Geo-Marine Letters, v. 1, p. 115–121.

Bouma, A. H., and W. R. Bryant, 1994, Physiographic features on the northern Gulf of Mexico continental slope: Geo-Marine Letters, v. 14, p. 252–263.

Bryant, W. R., 1986, Physiography and bathymetry, *in* A. Salvador, The Gulf of Mexico Basin: Geological Society of America, the geology of North America, v. J, p. 13–30.

Bryant, W. R., J. R. Bryant, M. H. Feeley, and G. R. Simmons, 1990, Physiographic and bathymetric characteristics of the continental slope, northern Gulf of Mexico: Geo-Marine Letters, v. 10, p. 182–199.

Bryant, W. R., J. Y. Liu, and J. Ponthier, 1995, The engineering and geological constraints of intraslope basins and submarine canyons of the northwestern Gulf of Mexico: Gulf Coast Association of Geological Societies Transactions, v. 45, p. 95–101.

Coleman, J. M., H. H. Roberts, and W. R. Bryant, 1986, Late Quaternary sedimentation, *in* A. Salvador, The Gulf of Mexico Basin: Geological Society of America, the geology of North America, v. J, p. 325-352.

Davies, D. K., 1972, Deep sea sediments and their sedimentation, Gulf of Mexico: AAPG Bulletin, v. 56, p. 2212–2239.

Ewing, M., and A. John, 1966, New seismic data concerning sediments and diapiric structures in Sigsbee Deep and continental slope, Gulf of Mexico: AAPG Bulletin, v. 50, p. 479–504.

Fisk, H. N., E. McFarlan, Jr., C. R. Kolb, and L. J. Wilbert, Jr., 1954, Sedimentary framework of the modern Mississippi Delta: Journal of Sedimentary Petrology, v. 24, p. 76–99.

Galloway, W. E., D. G. Bebout, W. L. Fisher, J. B. Dunlap, Jr., R. Cabrera-Castro, J. E. Lugo-Rivera, and T. M. Scott, 1991, Cenozoic, *in* A. Salvador, ed., The Gulf of Mexico Basin: Geological Society of America, The Geology of North America, v. J, p. 245–324.

GRASS, 1993, Geographic resources analysis support system: Champaign, Illinois, U.S. Army Construction Engineering Research Laboratories (USACERL), version 4.1.

Hardin, G. C., Jr., 1962, Notes on Cenozoic sedimentation in the Gulf Coast geosyncline, U.S.A.: Geology of the Gulf Coast and central Texas and guidebook of excursions; Geological Society America Annual Meeting and Houston Geological Society, p. 1–15.

Lee, G. H., 1990, Salt tectonics and seismic stratigraphy of the Keathley Canyon area and vicinity, northwestern Gulf of Mexico: Unpublished Ph.D. dissertation, Texas A&M University, Texas, 182 p.

Lehner, P., 1969, Salt tectonics and Pleistocene stratigraphy on continental slope of northern Gulf of Mexico: AAPG Bulletin, v. 53, p. 2431–2479.

Liu, J. Y., 1997, High resolution bathymetry in the Gulf of Mexico, covers 94 to 87.5°W longitude and

25.7–29.5°N latitude: AAPG Annual Convention Official Program, p. A70.

Liu, J. Y., and W. R. Bryant, 1999, Seafloor relief of northern Gulf of Mexico deep water, Texas Sea Grant College Program, Scale 1:1,000,000.

Martin, R. G., 1978, Northern and eastern Gulf of Mexico continental margin: Stratigraphic and structural framework, *in* A. H. Bouma, G. T. Moore, and J. M. Coleman, eds., Framework, facies, and oil-trapping characteristics of the upper continental margin: AAPG Studies in Geology 7, p. 21–42.

Martin, R. G., and A. H. Bouma, 1978, Physiography of the Gulf of Mexico, *in* A. H. Bouma, G. T. Moore, and J. M. Coleman, eds. Framework, facies, and oil-trapping characteristics of the upper continental margin: AAPG Studies in Geology 7, p. 3–19.

Martin, R. G., and A. H. Bouma, 1982, Active diapirism and slope steepening, Texas–Louisiana slope, northern Gulf of Mexico: Marine Geotechnology, v. 5, p. 63–91.

Middleton, G. V., and Hampton, M. A., 1976, Subaqueous sediment transport and deposition by sediment gravity flows, *in* D. J. Stanley and D. J. P. Swift, eds., Marine sediment transport and environmental management: John Wiley & Sons Course Notes No. 3, 2d ed., 401 p.

Moore, D. G., and P. G. Scruton, 1957, Minor internal structures of some recent unconsolidated sediments: AAPG Bulletin, v. 54, p. 326–333.

Moore, G. T., G. W. Starke, L. C. Bonham, and H. O. Woodbury, 1978, Mississippi Fan, Gulf of Mexico–physiography, stratigraphy, and sedimentational patterns: AAPG Bulletin, v. 62, p. 155–191.

Murray, G. E., 1952, Volume of Mesozoic and Cenozoic sediments in central Gulf Coastal Plain of United States: Geological Society of America Bulletin, v. 63, p. 1177–1192.

Peel, F. J., C. J. Travis, and J. R. Hossack, 1995, Genetic structural provinces and salt tectonics of the Cenozoic offshore U.S. Gulf of Mexico: a preliminary analysis, *in* M. P. A. Jackson, D. G. Roberts, and S. Snelson, eds., Salt tectonics: a global perspective: AAPG Memoir 65, p. 153–175.

Roberts, H. H., I. B. Singh, and J. M. Coleman, 1986, Distal shelf and upper slope sediments deposited during rising sea-level, north-central Gulf of Mexico: Gulf Coast Association of Geological Societies Transactions, v. 36, p. 541–551.

Salvador, A., 1991, Origin and development of the Gulf of Mexico Basin, *in* A. Salvador, ed., The Gulf of Mexico Basin: Geological Society of America, the geology of North America, v. J, p. 389–444.

Shinn, A. D., 1971, Possible future petroleum potential of upper Miocene and Pliocene, western Gulf basin, *in* I. H. Cram, Future petroleum provinces of the United States—their geology and potential: AAPG Memoir 15, v. 2, p. 824–835.

Simmons, G. R., 1992, The regional distribution of salt in the northwestern Gulf of Mexico: styles of emplacement and implications for early tectonic history: Unpublished Ph.D. dissertation, Texas A&M University, Texas, 180 p.

van den Bold, M. C., T. F. Moslow, and J. M. Coleman, 1987, Origin and timing of seafloor erosion on the Louisiana continental slope: Gulf Coast Association of Geological Societies Transactions, v. 37, p. 487–498.

Weimer, P., 1989, Sequence stratigraphy, facies geometry, and depositional history of the Mississippi Fan, Gulf of Mexico: AAPG Bulletin, v. 74, p. 425–453.

Woodbury, H. O., I. B. Murray, Jr., P. J. Pickford, and W. H. Akers, 1973, Pliocene and Pleistocene depocenters, outer continental shelf, Louisiana and Texas: AAPG Bulletin, v. 57, p. 2428–2439.

Scott, E. D., A. H. Bouma, and H. DeVille Wickens, 2000, Influence of tectonics on submarine fan deposition, Tanqua and Laingsburg subbasins, South Africa, *in* A. H. Bouma and C. G. Stone, eds., Fine-grained turbidite systems, AAPG Memoir 72/SEPM Special Publication No. 68, p. 47–56.

Chapter 5

Influence of Tectonics on Submarine Fan Deposition, Tanqua and Laingsburg Subbasins, South Africa

Erik D. Scott
BHP Petroleum
Houston, Texas, U.S.A.

Arnold H. Bouma
Department of Geology and Geophysics, Louisiana State University
Baton Rouge, Louisiana, U.S.A.

H. DeVille Wickens
Kuils River, South Africa

ABSTRACT

The Permian Tanqua and Laingsburg subbasins in the southwestern Karoo Basin, South Africa, had near-contemporaneous formation and filling. Five submarine fan systems are in the broad, shallow Tanqua subbasin and four in the more typical foredeep-style Laingsburg subbasin. Petrologic and microprobe analysis of the sandstones indicates a distant source area. Tectonic events led to varying sea-floor topography that directed sediment transport to the subbasins. Tectonic activity was low to nonexistent during deposition of any one or more of the submarine fans and indicates that the depositional cycle is much shorter than a tectonic cycle. The tectonic style of a basin may not always define the sedimentary characteristics of turbidite systems deposited in that basin.

INTRODUCTION

The Karoo Basin covers a major portion of present-day South Africa (see map in Wickens and Bouma, this volume). It developed during the late Carboniferous and early Permian as part of a series of foreland basins associated with subduction of the paleo-Pacific plate under the southern margin of Gondwana (de Wit and Ransome, 1992). The subduction caused the formation of a fold-thrust belt along the southern portion of Gondwana, which includes the Cape Fold Belt (CFB). Five tectonic episodes, associated with the formation of the CFB, were found to have occurred at 278 ± 2, 258 ± 2, 247 ± 3 , 230 ± 3, and 215 ± 2 Ma by Hälbich (1983) using ^{40}Ar–^{39}Ar step-wise heating. The CFB today is a continuous mountain chain that borders the southern (E–W-trending Southern Branch) and western (N–S-trending Western Branch) sides of the southwest Karoo Basin (de Beer, 1990). The two branches underwent different directions and amounts of compression that resulted in the northeast trending Baviaanshoek/Hex River anticlinorium, where the branches converge (de Beer, 1990, 1992; Wickens and Bouma, this volume). The anticlinorium was a basin-floor high that separated the

Tanqua subbasin from the Laingsburg subbasin and influenced sedimentation paths into both subbasins (Figure 1). A foredeep, the Tanqua and Laingsburg subbasins, developed in the southwest Karoo basin in association with the subduction zone, as separate, but geologically contemporaneous, subbasins that experienced similar, but individually distinct, histories (Bouma and Wickens, 1994; Wickens, 1994; Visser, 1995; Scott, 1997).

In the Tanqua subbasin, deposition of dilute, muddy turbidity currents and hemipelagic suspension sedimentation resulted in the shales of the Tierberg Formation (see stratigraphic column in Wickens and Bouma, this volume). This shale deposition was interrupted five times by deposition of fine-grained, basin-floor fans of the Skoorsteenberg Formation, designated Fan 1 through 5, from oldest to youngest. Individual fans vary in thickness from 20 to 60 m, with the intervening shales ranging from 20 to 75 m. Paleocurrent directions for Fans 1, 2, 3, and 5 range from the SSW to S and from W to WNW for Fan 4. Turbidites in the Laingsburg subbasin are found in the Vischkuil and Laingsburg formations. The shales and very fine- to fine-grained sandstones of the Vischkuil Formation were deposited from muddy, dilute turbidity currents and vary in thickness from 200 to 400 m. The Laingsburg Formation includes four submarine fans, designated Fan A to Fan D from oldest to youngest, ranging from mid-fan channel fill deposits to outer-fan lobe deposits (Wickens, 1994). Individual fan thickness in the Laingsburg subbasin exhibits a greater variation than in the Tanqua subbasin, ranging from 35 to 250 m, with the intervening shales ranging

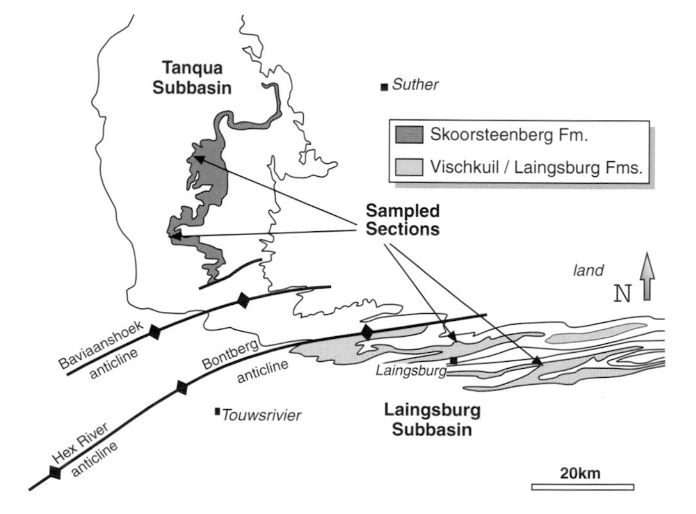

Figure 1—Outline of the outcrops of the Tanqua and Laingsburg subbasins showing relationship to the Hex River/Baviaanshoek anticlines. Shown are the locations of sampled sections through the Skoorsteenberg, Vischkuil, and Laingsburg formations. For study area location map, see Wickens and Bouma (this volume).

from 25 to 180 m in thickness. A detailed description of the stratigraphy in the subbasins can be found in Wickens and Bouma (this volume).

SEDIMENT SOURCE AREA

The majority of the previous studies based their suggestions for the source area of the submarine fan systems on paleoreconstructions, relying heavily on traditional sedimentation models of foreland basins (Johnson 1991; Scott, 1997 and references therein). Most of the studies concluded that the associated fold-thrust belt (CFB) was uplifted and eroded to shed sediment into the subbasins; others suggested that the source area was outside the CFB. The field and laboratory data gathered in this study, along with field observations, indicate that the sandstones of the submarine fan systems have a source area other than the CFB and underwent a longer sediment pathway than would normally be associated with a foreland basin model.

The sandstones of the Skoorsteenberg, Vischkuil, and Laingsburg formations are fine- to very fine-grained with grains ranging from 177 to 62µ, with the majority between 125 and 62µ. The majority of the sandstones are grain-supported with varying amounts of mud matrix and thin mud laminae. Mineral composition is similar throughout, with mono- and polycrystalline quartz as the major component. A number of feldspars in the sandstones are partially euhedral, exhibiting sharp angles. All the samples have both detrital muscovite and biotite, and contain a wide variety of heavy minerals including tourmaline, garnet, zircon, apatite, sphene, rutile, and epidote. The sandstones in the subbasins consist of immature sediments. They are well sorted, with the majority of the grains being subrounded to subangular, indicating shorter overall weathering time. Alteration of the micas (only minor vermiculitization shown by a lower K content of the biotites) and feldspars in the sandstones are minimal (Scott, 1997).

Petrologic and microprobe analyses of tourmaline, garnet, and biotite in the sandstones from both subbasins are consistent with an origin from a high-grade metamorphic terrain of amphibolite to lower granulite facies. Plotting detrital tourmaline compositions on a Ca-Fe-Mg ternary diagram, the tourmalines primarily fall in the Ca-poor metapelite-metapsammite and calc-silicate fields (Figure 2a). On the Al-Fe-Mg diagram, the majority of tourmalines are consistent with derivation from a metapelite and metapsammite (both Al-saturating and no Al-saturating phases) zone, from $Fe^{(3+)}$-rich quartz-tourmaline rocks, and from a calc-silicate zone (Figure 2b). The composition of garnets from the subbasins are predominately in the gneiss and schist fields of a Ca-Fe+Mg-Mn ternary diagram (Figure 2c). There are a number of garnets with a high Ca content, possibly derived from skarn zones (metasomatized carbonates) (Scott, 1997). Biotites found in the sandstones are enriched in titanium, as indicated by their reddish color and from microprobe analysis, again indicating formation under high-grade metamorphic conditions. The heavy mineral assemblage and the chemical composition of the biotites imply a provenance from the upper amphibolite to lower granulite facies (D. Henry, personal communication, 1996).

The petrography, chemistry, and the fine-grained, mud-rich texture of the deposits suggests a distant source area outside of the CFB with a sediment pathway that emphasizes deeper water transport over continental and inner shelf transport. Sediments shed from fold-thrust belts into the adjacent foreland basins are typically coarse-grained and sand-rich, whereas the sandstones in the Tanqua and Laingsburg subbasins are fine- to very fine-grained and mud-rich. The sandstones in the subbasins are sorted but contain a large number of angular and subangular grains with little apparent chemical alteration, particularly the feldspars and micas, and not well-weathered, rounded grains expected in reworked sediments. The grains in the sandstones indicate a relatively short time under continental/shelf transport conditions and implies a longer density flow pathway. The condition of the mica and feldspar grains in the samples does not lead to the conclusion that the sediments were from weathered and reworked older sediments of the CFB. In addition, previous studies indicate that the CFB only experienced metamorphism to greenschist facies (~300°C and 3 kbar) (Hälbich and Cornell, 1983) or to prehnite-pumpelyite facies (Martini, 1974). The present study strongly suggests that the high-grade nature of the source area, the immaturity of the sandstones, and the condition of the grains appear to negate the CFB as the main sediment source and therefore place the provenance for the sandstones outside of the fold-thrust belt.

Sediments appear to be sourced from an orogenic belt associated with the subduction zone on the western side of South America. Although it is very difficult to precisely pinpoint the location of the source area, most likely the sediments came from an area between the magmatic arc of the subduction zone and the rising, but submerged, fold-thrust belt. This may possibly be an uplifted area of older, metamorphosed sediments in the paleo-Patagonia region toward the southwest, approximately 200–500 km away.

CORRELATION AND TIMING OF SUBBASIN DEPOSITS

At present, correlation of the deposits between the Tanqua and Laingsburg subbasins can only be made on a formation level for the Dwyka, Prince Albert, Whitehill, and Collingham formations (Visser, 1993; Wickens, 1994). These formations were deposited before the anticlinorium separated the sediment transport into the subbasins. Above the Collingham Formation, the stratigraphic sections of the subbasins split until the formations of the Beaufort Group (see stratigraphic column in Wickens and Bouma, this volume).

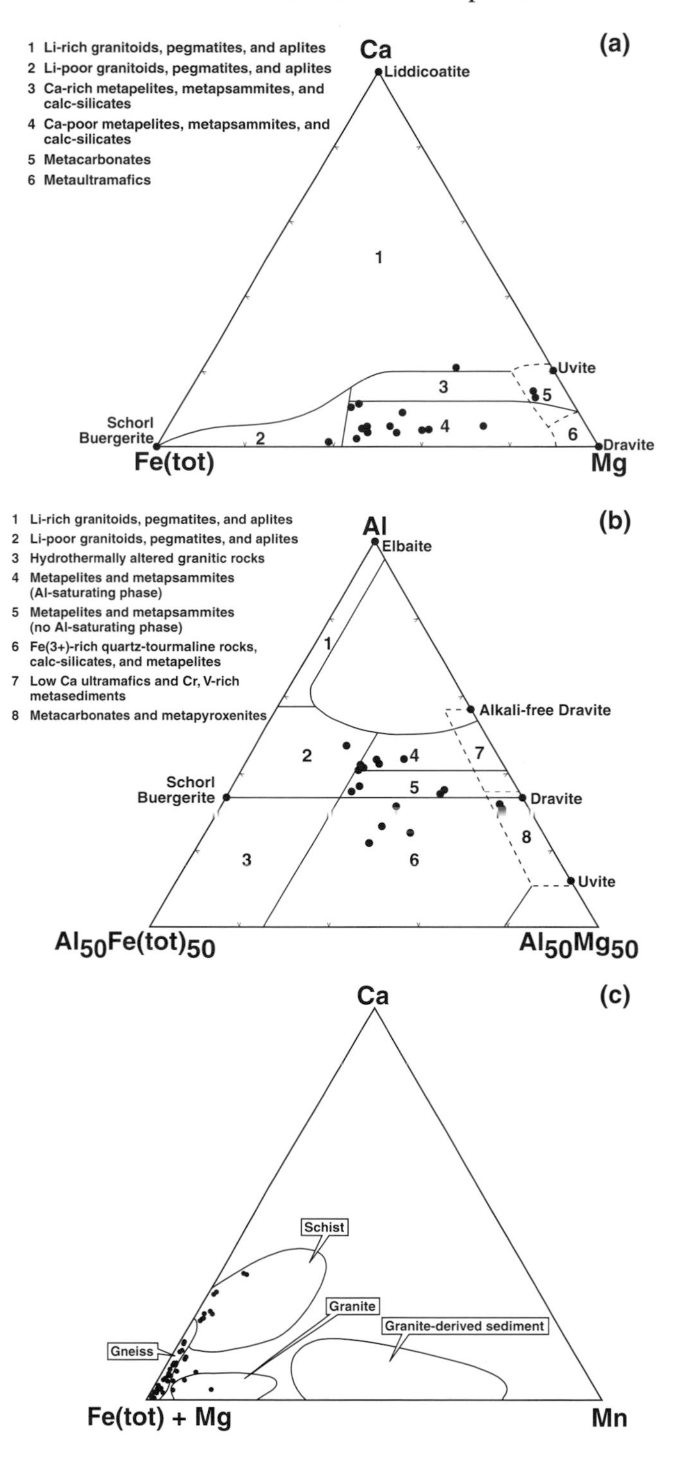

Figure 2—Source area ternary diagrams for tourmalines (after Henry and Dutrow, 1992) and garnets (after Smale and Morton, 1987). (a) Ca-Mg-Fe(tot) for tourmalines, (b) Al-Mg-Al-Fe(tot) for tourmalines, and (c) Ca-Mn-Fe(tot)+Mg for garnets.

Fine-scale correlation of the sand/shale packages of the Skoorsteenberg and Laingsburg Formations is difficult. Wickens (1994) initially correlated the five fan systems in the Tanqua subbasin with the four fan systems in the Laingsburg subbasins. However, direct correlation of the sand-rich sections of the Skoorsteenberg and Laingsburg formations would assume that active fan deposition occurred simultaneously in both subbasins. Field and laboratory data suggest there was a single sediment pathway and point source for both subbasins, and make the idea of concurrent sand deposition in two adjacent basins very unlikely. Most likely one subbasin was the site of active sand transport, while the other experienced hemipelagic sedimentation and/or deposition from very dilute muddy turbidity currents. Fine-scale correlation of the deposits of the Tanqua and Laingsburg subbasins is subject to (1) whether there was a single or multiple feeder system and (2) what the influence of sea-floor topography was on sedimentation pathways. As noted earlier, active submarine fan deposition probably only occurred in one subbasin at a time, making a very precise correlation of the sand-rich packages from one subbasin to the other virtually impossible. Additionally, the current data are not conclusive in assessing whether one subbasin filled completely and then the other or if sedimentation alternated between the two subbasins.

From Inductive Coupled Plasma–Atomic Emission Spectrum chemical analysis of bulk shale samples from sections through the Skoorsteenberg, Vischkuil, and Laingsburg formations in the subbasins, the barium, vanadium, strontium, titanium, and arsenic concentration profiles indicate possible correlations between the subbasins (Figure 3). Correlation of some of the sand/shale packages within an individual subbasin is evident, however, correlation between the Tanqua and Laingsburg subbasins is not as clear. The barium, vanadium, and strontium profiles suggest that deposition of Fans 1 and 2 in the Tanqua subbasin are time equivalent with the deposition of the lower section of the Vischkuil Formation in the Laingsburg subbasin (Figure 3a, b, c). Fan 4 in the Tanqua subbasin may be equivalent to the upper portion of the Vischkuil formation and the lower section of Fan A in the Laingsburg subbasin as indicated by the titanium and arsenic profiles (Figure 3d, e). From the profiles, tentative correlations can be made for the lower Skoorsteenberg Formation to the Vischkuil and lower Laingsburg formations (Figure 4). Correlation of the upper sections of the Skoorsteenberg and Laingsburg formations

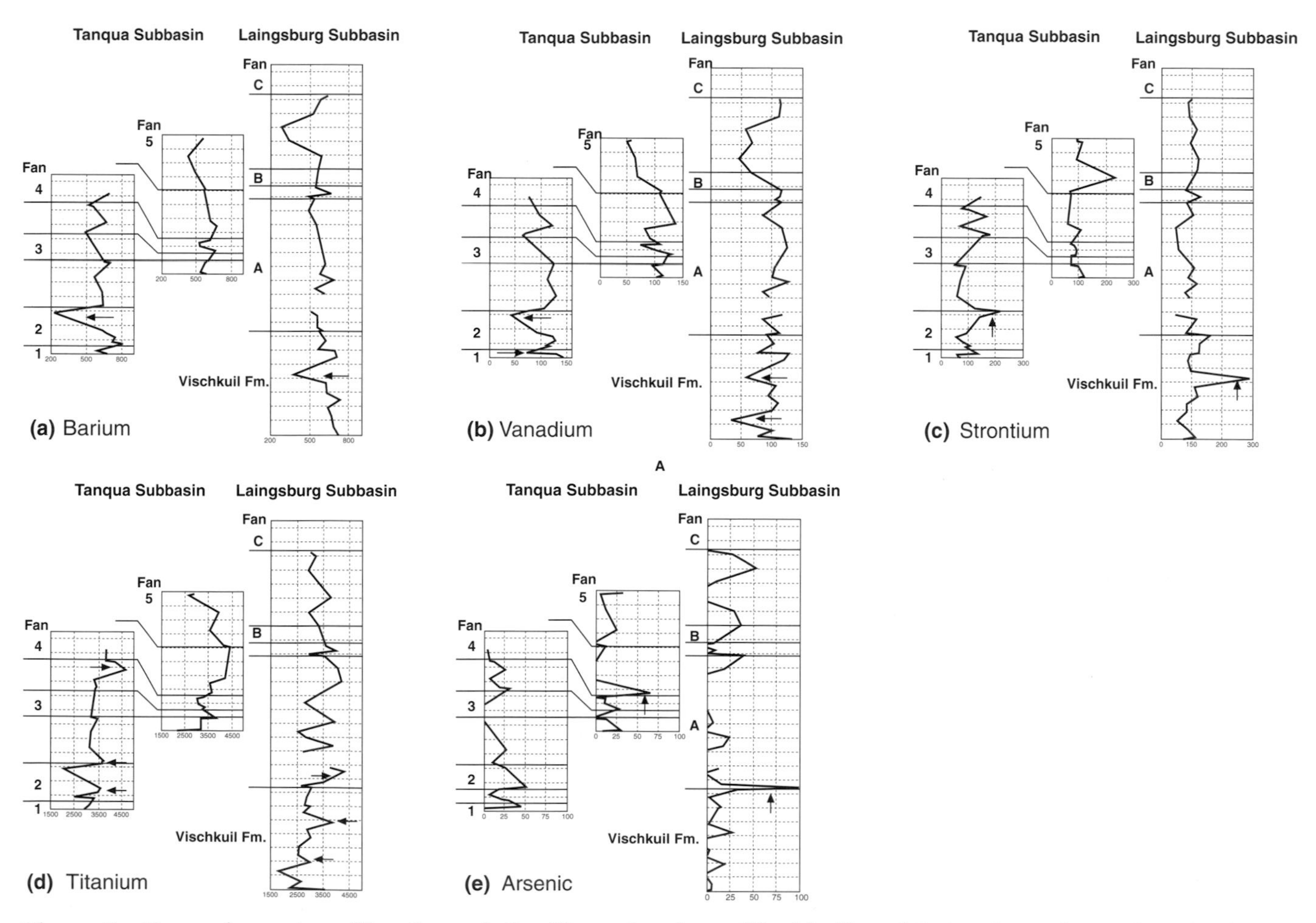

Figure 3—Trace element profiles through the Skoorsteenberg, Vischkuil, and Laingsburg formations in the subbasins. Arrows indicate possible correlation points. (a) barium, (b) vanadium, (c) strontium, (d) titanium, and (e) arsenic.

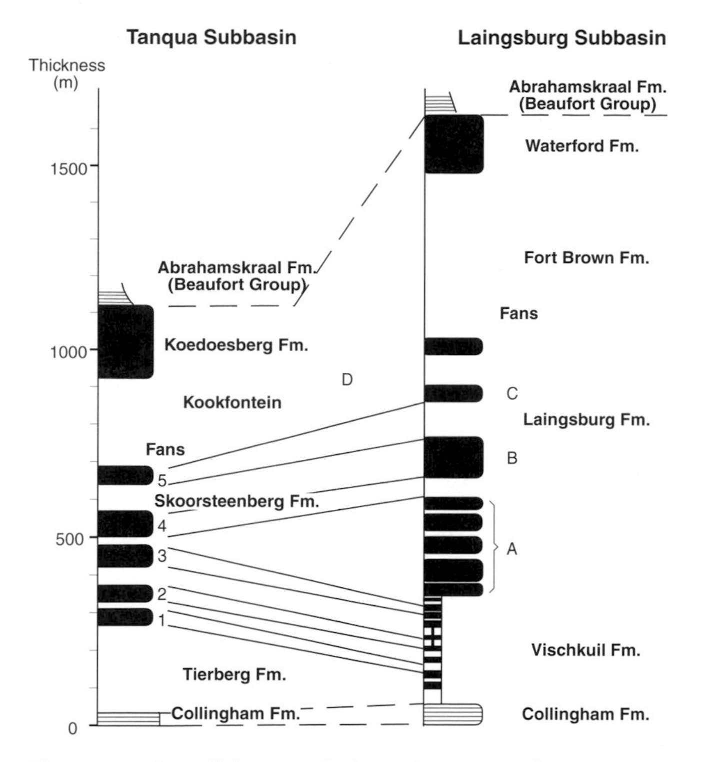

Figure 4—Possible correlations between the Tanqua and Laingsburg subbasins.

becomes questionable with numerous scenarios of alternated submarine fan building being equally valid.

TECTONIC INFLUENCE ON BASIN SHAPE AND SEDIMENTATION

The evolution of the Tanqua and Laingsburg subbasins is connected with the tectonic events and the resulting structures of the CFB. Differential compression on the two branches of the CFB led to distinct differences in the shape of the Tanqua and Laingsburg subbasins and their position in relationship to the branches of the CFB (de Beer, 1992). With less compression on the Western Branch of the CFB, the north–south trending Tanqua subbasin developed into a broad, open style of basin with unrestricted deposition (Wickens and Bouma, this volume). Four of the five submarine fan systems in the Tanqua subbasin parallel the Western Branch and show the subbasin's foreland style relative to the fold-thrust belt. Increased compression on the Southern Branch of the CFB caused the east-west-trending Laingsburg subbasin to evolve into a deeper, narrower basin of a more typical foredeep configuration with restricted sedimentation. Existing outcrops are highly folded, resulting in valley and ridge topography, and extend east–west for approximately 200 km (Wickens, 1994).

Tectonic events and the resulting structures associated with the CFB-influenced sedimentation in the Tanqua and Laingsburg subbasins, starting with the Whitehill Formation. Apparent thinning of the Whitehill and Collingham formations over the basin-floor high (Wickens, 1994) suggests that it was active in controlling gravity-controlled bottom flows into the adjacent subbasins since the earliest compressional event. The sudden change from the carbonaceous shale of the Whitehill formation to the turbidity flows and numerous ash fall deposits of the Collingham formation indicates a shift in the tectonic conditions of the region with basin margin instability and volcanic activity (Wickens, 1994). During deposition of the Tierberg, Skoorsteenberg, Laingsburg, and Vischkuil formations, the basin-floor high had significant sea-floor expression, separating the subbasins geographically as well as directing sediment away from the anticlinorium (de Beer, 1992; Bouma and Wickens, 1994; Wickens, 1994). Although the CFB did not have enough expression to shed sediments into the subbasins, it had developed sufficient elevation to effect sedimentation paths into the subbasins.

GEOLOGIC HISTORY OF THE TANQUA AND LAINGSBURG SUBBASINS

The relationship of the formation and sedimentation of the subbasins with different events of the CFB is an important aspect in the geologic history of the subbasins. With no age control available from the Skoorsteenberg, Vischkuil, and Laingsburg formations, timing of deposition of the deposits in these formations can only be inferred from the suggested ages of the under- and overlying formations. The earliest time for deposition of the Whitehill Formation has been suggested to be middle–early Permian (late Sakmarian) (Oelofsen and Araujo, 1987). The upper age limit for the boundary between the Ecca and Beaufort groups, set by tetrapod fauna, is early late Permian, possibly at the start of the lower late Permian (Kazanian) (Rubidge, 1987). From these ages, it appears the subbasins formed and filled during the 278- and 258-Ma tectonic episodes of the CFB.

Some previous studies associate the 278-Ma tectonic event of the CFB with the end of Dwyka glaciation (Visser, 1990; Cole, 1992). However, the mudstones and shales of the overlying Prince Albert Formation shows no influence of tectonics during deposition. This suggests that the CFB did not have enough expression at this time to affect sedimentation of the subbasins. Other studies suggest that the CFB experienced enough uplift to produce topography on the sea floor and influence sedimentation to some extent with the 278-Ma event (de Beer, 1990; Wickens, 1992, 1994). Some reconstructions show that not only did the CFB emerge with the 278-Ma event, but also that the complete Ecca Group was deposited before the next event at 258 Ma (Hälbich, 1992).

Collingham Formation

The transition from the dark, carbonaceous shales of the Whitehill Formation to numerous ash fall and siliciclastic turbidites of the Collingham Formation is most likely related to the 278-Ma tectonic episode (Figures 5, 6). The event caused subsidence of the two subbasins

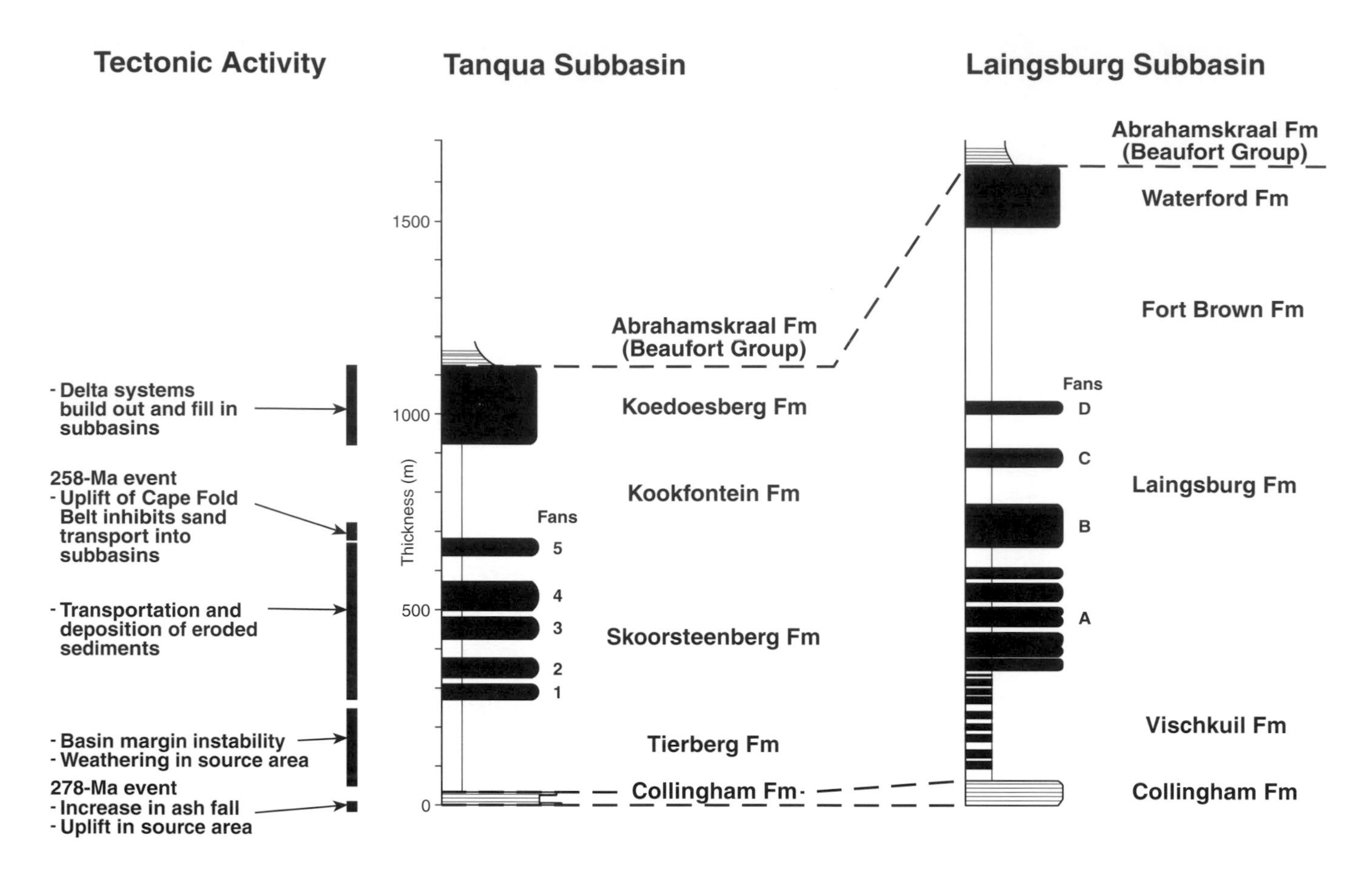

Figure 5—Relationship of the deposits in the Tanqua and Laingsburg subbasins with the tectonic events associated with the Cape Fold Belt.

against the branches of the CFB, bringing about basin margin instability and slope failure causing mainly low-density turbidity currents to travel into the developing basins. However, since there was not enough time for weathering of the uplifted orogenic belt to shed sufficient sediments to feed the density flows, along with the long transport distance between the source area and the subbasins, only infrequent transport of sediment into the subbasins occurred. The Collingham Formation thins over the anticlinorium area, indicating the continuing development of the basin floor high. Using U/Pb dating on zircons, an ash bed in the Collingham Formation was recently dated at 270 ± 1 Ma (Trouw and De Wit, 1999), suggesting the relationship of the age of the formation and the increased activity in the magmatic arc in conjunction with the 278 ± 2 Ma tectonic event.

Tierberg, Skoorsteenberg, Vischkuil, and Laingsburg Formations

After deposition of the Collingham Formation, the basin-floor high developed enough to effectively split the sedimentation paths into the subbasins. After the first major compressional event of the CFB at 278 Ma, the Laingsburg subbasin may have experienced greater subsidence than the Tanqua subbasin due to increased compression on the Southern Branch of the CFB. This led to basin margin instability and density flows in the Laingsburg subbasin during the early stages of subbasin development. This is seen in the lower Vischkuil Formation by the presence of distal turbidites and slump features. At the same stratigraphic level, the slower subsiding Tanqua subbasin underwent dilute, muddy turbidite and hemipelagic sedimentation, depositing the dark shales of the Tierberg Formation.

Previous studies have correlated onset of submarine fan deposition in the Tanqua and Laingsburg subbasins with the 258-Ma event (Cole, 1992; Visser, 1995). This correlation is based on short sediment transport paths of traditional foreland basin sedimentation models and assumes that the sandstones represent a syn-tectonic phase of sedimentation. This premise implies that the source area was uplifted, weathering occurred, sediments were transported by fluvial and shelf processes, and then sediments were transported by sediment gravity flows to the deeper basins, all within a very short time span. Sedimentation into the Tanqua and Laingsburg subbasins, with the longer transportation distance from the source area, most likely represents a post-tectonic depositional stage (occurring during a tectonic quiescence) rather than syn-tectonic deposition. After

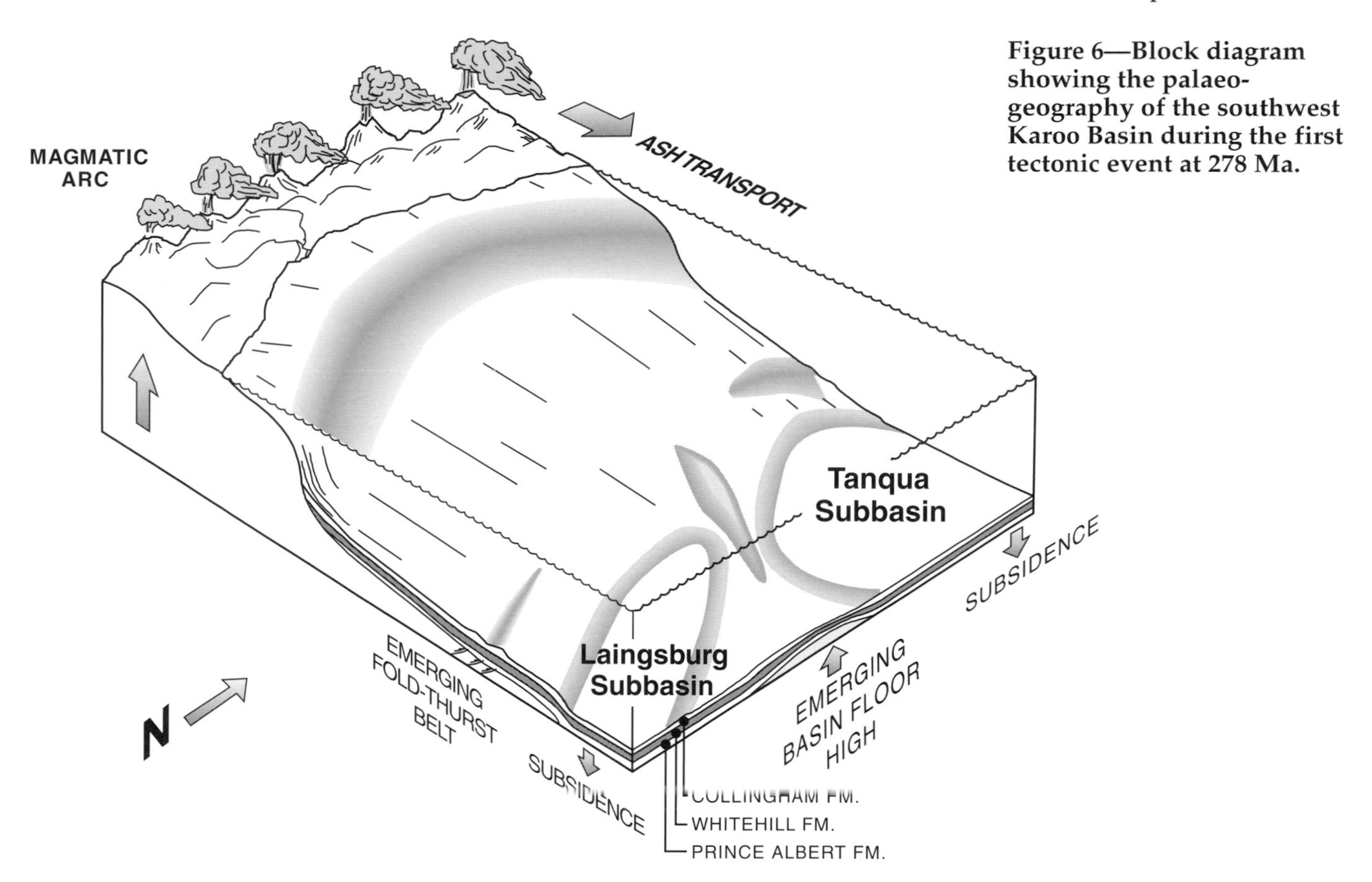

Figure 6—Block diagram showing the palaeogeography of the southwest Karoo Basin during the first tectonic event at 278 Ma.

the uplift of the orogenic belt with the 278-Ma event, sufficient time is needed to weather and transport sediments from the source area to the subbasins (Figures 5, 7).

Submarine fan deposition most likely alternated in some respect between the subbasins, with many different interpretations likely being equally valid. Geochemical analysis suggests Fans 1 and 2 are roughly

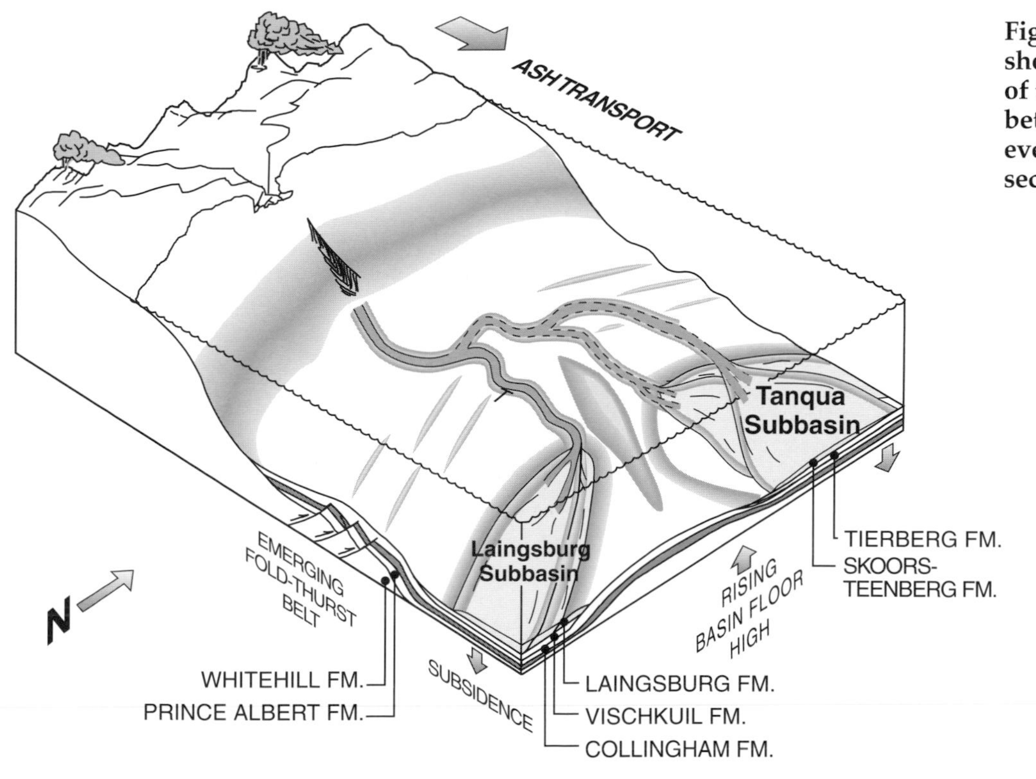

Figure 7—Block diagram showing the palaeogeography of the southwest Karoo Basin between the first tectonic event at 278 Ma and the second at 258 Ma.

equivalent to the lower half of the Vischkuil Formation and that Fan 3 may correlate with the upper part of the Vischkuil Formation and the lower portion of Fan A. During deposition of the fan systems it may be possible for sand input to switch from one subbasin to the other during deposition of one of the larger sand-rich packages. Fan A contains a number of shale breaks inside the overall larger sand-rich sequence, and Fan 2 consists of three sand-rich packages separated by shales (Figure 4). These breaks may indicate the change or temporary interruption of the sediment path from one subbasin to the other, possibly resulting from short, still stands and relative sea level or tectonic activity. A small change in seafloor topography, a rise or depression, could lead to major change in the sediment transport path to the subbasins.

Kookfontein/Fort Brown and Koedoesberg/Waterford Formations

The second compressional event at 258 Ma uplifted the CFB enough to cut off sedimentation paths and stopped submarine fan deposition in the subbasins (Figures 5, 8). With the sand now trapped between the source area and the CFB, the subbasins experienced predominantly hemipelagic and muddy turbidite deposition. Containing upward thickening and coarsening cycles, starting with dark-gray shale and siltstone followed by alternating siltstone, shale, and sandstones, the Kookfontein and Fort Brown formations represent progressively shallowing conditions (Wickens, 1994). Deltaic systems of the Koedoesberg/Waterford formations built out over the fold-thrust belt from the west and southwest and filled in the rest of the subbasins.

IMPLICATIONS FOR THE SEDIMENTOLOGY OF SUBMARINE FAN DEPOSITS

The submarine fan systems in the Tanqua and Laingsburg subbasins were deposited in a foreland basin setting in relation to the subduction of the paleo-Pacific plate under the southern edge of Gondwana, an active margin setting. The fan systems, however, consist of fine- to very fine-grained, mud-rich sandstones of the style and architecture normally associated with a passive margin. Submarine fan systems associated with an active margin typically are assumed to be coarse-grained and sand-rich, whereas submarine fan systems associated with a passive margin are assumed to be fine-grained with a considerably lower sand/shale ratio. The regional geologic setting of southwestern Gondwana, with the long distance between the subduction zone and its foreland basin, created a longer transportation distance for the sediments, which led to the "passive margin style" of deposition of the submarine fan systems in the Tanqua and Laingsburg subbasins. Consequently, the distinction of submarine fans being associated either with an active or a passive margin as a classification and genetic tool is not always accurate. It appears that a better classification of submarine fans would be in more descriptive terms between the end members of coarse-grained, sand-rich systems and fine-grained, mud-rich systems.

Submarine fan systems in the Tanqua and Laingsburg subbasins appear to have been deposited during a tectonic quiescence. The fan systems probably were not

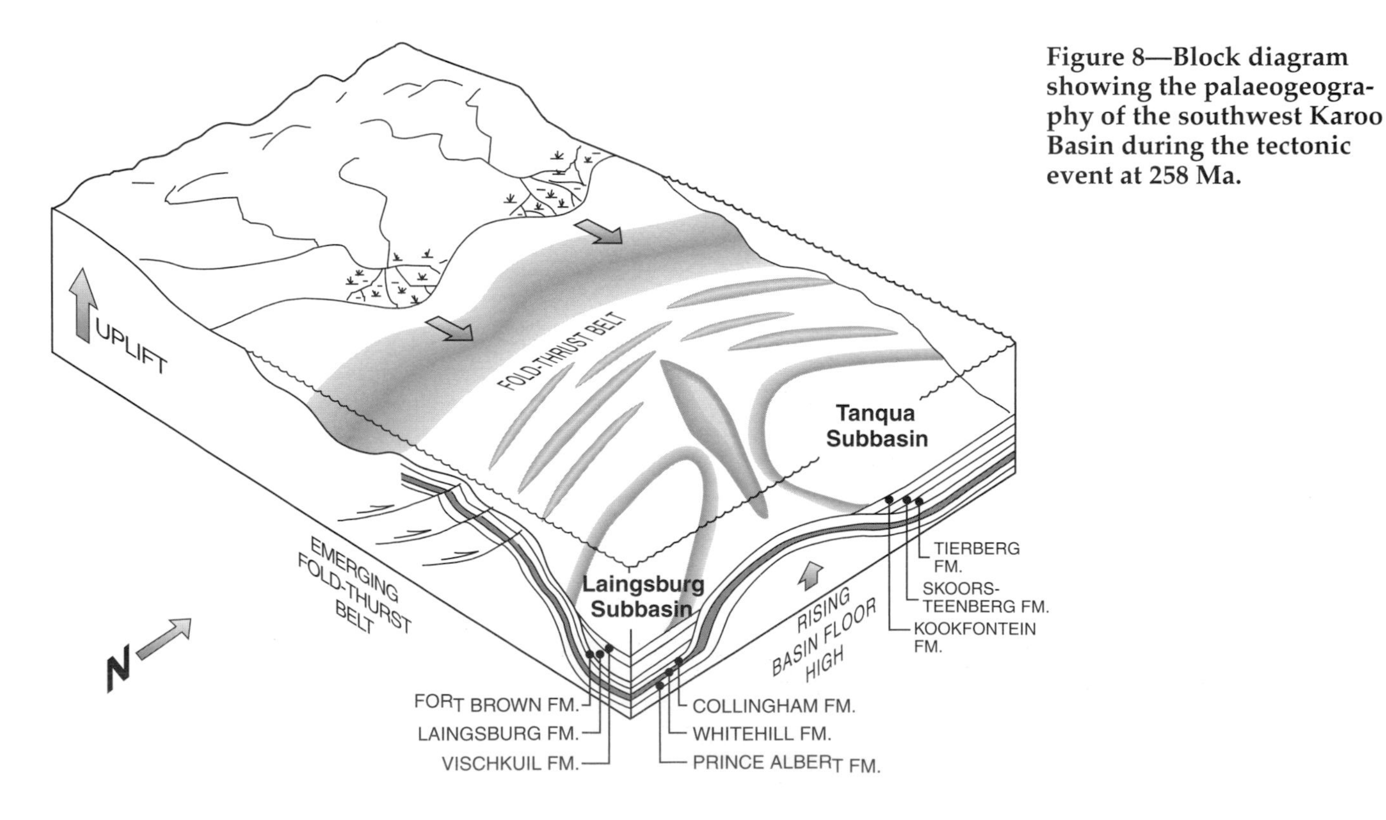

Figure 8—Block diagram showing the palaeogeography of the southwest Karoo Basin during the tectonic event at 258 Ma.

deposited concurrently or immediately after a tectonic event but, instead, represent the subsequent weathering and erosion of the uplifted orogenic belt a substantial distance away. Even though the Laingsburg subbasin overall experienced greater compression and has a larger tectonic overprint than the Tanqua subbasin, the depositional processes that built the submarine fan systems in both subbasins were essentially the same. With comparable sedimentary processes occurring, the architecture of the deposits in the two subbasins is similar in nature. This similarity would make it difficult to distinguish between the open basin setting and the restricted basin setting using typical subsurface data. This study strongly suggests that the irregular cyclicity of tectonic activities and submarine fan deposition differ sufficiently to allow construction of a fan during a tectonic window of inactivity.

Small changes in topography of the sea floor, especially nearer the deltaic sediment supply system, could lead to a large change in the sediment transport paths to their depositional basins. The alternation of sand-rich deposition between the subbasins and the switching of paleocurrent directions between the five fans in the Tanqua subbasin likely is related to either or both, minor tectonic activities causing changes in the sea floor topography and deltaic switching.

REFERENCES CITED

Bouma, A. H., and H. deV. Wickens, 1994, Tanqua Karoo, ancient analog for fine-grained submarine fans, *in* P. Weimer, A. H. Bouma, and B. F. Perkins, eds., Submarine fans and turbidite systems: sequence stratigraphy, reservoir architecture and production characteristics: Gulf Coast Section of the SEPM 15th Research Conference Proceedings, p. 23–34.

Cole, D. I., 1992, Evolution and development of the Karoo Basin, *in* M. J. de Wit and I. G. D. Ransome, eds., Inversion tectonics of the Cape Fold Belt, Karoo and Cretaceous basins of Southern Africa: Rotterdam, A. A. Balkema, p. 87–100.

De Beer, C. H., 1990, Simultaneous folding in the western and southern branches of the Cape Fold Belt: South African Journal of Geology, v. 93, p. 583–591.

De Beer, C. H., 1992, Structural evolution of the Cape Fold Belt syntaxis and its influence on syntectonic sedimentation in the SW Karoo Basin, *in* M. J. de Wit and I. G. D. Ransome, eds., Inversion tectonics of the Cape Fold Belt, Karoo and Cretaceous basins of Southern Africa: Rotterdam, A. A. Balkema, p. 197–206.

de Wit, M. J., and I. G. D. Ransome, 1992, Regional inversion tectonics along the southern margin of Gondwana, *in* M. J. de Wit and I. G. D. Ransome, eds., Inversion tectonics of the Cape Fold Belt, Karoo and Cretaceous basins of Southern Africa: Rotterdam, A. A. Balkema, p. 15–21.

Hälbich, I. W., and D. H. Cornell, 1983, Metamorphic history of the Cape Fold Belt: *in* A. P. G. Söhnge and I. W. Hälbich, eds., Geodynamics of the Cape Fold Belt: Geological Society of South Africa Special Publication 12, p. 149–164.

Hälbich, I. W., 1983, A tectonogenesis of the Cape Fold Belt (CFB), *in* A. P. G. Söhnge and I. W. Hälbich, eds., Geodynamics of the Cape Fold Belt: Special Publication of the Geological Society of South Africa, v. 12, p. 165–175.

Hälbich, I. W., 1992, The Cape Fold Belt Orogeny: state of the art 1970s–1980s, *in* M. J. de Wit and I. G. D. Ransome, eds., Inversion tectonics of the Cape Fold Belt, Karoo and Cretaceous basins of Southern Africa: Rotterdam, A. A. Balkema, p. 141–158.

Henry, D. J., and B. L. Dutrow, 1992, Tourmaline in a low grade clastic metasedimentary rock: an example of the protogenetic potential of tourmaline: Contributions to Mineralogy and Petrology, v. 112, p. 203–218.

Johnson, M. R., 1991, Sandstone petrography, provenance and plate tectonic setting in Gondwana context of the southeastern Cape–Karoo Basin: South African Journal of Geology, v. 94, p. 137–154.

Martini, J. E. J., 1974, On the presence of ash beds and volcanic fragments in graywackes of the Karoo system in the southern Cape Province: Transactions of the Geological Society of South Africa, v. 77, p. 113–116.

Oelofsen, B. W., and D. C. Araujo, 1987, *Mesosaurus tenuidens* and *Stereosternum tumidum* from the Permian Gondwana of both Southern Africa and South America: South African Journal of Science, v. 83, p. 370–372.

Rubridge, B. S., 1987, South Africa's oldest land-living reptiles form the Ecca–Beaufort transition in the southern Karoo: South African Journal of Science, v. 83, p. 165–166.

Scott, E. D., 1997, Tectonics and sedimentation: evolution, tectonic influences and correlation of the Tanqua and Laingsburg subbasins, Southwest Karoo Basin, South Africa: Unpublished Ph.D. dissertation, Louisiana State University, 253 p.

Smale, D., and Morton, A. C. 1987. Heavy mineral suites of core samples from the McKee Formation (Eocene–lower Oligocene), Taranaki: implication for provenance and diagenesis: New Zealand Journal of Geology and Geophysics v. 30, p. 299–306.

Trouw, R. A. J., and de Wit, M. J., 1999, Relation between the Gondwanide Orogen and contemporaneous intracratonic deformation: Journal of African Earth Sciences, v. 28, no. 1, p. 203–213.

Visser, J. N. J., 1990, The age of the late Paleozoic glaciogene deposits in southern Africa: South African Journal of Geology, v. 93, p. 366–375.

Visser, J. N. J., 1993, Sea-level changes in a back-arc-foreland transition: The late Carboniferous-Permian Karoo Basin of South Africa: Sedimentary Geology, v. 83, 115–131.

Visser, J. N. J., 1995, Post-glacial Permian stratigraphy and geography of southern and central Africa: boundary conditions for climatic modeling: Palaeogeography, Palaeoclimatology, Palaeoecology, v. 118, p. 213–243.

Wickens, H. DeV., 1992, Submarine fans of the Permian Ecca group in the SW Karoo Basin: their origin and reflection on the tectonic evolution of the basin and its source areas, *in* M. J. de Wit and I. G. D. Ransome, eds., Inversion tectonics of the Cape Fold Belt, Karoo

and Cretaceous basins of Southern Africa: Rotterdam, A. A. Balkema, p. 117–126.
Wickens, H. DeV., 1994, Basin floor building turbidites of the southwestern Karoo Basin, Permian Ecca Group, South Africa: Unpublished Ph.D. dissertation, University of Port Elisabeth, South Africa, 233 p.

Pratson, L. F., J. Imran, G. Parker, J. P. M. Syvitski, and E. Hutton, 2000, Debris flows versus turbidity currents: a modeling comparison of their dynamics and deposits, *in* A. H. Bouma and C. G. Stone, eds., Fine-grained turbidite systems, AAPG Memoir 72/ SEPM Special Publication No. 68, p. 57–72.

Chapter 6

Debris Flows vs. Turbidity Currents: a Modeling Comparison of Their Dynamics and Deposits

Lincoln F. Pratson
Division of Earth and Ocean Sciences, Duke University
Durham, North Carolina, U.S.A.

Jasim Imran
Department of Civil Engineering, University of South Carolina
Columbus, South Carolina, U.S.A.

Gary Parker
St. Anthony Falls Laboratory, University of Minnesota
Minneapolis, Minnesota, U.S.A.

James P. M. Syvitski
Eric Hutton
Institute of Arctic and Alpine Research, University of Colorado
Boulder, Colorado, U.S.A.

ABSTRACT

Debris flows tend to conserve their density, whereas turbidity currents constantly change theirs through erosion, deposition, and entrainment. Numerical models illustrate how this distinction leads to fundamental differences in the behaviors of debris flows and turbidity currents and the deposits they produce. The models predict that when begun on a slope that extends onto a basin floor, a debris flow will form a deposit that begins near its point of origin and gradually thickens basinward, ending abruptly at its head. By contrast, deposition from an ignitive turbidity current (i.e., one that causes significant erosion) will largely be restricted to the basin floor and will be separated from its origin on the slope by a zone of erosion. Furthermore, the turbidite will be thickest just beyond the slope base and thin basinward. These contrasting styles of deposition are accentuated when debris flows and turbidites are stacked.

INTRODUCTION

Subaqueous debris flows and turbidity currents are separate processes that generally produce distinctive deposits. What makes the two types of flows different is density and clay content. Debris flows are dense and have enough clay so that a mud matrix supports coarser-grained material as the flow moves (Middleton and Hampton, 1976). Turbidity currents are dilute and carry clay like other grain sizes in a solution of water and dispersed sediments, the latter of which are kept in suspension by fluid turbulence (Middleton and Hampton, 1976). This difference explains why debris flow deposits tend to be massive and poorly sorted, and turbidites are normally graded. When a debris flow stops, the mud matrix inhibits the settling of coarser grains jumbled during transport, and the mixture freezes. When a turbidity current stops, all grain sizes are free to settle. The heaviest, generally coarsest grains settle first and the finest, lightest grains settle last, leading to a deposit that fines upward.

What is not obvious is how density and clay content affect the behavior of debris flows and turbidity currents, and, consequently, the geometries of their deposits. This is because although there are differences between the two types of flows, there are also similarities. Principal among these is that both are propelled downslope by gravity and fluid pressure while being retarded by friction.

The purpose of this chapter is to compare and contrast the dynamics and deposits of debris flows and turbidity currents through numerical modeling. The numerical models are mathematical translations of the standard definitions for a debris flow and turbidity current. They encapsulate the fundamental physics of the flows as observed in the field and laboratory, and as interpreted from sedimentologic and stratigraphic data. The models are used to: (1) illustrate the major influences that drive debris flows and turbidity currents; (2) predict how divergences in the behavior of these flows lead to differences in the geometries of their deposits; and (3) anticipate the stratal patterns that would be produced by the stacking of debris flows versus turbidites.

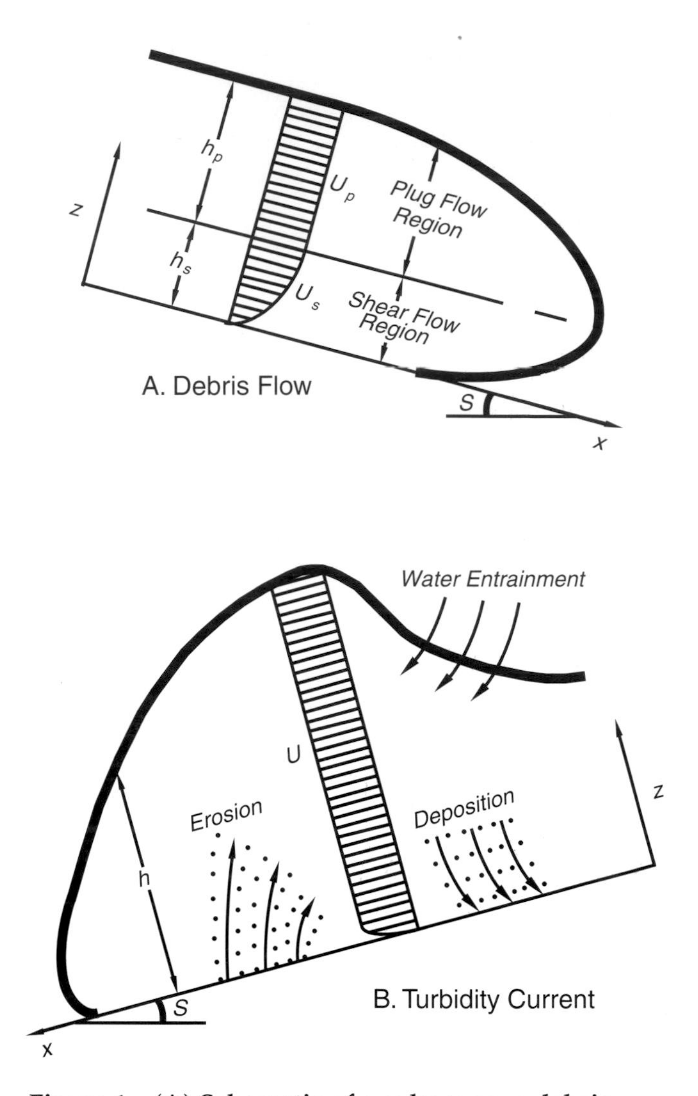

Figure 1—(A) Schematic of a subaqueous debris flow. Flow movement is divided into two regions: a lower shear layer, h_s, in which velocity, U_s, increases with height above the bed due to shearing; and a plug layer, h_p, in which velocity, U_p, is uniform. (B) Schematic of a turbidity current indicating the sources and sinks of mass and momentum to the current. *S* is sea-floor slope, *h* is current thickness, and *U* is current velocity.

MATHEMATICAL MODELS

Conceptual Framework

A debris flow is a relatively dense, viscous, sediment–water mixture that moves downslope like wet concrete (Middleton and Hampton, 1976). Its viscosity stems from the cohesion of the wet clay that forms the flow matrix. This cohesion gives the matrix a shear strength. Shear stresses in the bottom of the flow generated by its movement exceed the matrix shear strength and cause the flow to shear. This shearing decreases upward through the flow as stresses drop and velocity increases (Figure 1A). At the height that the shear stress becomes less than the matrix shear strength, shearing stops and the flow moves as a plug with uniform velocity (Figure 1A). Thus a debris flow can be subdivided into a shear layer overlain by a plug layer (Lin and Leblond, 1993).

Laboratory experiments show that a subaqueous debris flow causes negligible sea-floor erosion (Mohrig et al., 1998, 1999). This is in part because the head of the flow will often hydroplane, but it is also because the cohesive mud matrix of a debris flow inhibits the exchange of water and sediment across its surface. As a consequence, the volume of sediment and water in a subaqueous debris flow does not change significantly between when it starts and ends.

This is not true for a turbidity current. Although dilute, the current has a higher density than the water surrounding it. This gives rise to fluid pressure forces that along with the weight of the current when on a slope cause it to move (Figure 1B). The movement induces turbulent mixing of ambient water into the current, making it grow in size. At the same time, the current loses sediment through deposition. If moving fast enough, it can also gain sediment by eroding the bed (Figure 1B). Thus unlike a debris flow, the volume of sediment and water in a turbidity current changes continuously.

Governing Equations

Three equations describe the basic physics of a subaqueous debris flow (Lin and Leblond, 1993).

$$\underbrace{\frac{\delta h}{\delta t}}_{(a)} + \underbrace{\frac{\delta}{\delta x}\left[U_p h_p + \tfrac{2}{3} U_p h_s\right]}_{(b)} = 0 \quad (1)$$

$$\underbrace{\tfrac{2}{3}\frac{\delta}{\delta t}\left(U_p h_s\right) - U_p \frac{\delta h_s}{\delta t} + \tfrac{8}{15}\frac{\delta}{\delta x}\left(U_p^2 h_s\right) - \tfrac{2}{3} U_p \frac{\delta}{\delta x}\left(U_p h_s\right)}_{(a)}$$
$$= \underbrace{h_s g\left(1 - \frac{\rho_w}{\rho_m}\right) S}_{(b)} - \underbrace{h_s g\left(1 - \frac{\rho_w}{\rho_m}\right)\frac{\delta h}{\delta x}}_{(c)} - \underbrace{2\frac{\mu_m}{\rho_m}\frac{U_p}{h_s}}_{(d)} \quad (2)$$

$$\underbrace{\frac{\delta}{\delta t}\left(U_p h_p\right) + \frac{\delta}{\delta x}\left(U_p^2 h_p\right) + U_p \frac{\delta h_s}{\delta t} + \tfrac{2}{3} U_p \frac{\delta}{\delta x}\left(U_p h_s\right)}_{(a)}$$
$$= \underbrace{h_p g\left(1 - \frac{\rho_w}{\rho_m}\right) S}_{(b)} - \underbrace{h_p g\left(1 - \frac{\rho_w}{\rho_m}\right)\frac{\delta h}{\delta x}}_{(c)} - \underbrace{\frac{\tau_m}{\rho_m}}_{(d)} \quad (3)$$

These 1-D equations yield the thickness *(h)* and depth-averaged velocity *(U)* of the flow at any time *(t)* and downslope location *(x)* (additional variables defined in appendix).

Equation (1) is the continuity equation for a debris flow, and expresses the conservation of sediment and water combined in the flow. It states that the height of the flow (1a) is inversely proportional to its velocity (1b). So if the flow slows, it will thicken, and if it speeds up, it will thin. Note that there are no other controls on flow thickness. This is a critical assumption that does not allow for the possible evolution of a debris flow into a turbidity current. However, it is appropriate for this study, because its purpose is to compare debris flows and turbidity currents.

Equations (2) and (3) are the momentum equations for the shear and plug layers, respectively, in a debris flow. Separate equations are needed because of the different ways the two layers move. However, the same forces govern the change in momentum of both layers: (1) the weight of the flow scaled by the sea-floor slope (2b and 3b); (2) fluid pressure forces produced by lateral variations in flow height (2c and 3c); and (3) frictional forces (2d and 3d). For the shear layer, the frictional force relates to the viscosity of the flow matrix, whereas for the plug layer, it derives from the yield strength of the matrix.

Four 1-D equations capture the basic physics of a turbidity current (Parker et al., 1986).

$$\underbrace{\frac{\delta h}{\delta t}}_{(a)} + \underbrace{\frac{\delta Uh}{\delta x}}_{(b)} = \underbrace{e_w U}_{(c)} \quad (4)$$

$$\underbrace{\frac{\delta Ch}{\delta t}}_{(a)} + \underbrace{\frac{\delta UCh}{\delta x}}_{(b)} = \underbrace{v_s\left(E_s - r_o C\right)}_{(c)} \quad (5)$$

$$\underbrace{\frac{\delta Uh}{\delta t} + \frac{\delta U^2 h}{\delta x}}_{(a)} = \underbrace{RgChS}_{(b)} - \underbrace{\tfrac{1}{2} Rg \frac{\delta Ch^2}{\delta x}}_{(c)} - \underbrace{u_*^2}_{(d)} \quad (6)$$

$$\underbrace{\frac{\delta Kh}{\delta t} + \frac{\delta UKh}{\delta x}}_{(a)} = \underbrace{u_*^2 U + \tfrac{1}{2} U^3 e_w}_{(b)} - \underbrace{\varepsilon_0 h}_{(c)} \quad (7)$$
$$\underbrace{-Rg v_s Ch}_{(d)} - \underbrace{\tfrac{1}{2} RgChUe_w}_{(e)} - \underbrace{\tfrac{1}{2} Rghv_s\left(E_s - r_o C\right)}_{(f)}$$

In addition to the thickness and depth-averaged velocity of the current, these equations also yield its depth-averaged bulk sediment concentration *(C)* at any time or downslope location (additional variables defined in appendix).

Equations (4) and (5) are the continuity equations for the water and sediment, respectively, in the current. They are similar to the debris-flow continuity [equation (1)] in that the thickness of water in the current *(h)* and the thickness of sediment suspended in that water *(Ch)* are inversely proportional to the current velocity. However, each variable also depends on an additional factor, which is why they require separate continuity equations. The water in the current is increased by entrainment (4c). The sediment in the current can increase or decrease depending on the net rate of erosion (5c), which is positive if erosion exceeds deposition and negative if vice versa.

Only a single equation, (6), is needed to describe the momentum of a turbidity current because it is turbulent throughout. Whereas it is different from the momentum equations for the shear and plug layers in a debris flow, (6) contains the same three sources and sinks of momentum: the weight of the current (6b), fluid pressures within it (6c), and friction (6d). In this case, the friction is that generated along the base of the current as it moves over the sea bed.

Finally, the balance of turbulent kinetic energy (7) is used to constrain the erosion caused by a turbidity current. Turbulent kinetic energy is created by the current's velocity (7b); the faster it moves, the more energy it generates. At the same time, this energy: (1) is dissipated by the current viscosity (7c); (2) goes into keeping sediments in suspension (7d); and (3) is used to entrain water (7e). Any remaining energy is available for eroding the bed (7f). So if no energy remains, the current has reached its carrying capacity and can not erode.

Numerical Solution

The debris flow model (equations 1–3) and the turbidity current model (equations 4–7) are solved numerically in a Lagrangian reference frame using a staggered-grid, finite-difference method. The details of this method are being published (Pratson, et al., in press; and Imran, et al., in press). All model runs start with the same initial conditions. A parabolic-shaped flow 1 km long and with a maximum height of 4 m is begun from rest on a slope that dips 3° for 5 km before becoming flat (Figure 2A.1).

The only initial condition that is different between the models is the starting flow density. Subaqueous debris flows typically range in density between 1200 and 2000 kg/m^3, and a moderate density of 1500 kg/m^3 is used here. Turbidity currents on the other hand have much lower densities. Although there has been considerable discussion of "high-density" turbidity currents, a bulk concentration *(C)* of 10% is generally believed to be the maximum a turbidity current can reach. This is because when concentrations exceed 10%, sediments come in constant contact with one another, which is characteristic of a debris flow and not a turbid flow. In the turbidity current model, the flow is started with a bulk concentration of 5%, which, in sea water with a density of 1030 kg/m^3, equates to a current density of 1111 kg/m^3.

COMPARISON OF SIMULATED DYNAMICS

Similarities between Debris Flows and Turbidity Currents

As pointed out above, weight, fluid pressure forces, and friction forces govern the momentum of both debris flows and turbidity currents. Consequently, these forces produce common behaviors in both types of flows. The impact and interactions of these forces are illustrated by introducing them into the debris flow model one at a time.

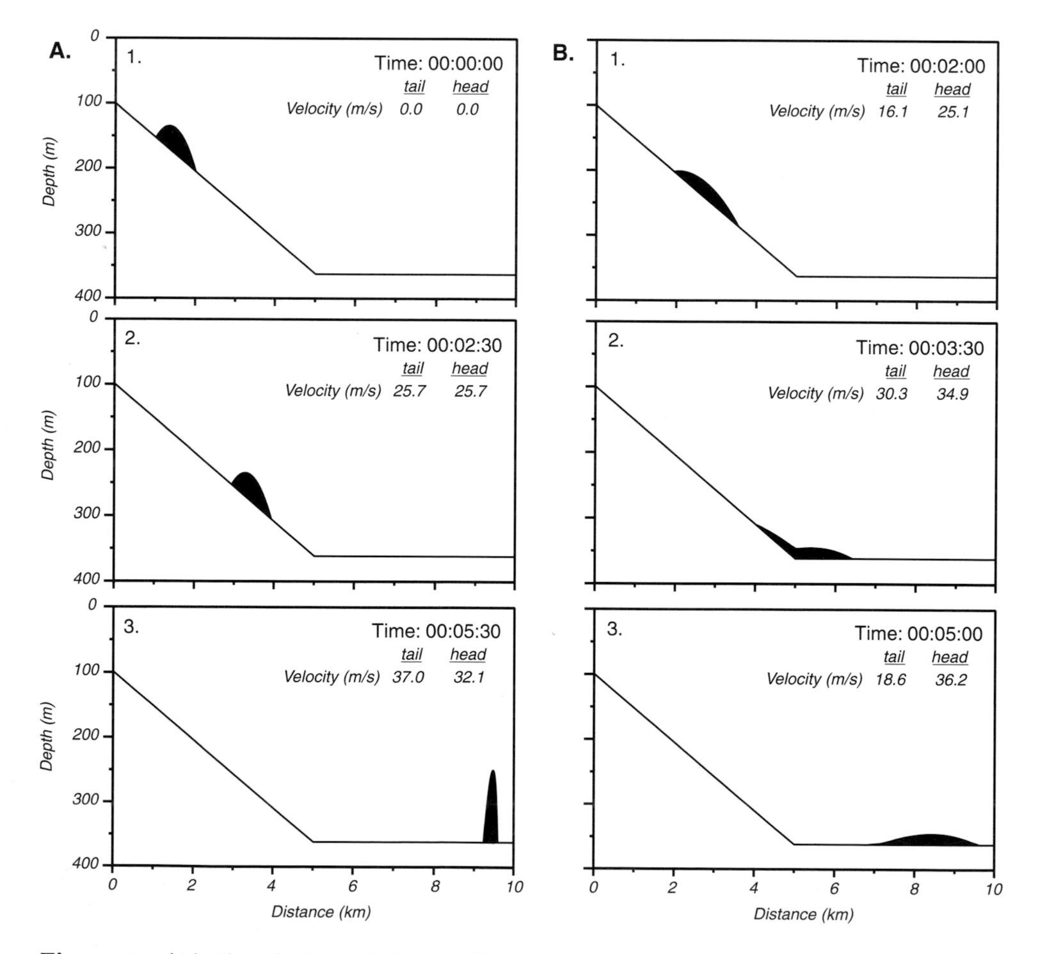

Figure 2—(A) Simulation of debris flow movement due only to the downslope component of flow weight. (B) Debris flow movement when fluid pressure is added to the simulation. Flow thickness in both simulations has been exaggerated 5X.

Flow Weight

The contribution of flow weight to the movement of a debris flow/turbidity current is demonstrated by using terms (2b) and (3b) as the only sources of momentum in the debris flow model (i.e., 2c = 3c = 2d = 3d = 0). On the slope, the weight of the flow causes it to accelerate from rest (Figure 2A.2). The greater the slope angle and/or the mass of the flow, the quicker it accelerates. Note that the flow does not change form as it moves down the slope. Since the slope is of constant dip, all parts of the flow accelerate at the same rate (Figure 2A.2).

This changes when the flow passes out onto the flat seafloor at the base of slope. Here the surface gradient is zero and the weight of the flow adds no further momentum (i.e., 2b = 3b = 0). The flow stops accelerating and moves at the velocity it attained at the base of the slope. Because the tail of the flow traveled the greatest distance over the slope, it has the highest velocity, and begins to catch up to the head (Figure 2A.3). This causes the flow to contract and, since volume is conserved (1), thicken (Figure 2A.3).

Fluid Pressure

Figure 2B shows the result when fluid pressure forces (terms 2c and 3c) are added the model. As the flow begins to move down the slope, it now also starts to collapse and spread (Figure 2B.1). This is due to the migration of mass from the interior of the flow where it is thickest and fluid pressures are highest, to the margins of the flow where it is thinnest and fluid pressures are lowest. Toward the head of the flow, the spreading adds to the acceleration caused by the flow weight (Figure 2B.2). But toward the tail of the flow, the spreading decreases the acceleration. Note that the flow continues to spread even after it has passed out onto the flat sea floor where there is no further change in momentum due to flow weight (Figure 2B.3). This is because the varying thickness of the flow, although greatly diminished, maintains a pressure gradient toward the flow margins.

Friction

The complete model involves friction (2d and 3d), which slows and eventually stops a debris flow/turbidity current by retarding its movement in any direction. Toward the rear of the flow, fluid pressure and flow weight oppose one another, which in the presence of friction causes the momentum to rapidly fall below the threshold for movement. As a result, the tail does not travel far before stopping on the slope (Figure 3A). But toward the front of the flow, fluid pressure and flow weight act in concert to drive the flow downslope (Figure 3A). Here the impact of friction is minimized until the flow passes out onto the flat sea floor and spreading causes fluid pressures to fall below the level of friction (Figure 3B). The last part of the flow to be halted by this is the head (Figure 3C). So the effect of friction on the overall shape of the flow is to exaggerate the spreading induced by fluid-pressure gradients.

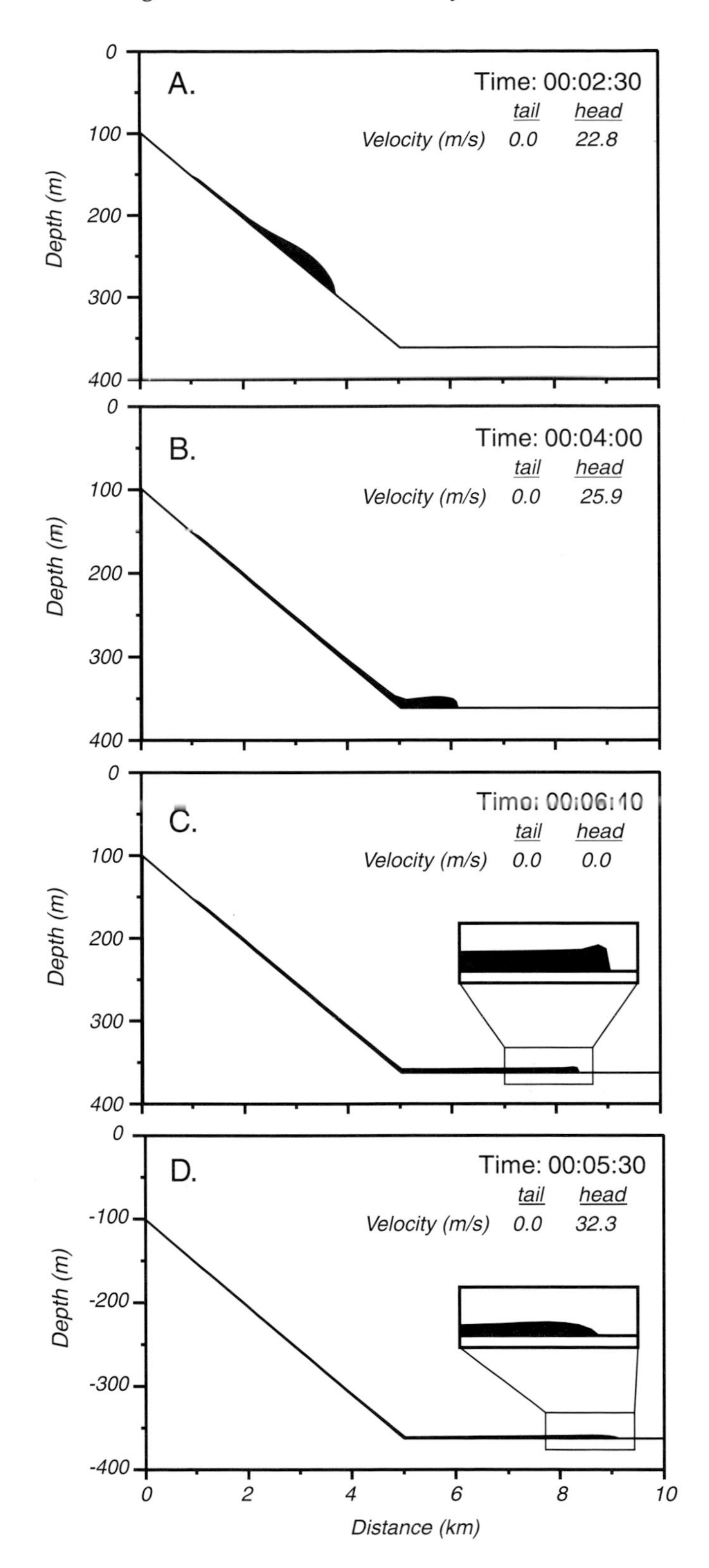

Figure 3—(A)–(C) Debris flow movement when friction is added to the simulation. (D) Simulation of turbidity current produced when there is no entrainment and no deposition or erosion. Because of the low friction in the model, the current is still moving. Flow thickness in both simulations has been exaggerated 5X. Insets show a blowup of the heads of the deposits. Note the similarity in their overall shape.

Relationship to the Turbidity Current Model

The debris flow model has been used to demonstrate the effects of flow weight, fluid pressure, and friction on the behavior of both a debris flow and a turbidity current. The same general results can be obtained using the turbidity current model when its continuity equations (4 and 5) are set to be conservative such as (3) in the debris flow model (i.e., when 4c = 5c = 0). This is demonstrated by Figure 3D, which shows the deposit produced by the turbidity current model when there is no entrainment and no deposition or erosion. The result is not identical to that generated by the debris flow model (Figure 3C) because of the variables that govern the weight, fluid pressure force, and friction force in the turbidity current model are different [e.g., shear velocity (6d) versus viscosity (2d) and yield strength (3d)]. But the result does verify that these three forces have the same fundamental effect on the relative motion and shape of the flows despite the different forms of the momentum equations in both models.

Differences between Debris Flows and Turbidity Currents

From a dynamic standpoint, the most significant difference between a debris flow and a turbidity current is that while the mud matrix of a debris flow inhibits the exchange of water and sediment across its surface, the dilute suspension of a turbidity current allows for it. The impact of this exchange on the behavior of the latter process is illustrated using the turbidity current model in the same fashion as the debris flow model above. However, in the turbidity current model, flow weight, fluid pressure, and friction are all accounted for. What is successively added to the model are the effects of entrainment, erosion and deposition, the conservation of turbulent kinetic energy, and the frictional resistance of ambient water.

Entrainment

Entrainment of ambient water into a turbidity current increases its overall thickness while diluting it. Entrainment also generates friction along the upper surface of the current, which slows it down dramatically. Since the rate of entrainment (term 4c) is a function of velocity, it is not uniform over the length of the current. Entrainment is most rapid at the head where the current is moving fastest (Figure 4a). This thickens the head, which in the model becomes a peak with a sharp front. The peak is the result of the model only simulating the depth-averaged velocity and density along the current. In a real turbidity current, velocity and density decrease with height above the sea floor, and the head is swept back over the trailing body. Note in the model that as the head slows, the faster moving body immediately behind it catches up, contracts, and thickens. Toward the rear, the body moves slower and thus thins, tapering out at the tail (Figure 4a).

Erosion and Deposition

Dilution due to entrainment is abetted by deposition and counterbalanced by erosion. If erosion does not exceed deposition soon after the current begins to move,

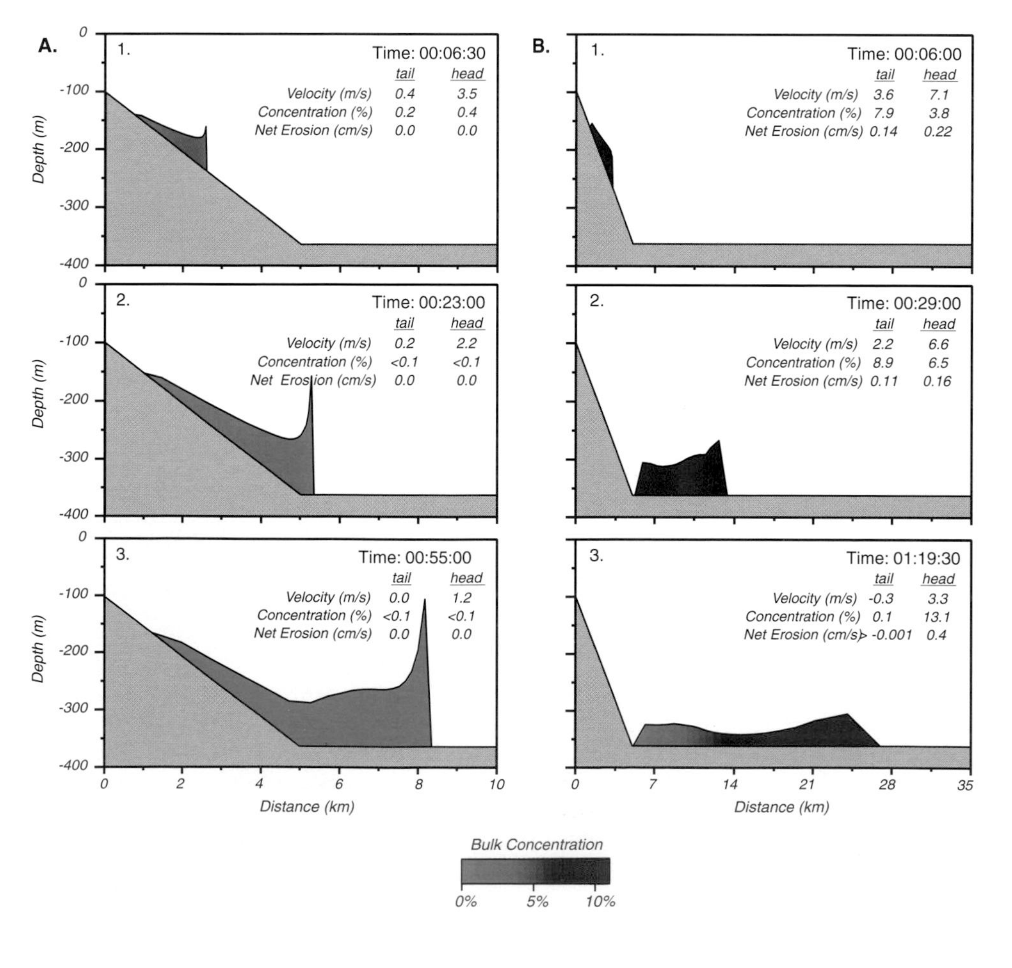

Figure 4—(A) Simulation of a turbidity current movement due to weight, fluid pressure, friction, and entrainment. (B) Turbidity current movement when erosion and deposition are added to the simulation. Flow thickness in both simulations has been exaggerated 5X.

it becomes too dilute and "dies." This occurs when the initial density of the current is too low, the sea-floor slope is too gradual, and/or the sea bed is not easy enough to erode. However, if net erosion is achieved quickly, the current will begin to take off. The eroded sediments increase the mass of the current, fueling its acceleration and further erosion in a positive feedback.

This latter case is illustrated in Figure 4B, which shows the results when erosion and deposition (term 5c) are factored into the turbidity current model. As the current moves downslope and thickens through entrainment, it maintains a relatively high density through the net erosion. This is true not only in the head of the current, but over its entire body, including the tail (Figure 4B.1). As a result, the current moves downslope en masse, and does not become significantly stretched by fluid pressures and friction (compare Figure 4, A, B).

Thickening of the current is also dampened (compare Figure 4, A, B). This is because the relatively high sediment concentrations depress mixing across the surface of the current (see appendix), signficantly reducing the entrainment, even in the fast-moving head.

By the time the tail passes out onto the flat seafloor, the overall concentration of the current exceeds its initial concentration of 5% (Figure 4B.2). In the central body of the current, concentrations exceed 10%. This region contracts and thickens as the tail joins the body stacked up behind the decelerating head (Figure 4B.2). The thickening leads to high fluid pressures that cause the body to then collapse. The rear of the current is driven back toward the slope, draining its momentum and forcing it to deposit sediments on the sea floor that it had previously eroded. Simultaneously, the head is accelerated basinward, leading to increased erosion and the generation of concentrations in this region that are >10% (Figure 4B.3). In between the two, the body essentially stops; erosion turns to deposition, and the current is diluted.

Turbulent Kinetic Energy

Turbidity currents cause erosion through the shear stress they exert on the sea bed as they move over it. Parker et al. (1986) propose that the magnitude of this shear stress depends on the balance of turbulent kinetic energy in the current given in equation (7). Figure 5A illustrates the result when this balance is included in the turbidity current model. As the current moves downslope, the erosion it causes is now constrained. At no point does the current exceed the critical concentration of 10%. Instead, concentrations in the current briefly reach a maximum of 7%, while the current is at its peak velocity and then drop to 5% or less for the remainder of the simulation (Figure 5A).

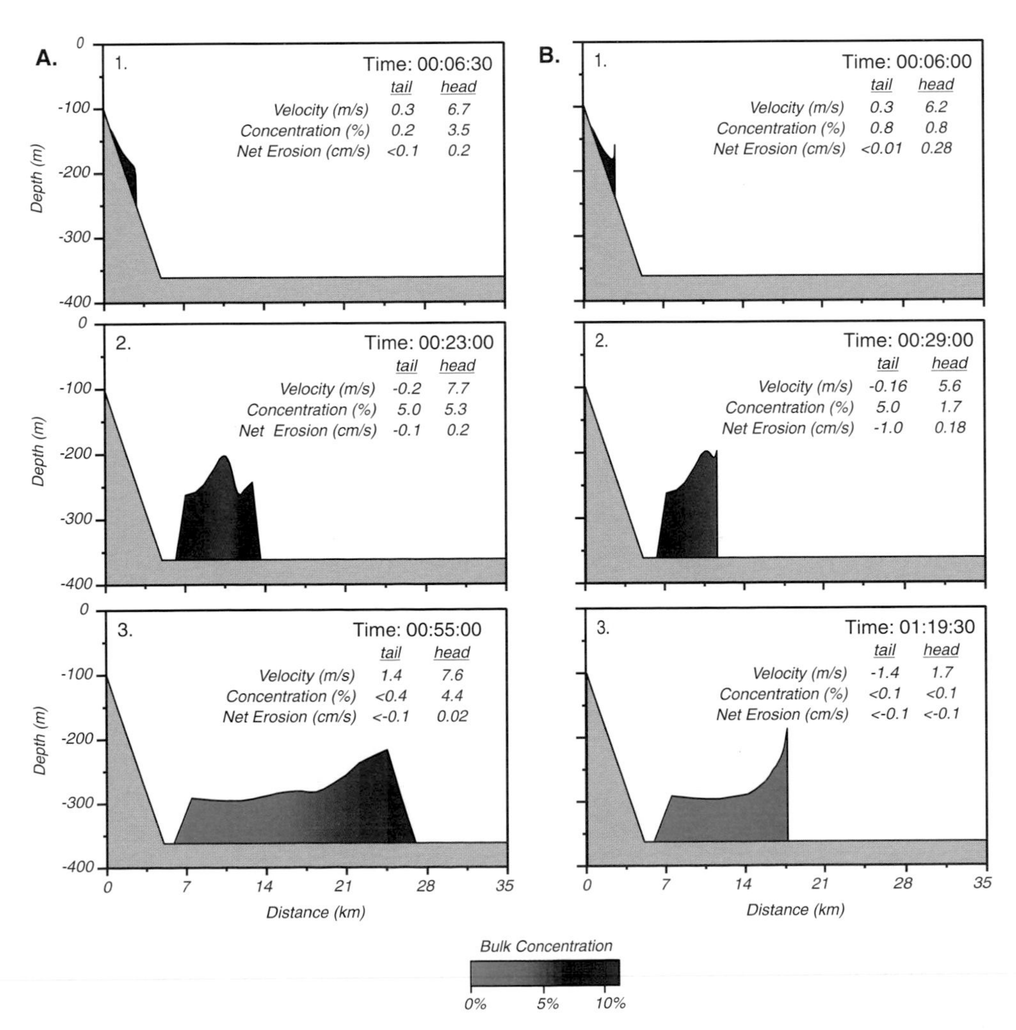

Figure 5—(A) Turbidity current movement when the balance of turbulent kinetic energy is added to the simulation. (B) Turbidity current movement when the resistance of ambient water is added to the simulation. Flow thickness in both simulations has been exaggerated 5X.

Resistance of Ambient Water

To move, a turbidity current must push ambient water out of its way. The resistance of the water causes the current to develop an abrupt front similar to a shock wave. The velocity of this front, U_f, can be approximated using the empirical relationship of Britter and Linden (1980)

$$U_f = 1.5RgCh_f \qquad (8)$$

where h_f is the thickness of the head of the current. When this equation is included in the turbidity current model, the head of the current moves more slowly, advancing as a wall through the surrounding water (Figure 5B). The body of the current behind the head is forced to slow as well, which thickens the current front, drops its speed, and reduces its rate of erosion.

DEPOSIT SIMULATIONS

In this section, the debris flow and turbidity current models are used to predict how the distinctive behaviors of these two flows affect the geometries of their deposits. Since the initial attributes of the flow and the form of the sea bed can also influence deposit geometry, the models are run using a range of initial flow sizes and densities, and slope lengths and dips.

The debris flow deposits have the same density as the flow that produced them, so, in effect, they are the deposits formed immediately after the flow has stopped. The turbidity current deposits, on the other hand, have zero porosity. This is because of a simplification in the turbidity current model in which pore volume was not accounted for in either erosion of the seabed or deposition on it. If included, pore volume would have increased the depth of erosion and thickness of deposition; otherwise, the results would be similar.

For this and other reasons, such as uncertainties concerning the scaling of laboratory-derived relationships for entrainment and erosion up to natural turbidity currents, the comparisons of the simulated deposits are qualitative. They focus on predicted differences in the relative extents, locations, and shapes of debris flow and turbidity current deposits. As is shown, these differences are repeated under a variety of initial and boundary conditions. If they exist in nature, they should be observable in seismic reflection data and outcrops.

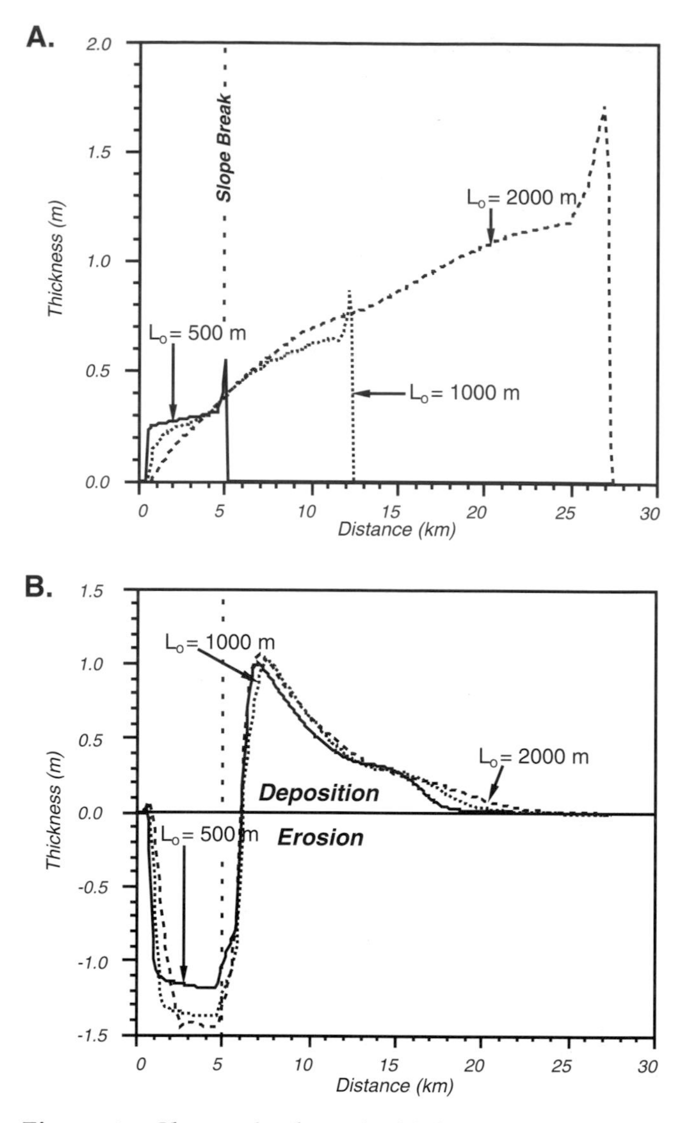

Figure 6—Change in deposit thickness and length with change in initial size of a debris flow and turbidity current. Initial lengths and heights of the flow/current were 500 × 2 m, 1,000 × 4 m, and 2,000 × 8 m. Vertical axis is thickness above or below the initial sea floor. Horizontal axis is distance along the sea floor. All flows/currents were begun at a distance of 500 m. (A) Debris flow deposits. L_o indicates original flow length. (B) Turbidite deposits. Negative heights indicate erosion and positive heights indicate deposition.

Initial Conditions

Initial Flow Size

As the initial size of a debris flow increases, so too does the length and thickness of the deposit it produces (Figure 6A). A somewhat similar result is obtained with the turbidity current model. Increasing the initial size of the current leads to a correspondingly larger deposit, and in this case, one that grows more in length than in height (Figure 6B).

Beyond this similarity, there are several key differences between the debris flow and turbidity current deposits. One is that the debris flows extend from the basin floor back up the slope all the way to near their point of origin. The turbidites, on the other hand, are confined to the basin floor and are separated from their point of origin by a zone of erosion that extends to the base of the slope. Second, the debris flow deposits thicken from their tails to their heads, which end relatively abruptly (Figure 6A). In contrast, the turbidites are thickest near their tail and thin toward their head (Figure 6B). Finally, for initial lengths of 500–1000 m, the simulated turbidites extend farther seaward than the debris flows. This is in part due to the greater mobility of the currents, but also due to their ability to exchange water and sediment through entrainment, deposition, and erosion.

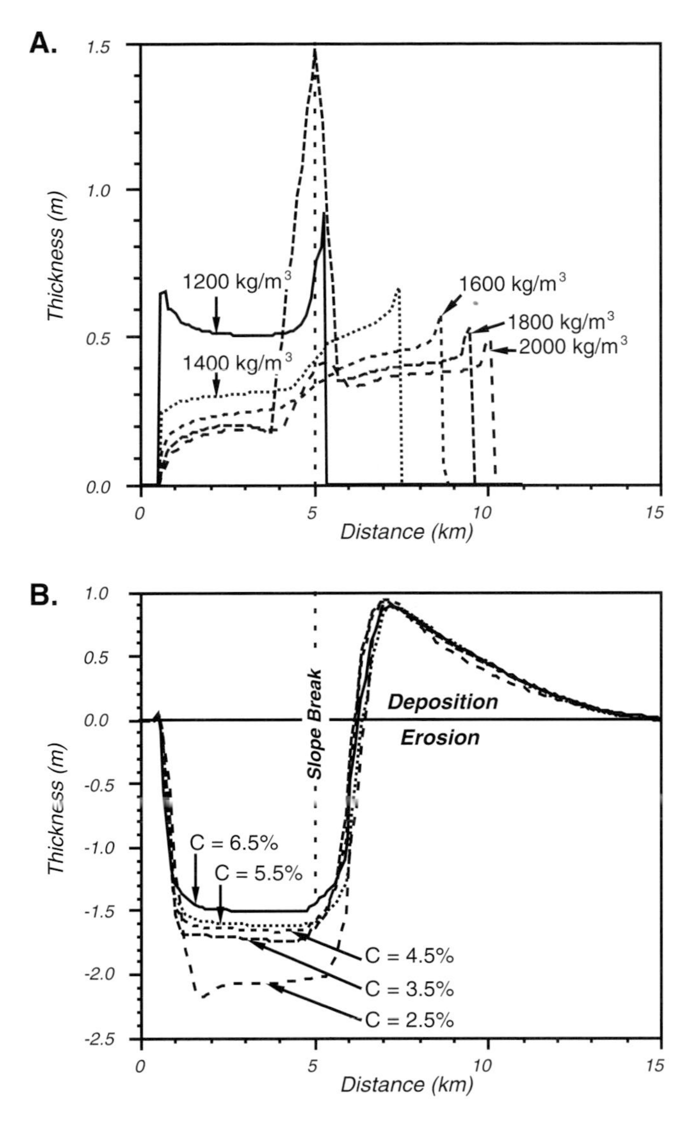

Figure 7—Change in deposit thickness and length with change in initial density of a debris flow and turbidity current. (A) Debris flow deposits. Density is expressed in terms of kilograms per cubic meters. Large spike in 1600 kg/m^3 deposit due to flow thickening at the slope base that was still present when the simulation was stopped. (B) Turbidity current deposits. Density is expressed in terms of bulk concentration (%). Note that increasing initial density results in progressively less erosion of the slope and similar-sized basin floor deposits. See text for further discussion.

Initial Flow Density

In the debris flow model, higher initial densities lead to successively longer and thinner debris flow deposits (Figure 7A). This is because higher densities impart more momentum to a flow, which drives it further downslope.

Increasing the initial density in the turbidity current model, or the bulk sediment concentration as expressed in Figure 7B, leads to a different result. Turbidity currents cause progressively less erosion of the slope yet produce basin floor deposits that are almost identical in size (Figure 7B). This appears to be due to the effect of the initial density of the current on its subsequent entrainment of ambient water and dispursement of turbulent kinetic energy. Higher initial densities require a greater expenditure of turbulent kinetic energy in keeping the sediments in suspension. This in turn reduces the energy available for erosion. The higher initial densities also dampen entrainment, reducing the growth rate of the current and thus the space it has for accommodating new sediments gained through erosion. The combined effect is a reduction in the erosion rate. For the case shown, this leads to similar-sized turbidites. The result is intriguing, because it suggests that a relationship may exist between the size of a turbidite and the initial size of the turbidity current that generated it.

Boundary Conditions

Slope Length

Figure 8 shows the deposits produced by the models when the length of the slope is increased

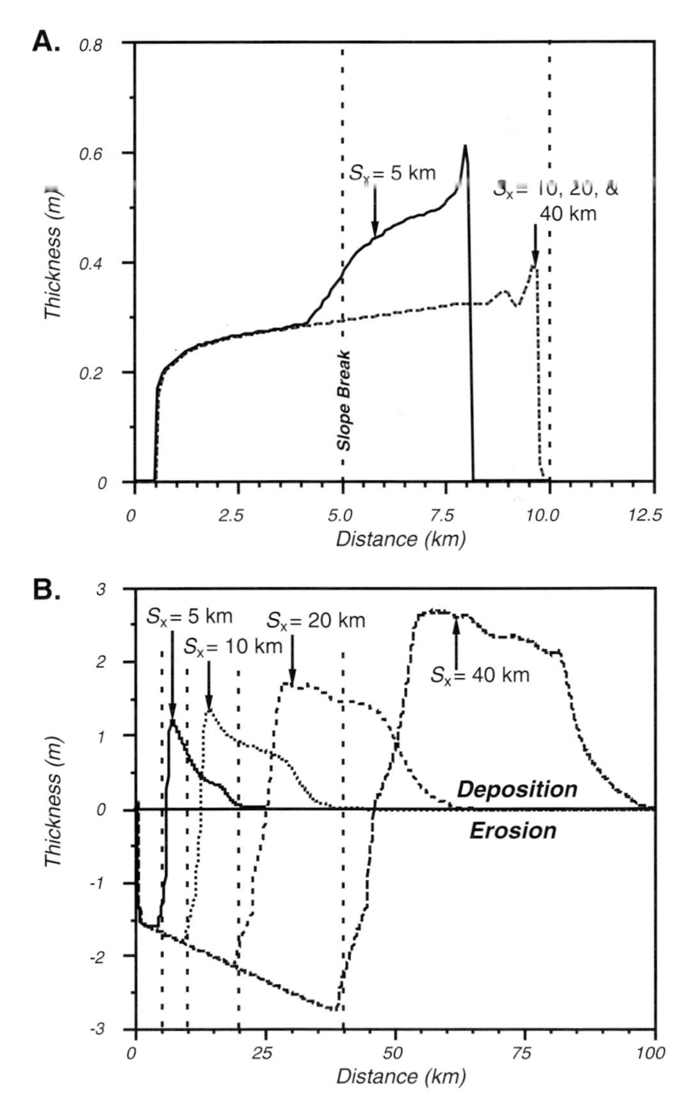

Figure 8—Change in deposit thickness and length with change in slope length, S_x. (A) Debris flow deposits. Note that for slope lengths of 10, 20, and 40 km, the deposit is the same. (B) Turbidity current deposits.

5–40 km. From slope lengths of 5–10 km, the debris flow deposits become longer and thinner (Figure 8A). However, longer slopes produce no further change in the final deposit. This is because a debris flow is unable to erode and increase mass, so it can only travel a finite distance down an incline before becoming too thin to move.

Turbidity currents are the opposite. As slope length is increased, the turbidity current model generates flows that cause more erosion of the slope, travel a greater distance basinward, and create a larger deposit on the basin floor (Figure 8B). These differences are a direct consequence of the current's being able to add sediment through erosion and accommodate this additional sediment by increasing size through entrainment.

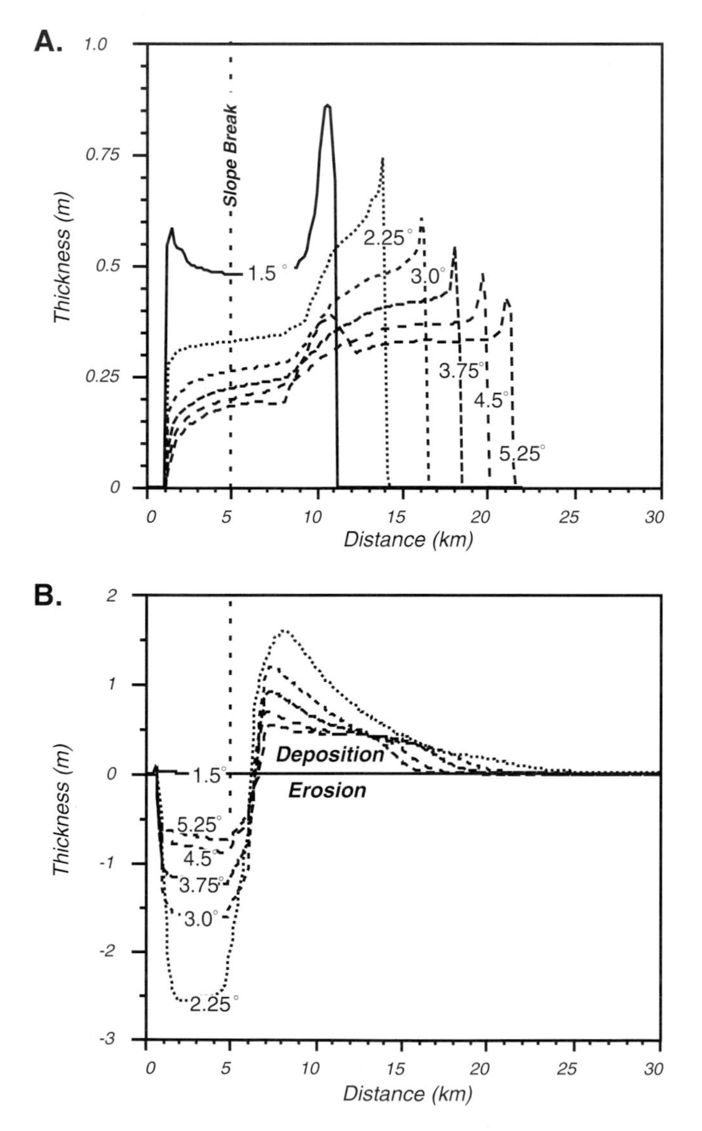

Figure 9—Change in deposit thickness and length with change in slope gradient (°). (A) Debris flow deposits. (B) Turbidity current deposits. Deposition is restricted to the slope when the slope gradient is 1.5°. At gradients above 2.5°, slope erosion and basin floor deposition decrease. See text for further discussion.

Slope Gradient

Increasing the slope angle increases the runout distance of debris flow deposits (Figure 9A). Steeper slopes impart more momentum to the flows, propelling them farther basinward. But as before, the unchanging volume of the flows limits the extent of their deposits. The flows stop when their heads become too thin to move.

The impact of slope gradient on turbidite deposition is different. As noted above, when a turbidity current is initiated, it must quickly gain mass through erosion in order to offset being diluted by entrainment and deposition. At too low a slope (e.g., 1.5° in Figure 9B), the turbidity current does not accelerate fast enough to achieve this erosion and dies, depositing its sediment load on the slope near where it began (Figure 9B). However, as Parker et al. (1986) have shown, there is a threshold gradient at or above which the turbidity current will quickly gain mass by erosion shortly after initiation (~2.25° in Figure 9B). When this occurs, the turbidity current is referred to as being ignitive (Parker et al., 1986). In Figure 9B, such currents erode to just beyond the slope base before depositing their sediment load out on the flat basin floor.

Somewhat surprisingly, higher slope gradients do not lead to increased slope erosion and larger basin floor deposits such as resulted from increasing the slope length. Instead, erosion is reduced and the basin floor deposits get smaller (e.g., ≥ 3° in Figure 9B). The principal causes for this appear to be entrainment and the turbulent kinetic energy available for erosion. As the sea floor is steepened, the current accelerates quicker and erodes more in a shorter time following its initiation. However, this increased erosion quickly dampens entrainment. With a slower rate of thickening, the rate at which the current can accommodate and transport eroded sediments slows as well, translating into a slower erosion rate, and ultimately smaller basin floor deposits.

Thus the turbidity current model predicts that a turbidity current of a given initial size and density will cause the most erosion and create the largest basin floor deposit when it travels across a slope that is inclined at or just above the threshold angle for ignition. If the slope gradient lies below this angle, then the turbidity current will die, if it lies too far above this angle, it will erode less and create a smaller basin floor deposit.

STRATIGRAPHIC SIMULATIONS

The debris flow and turbidity current models predict there should be fundamental differences in the relative extent, location, and shape of a debris flow deposit versus a turbidite. But because of the limited exposure of outcrops and the finite resolution of seismic reflection data, it is often not possible to resolve the areal geometry of individual deposits. More easily observed in these data is the stacking of multiple deposits. Stacking patterns of debris flows and turbidites are simulated for the simple slope and basin floor configuration used in the analyses above. This geometry approximates the slope-rise transition across many passive continental margins. In the models, the debris flows and turbidity currents are begun at random locations on the sea floor between 100

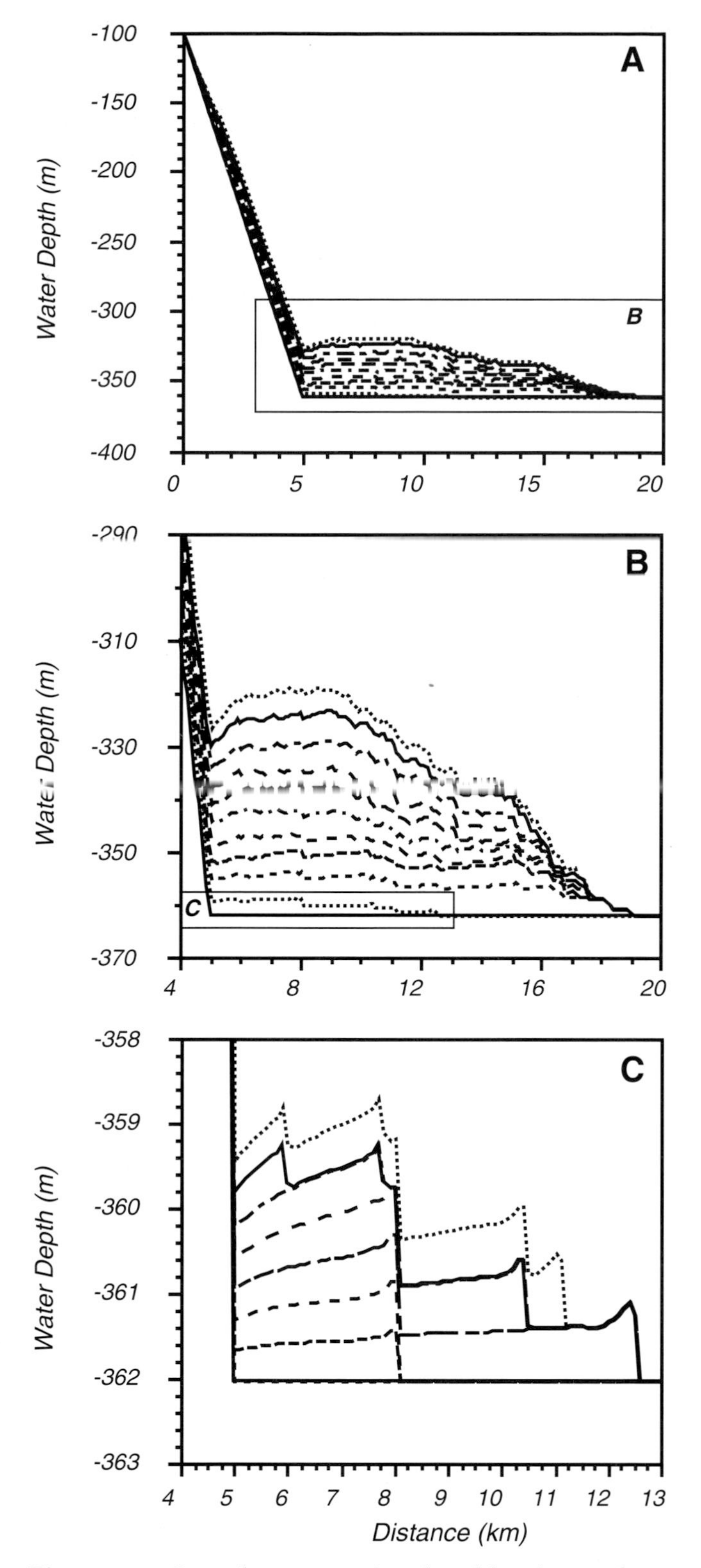

Figure 10—Stratal patterns simulated by the stacking 100 debris flows. (A) Overall stacking pattern. Lines represent the surface of every tenth deposit. (B) Close-up of slope-basin floor transition. (C) Further close-up of the first ten deposits.

and 250 m water depth. The initial size of the flows is also varied randomly, with lengths and maximum heights ranging from 500 × 2 m to 2000 × 8 m, respectively.

Stacked Debris Flows

Stacking of debris flows produces a sequence in which successive deposits are progressively rotated backwards (Figure 10). This stacking pattern is a direct consequence of the basinward thickening of the individual deposits. The first debris flow tilts the basin floor at the base of the slope back toward the slope (Figure 10C). This dip then becomes accentuated as additional debris flows are stacked on the first (Figure 10A, B).

The stacking pattern is complicated by the varying size of the debris flows. Small debris flows come to rest on the backs of the flows that preceded them; large debris flows cascade over the previous flows even farther out onto the basin floor (Figure 10C). This extends the sequence seaward even as it grows vertically and becomes more tilted (Figure 10A).

Importantly, at the same time the basin floor is aggrading and prograding, so too is the slope (Figure 10A, B). This is because each debris flow deposit begins near its origin, and extends down the slope out onto the basin floor. And since these deposits thicken down the slope, successive deposits gradually reduce the lower slope gradient.

Stacked Turbidites

When turbidites are stacked on the simple slope and basin floor, a different stratigraphic pattern develops. Successive turbidity currents downcut rather than build up the slope (Figure 11A, B). Initially, the greatest erosion is at the slope base, where the currents expend their energy before moving out onto the low gradient basin floor (Figure 11C). This leads to the formation of an erosional trough similar to that found at the foot of waterfalls [e.g., Niagara Falls (Van Diver, 1985)].

Seaward of the trough, the turbidity currents deposit their load. The first turbidite is thickest within a short distance of the trough and tapers off basinward (Figure 11C). This creates a depositional bulge that impedes the flow of subsequent turbidity currents. As they cross over the backside of the bulge, the later currents are slowed and forced to deposit much of their sediment load. The backside of the bulge builds up preferentially, evolving into a sedimentary sequence that migrates updip (Figure 11B, C).

As it migrates updip, the sequence progrades into the trough at the slope base (Figure 11B). The axis of the trough now lies between an erosional slope on one side and a depositional slope on the other. With further turbidity current deposition, the sedimentary sequence completely infills the original erosional trough (Figure 11B). The sequence begins to onlap the slope unconformity and pushes the existing trough up dip.

Interestingly, troughs similar to that produced by the turbidity current model have been discovered at the base of the North Carolina continental slope (Bunn and McGregor, 1980), the Malta Escarpment in the Eastern Mediterranean Sea (Cita et al., 1982; Biju-Duval et al., 1983), and the New Jersey continental slope (Farre and Ryan, 1987). Along the foot of the New Jersey continental slope, the troughs are 4–7 km

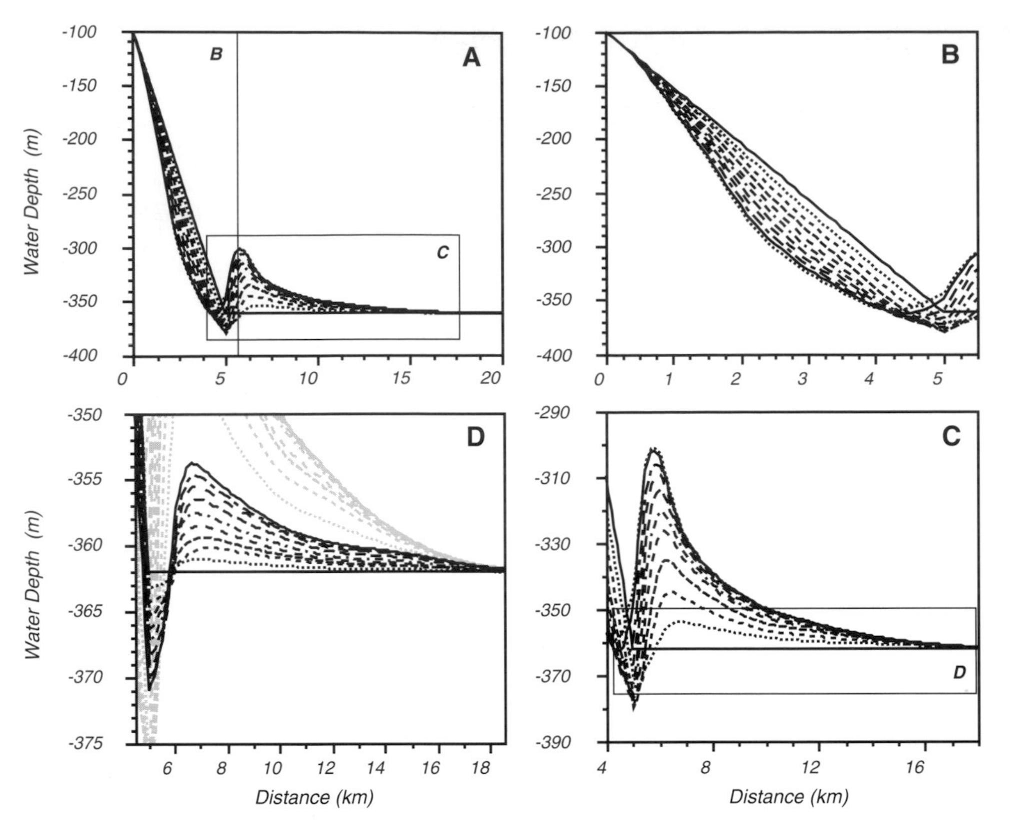

Figure 11—Stratal patterns simulated by the stacking 100 turbidites. (A) Overall stacking pattern. Lines represent the surface of every tenth deposit. (B) Close-up of slope erosion. (C) Close-up of basin floor. (D) Further close-up of the first ten turbidites (black lines).

long, 1–2 km wide, and 30–70 m deep (Figure 12A). A seismic profile across one of these troughs is shown in Figure 12B. Nearby Deep Sea Drilling Project (DSDP) boreholes indicate that on its continental slope side, the trough is walled by an outcrop of Eocene sediments (Farre and Ryan, 1987). On its continental rise side, the trough is walled by Pleistocene sediments that onlap the slope. Furthermore, these Pleistocene sediments are skewed in thickness, so they form a ridge that runs along the base of the slope (Figure 12B).

The mounded geometry of the Pleistocene sequence and the way it onlaps the erosional slope unconformity parallels the stratigraphy produced by the turbidity current model (Figure 11). The similarity suggests that the Pleistocene sequence may be largely composed of turbidites. Unfortunately, further evidence to support this is lacking. The reflectors within the sequence are poorly defined and do not indicate a simple updip migration like that predicted by the model. More importantly, the deposits that compose the Pleistocene sequence are ill defined. Core recovery at the DSDP sites was poor and all that could be interpreted is that the sequence consists of "mass flows" (Farre and Ryan, 1987; Poag, 1987), a term that lumps turbidites and debris flows together. However, the model results are intriguing and suggest that turbidity currents may be the cause for the troughs along the slope–rise transition offshore New Jersey and possibly elsewhere.

CONCLUSIONS

(1) The main similarity between the dynamics of a subaqueous debris flow and turbidity current is that their momentum is dictated by flow/current weight, fluid pressures within the flow/current, and friction. The numerical models used in this study show that: (1) weight drives a flow/current downslope; (2) fluid pressures cause it to spread; and (3) friction elongates it further by slowing and eventually stopping the flow/current where the net momentum imparted by weight and fluid pressure fall below the frictional threshold for movement.

(2) The main difference between the dynamics of a subaqueous debris flow and turbidity current is that a debris flow conserves mass, whereas a turbidity current changes its mass through the entrainment of ambient water and the deposition and erosion of sediment. The numerical models illustrate that entrainment thickens and dilutes a current while significantly slowing it down. Deposition and erosion also alter current density. When erosion does not offset the dilution caused by entrainment and deposition, the current will wane. But when it does, the current accelerates, creating a positive feedback that leads to more erosion and further acceleration. This feedback is regulated by the store of turbulent kinetic energy in the current, which places a dynamic threshold on the maximum amount of sediment the current can carry.

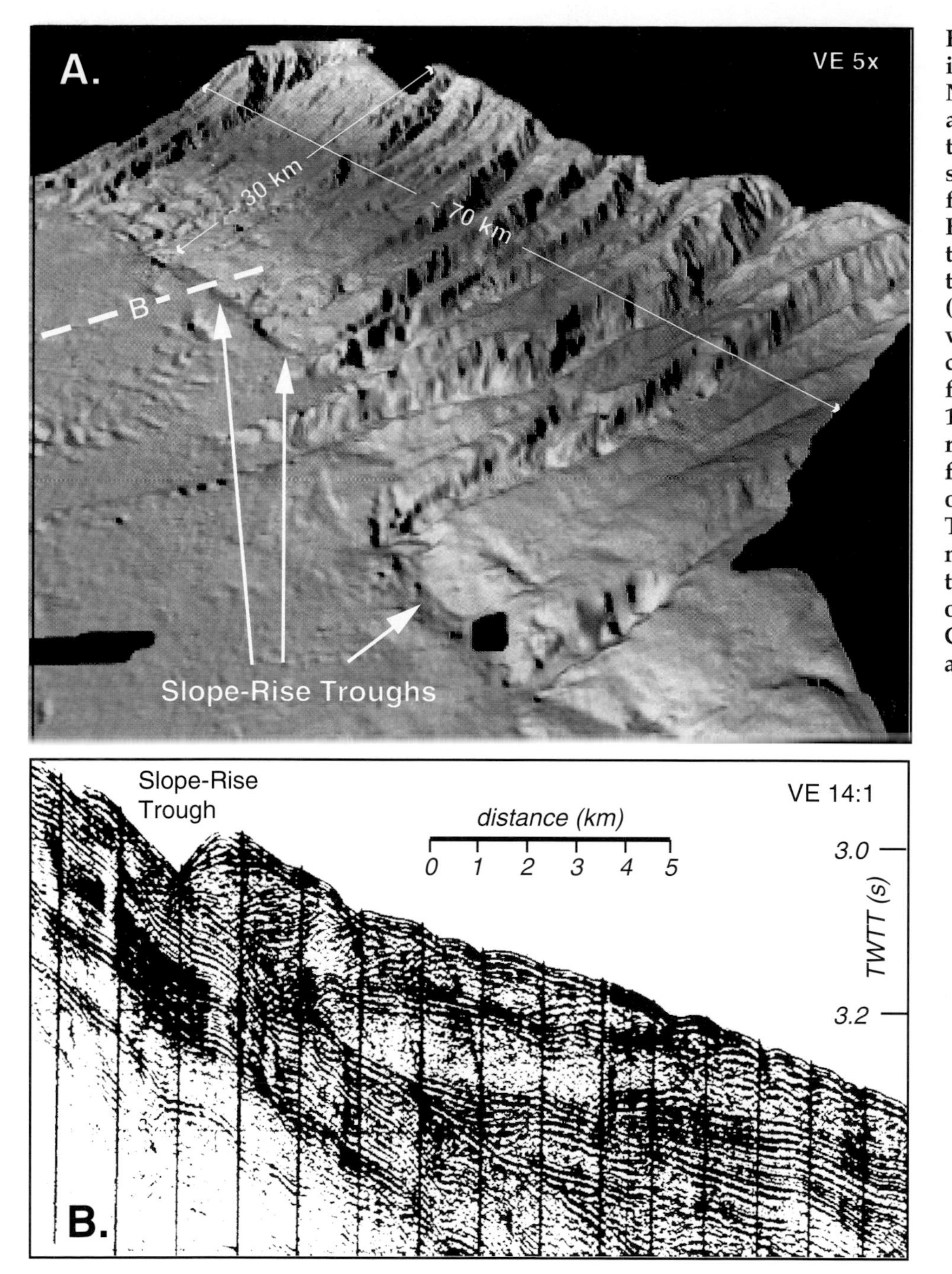

Figure 12—(A) Perspective image of bathymetry of the New Jersey continental slope and upper rise showing the troughs that exist along the slope–rise transition (modified from Pratson and Haxby, 1996). Dashed line is the approximate location of the seismic profile shown in (B). (B) U.S. Geological Survey airgun profile 176, which crosses the trough (modified from Poag and Mountain, 1987). Profile shows that the rise-side of the trough is formed by a mounded ridge of Pleistocene sediments. These onlap Eocene sediments exposed at the base of the slope, which form the opposite wall of the trough. Compare with Figure 11(A) and (B).

(3) The different way mass is managed in subaqueous debris flows and turbidity currents has a direct impact on the relative extent and shape of their deposits. The numerical models show that a debris flow, when begun on a slope that leads to a basin floor, will produce a deposit that starts near its point of origin and extends basinward as far as the momentum of its head can drive it. The deposit will be thinnest near its tail and will gradually thicken toward its head, where it will end rather abruptly. On the same sea floor, an ignitive turbidity current will produce a different style of deposit. This deposit will begin on the basin floor near the base of slope and will be separated from its point of origin by a zone of erosion. The deposit will be thickest near the slope base and will taper seaward.

(4) A debris flow will extend farther basinward if its size is increased, its density is increased, and/or the slope gradient is increased. Similarly, a turbidity current will erode more of the slope and produce a larger, more extensive basin deposit if its size is increased or the slope length is increased. However, this is not true when the slope gradient or initial density of the current is increased. Under

these circumstances, there appears to be an optimal slope gradient and maybe an optimal initial density that will lead a current of a given initial size to cause its greatest possible erosion and generate its largest possible deposit.

(5) The models predict that stacking of debris flows will form a sedimentary sequence that builds out the slope, lowers the slope gradient, raises the basin floor, and progressively tilts it back toward the slope. In contrast, successive turbidity currents will cut back the slope and form a trough at its base. The slope side of the trough will be erosional, and the basinward side will be a depositional bulge formed of turbidites. Preferential deposition on the backside of this bulge by successive turbidity currents will cause it to migrate updip and onlap the eroded slope. Troughs and depositional bulges similar to those produced by the turbidity current model have been observed along the bases of a number of continental slopes. The parallels suggest that turbidity currents and debris flows may indeed produce distinctive stacking geometries that can be identified in seismic-reflection data and possibly outcrops where the nature of the deposits are in debate. The relationship between stacking pattern and deposit type warrants further investigation.

ACKNOWLEDGMENTS

This study was made possible by funding from the Office of Naval Research (ONR Grant Nos. N00014-99-0044 (LP), N00014-93-1-0300 (GP), andN00014-95-1-1281 (JS). Special thanks are extended to J. Kravitz for supporting the work as part of STRATAFORM. The authors thank those who reviewed the manuscript and helped improve it: J. Southard, G. Kuecher, and A.H. Bouma, who even though he did not completely believe our results, was still willing to publish them as part of this volume.

REFERENCES CITED

Biju-Duval, B., et al., 1983, Dépressions circulariés au pied de l'éscarpment de Malte et morphologies des éscarpments sous-marins: Problems d'interprétation: Revue Institute Fran cais du Petrole, v. 38, p. 605–619.

Britter, R. E., and P. F. Linden, 1980, The motion of the front of gravity current traveling down an incline: Journal of Fluid Mechanics, v. 99, p. 531–543.

Bunn, A. R., and B. A. McGregor, 1980, Morphology of the North Carolina continental slope, Western North Atlantic, shaped by deltaic sedimentation and slumping: Marine Geology, v. 37, p. 253–266.

Cita, M. B., et al., 1982, Unusual debris flow deposits at the base of the Malta Escarpment (eastern Mediterranean), *in* S. Saxon and J.K. Nieuwenhuis, eds., Marine slides and other mass movements: New York, Plenum, p. 305–322.

Farre, J. A., and W. B. F. Ryan, 1987, Surficial geology of the continental margin offshore New Jersey in the vicinity of Deep Sea Drilling Project Sites 612 and 613, *in* C. W. Poag, et al., Initial Reports of the Deep Sea Drilling Project: Washington D.C., U.S. Government Printing Office, v. 95, p. 725–759.

Imran, J., G. Parker, and P. Harff, A numerical model of submarine debris flow with praphical user interface, *in* Computers and Geosciences.

Lin, J., and P. H. Leblond, 1993, Numerical modeling of an underwater bingham plastic mudslide and the waves which it generates: Journal of Geophysical Research, v. 98, no. C6, p. 10,303–10,317.

Middleton, G.V., and M.A. Hampton, 1976, Subaqueous sediment transport and deposition by sediment gravity flows, *in* D.J. Stanley and D.J.P. Swift, eds., Marine Sediment Transport and Environmental Management: New York, Wiley, p. 197–218.

Mohrig, D., K. X. Whipple, M. Hondzo, C. Ellis, and G. Parker, 1998, Hydroplaning of subaqueous debris flows: Bulletin of the Geological Society of America, v. 110, p. 387–394.

Mohrig, D., A. Elverhoi, and G. Parker, 1999, Experiments on the relative mobility of muddy subaqueous and subaerial debris flows, and their capacity to remobilize antecedent deposits: Marine Geology, v. 154, p. 117–129.

Parker, G., Y. Fukushima, and H. M. Pantin, 1986, Self-accelerating turbidity currents: Journal of Fluid Mechanics, v. 171, p. 145–181.

Poag, C. W., 1987, The New Jersey Transect: stratigraphic framework and depositional history of a sediment-rich passive margin, in C. W. Poag, et al., eds., Initial Reports of the Deep Sea Drilling Project: Washington D.C., U.S. Government Printing Office, v. 95, p. 763–817.

Poag, C. W., and G. S. Mountain, 1987, Late Cretaceous and Cenozoic evolution of the New Jersey continental slope and upper rise: an integration of borehole data with seismic reflection profiles, *in* C. W. Poag, et al., Initial Reports of the Deep Sea Drilling Project: Washington, D.C., U.S. Government Printing Office, v. 95, p. 673–724.

Pratson, L.F., and W. Haxby, 1996, What is the slope of the U.S. continental slope?: Geology, v. 24, p. 3–6.

Pratson, L. F., J. Imran, E. Hutton, G. Parker, and J. P. M. Syvitski, in press, BANG1D: a one-dimensional Lagrangian model of subaqueous turbid clouds *in* Computers and Geosciences.

Van Diver, B. B., 1985, Roadside Geology of New York: New York, Mountain Press, 411 p.

Appendix: Closure Equations for the Turbidity Current Model

Additional relationships are needed to solve the governing equations for a turbidity current. These relationships are discussed by Parker et al. (1986) and are simply repeated here for completeness.

The equation for the entrainment coefficient is

$$e_w = \frac{0.00153}{0.0204 + Ri} \qquad (8)$$

where

$$Ri = \frac{RgCh}{U^2} \tag{9}$$

The equation for the shear velocity is

$$u_*^2 = \alpha K$$

The equation for the coefficient of erosion given by the empirical relation

$$E_s = \frac{1.3 \times 10^{-7} Z^5}{1 + \frac{1.3 \times 10^{-7}}{0.3} Z^5} \tag{10}$$

where

$$Z = \frac{\sqrt{C_d} U}{v_s} \left(\frac{\sqrt{RgD^3}}{\nu} \right)^{0.6}$$

Finally, the average rate that turbulent energy is dissipated by the viscosity of the turbidity current, ε_0 is given by the equation

$$\varepsilon_0 = \beta \frac{K^{\frac{3}{2}}}{h}$$

where β is

$$\beta = \frac{\frac{1}{2} e_w \left(1 - Ri - 2 \frac{C_d}{\alpha} \right) + C_d}{\left(\frac{C_d}{\alpha} \right)^{\frac{3}{2}}}$$

Notations

The following symbols are used in this paper (meter-kilogram-second units in parentheses):

- C vertically averaged bulk concentration of a turbidity current
- C_d nondimensional coefficient for the drag at the base of the flow
- E_s nondimensional coefficient of erosion
- e_w nondimensional coefficient of water entrainment
- g gravitational acceleration (m/s^2)
- h flow/current thickness perpendicular to the sea floor (m)
- h_p plug layer thickness (m)
- h_s shear layer thickness (m)
- K turbulent kinetic energy in a turbidity current (m^2/s^2)
- R nondimensional submerged specific gravity of the sediment
- Ri Richardson number
- r_0 nondimensional coefficient for the bulk concentration at the base of a turbidity current
- S sea-floor slope (radians)
- t time *(s)*
- U vertically averaged flow/current velocity (m/s)
- Up vertically averaged velocity of the plug layer (m/s)
- u* shear velocity at the base of a turbidity current (m/s)
- x distance parallel to the sea floor (m)
- α nondimensional coefficient relating shear velocity to turbulent kinetic energy
- β nondimensional coefficient relating viscosity to turbulent kinetic energy
- ε_0 vertically averaged viscosity of a turbidity current (m^2/s^2)
- μ_m kinematic viscosity of a debris flow (m^2/s)
- ρ_m debris flow bulk density (kg/m^3)
- ρ_w ambient water density (kg/m^3)
- τ_m yield strength of debris flow matrix (N/m^2)
- v_s settling velocity of the suspended sediments (m/s)

Peakall, J., W. D. McCaffrey, B. C. Kneller, C. E. Stelting, T. R. McHargue, and W. J. Schweller, 2000, A process model for the evolution of submarine fan channels: implications for sedimentary architecture, *in* A. H. Bouma and C. G. Stone, eds., Fine-grained turbidite systems, AAPG Memoir 72/ SEPM Special Publication No. 68, p. 73–88.

Chapter 7

A Process Model for the Evolution of Submarine Fan Channels: Implications for Sedimentary Architecture

J. Peakall
W. D. McCaffrey
B. C. Kneller
School of Earth Sciences, University of Leeds
Leeds, West Yorkshire, United Kingdom

C. E. Stelting
Chevron U.S.A. Production Company
New Orleans, Louisiana, U.S.A.

T. R. McHargue
Chevron Petroleum Technology Company
San Ramon, California, U.S.A.

W. J. Schweller
Chevron Petroleum Technology Company
La Habra, California, U.S.A.

ABSTRACT

Medium- to high-sinuosity, aggradational submarine channels have frequently been considered analogous to subaerial channels. However, planform evolution and resulting architecture in these submarine channels are characterized by absence of downstream migration, eventual cessation of movement, and ribbon geometries. In contrast, alluvial rivers undergo continuous downstream and lateral movement to form tabular, sheetlike bodies. A simple process model of flow structure and evolution is described for these submarine channels. Flows are predicted to be highly stratified, have significant supra-levee thicknesses, and form broad overbank wedges of low-concentration fluid. The model, for the first time, provides a coherent set of process explanations for the primary observations of submarine channels.

INTRODUCTION

A close analogy has been drawn between subaerial channels and medium- to high-sinuosity, aggradational, leveed submarine channels, both in terms of morphology and flow processes. This analogy has been based in part on qualitative comparisons (Figure 1), in which many features of fluvial type have been observed, including bend cutoffs, high channel sinuosity, crevasse-splays, point bars, scroll bars, meander belts, and chute channels and pools, (Klaucke and Hesse, 1996), and in part on quantitative analysis of planform characteristics (Flood and Damuth, 1987). Paradoxically, processes such as flow stripping have been considered to be unique to submarine channels, as has the development of nested mounds on the outside of channel bends. However, the basic fluid dynamics of open-channel

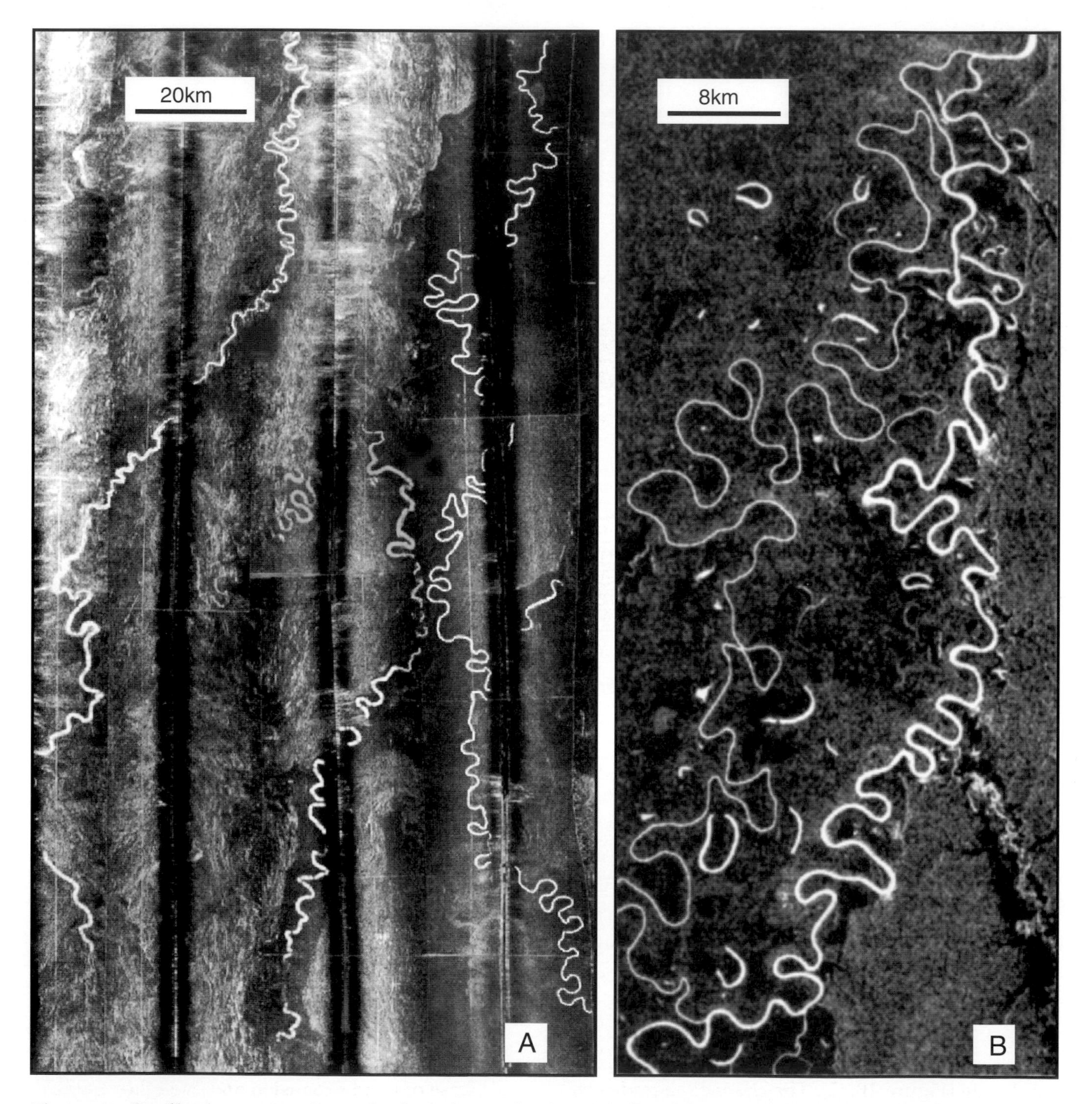

Figure 1—Qualitative comparison of submarine and subaerial channels. (A) GLORIA image of the Indus submarine channels (from Kenyon et al., 1995, with permission from Klüwer Academic Publishers). (B) Synthetic aperture radar image (inverted) of the Parana Breu (left) and Juruá rivers (right), both tributaries of the Amazon River (image 31 km × 70 km) (adapted from Stølum, 1998), © NASDA/MITI.

and gravity flows are sufficiently different to suggest that there should be significant differences in the structure of each type of flow and their deposits. For instance, the effective gravity acting on overbank submarine flows is a small fraction of that acting on subaerial flows due to the far smaller density difference with the surrounding medium. Additionally, turbidity currents have significant friction at the upper boundary in contrast to subaerial flows, and entrain fluid at the upper boundary and beneath the head, thereby modifying the flow. These differences in fluid mechanics suggest that the analogy between submarine channels and rivers is imperfect and that submarine channel processes and evolution may be significantly different. Despite this, fluvial models are routinely applied to predict the architecture of submarine systems. This paper builds on a process model of channelized turbidity currents presented by Peakall et al. (in press) and goes on to investigate the implications of the model for the development of submarine channel architecture.

CHANNEL PLANFORM EVOLUTION AND SEDIMENTARY ARCHITECTURE

Submarine channels at first appear very similar to their subaerial counterparts. However, a number of characteristics suggest that the analogy is only superficial. Probably the most conspicuous example is that submarine channels have a very small number of bend cutoffs relative to many subaerial channels (Figure 1A, B). Although individual cutoffs have been examined in detail (Lonsdale and Hollister, 1979), their frequency is low in comparison with subaerial meandering channels. For instance, no cutoffs have been recognized on the youngest Mississippi Fan channel (Garrison et al. 1982; EEZ-SCAN, 1987), and only nine cutoffs have been recognized in a more than 800-km reach of the Amazon Channel (Pirmez and Flood, 1995). By way of contrast, many high-sinuosity meandering alluvial reaches display hundreds of bend cutoffs, suggesting a difference in frequency of occurrence of one to two orders of magnitude.

Bend Development in the Youngest Mississippi Fan Channel

The most complete history of submarine bend development and migration is provided by the youngest mid-fan channel of the Mississippi Fan (Kastens and Shor, 1985; Stelting et al., 1985a). Isopachs derived from seismic-reflection profiles allow three stages of channel development to be determined (Stelting et al., 1985a) and two further stages can be identified from seismic cross sections. In stage 1, the channel is almost straight but contains a sinuous thalweg (Figure 2A). In subsequent stages 2 and 3, a prominent bend begins to develop and grows in amplitude (Figure 2B, C). The channel aggrades by approximately 175 m during stages 1 to 3 indicating that the bend growth rate is very slow compared with the aggradation rate. As the channel evolves, the channel thalweg width decreases from approximately 5 km during stage 1 to just 2 km at the end of stage 3 (Stelting et al., 1985a), and the levee-to-levee width from approximately 10 km to 3 km (Figure 2). A further 90–100 m of aggradation occurred during stage 4, with no change in bend amplitude, wavelength, or position (Stelting et al., 1985b). The channel was subsequently plugged by approximately 90 m of debris flow material during stage 5. After the initial bend growth of stages 1–3 (Figure 2), the absence of channel movement in stage 4 suggests that the channel reached an equilibrium planform state.

Ridge and Swale Topography on the Mississippi Fan

Fluvial point bars may be characterized by ridge and swale topography, marking the incremental movement of the channel through time (Figure 3). Ridge and swale topography has been described from an area surrounding the same bend of the youngest Mississippi Fan channel (Stelting et al., 1985b; Pickering et al., 1986), suggesting that substantial channel movement has occurred. Such an interpretation is incompatible with the seismically derived channel history described previously. This Mississippi Fan example has been used to suggest that ridge and swale topography is a general feature of sinuous submarine channels (Clark and Pickering, 1996a,b). There is, however, evidence that indicates the ridges associated with the Mississippi Fan submarine channel are not examples of ridge and swale topography:

(1) The swales have maximum subsurface depths of between 15 and 50 m (Stelting et al., 1985b; Pickering et al., 1986). However, since the channel aggraded by 180–190 m in the absence of bend movement, these features cannot be related to the history of point bar growth, and therefore cannot be genuine ridge and swale deposits.
(2) Ridges occur in the channel thalweg, crevasse-splay deposits, and beyond the mapped channel limits. True ridge and swale deposits would be only found on the inner bend (Figure 3). These features have instead been interpreted as deformation structures on the upper surface of debris flow deposits (Kastens and Shor, 1985).
(3) The ridges and swales exhibit different genetic relationships to fluvial examples. Fluvial ridges are formed by lateral accretion surfaces or from sedimentation onto lateral accretion surfaces (Figure 3). However, seismic cross sections (Prior et al. 1983; Stelting et al., 1985b; Pickering et al., 1986) show that ridge and swales are not associated with lateral accretion surfaces in this submarine example.

If scroll bars (ridges) do form in high-sinuosity submarine point bars, they will only do so during the initial phase of bend expansion and even then because the rate of bend growth is slow compared with the rate of channel aggradation each will be rapidly buried, suppressing the development of ridge and swale topography.

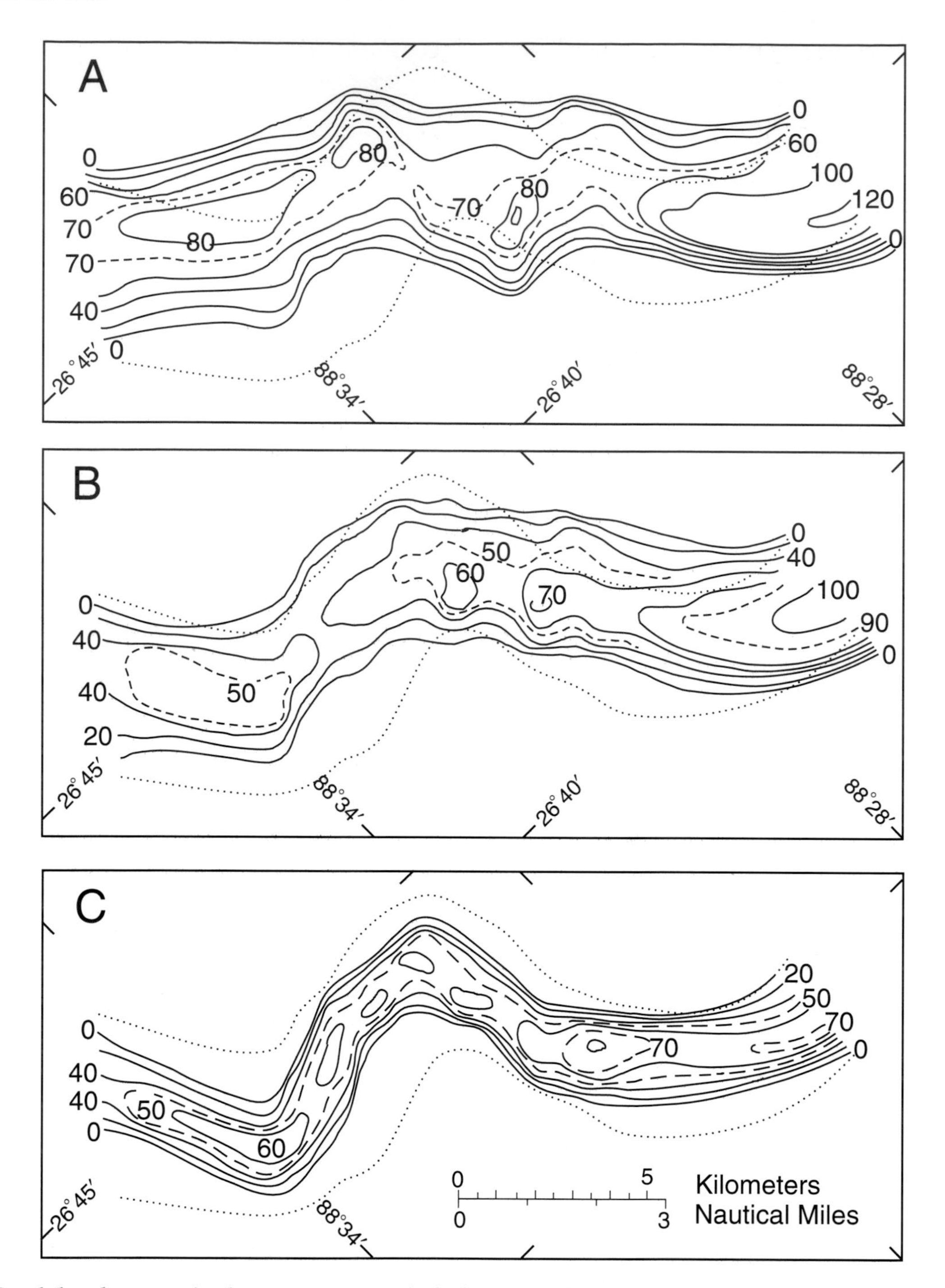

Figure 2—Bend development in the most recent Mississippi Fan channel, as illustrated by isopach maps of the channel-lag deposits (acoustically high-amplitude zones) for successive time increments. Contours are in milliseconds; 20 ms is approximately equivalent to 15 m in depth. Adapted from Stelting et al. (1985a). (A) stage 1, lowermost ~60 m of sediment; (B) stage 2, ~60 m thick section of overlying stage 1; (C) stage 3, ~55 m thick section overlying stage 2.

Bend Development in a Sinuous Passive Margin Channel

A subsurface example of a sinuous single-thread submarine channel is shown in Figure 4. This is the uppermost section (~40 m) of a 120 m (±20 m) thick aggradational channel-levee stack that infills the upper part of a fan valley on the west African margin. The illustrated reach (Figure 4) is 12 km long, has a maximum width of 1.5 km, and is approximately 40 m (130 ft) thick; flow was from right to left. Sequential thalweg positions were determined from a 3-D seismic volume by making horizon slices through the amplitude data at 8 ms (~10 m; 33 ft) intervals through the upper 40 m of the channel unit. Channel thalweg maps become increasingly ambiguous for horizons below

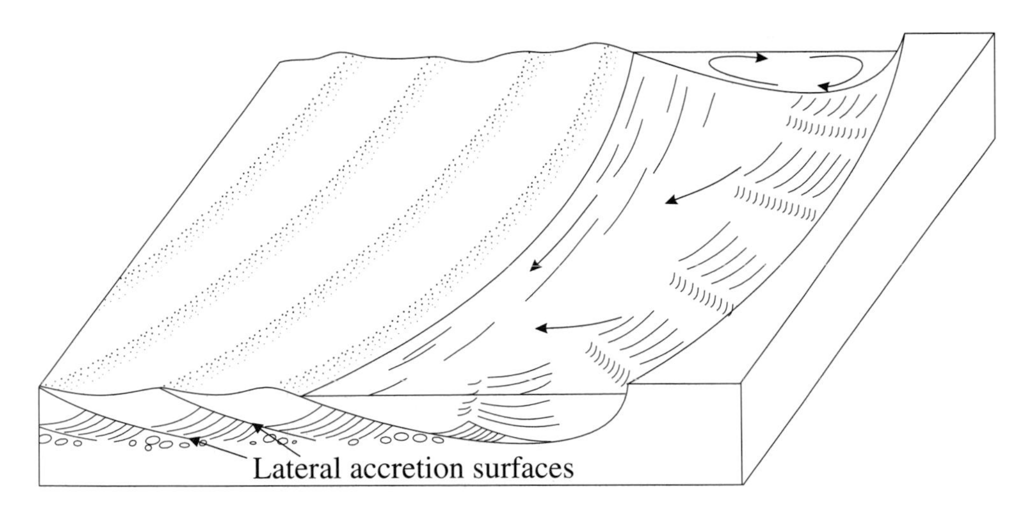

Figure 3—Morphological and genetic relationships of ridge and swale deposits in fluvial systems. After Gibling and Rust (1993).

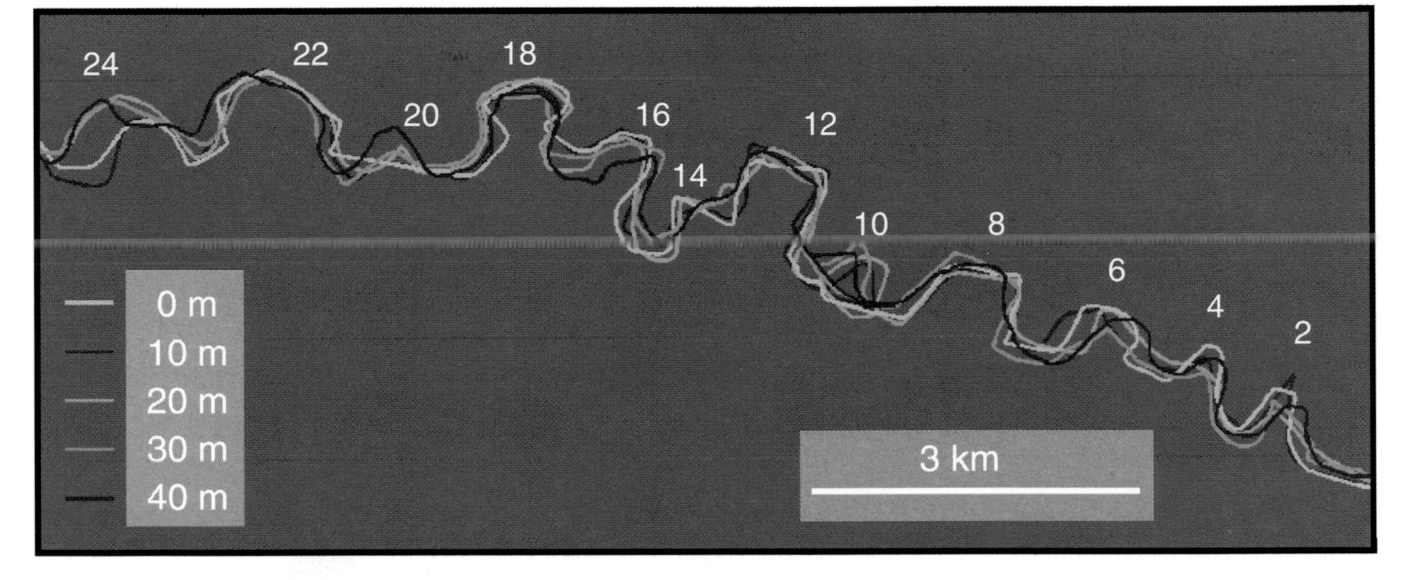

Figure 4—Bend development in a subsurface example of a sinuous submarine channel. Successive positions of the channel thalweg are shown at aggradation intervals of 10 m, and were picked from amplitudes on successive horizon slices; thalweg width is ~60 m, channel width is ~400 m. Flow is from right to left. ***Note:*** **Poor seismic resolution at bend 10 precludes a precise delineation of channel course.**

the uppermost 40 m and are therefore not illustrated in Figure 4. Nevertheless, seismic resolution in the lower part of the channel levee stack (~80 m) is sufficient to indicate that the bend nodes are located in virtually the same positions as those illustrated in Figure 4; consequently, no significant downstream bend migration (translation) occurs at any stage in the channel history. The lower part of the channel-levee stack is characterized by a change from a relatively broad, low-sinuosity channel at the base, to a narrower, higher sinuosity channel at the top (lowermost unit in Figure 4), with progressive upward increases in sinuosity and bend amplitude, and a reduction in channel width.

In the uppermost section (Figure 4), of the 24 bends, only five of the bends (6, 15, 16, 20, and 24 on Figure 4) show any significant lateral movement during aggradation of the channel. In these cases, the movement of the channel thalweg does not always cause a systematic increase in channel sinuosity. In bends 6 and 20, for example, the last phase of channel evolution tends to return the planform toward an earlier configuration. Very few bends display significant lateral movement, and maximum movement (~400 m in one example) is approximately equivalent to channel width. Lateral movement *(M)* is also low as a function of vertical aggradation *(A)* with ratios of *M/A* of 7–11 for the most active bends; one to four orders of magnitude lower than in river channels (Peakall et al., in press). The stability of channel bends during this 40 m of aggradation, and throughout the whole 120 (±20 m), suggests that the channel planform is essentially in equilibrium. It should be noted that at least five periods of channel incision punctuate the stacked channel sequence, presumably the result of some external (possibly tectonic) forcing. However, the incisional phases are relatively shallow and do not appear to influence the planform stability of the channel sequence significantly.

Comparative Bend Evolution in Submarine and Subaerial Channels

The evolution of the youngest Mississippi Fan channel and the channel from the west African margin can be contrasted with the channel development of a typical high-sinuosity alluvial river (Figure 5). A single-thread straight river would also undergo rapid bend expansion with bend amplitude increasing and channel wavelength decreasing. However, fluvial meander bends also progressively translate downstream (sweep) and migrate laterally (swing) across their flood plains (Figure 5) and consequently meander bends never reach equilibrium. Individual bends translate at different rates because of floodplain heterogeneities, leading to bend cutoff and a repetition of the bend development sequence. River bend development, channel sweep, and swing are all very fast compared with net channel aggradation; river channels also maintain a fairly constant width during channel evolution and aggradation. The comparative patterns of channel evolution are, therefore, very different,

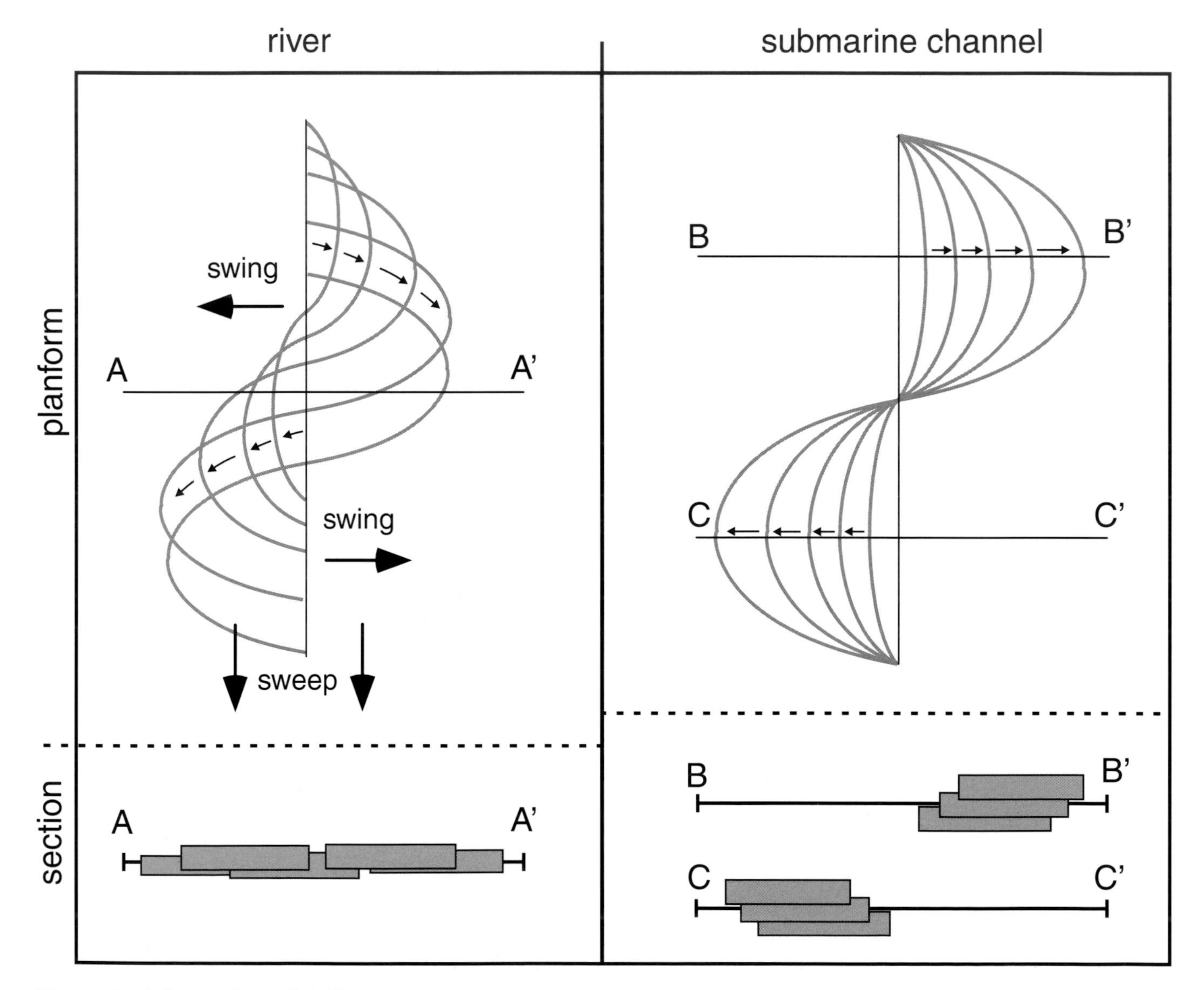

Figure 5—Schematic model illustrating the comparative evolution of submarine and subaerial channels. Bend growth from an initially straight channel (central black line) is shown in both cases. Submarine channels increase in bend amplitude before reaching an equilibrium planform characterized by a cessation of movement. At no point is there any significant downstream bend translation (sweep). Alluvial river channels also increase in bend amplitude but undergo continuous sweep and swing across their flood plains, forming tabular, sheetlike bodies referred to as a channel belts. In contrast, submarine channels produce ribbon geometries, with no reworked, homogenized channel sediments between bends.

with three points particularly prominent: (1) very slow bend formation in the submarine channels relative to aggradation rate; (2) absence of significant sweep in the submarine channels, and no indication of swing after initial bend growth; and (3) the cessation of bend movement and attainment of planform equilibrium in the submarine channel, after initial bend formation or development of initial sinuosity.

Three-Stage Model of Submarine Channel Architecture

There are a number of lines of evidence to support the model of submarine planform development postulated above. The lack of significant bend translation, coupled with the attainment of planform equilibrium, is consistent with the limited bend cutoff observed in such systems. High-resolution seismic sections of modern systems generally show a single steeply aggrading ribbon of channel axis deposits (Kenyon et al., 1995). This suggests that the channels do not undergo significant swing or sweep, otherwise there would be a meander belt underlying the channel that would be broader than the width of an individual channel. Indus Fan channels show a similar pattern, characterized by a progressive decrease in lateral movement, an increase in aggradation, and a narrowing of thalweg deposits with time (McHargue, 1991). Further examples of comparable channel evolution include the lower-sinuosity Einstein channel from the Gulf of Mexico and the most recent Amazon Fan channels (Hackbarth and Shew, 1994; Pirmez, 1994; Peakall et al., in press). In each of these cases, channel thalweg aggradation and levee growth appear to be broadly contemporaneous.

In cross section, three stages can be identified in the development of high-sinuosity, aggradational submarine channels (Figure 6): (1) point bar development associated with bend growth and the possible preservation of well-defined lateral-accretion surfaces due to the lack of sweep; (2) equilibrium phase (stable planform geometry), where the channel is dominantly acting as a bypass zone, with near-vertical aggradation; and (3) a channel abandonment phase, in which the channel fill is either broadly fining-up as the magnitude of successive flows decreases, or the channel is plugged by debris flow deposits, or the channel is left open and draped by hemipelagic material.

FLOW PROCESSES

Flow Stripping and Nested Mound Formation

Flow stripping is the most frequently identified process unique to submarine channels and provides a good starting point for any consideration of flow processes in submarine channels. Flow stripping [Piper and Normark (1983)] predicts that turbidity currents that are overbanking will be unable to negotiate sharp channel bends while staying intact (Figure 7). Thus the flow splits into two distinct parts with the upper portion of the flow moving tangentially to the main channel, down the back side of the outer levee, while the lower flow section continues down the channel (Figure 7). Overbank spill is predicted to lead to a loss of momentum in the channel and the deposition of sand just beyond the channel bend. Channelized nested mounds, which stack from the outside to the inside of bends, were first reported from dipmeter interpretations of the Permian Indian Draw Field, New Mexico (Phillips, 1987),

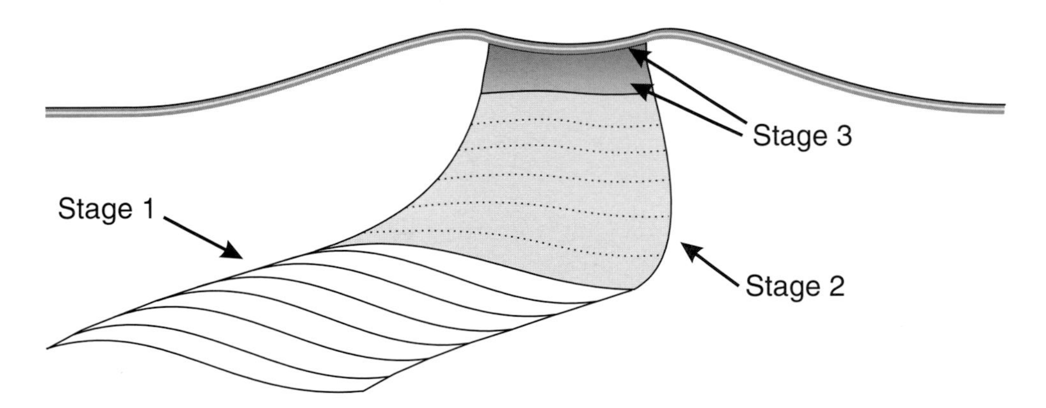

Figure 6—Schematic cross section, illustrating the three stages of submarine channel development: (1) Climbing clinoforms as channel bends broaden; (2) equilibrium phase aggradation when the channel is dominantly acting as a zone of bypass while systematically aggrading both the thalweg and the levees, and (3) a channel abandonment stage identified by a broadly fining-up sequence as mean channel velocities gradually decline, followed by the deposition of a hemipelagic drape. There are, however, two other possible abandonment facies to that shown here—infill by a large debris flow(s) or instantaneous channel abandonment followed by hemipelagic deposition.

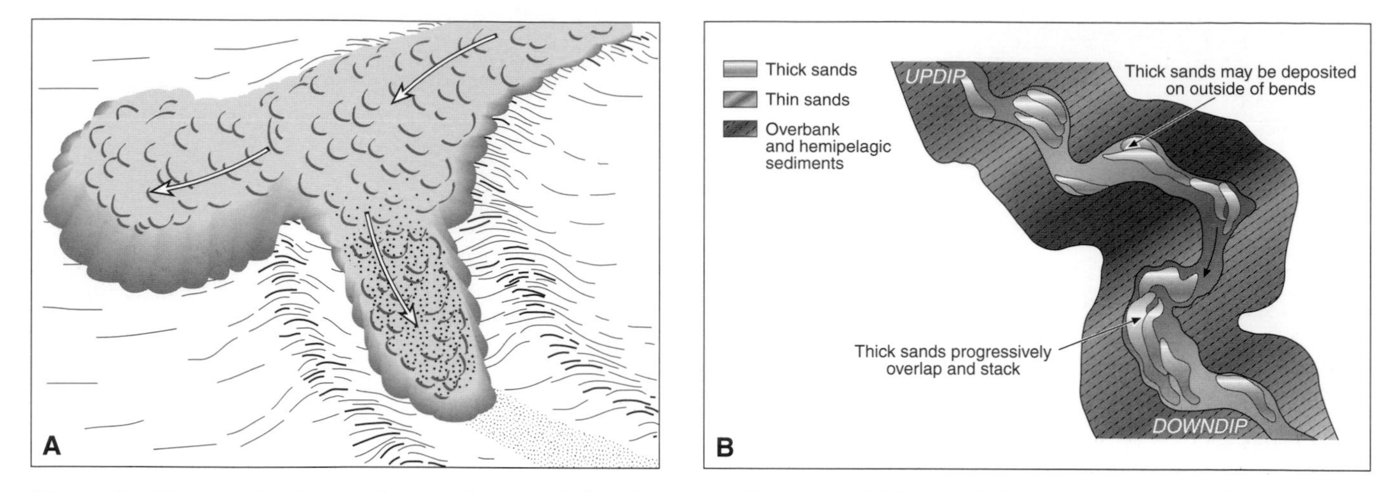

Figure 7—Flow stripping and nested mound development [Piper and Normark (1983) and Timbrell (1993)]. (A) Flow is unable to negotiate a bend of high-channel curvature and splits into two parts. Overbank spill leads to a loss of momentum in the channel and the deposition of sand just beyond the channel bend. (B) Schematic view of planform development of nested mounds. Adapted from Timbrell (1993).

and have subsequently been interpreted as the result of flow-stripping–induced sand dumping (Clark and Pickering, 1996a). Other channelized podlike geometries have been similarly interpreted: seismically identified features of the Upper Paleocene Balder Formation (Timbrell, 1993) and Oligocene Hackberry Formation (Cossey and Jacobs, 1992), and high back scatter elements in sonar images of channel bends in the Umnak channels of the Bering Sea (Clark and Pickering, 1996a). On this basis, nested mounds have been considered to be a significant channel fill architectural element and have been incorporated into schematic facies models of submarine channels (Figure 7B; Timbrell, 1993; Clark and Pickering, 1996a).

Flow Stripping Theory

Flow stripping has been modeled using Komar's equation for the depth-integrated velocity of a turbidity current (Bowen et al., 1984), which can be simplified to:

$$U^2 = kCh$$

where k is taken as a constant, C is the volume concentration, U the velocity, and h the height of the main body of the flow. It follows that since the volume concentration is taken to be constant, a decrease in the flow height due to flow stripping inevitably leads to a rapid decrease in the velocity of the turbidity current and increased potential for deposition at the base of the flow. Such an analysis assumes that channelized turbidity currents are unstratified. However, strong flow stratification in submarine turbidity currents is indicated by the general fining-up character of levee sequences, the absence of marked breaks in levee grain size or net-to-gross ratios, quantitative experimental studies (Altinaker et al., 1996), field studies (Chikita, 1990), and theoretical models (Stacey and Bowen, 1988). It is, therefore, apparent that the assumption of a depth-averaged value for concentration, as used by Bowen et al. (1984), is inappropriate for the description of flow stripping.

We reassess the flow stripping model by calculating the in-channel depth-averaged velocity before and after flow stripping, while considering the effects of a concentration gradient (Figure 8). The proportional reduction in velocity decreases as flow stratification increases; consequently, stripping of a high proportion of the upper part of a well-stratified flow makes relatively little difference to the in-channel velocity (in Figure 8, for example, the loss of the upper 50% of the flow may result in a velocity reduction of only 5%), suggesting that flow stripping need not entail catastrophic changes to the residual in-channel flow.

Flow Stripping and the Navy Fan

The initial concept and modeling of flow stripping arose from studies of the Navy Fan in the California Borderland (Piper and Normark 1983; Bowen et al., 1984). Here, the field and theoretical studies are in excellent agreement, despite the modeling limitations discussed above. However, a number of factors in the Navy Fan militate against the wider use of the flow stripping hypothesis. First, the bend angles are abrupt, at approximately 60°, second, there is a partially breached right-hand levee on the first left-hand bend; and finally, stripping of 85–90% of the sediment from a single flow occurred, with less than 10% of the deposition taking place downstream of the same bend (Piper and Normark, 1983; Bowen et al., 1984). The loss of flow containing 80–90% of the sediment, as apparently occurred on the Navy Fan, is likely to cause rapid deposition of sediment within the channel downstream of the bend, regardless of the degree of flow stratification (Figure 8). However, this type of flow

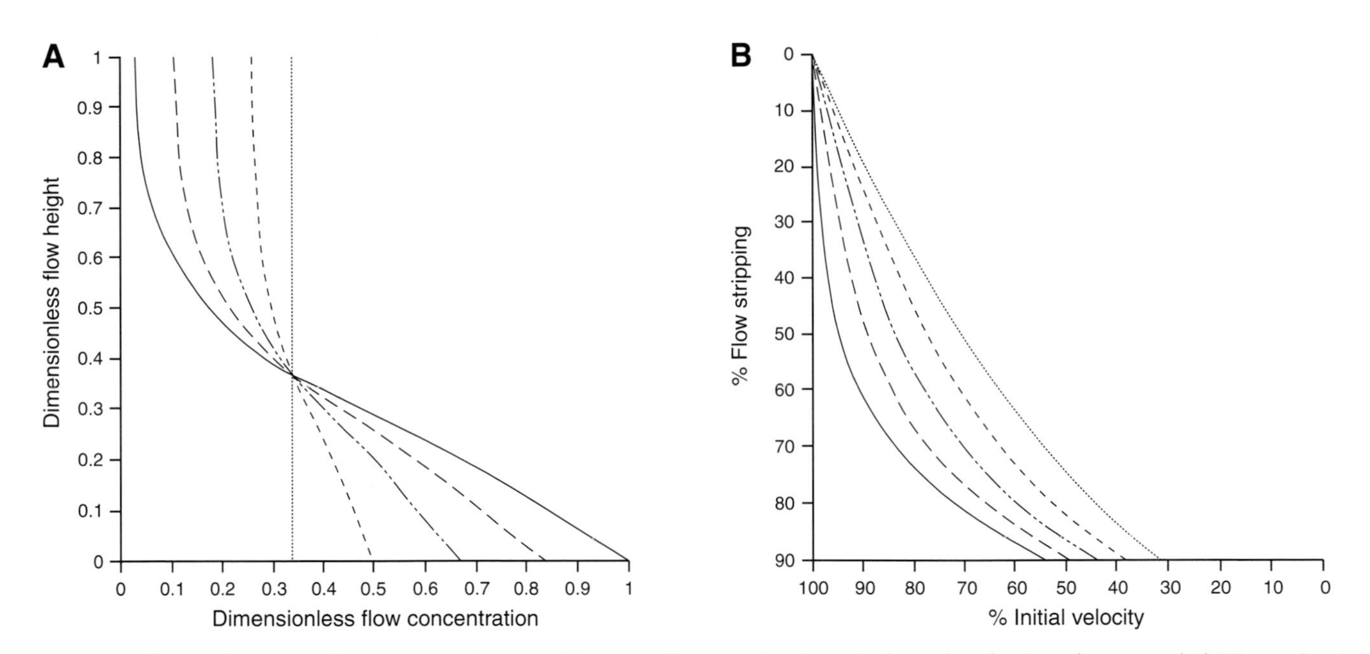

Figure 8—The influence of concentration gradient on flow-stripping–induced velocity changes. (A) Hypothetical concentration gradients varying from a stratified experimental distribution (Altinaker et al., 1996) to an unstratified flow. (B) Reduction in flow velocity as a function of flow stripping and stratification (shown by line type, see part A).

stripping must be an anomalous form of overbank flow; otherwise, turbidity currents of significant size would never travel long distances through sinuous channels.

The Nature of Overspill

Flow stripping produces localized overbank flow across the outside of channel bends. However, the formation of continuous levees along submarine channels requires a more general process of overbank flow, referred to here as overspill, that also accounts for overbank flow across inner-bend levees and bend inflection points. The flow stripping model (Bowen et al., 1984) is a generalized description of channelized turbidity current flow; consequently, the previous discussion and reevaluation of this process may be equally well applied to the overspill process. We consider the general case of a flow that is initially at the height of the levee crest, exhibiting a strong vertical density stratification, and flowing through a channel without breached levees. For simplicity, inherited fluid momentum, the Coriolis Force and any slope on the upper surface of the flow are ignored. Therefore, the initial driving force (per unit area) of the first parcel of supra-levee fluid (Figure 9A) is the density difference between the supra-levee and ambient fluids, multiplied by the component of gravity in the downslope (levee backslope) direction. In a highly stratified flow, the upper part will consist of relatively fine-grained material. Therefore, for a turbidity current where the interstitial fluid is seawater and where the surrounding medium is also seawater, the density difference will solely be a function of the fine-grained material. As the parcel of fluid moves down the levee backslope, it may deposit part of its sedimentary load and thereby decrease the mean density difference with the ambient fluid. Succeeding parcels of overbank fluid (Figure 9B) have a driving force that is a function of the density difference between themselves and the supra-levee fluid, which has already undergone partial sedimentation. As a consequence, the excess gravity-driving force decreases both spatially and temporally.

Long-lived flows will, therefore, gradually form a body of low-concentration overbank fluid that will increase in dimensions with time, as the excess gravity-driving force decreases (Figure 10). In planform this fluid may take the form of a wedge (Figure 10A), which is wider upstream than at the downstream end (Peakall et al., in press). With time, the overbank fluid wedge will acquire further momentum due to flowing down the same regional gradient as the channel and from shearing of the underlying channelized flow. Momentum will also be transferred to the overlying ambient fluid, creating a broadening zone of moving ambient and low-concentration turbid fluid, surrounding the main channelized flow.

This model predicts that overbank flows will be spatially and temporally extensive, deep, and consist of low-concentration, low-velocity fluid. There is abundant evidence for such flows, which is in conflict with models that predict thin overbank flows (Hiscott et al., 1997a). This evidence can be subdivided into six broad themes:

(1) Acoustic mapping of turbidity currents (Hay et al., 1982)
(2) Correlation of individual sediment layers on levees over large distances (≤300 km in the NAMOC submarine system: Hesse, 1995),
(3) Large-scale levee erosion and breaching (e.g., Masson et al., 1995)

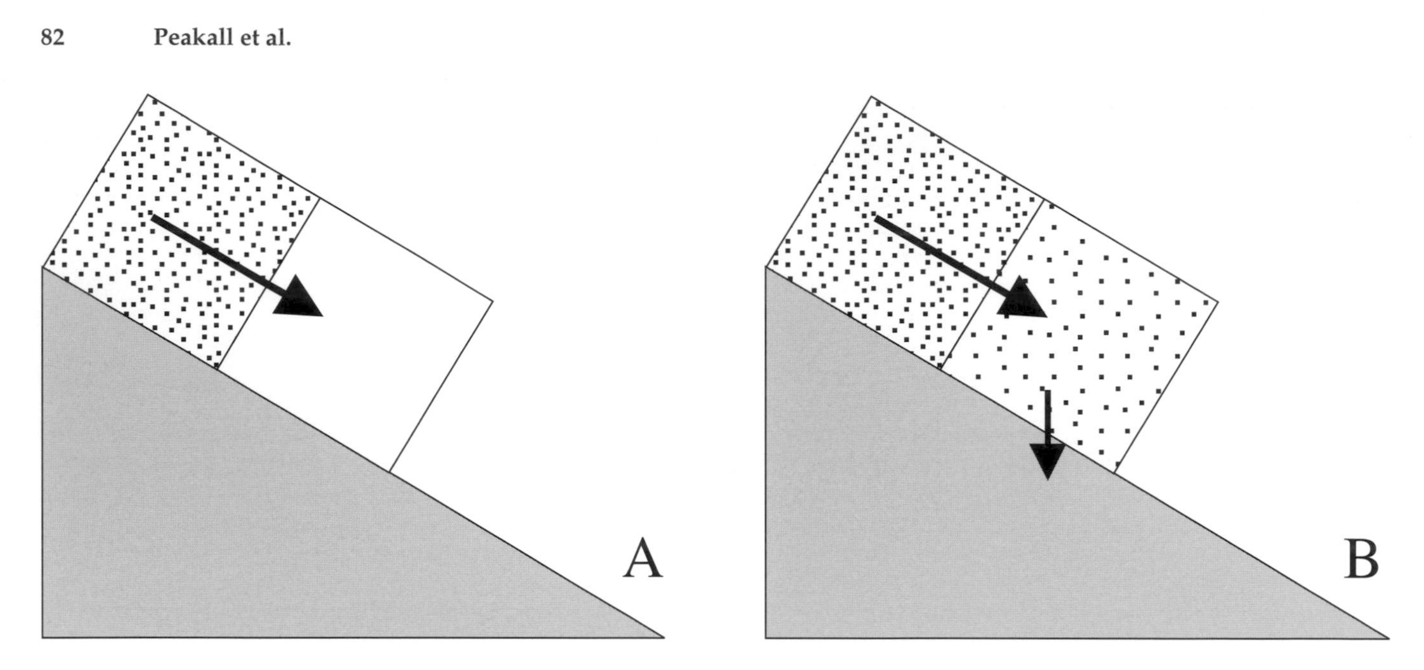

Figure 9—Simplified schematic description of the driving forces affecting successive parcels of supra-levee fluid on the backslope of a levee. The main channel is to the left of each figure. (A) First parcel of supra-levee fluid, (B) second parcel of overbank fluid. See text for explanation.

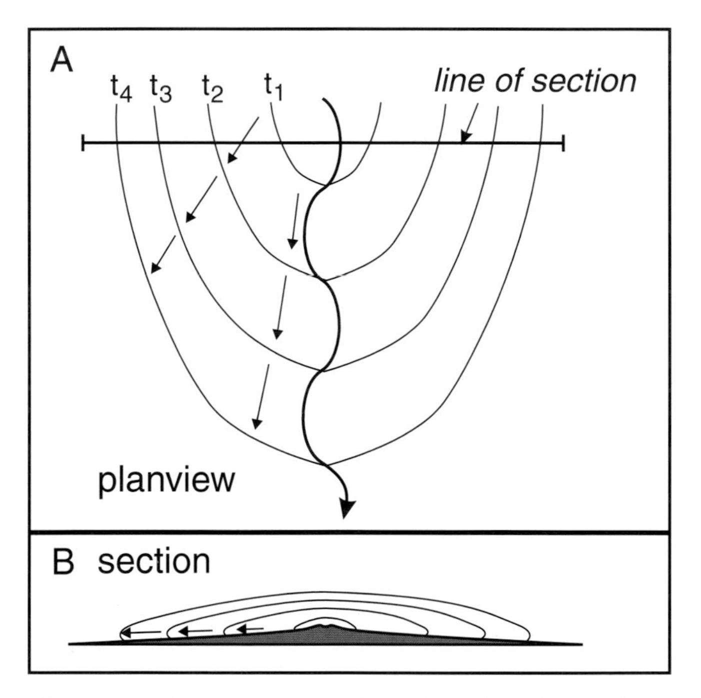

Figure 10—Development of a submarine channel overbank fluid wedge with time. (A) Planform view and (B) cross-section view through successive time increments ($t_1 - t_4$).

(4) Flow estimates for the formation of sediment waves on channel levees that predict thick (100–800 m), low-velocity (~0.1 ms^{-1}), very low-concentration (~2.5 mg l^{-1}) overbank flows (Normark et al., 1980)

(5) Thin, laminated interchannel deposits of the Laurentian Fan that were calculated to have been deposited by low concentration (~2500 mg l^{-1}), thick (100s of meters), slow-moving (0.1–0.2 ms^{-1}) currents, produced by gradual overspill (Stow and Bowen, 1980)

(6) Downstream decline in cross-sectional area of submarine channels. Reduction in cross-sectional area by a factor of up to 30, coupled with declining slopes (and therefore flow velocity), reduces the discharge capacity of the channel and suggests that flow must be lost overbank on at least a semicontinuous basis (Hiscott et al., 1997a).

Effects of Overspill on Flow Density and Grain-Size Stratification

Flows exhibiting continuous overspill will experience further entrainment of ambient fluid at the upper boundary together with re-equilibration of the sediment concentration profile. The finest grade material, frequently referred to as washload, is essentially evenly distributed within the flow after each increment of overspill and re-equilibration; therefore, the mean concentration of the finest material (clays and silts) decreases with time. The transport efficiency of sand is greater in flows that have large quantities of fines; therefore, mud-rich currents that experience this progressive sorting should be able to transport sand-grade sediment much farther downsystem than similar flows that are initially sand-rich. This continuous-sorting mechanism may be an important process controlling the deposition of clean sands in submarine channels and on the lobes and lower fan downstream of channel mouths (Hiscott et al., 1997a). The concentration of sand increases as clay is progressively removed from the flow until the clay content falls to a point where the flow is no longer sufficiently efficient to carry the basal sediment load, whereupon deposition of clean sand with very low interstitial clay content takes place. Any remaining fine-grained upper parts of the flow may bypass the zone of sand deposition and be deposited farther downstream. Such a mechanism is relevant to the sandy debris flow versus turbidity current debate (Shanmugam et al., 1995, 1997; Hiscott et al., 1997b), as it suggests that although sediment may be transported by a turbidity current (for up to hundreds of kilometers) it can be rapidly deposited once sufficient clay has been removed and may form a deposit that would be identified as that of a sandy debris flow, using the criteria of Shanmugam et al. (1995, 1997). This process may be particularly important in generating sandy deposits in mud-rich systems such as the Mississippi and Amazon fans (Gibbs, 1967). Such a mechanism offers an alternative process explanation to the bimodal flow concept of thick, slow-moving muddy turbidity currents and thin, fast-moving sandy flows (also based on the depth-averaged Komar equation; Piper and Normark, 1983; Bowen et al., 1984), although the two processes need not be mutually exclusive.

COMPARISON OF PROCESS MODEL WITH TURBIDITY CURRENTS IN BUTE INLET

The sinuous channel system in Bute Inlet, British Columbia, has been intensively studied using a variety of techniques including sonar imaging, coring, current meters, deflection vanes, a transmissometer-based turbidity event detector, and the use of suspended sediment traps (Prior et al., 1986, 1987; Bornhold et al., 1994). These studies provide the most complete field dataset with which to test directly the proposed flow process model.

Turbidity currents in the inlet occur from May to September and are related to floods in the two rivers that enter the inlet (Prior et al., 1987). The lower parts of the turbidity currents move sand-sized material and have caused extensive damage to in-situ equipment (Prior et al., 1986). Higher parts of the currents are much finer grained, as indicated by suspended sediment traps 6–7 m above the floor, suggesting that flows are highly stratified (Zeng et al., 1991). Data from a transmissometer moored above the flank of the channel recorded upper, unchannelized fine-grained portions of the flow (Bornhold et al., 1994). During a series of closely spaced turbidity current events, the background turbidity recorded by the transmissometer increased with time (Bornhold et al., 1994, their Figure 5), consistent with the development of an overbank wedge of fluid with more dilute fluid moving progressively further away from the transmissometer (Figure 10). Large-scale advection of fine-grained material was also indicated by accumulation rates of muddy sediment up to 220 times faster than the background rate after the passage of turbidity currents (Zeng et al., 1991). A channel-lobe complex consisting of well-sorted, fine- to medium-grained, predominantly massive sands is present at the end of the channel system. Distal splays beyond the channel-lobe complex consist of fine-grained sands with some capping mud layers (Zeng et al., 1991). Taken together, the sediments in the channel-lobe complex and the distal splays suggest that sufficient mud is advected by the time currents reach the end of the channel that clean massive sands are deposited at the channel mouth. Any remaining advected material appears to dominantly bypass the lobes and is deposited in the distal splays, along with some fine sand-grade material from the very largest currents.

Although more detailed field datasets are required to thoroughly test turbidity current models, the studies in Bute Inlet are broadly consistent with the proposed model of submarine channel flow structure and flow evolution. Further field studies using two or more transmissometers in such a setting would hold great promise for elucidating both the instantaneous and temporal behavior of overbank turbidity flows.

IMPLICATIONS FOR SEDIMENTARY ARCHITECTURE

Nested Mounds

This reappraisal of the flow stripping theory has several major implications for nested mound formation. In normal circumstances, the within-channel velocity deceleration related to flow stripping at a bend is generally too small for large-scale deposition of coarse sediment. The new theory predicts that extreme disequilibrium flows are required for nested mounds to form through flow stripping. Furthermore, flow-stripping–induced nested mounds would be concentrated in specific bends (i.e., the first sharp bend in a channel system, or a bend with extreme sinuosity) and would not be a general feature of most submarine channel bends as predicted in existing submarine channel architecture models (Figure 7B). Alternative mechanisms are required to explain the presence of the majority of sand mounds on the outside of channel bends. One possible explanation is the frequent observation of channel-bank failures (Prior et al., 1986; Ren et al., 1996), which may be concentrated on the outer bank due to higher shear stresses, steeper slopes, and large-scale bank erosion during initial bend evolution. Temporary deposition of such levee material in the channel would create a localized topographic feature, and flow deceleration over and around the obstacle would lead to deposition of coarse-grained material from the base of the turbidity current (Kneller, 1995). Such a process would explain both the presence of nested mounds on extensive sequences of bends (rather than the first bend at which flow was stripped) and the production of stacked mounds, which could be formed by successive slumping events. Steep slopes, higher shear stresses, and large-scale bank erosion may lead to a high proportion of nested mounds forming on the outside of channel bends. However, slumping-related nested mounds would not be located exclusively on the outer bank of channel bends as the current flow stripping model predicts, but would also be present on channel crossovers and even on the inner channel bend.

Levee Deposits

The flow model outlined above suggests that the overbank portion of many flows will be thick, spatially and temporally extensive, and consist of low-concentration, low-velocity fluid. The resulting deposits will dominantly consist of thin, yet spatially extensive beds of sediment. Such a model of turbidity current deposition is supported by the increasing recognition of turbiditic sands in levees rather than sands of contourite origin (Shew et al., 1994). Ignoring, for the moment, the effects of levee instability, the model suggests that many levees may contain large volumes of sand, disposed in thin, widespread beds, and may therefore be more prospective from a hydrocarbon perspective than has previously been recognized. There is some evidence from modern systems that levee sediments are widespread: individual sediment layers on the NAMOC system have been traced for more than 300 km (Hesse, 1995). There is also increasing evidence that levee production can exceed expectations (Kendrick, 1998), undermining the view that laminated sand beds are unlikely to be widespread. Gas production rates from the Ram Powell Field in the Gulf of Mexico have outperformed predictions, confirming that thin-bedded levee-overbank reservoirs can have better connectivity than logs and cores would suggest (Clemenceau et al., 1998). Additionally, pressure decline histories indicate that thin-bedded levee sands are apparently very continuous in the Tahoe and Popeye prospects (Kendrick, 1998), with aquifer pressure support suggesting that the thin beds are continuous some distance downstream as well as laterally; in these prospects the area of reservoir compartments may exceed 1000 acres (2470 hectares). The existence of good lateral reservoir continuity within levee-hosted sheet sands has also been noted by Kolla et al. (1998).

The view that levee facies are gravitationally unstable, and prone to slumping and sliding, at least in the zone around the levee crest (Cossey, 1994, and references therein; Clark and Pickering, 1996a), is in apparent contradiction to the evidence cited above that thin-bedded levee facies can be laterally extensive and well connected, not only at deposition but after burial. The explanation may lie in the fact that most outcrops identified as levee analogs do not provide good 3-D control. Although slumps and slides may disrupt apparent connectivity in 2-D sections, they appear to be of a much smaller areal extent than individual levee beds, and so in three dimensions they are more likely to form baffles than barriers to flow.

Channel Thalweg Deposits

The schematic submarine channel planform and cross-sectional models (Figures 5, 6) can be combined to produce a 3-D model of the resultant channel architecture (Figure 11). Initially, point bar growth produces a single-thread sandbody that migrates laterally at the channel bends, maintaining a constant thickness. Eventually an equilibrium planform geometry is reached and the channel aggrades without further bend movement (Figure 11). The resultant 3-D geometry is predictable and suggests that facies sequences vary as a function of bend position. The thalweg deposits at crossover points are near-vertical; the bend apex is characterized by a lower sequence extending obliquely outward from the inner part of the bend and a vertical upper sequence (Figure 11). The issue of connectivity between levee and channel-axis facies is not yet satisfactorily resolved (Clark and Pickering, 1996a and references therein), although there is evidence to suggest that channel axis facies can be isolated from the levee facies, due perhaps to mud draping or filling the channel (Kendrick, 1998).

Generality of the Architectural Model

The proposed architectural model is for systems that are unaffected by tectonism (Nakajima et al., 1998). Although a detailed discussion of the affects of tectonism is beyond the scope of this paper, it is worth noting, for example, that local tectonic or synsedimentary faulting may affect channel development by changing local gradients, by producing barriers to flow, and by pinning individual channel segments to particular incision points through fault planes. A consequence of this control is that channels may be unable

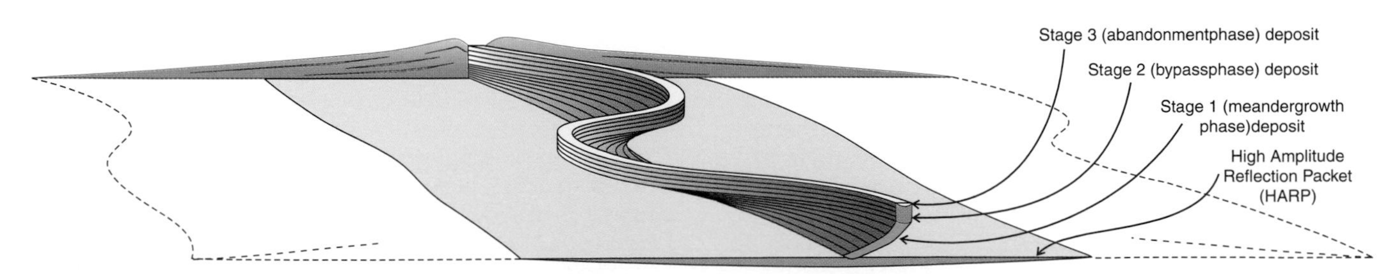

Figure 11—Three-dimensional block model of bend evolution and sedimentary architecture in a typical aggradational, high-sinuosity, sinuous submarine channel (not to scale). For simplicity, the channel axis width is constant in the figure, although there is evidence for progressive reduction of channel width in some submarine channels.

to evolve to equilibrium sinuosities, and may change from aggradational to incisional styles, evolve into sheet systems and back again, and swing downstream of a point of lateral confinement. None of these possibilities can be predicted by application of the model described above, and caution must be urged when attempting to make predictions regarding channel style and fill in such areas.

CONCLUSIONS

A simple process model of flow structure and flow evolution can be applied to medium- to high-sinuosity, aggradational submarine channels. The model predicts that flows in such channels are highly stratified, have significant supra-levee thicknesses, and form broad overbank wedges of low-concentration fluid. The process model is in good agreement with extensive field studies undertaken in Bute Inlet, British Columbia (Bornhold et al., 1994 and references therein), and for the first time, provides a coherent set of process explanations for the primary sedimentary observations of submarine channels, including fining-up of channel levees; long-distance correlation of levee beds; morphological evidence for large-scale overbank flows; evidence for deposition of low-velocity, low-concentration flows across broad areas of submarine fans; and transport and deposition of clean sand at and beyond channel mouths.

This study has important implications for two particular aspects of deep marine sedimentology. First, true ridge and swale topography has yet to be recognized in submarine channel systems. Second, the incorporation of stratification into the flow stripping concept suggests that gradual stripping of large portions of the upper flow may occur without dramatically affecting basal flow processes. Consequently, nested mound formation is probably not the result of flow stripping except in a few special circumstances. We propose that the majority of nested mounds may instead form as a result of deceleration around topography formed by slumped levee material.

The model proposed here also has a number of implications for sedimentary architecture in submarine channel systems, including (1) the architecture of submarine channels is likely to be significantly different from fluvial channels and extreme caution should be used if applying fluvial models to submarine channels; (2) submarine channel deposits are characterized by ribbon geometries; (3) the broad 3-D geometry of the channel thalweg deposits is predictable with significant variations between bend apex and bend crossover positions; (4) because bends reach an equilibrium planform geometry, point bars and associated lateral accretion deposits (if present) are not reworked extensively and will be well preserved; (5) thin, spatially extensive levee beds with high initial connectivity will be commonly deposited; (6) because slumping and sliding of outer levee deposits occur on a smaller areal scale than that of individual levee beds, these features will therefore tend not to produce barriers to 3-D bed connectivity; (7) core analysis and 2-D field analog studies are likely to underestimate levee facies connectivity; (8) the composite levee facies model (points 5 and 6) suggests that discoveries made in levee deposits to date are not anomalous and that many submarine levee deposits may be prospective, and (9) although slumping-related intrachannel nested mounds will be concentrated on the outer bank of channel bends they may also be present in bend crossover zones and on the inner edge of channel bends.

It should be borne in mind that this channel model applies solely to medium- to high-sinuosity channels characterized by broadly contemporaneous channel thalweg aggradation and levee growth and unaffected by tectonism. Caution is urged if applying this model to other submarine channel types and in areas where tectonic controls are a significant factor.

ACKNOWLEDGMENTS

We thank reviewers Senira Kattah and Bradford Macurda and editor Arnold Bouma for input to this paper. This study was funded by a consortium of oil companies, including Amerada Hess, Amoco, ARCO, BG, BHP, BP, Chevron, Conoco, Elf, Enterprise, Fina, Mobil, Shell, and Texaco.

REFERENCES CITED

Altinakar, M. S., W. H. Graf, and E. J. Hopfinger, 1996, Flow structure in turbidity currents: Journal of Hydraulic Research, v. 34, p. 713–718.

Bornhold, B. D., P. Ren, and D. B. Prior, 1994, High-frequency turbidity currents in British Columbia fjords: Geo-Marine Letters, v. 14, p. 238–243.

Bowen, A. J., W. R. Normark, and D. J. W. Piper, 1984, Modelling of turbidity currents on Navy submarine fan, California continental borderland: Sedimentology, v. 31, p. 169–186.

Chikita, K., 1990, Sedimentation by river-induced turbidity currents: field measurements and interpretation: Sedimentology, v. 37, p. 891–905.

Clark, J. D., and K. T. Pickering, 1996a, Submarine channels: processes and architecture: London, Vallis Press, 231 p.

Clark, J. D., and K. T. Pickering, 1996b, Architectural elements and growth patterns of submarine channels: applications to hydrocarbon exploration: AAPG Bulletin, v. 80, p. 194–221.

Clemenceau, G., J. Colbert, F. Lockett, and B. Musso, 1998, Development drilling results from selected deepwater Gulf of Mexico fields—implications on turbidite reservoir characterization and production performance prediction (abs.) *in* Developing and managing turbidite reservoirs: case histories and experiences: EAGE/AAPG 3d research symposium, Almeria, Spain, October 3–9,1998.

Cossey, S. P. J., and Jacobs, R. E., 1992, Oligocene Hackberry Formation of southwest Louisiana: sequence stratigraphy, sedimentology, and hydrocarbon potential: AAPG, Bulletin, v. 76, p. 589–606.

Cossey, S. P. J., 1994, Reservoir modeling of deepwater clastic sequences: mesoscale architectural elements, aspect ratios and producibility, *in* P. Weimer, A. H. Bouma, and B. F. Perkins, eds., Submarine fans and turbidite systems: sequence stratigraphy, reservoir architecture and production characteristics: Gulf Coast Section SEPM 15th Annual Research Conference Proceedings, p. 83–93.

EEZ-SCAN 85 Scientific Staff, 1987, Atlas of the Exclusive Economic Zone, Gulf of Mexico: U.S. Geological Survey Miscellaneous Investigations Series I-1864-A, scale 1:500,000, Reston, Virginia, 104 p.

Flood, R. D., and J. E. Damuth, 1987, Quantitative characteristics of sinuous distributary channels on the Amazon Deep-Sea Fan: Geological Society of America Bulletin, v. 98, p. 728–738.

Garrison, L. E., N. H. Kenyon, and A. H. Bouma, 1982, Channel systems and lobe construction in the Mississippi Fan: Geo-Marine Letters, v. 2, p. 31–39.

Gibbs, R. J., 1967, The geochemistry of the Amazon River system: Part I. The factors that control the salinity and the composition and concentration of the suspended solids: Geological Society of America Bulletin, v. 78, p. 1203–1232.

Gibling, M. R., and B. R. Rust, 1993, Alluvial ridge-and-swale topography: a case study from the Morien Group of Atlantic Canada, *in* M. Marzo and C. Puigdefábregas, eds., Alluvial sedimentation: International Association of Sedimentologists Special Publication 17, p. 133–150.

Hackbarth, C. J., and R. D. Shew, 1994, Morphology and stratigraphy of a mid-Pleistocene turbidite leveed channel from seismic, core and log data, northeastern Gulf of Mexico, *in* P. Weimer, A. H. Bouma, and B. F. Perkins, eds., Submarine fans and turbidite systems: sequence stratigraphy, reservoir architecture and production characteristics: Gulf Coast Section SEPM 15th Annual Research Conference Proceedings, p. 127–133.

Hay, A. E., R. W. Burling, and J. W. Murray, 1982, Remote acoustic detection of a turbidity current surge: Science, v. 217, p. 833–835.

Hesse, R., 1995, Long-distance correlation of spill-over turbidites on the western levee of the Northwest Atlantic Mid-Ocean Channel (NAMOC), Labrador Sea, *in* K. T. Pickering, R. N. Hiscott, N. H. Kenyon, F. Ricci Lucchi, and R. D. A. Smith, eds., Atlas of deep water environments: architectural style in turbidite systems: London, Chapman and Hall, p. 276–281.

Hiscott, R. N., F. R. Hall., and C. Pirmez, 1997a, Turbidity-current overspill from the Amazon channel: texture of the silt/sand load, paleoflow from anisotropy of magnetic susceptibility and implications for flow processes, *in* R. D. Flood, D.J.W. Piper, A. Klaus, and L.C. Peterson, eds., Proceedings of the ODP, scientific results 155: College Station, TX, Ocean Drilling Program, p. 53–78.

Hiscott, R. N., K. T. Pickering, A. H. Bouma, B. M. Hand, B. C. Kneller, and G. Postma, 1997b, Basin-floor fans in the North Sea: sequence stratigraphic models vs. sedimentary facies, discussion: AAPG Bulletin, v. 81, p. 662–665.

Kastens, K. A., and A. N. Shor, 1985, Depositional processes of a meandering channel on Mississippi Fan: AAPG Bulletin, v. 69, p. 190–202.

Kendrick, J. W., 1998, Turbidite reservoir architecture in the Gulf of Mexico—insights from field development *in* Developing and managing turbidite reservoirs: case histories and experiences (abs.): EAGE/AAPG 3rd research symposium, Almeria, Spain, October 3–9,1998.

Kenyon, N. H., A. Amir, and A. Cramp, 1995, Geometry of the younger sediment bodies of the Indus Fan, *in* K. T. Pickering, R. N. Hiscott, N. H. Kenyon, F. Ricci Lucchi, and R. D. A. Smith, eds., Atlas of deep water environments: architectural style in turbidite systems: London, Chapman and Hall, p. 89–93.

Klaucke, I., and R. Hesse, 1996, Fluvial features in the deep-sea: new insights from the glacigenic submarine drainage system of the northwest Atlantic mid-ocean channel in the Labrador Sea: Sedimentary Geology, v. 106, p. 223–234.

Kneller, B. C., 1995, Beyond the turbidite paradigm: physical models for deposition of turbidites and their implications for reservoir prediction, *in* A. J. Hartley and D. J. Prosser, eds., Characterization of deep marine clastic systems: Geological Society of London Special Publication 94, p. 31–49.

Kolla, V., P. Bourges, J. M. Urruty, D. Claude, M. Morice, E. Durand, and N. H. Kenyon, 1998, Reservoir architecture in recent and subsurface, deepwater meandri-channel and related depositional forms (abs), *in* Developing and managing turbidite reservoirs: case histories and experiences: EAGE/AAPG 3rd research symposium, Almeria, Spain, October 3–9,1998.

Lonsdale, P., and C. D. Hollister, 1979, Cut-off at an abyssal meander south of Iceland: Geology, v. 7, p. 597–601.

McHargue, T. R., 1991, Seismic facies, processes, and evolution of Miocene inner fan channels, Indus submarine fan, *in* P. Weimer and M. H. Link, eds., Seismic facies and sedimentary processes of submarine fans and turbidite systems: New York, Springer-Verlag, p. 403–413.

Masson, D. G., N. H. Kenyon, J. V. Gardner, and M. E. Field, 1995, Monterey Fan: channel and overbank morphology, *in* K. T. Pickering, R. N. Hiscott, N. H. Kenyon, F. Ricci Lucchi, and R. D. A. Smith, eds., Atlas of deep water environments: architectural style in turbidite systems: London, Chapman and Hall, p. 74–79.

Nakajima, T., M. Satoh, and Y. Okamura, 1998, Channel-levee complexes, terminal deep-sea fan and sediment wave fields associated with the Toyama deep-sea channel system in the Japan Sea: Marine Geology, v. 147, p. 25–41.

Normark, W. R., G. R. Hess, D. A. V. Stow, and A. J. Bowen, 1980, Sediment waves on the Monterey Fan levee: a preliminary physical interpretation: Marine Geology, v. 37, p. 1–18.

Peakall, J., W. D. McCaffrey, and B. K. Kneller, (in press), A process model for the evolution, morphology and architecture of sinuous submarine channels: Journal of Sedimentary Research.

Phillips, S., 1987, Dipmeter interpretation of turbidite-channel reservoir sandstones, Indian Draw Field, New Mexico, *in* R. W. Tillman and K. J. Weber, eds., Reservoir sedimentology: SEPM Special Publication 40, p. 113–128.

Pickering, K. T., J. Coleman, M. Cremer, L. Droz, B. Kohl, W. R. Normark, S. O'Connell, D. A. V. Stow, and A. Meyer-Wright, 1986, A high-sinuosity, laterally migrating submarine fan channel-levee-overbank: results from DSDP Leg 96 on the Mississippi Fan, Gulf of Mexico: Marine and Petroleum Geology, v. 3, p. 3–18.

Piper, D. J. W., and W. R. Normark, 1983, Turbidite depositional patterns and flow characteristics, Navy submarine fan, California Borderland: Sedimentology, v. 30, p. 681–694.

Pirmez, C., 1994, Growth of a submarine meandering channel-levee system on the Amazon Fan: Unpublished Ph.D. dissertation: New York, Columbia University, 621 p.

Pirmez, C., and R. D. Flood, 1995, Morphology and structure of Amazon Channel: *in* R. D. Flood, D. J. W. Piper, A. Klaus, and L. C. Peterson, eds., Proceedings of the ODP, scientific results, 155: College Station, TX, Ocean Drilling Program, p. 23–45.

Prior, D. B., Adams, C. E., and Coleman, J. M., 1983, Characteristics of a deep-sea channel on the middle Mississippi Fan as revealed by a high-resolution survey: Gulf Coast Association of Geological Societies Transactions, v. 33, p. 389–394.

Prior, D. B., B. D. Bornhold, and M. W. Johns, 1986, Active sand transport along a fjord-bottom channel, Bute Inlet, British Columbia: Geology, v. 14, p. 581–584.

Prior, D. B., B. D. Bornhold, W. J. Wiseman, Jr., and D. R. Lowe, 1987, Turbidity current activity in a British Columbia Fjord: Science, v. 237, p. 1330–1333.

Ren, P., B. D., Bornhold, and D. B. Prior, 1996, Seafloor morphology and sedimentary processes, Knight Inlet, British Columbia: Sedimentary Geology, v. 103, p. 201–228.

Shanmugam, G., R. B. Bloch, S. M. Mitchell, G. W. J. Beamish, R. J. Hodgkinson, J. E. Damuth, T. Straume, S. E. Syvertsen, and K. E. Shields, 1995, Basin-floor fans in the North Sea: sequence stratigraphic models vs. sedimentary facies: AAPG Bulletin, v. 79, p. 477–512.

Shanmugam, G., R. B. Bloch, J. E. Damuth, and R. J. Hodgkinson, 1997, Basin-floor fans in the North Sea: sequence stratigraphic models vs. sedimentary facies, reply: AAPG Bulletin, v. 81, p. 666–672.

Shew, R. D., D. R. Rollins, G. M. Tiller, C. J. Hackbarth, and C.D. White, 1994, Characterization and modeling of thin-bedded turbidite deposits from the Gulf of Mexico using detailed subsurface and analog data, *in* P. Weimer, A. H. Bouma, and B. F. Perkins, eds., Submarine fans and turbidite systems: sequence stratigraphy, reservoir architecture and production characteristics: Gulf Coast Section SEPM 15th Annual Research Conference Proceedings, p. 327–334.

Stacey, M. W., and A. J. Bowen, 1988, The vertical structure of density and turbidity currents: theory and observations: Journal of Geophysical Research, v. 93, p. 3528–3542.

Stelting, C. E., et al., 1985a, Migratory characteristics of a mid-fan meander belt, Mississippi Fan, *in* A. H. Bouma, W. R. Normark, and N. E. Barnes, eds., Submarine fans and related turbidite sequences: New York, Springer-Verlag, p. 283–290.

Stelting, C. E., et al., 1985b, Drilling results on the Middle Mississippi Fan, *in* A. H. Bouma, W. R. Normark, and N. E. Barnes, eds., Submarine fans and related turbidite sequences: New York, Springer-Verlag, p. 275–282.

Stølum, H. H., 1998, Planform geometry and dynamics of meandering rivers: Geological Society of America Bulletin, v. 110, p. 1485–1498.

Stow, D. A. V., and A. J. Bowen, 1980, A physical model for the transport and sorting of fine-grained sediment by turbidity currents: Sedimentology, v. 27, p. 31–46.

Timbrell, G., 1993, Sandstone architecture of the Balder Formation depositional system, UK Quadrant 9 and adjacent areas, *in* J. R. Parker, ed., Petroleum geology of northwest Europe: Proceedings of the 4th Conference, Geological Society of London, p. 107–121.

Zeng, J., D. R. Lowe, D. B. Prior, W. J. Wiseman, Jr., and B. D. Bornhold, 1991, Flow properties of turbidity currents in Bute Inlet, British Columbia: Sedimentology, v. 38, p. 975–996.

Weimer, P., 2000, Interpreting turbidite systems with 2-D and 3-D seismic data: an overview, *in* A. H. Bouma and C. G. Stone, eds., Fine-grained turbidite systems, AAPG Memoir 72/SEPM Special Publication No. 68, p. 89–92.

Chapter 8

Interpreting Turbidite Systems with 2-D and 3-D Seismic Data: An Overview

Paul Weimer
Energy and Minerals Applied Research Center
Department of Geological Sciences, University of Colorado
Boulder, Colorado, U.S.A.

ABSTRACT

A significant portion of the geologic community's understanding of turbidite systems is derived from the interpretation of 2-D and 3-D seismic data. The images generated from 3-D seismic data, in particular, have caused the re-examination and reinterpretation of deep-water sedimentary processes. In the future, the integration of 3-D seismic data with other data sets and technologies will continue to increase our understanding these systems, and specifically for improved reservoir management in turbidite reservoirs. Some of the new areas of investigation will include: increased study of modern fans and intraslope turbidite systems by integrating multiple data sets (deep cores, sidscan images) with 3-D data; the use of 4-D seismic and reservoir monitoring; repeat 3-D seismic surveys integrated with reservoir simulation; time lapse, multicomponent (4-D, 3-D); artificial intelligence, viz, neutral networks for more accurate lithologic prediction; and the use of large-scale visualization rooms in enhancing interpretation.

INTRODUCTION

Seismic reflection data have been an essential tool to the geologic community's initial understanding, and more recently, significantly improving our understanding of turbidite systems. The development of 3-D data from 2-D data has made significant contributions to the understanding of these systems. Today, 3-D seismic data, when integrated with other disciplines and technology, provide remarkably accurate renderings of turbidite systems in the subsurface. The new images that are being generated from many continental margins from the world have challenged many of our previous concepts of sedimentary processes and facies distributions in turbidite systems (Mayall et al., 1998).

Two-dimensional seismic data allow for the general recognition of key bounding horizons, lateral and vertical distribution of seismic facies, and helping define a sequence stratigraphic framework for turbidite systems. General turbidite elements can be identified, correlated through a grid of data, and mapped. These seismic data were essential for the discovery of turbidite reservoirs in many sedimentary basins. The greatest shortcoming to 2-D seismic data concerns the way the seismic facies (turbidite elements) change between profiles, which can often be quite surprising to geologists, because the change is on a scale less than the 2-D line spacing.

Modern submarine fans were initially studied with sparker and small water gun data (Bouma et al., 1985a). With more extensive 2-D multifold seismic data, modern fans have been studied in greater detail by resolving a variety of seismic facies that may not appear in surficial sediments, and mapping their distributions.

The real growth in the use and application of 3-D seismic data has happened at a time when there have been increased exploration activities in many deepwater margins around the world, and these data are the essential tool for prospect generation. The addition of 3-D seismic data has been essential to helping bridge the scale gap of different datasets in turbidite systems first identified by Bouma et al. (1985b). Three-dimensional seismic data migration is a major improvement over 2-D migration, and allows for better imaging of the 3-D geometry, by way of improved lateral resolution. Key turbidite elements can be often resolved with 3-D data, such as a sinuous channel over long distances (>50 km). Variations in the elements can be identified and mapped in three dimensions. The integration of 3-D seismic data with other datasets, such as cores, wireline logs, and FMI images, provides for considerably more robust images of turbidite elements than was possible with 2-D seismic data.

Still, seismic reflection data can only resolve turbidite elements of larger scale; smaller features cannot be resolved. With seismic data, the interpreter will always be faced with tuning problems (1/4 seismic wavelet) and resolution (1/8 seismic wavelet). Outcrop studies are key in understanding the issues of scale in the study of turbidite systems (Slatt, this volume).

At a regional scale, closely spaced 3-D horizon slices (expressed as amplitude, coherency, and total energy maps, etc.) can allow for the interpretation of an entire facies distribution and evolution of a turbidite system along a continental margin, provided the area is not structurally disrupted. The extent of entire channel systems can be identified relatively quickly using a voxel (3-D pixel) product for visualization, in which a seed is placed in the data and the element can be traced over long distances in a matter of a few hours (Marotta et al., 1998). As always, the morphology of the 3-D image is presumed to be correct. The technology has evolved so quickly that it is forcing us to reconsider our understanding of depositional processes and facies relationships. What is impossible to accomplish at an outcrop due to limitations in the size of the outcrop can be accomplished rather quickly in the subsurface using 3-D seismic data.

The general architectural elements (channels, lobes, overbank, mass transport–related deposits filling in erosional surfaces) can be defined with great accuracy with 3-D seismic data, and are quite comparable to the scale of elements seen in modern fans using side-looking sonar systems [Armentrout et al., this volume; Pickering et al. (1995) with 3-D seismic data].

At the reservoir level, 3-D seismic data allow for the clear definition of reservoir interval, facies changes, fluid contact, and definition of possible reservoir compartments. When integrated with wireline logs, cores, petrophysics, and production information, 3-D seismic data are an essential in reservoir model simulation (Jack, 1998; Brown, 1999). Three-dimensional seismic data can also be helpful for delineating subsurface fluids (hydrocarbon versus non-hydrocarbon) when integrated with amplitude vs. offset analysis (Rudolph et al., 1998). Seismic attributes can be helpful in defining porosity and fluid content.

FUTURE AREAS

The future use of 3-D seismic data in interpreting turbidite systems will come with extensive integration with other data sets.

(1) Three-dimensional seismic data have been collected across only small portions of a few modern submarine fans. Additional 3-D seismic data, accompanied with coring and logging, will help improve our understanding of facies, architecture and depositional processes in modern submarine fans and intraslope turbidites. This will build on the databases from those fans that have already been studied with relatively deep cores, i.e., Mississippi and Amazon fans.

(2) Four-dimensional seismic (or time-lapse) data are now becoming more common with companies as an essential tool in reservoir management (Jack, 1998). Permanent seismic monitoring is beginning to become more common with many offshore fields where geophones are embedded in the sea floor, and new seismic data are being constantly collected to monitor changes in fluid saturation (i.e., movement) during reservoir development. For example, the turbidite fields west of the Shetland fields (Schielallion, Foinaven) are monitored using this technology (Cooper et al., 1999).

(3) Another technique is the use of repeated 3-D seismic data in reservoir simulation (Johnston, 1998). Two 3-D seismic surveys across the Fulmar field in the North Sea were summed and the differences noted. These results were then integrated with the reservoir simulation model to explain the modeled movement of fluid. In the future, this technique will have tremendous application for influencing the development of deepwater fields.

(4) Time-lapse, multicomponent seismic (4-D, 3-C) data have a tremendous potential for the delineation of lithology and fractures in deepwater reservoirs (Davis et al., 1998). The methodology involves obtaining the velocity of P and S waves in an interval, and dividing the two isochrons to establish a relationship between two velocities. When calibrated with the petrophysics, Vp/Vs can be used as fracture, lithology, porosity, and fluid discriminators. Differences between S1 and S2 waves can help determine the main directions of fractures (anisotropy analysis). Several recent papers summarize the status of multicomponent marine seismic data (Coldwell, 1999; MacLeod

et al., 1999; Zhu et al., 1999). The process is still a major research subject at select universities, and may provide valuable information that will be applied to all geophysics, but specifically to deepwater reservoirs. This technique will allow interpreters to see the changes in the dynamic properties of the reservoir during development.

(5) Artificial intelligence, specifically neural networks, will become increasingly useful for developing criteria for potential pay in lightly explored areas for both 2-D and 3-D seismic data. A recent successful application of this technique was in the development of the Zafiro Complex in Equatorial Guinea (Humphreys et al., 1999). The discovery was a fast-track development; first production was accomplished 17 months after the initial discovery of the field. Neural networks were used to help in lithology and porosity prediction by first determining the relationship between the wireline logs and the seismic attributes, and verification of derived relationships. A series of displays were produced that gave a more accurate rendering of the porosity and lithology relationships than routine 3-D prediction. These displays included V(shale), resistivity prediction, and density-neutron separation prediction.

(6) Large-scale visualization room, termed "interpretation tunnels or caves" (*Oil & Gas Journal*, 1998), is another approach that is used in most major oil companies, and will be used routinely in the future, especially as the cost for the systems declines. In the future, interpreters will be able to walk through a modeled turbidite system in a cave using virtual reality. Automated interpretation techniques with workstations continue to evolve and expedite interpretation of large 3-D datasets. Reality centers will help by assembling integrated teams to look at larger 3-D datasets and help develop a more integrated understanding and interpretation.

ACKNOWLEDGMENTS

I thank Arnold Bouma, Renato Fonseca, Gerald Kuecher, Jack Thomas, and Peter Varnai for their input and careful reviews that greatly improved this manuscript.

REFERENCES CITED

Bouma, A. H., W. R. Normark, and N. E. Barnes, eds., 1985a, Submarine fans and related turbidite systems: Springer-Verlag, New York, 350 p.

Bouma, A. H., W. R. Normark, and N. E. Barnes, 1985b, COMFAN: Needs and initial results, *in* A. H. Bouma, W. R. Normark, and N. E. Barnes, eds., Submarine fans and related turbidite systems: Springer-Verlag, New York, p. 7–11.

Brown, A. R., 1996, Interpretation of three-dimensional seismic data: AAPG Memoir 42, 5th edition, 514 p.

Caldwell, J., 1999, Marine multicomponent seismology: The Leading Edge, v. 18, p. 1274–1282.

Cooper, M., A. O'Donovan, A. Los, R. Parr, and G. Neville, 1999, 4-D seismic: active reservoir management of the Foinaven and Schielallion fields, West of Shetland (abs.): AAPG International Congress, Birmingham, England.

Davis, T. L., R. D. Benson, S. L. Roche, and D. J. Talley, 1998, 4D, 3C seismology and dynamic reservoir characterization—a geophysical renaissance (abs.): EAGE 60th Annual Conference.

Humphreys, N. V., T. A. Williams, G. D. Monson, and L. C. Blundell, 1999, Technology application as an enabler for rapid development of the Zafiro complex, Equatorial Guinea (abs.): AAPG International Congress, Birmingham, England, p. 246.

Jack, I., 1998, Time-lapse seismic in reservoir management: SEG Distinguished Instructor Short Course Notes, 200 p.

Johnston, D. H., 1998, Time-lapse seismic analysis of the North Sea Fulmar Field, *in* Time-lapse seismic in reservoir management: SEG Distinguished Instructor Short Course Notes, p. 9-68–9-89.

MacLeod, M. K., R. A. Hanson, C. R. Bell, 1999, The Alba Field ocen bottom cable seismic survey: impact on development: The Leading Edge, v. 18, p. 1306–1312.

Marotta, D., C. S. Alexander, K. Cotterill, K. Hartman, M. Pasley, T. C. Stitelar, and G. Tari, 1998, The use of 3D visualization for understanding Tertiary deep-water clastic systems: a West Africa example (abs.): AAPG International Congress, Rio de Janeiro, Brazil, p. 746–747.

Mayall, M., I. Stewart, and P. Ventris, 1998, The architecture of turbidite slope channels (abs.): AAPG International Congress, Rio de Janeiro, Brazil, p.384.

Oil & Gas Journal, 1998, Immersive visualization provides an insider's view of the subsurface, v. 96, no. 22, p. 41–47.

Pickering, K. T., et al., eds., 1995, Atlas of deep water environments: Chapman & Hall, London, 333 p.

Rudolph, K., W. Fahmy, and J. Stober, 1998, Direct hydrocarbon indicators: Exxon's worldwide experience (abs.): AAPG International Congress, Rio de Janeiro, Brazil, p. 942.

Zhu, X., S. Altan, and J. Li, 1999, Recent advances in multicomponent processing: The Leading Edge, v. 18, p. 1283–1288.

Armentrout, J. M., K. A. Kanschat, K. E. Meisling, J. J. Tsakma, L. Antrim, and D. R. McConnell, 2000, Neogene turbidite systems of the Gulf of Guinea continental margin slope, Offshore Nigeria, *in* A. H. Bouma and C. G. Stone, eds., Fine-grained turbidite systems, AAPG Memoir 72/SEPM Special Publication No. 68, p. 93–108.

Chapter 9

Neogene Turbidite Systems of the Gulf of Guinea Continental Margin Slope, Offshore Nigeria

John M. Armentrout
Katherine A. Kanschat
Kristian E. Meisling[1]
Mobil Technology Company, Inc.
Dallas, Texas, U.S.A.

Jerome J. Tsakma
Mobil Producing Nigeria
Lagos, Nigeria

Lisa Antrim
Dennis R. McConnell[2]
Amoco Production Company[3]
Houston, Texas, U.S.A.

ABSTRACT

In the study area of the eastern Gulf of Guinea continental margin slope, offshore Nigeria, turbidite depositional systems are confined to slope-valley and slope-basin bathymetric lows bounded by densely faulted, structurally complex zones. Each depositional system consists of three architectural segments (in order): (1) upper slope, small-scale channel elements converging downslope, (2) single channel and nested channel elements with linear to sinuous map patterns, grading further downslope (3) slope-basin lobe and sheet elements. Incised channels and constructional levees indicate transport by turbulent flow. Comparison of the mapped seismic amplitude patterns of different sequences suggests switching of the inferred sand-prone turbidite systems from one slope valley to another through time. This is interpreted to reflect both the lateral shifting of the fluvial sediment supply on the shelf, and the local tectonic modification of slope-valley geometry.

[1]Presently with Arco Oil and Gas, Plano, Texas
[2]Presently, Geological Consultant, Houston, Texas
[3]Now BP-Amoco

INTRODUCTION

The increased availability of large volumes of 3-D seismic reflection surveys provides an opportunity to map gravity-flow depositional systems in unprecedented detail. Deepwater sands are currently among the most promising worldwide exploration targets. Because gravity-flow processes govern transport of sand into deepwater basins, the sand-prone facies occur within the bathymetric lows of the sea floor, and potential reservoir sands commonly do not occur on the crests of structural highs that existed during deposition of the sand. Overlaying a gravity-flow depositional systems map with a present-day structural contour map for the same stratigraphic interval results in superpositioning of potential reservoir and trap, providing the petroleum explorationist with a clear definition of potential drilling targets.

Study Area and Data

This chapter illustrates a successful sequence stratigraphic methodology for mapping deepwater depositional systems using a 3-D seismic reflection volume. The study area is located on the upper slope of the northeastern Gulf of Guinea continental margin slope, south of the Niger delta and Port Harcourt (Figure 1)(CD Figure 1). The study area covers 2200 km^2 (840 mi^2) and extends from outer shelf water depths of less than 200 m (600 ft) to bathyal water depths of more than 1000 m (3000 ft) in the southeast corner. Structurally, the study area spans the growth-fault-dominated upper slope and extends downslope into the toe-of-slope thrust fault zone. Twenty-three sequences have been recognized within the data volume and 12 have been mapped and interpreted for depositional facies patterns.

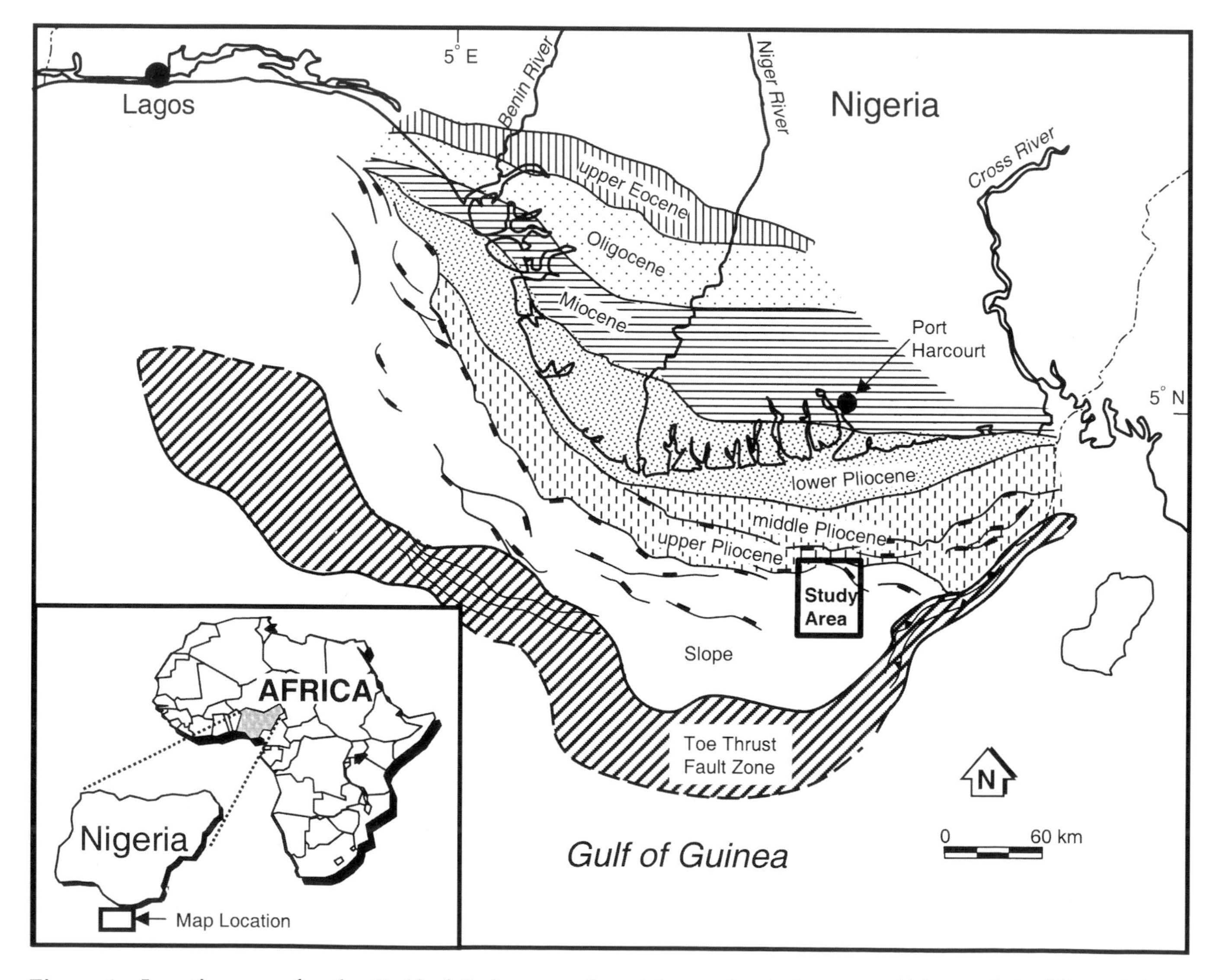

Figure 1—Location map for the Gulf of Guinea continental margin study area, offshore of the Niger delta. The pattern of basinward prograding marginal marine depocenters is taken from Doust and Omatsola (1990). Basinward collapse along outer shelf and upper slope growth faults (hachured lines) translates downslope into a toe-of-slope thrust fault zone (Damuth, 1994).

Aspects of three of these sequences are included in this chapter.

The 3-D seismic survey used for this study was acquired by Geco in 1994 and processed by Digicon UK under the direction of Mobil Producing Nigeria and Mobil Exploration and Producing Technical Center. The survey covers an area of approximately 2200 km^2 (840 mi^2). Nominal coverage is 42-fold, with 4200-m (13,775-ft) streamers and a final bin size of 12.5 m × 12.5 m (41 ft × 41 ft). The processing included 3-D DMO and a steep-dip migration algorithm. No demultiple was added to these data. For regional amplitude work 1500 ms automatic gain control (AGC) was applied to the data to compensate for the effects of a rapidly changing water bottom. Considerable effort was directed toward successfully displaying the data with zero phase.

The data volume was reprocessed for coherency (Bahorich and Farmer, 1995). By calculating localized waveform similarity in both in-line and cross-line direction, a quantitative measure of 3-D seismic coherence is obtained. Small regions of seismic traces cut by a fault or stratigraphic boundary generally have different seismic reflection character than corresponding regions of neighboring traces. This processing results in a sharp discontinuity in local trace-to-trace coherence. Calculating coherence for each grid point along a time slice results in lineaments of low coherence along faults and sharp stratigraphic boundaries. Features enhanced by coherency processing of the Gulf of Guinea continental slope seismic volume include channel forms, mud volcanoes, gas escape craters, and curvilinear faults. Within the 3-D seismic reflection volume of the Niger slope study area, coherency processed data facilitates observation of these features more than 2 s [two-way travel time (TWT)] below the seafloor.

The analysis of the seismic data predated drilling of the study area, necessitating rock-physics modeling for lithofacies prediction using well data from the Joint Venture areas approximately 50 km (30 mi) to the northeast. Subsequently, results from four exploration wells confirmed most of the initial facies interpretations. (Well data are not reported in this chapter).

Interpretation Methods

The depositional systems illustrated in this chapter were identified in a 3-D seismic volume using sequence stratigraphic interpretation methods (Vail, 1987). The sequence stratigraphic model proposed by Mitchum et al. (1977a, b), Posamentier and Vail (1988), and Van Wagoner et al. (1990) is useful in interpreting seismic reflection geometry for depositional environments. Particularly relevant to deepwater depositional systems are papers by Mitchum (1985) for 2-D seismic profiles, and by Prather et al. (1998), Risch et al. (1996), and Weimer et al. (1998) for interpretation of 3-D seismic volumes. In seismic sequence stratigraphic interpretation, the initial step is to identify patterns of reflection terminations on seismic reflection profiles. In basinal deepwater settings, the observed pattern is most often concordant reflections with local patterns of truncation, onlap, and downlap. Once these lapout patterns of reflection termination are noted and candidate sequence boundaries identified, the sequence boundaries are correlated throughout the seismic grid/volume to define seismic horizons that separate regionally significant patterns of underlying truncation from patterns of overlying onlap and downlap (Vail, 1987) (Figure 2) (CD Figure 2).

In deepwater settings, the sequence boundary converges toward the underlying maximum flooding surface, which occurs within an interval of condensed sedimentation, such as a fossil-rich claystone (Loutit et al., 1988). The condensed section interval is often expressed as a regionally continuous, relatively uniform amplitude reflection, affording an excellent mapping horizon (Armentrout, 1996; Prather et al., 1998). Once the condensed sections have been correlated, the reflection profiles are examined for the locally expressed lapout geometry of candidate sequence boundaries and these are correlated throughout the seismic data volume. Local discontinuities, usually associated with faults or deposits of mass wasting events, must be carefully identified and separated from the truly regional sequence boundaries.

Paleoshelf edges are often identified by clinoform topset/foreset inflections. The patterns of shelf-ward toplap vs. basinward downlap also define candidate surfaces for depositional sequence boundaries (Vail, 1987) (Figure 3) (CD Figure 3). Correlation of the paleoshelf-edge sequence boundary candidates at the base of shelf-edge progradational complexes (Figure 3) (CD Figure 3), and the basinal sequence boundary candidates (Figure 2) (CD Figure 2), provides a test of the regional significance of each candidate sequence boundary. Once the numerous sequence boundaries within a data volume have been identified and carefully correlated, and their coeval shelf edges mapped, they provide a stratigraphic framework for interpreting the genetically linked facies of depositional systems within each sequence.

The seismic sequence stratigraphic depositional model predicts siliciclastic turbidite systems in the lowstand systems tract directly overlying the sequence boundary in slope and basinal areas (Vail, 1987; Posamentier and Vail, 1988). In this study, sequence boundaries were correlated throughout the 3-D seismic volume, and RMS (root mean square) amplitude extractions were made from the entire seismic volume between the sequence boundaries. Where the amplitude extractions showed patterns suggestive of gravity-flow depositional elements, amplitude extractions were made of narrower windows within the 3-D volume to more clearly define the stratigraphic position and geometric character of the depositional systems. This analysis of the Gulf of Guinea 3-D seismic volume provides a test of the sequence stratigraphic model.

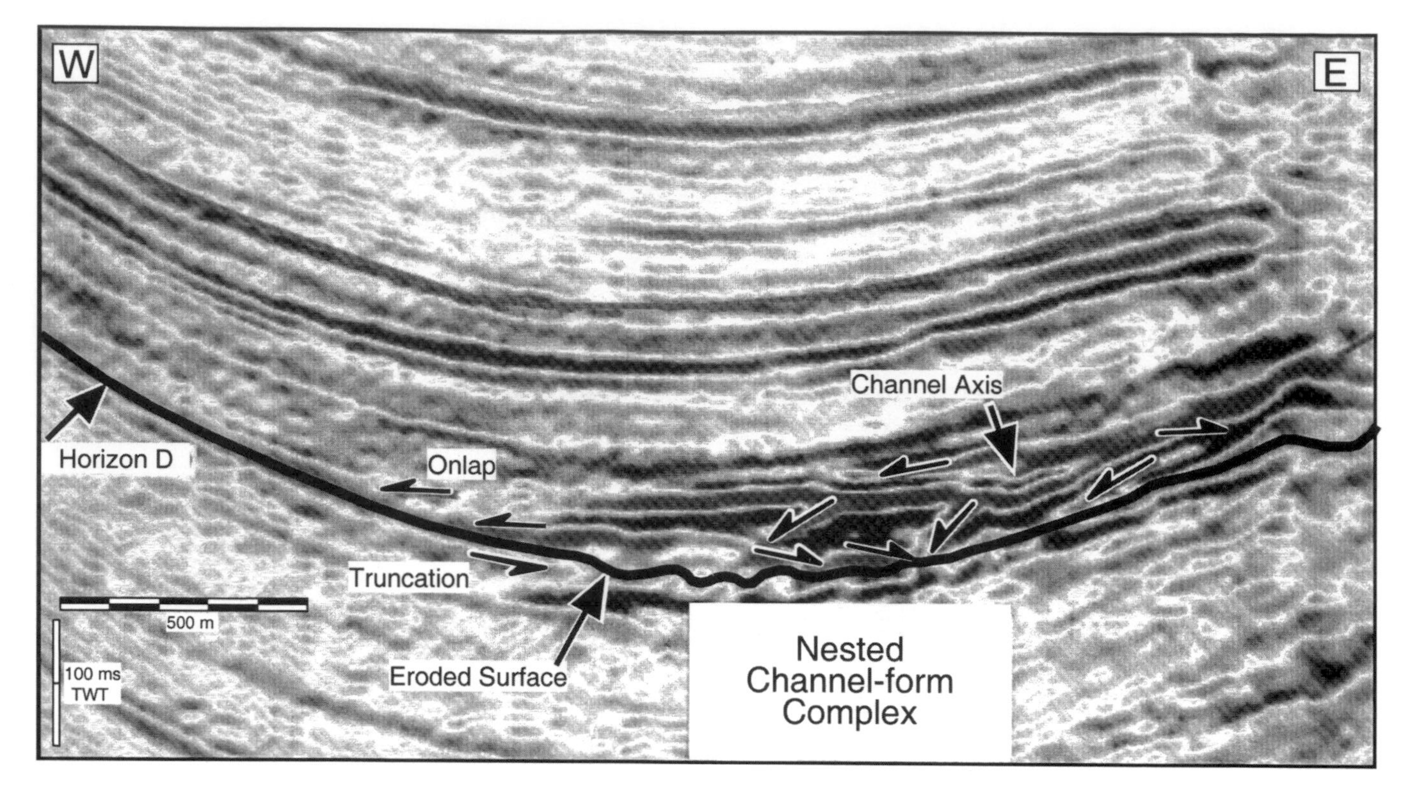

Figure 2—Seismic reflection profile from the 3-D volume showing observational patterns defining a candidate sequence boundary (location shown on Figure 4). The reflection profile is a depositional-strike view through a synclinal-valley isochron. The pattern of reflection terminations is interpreted to reflect erosional truncation below a sequence boundary and bidirectional lapouts and onlaps above within a nested channel complex.

TECTONOSTRATIGRAPHIC SETTING

Understanding the turbidite systems of the Gulf of Guinea continental slope requires an understanding of the tectonic setting into which they were deposited. The 3-D survey interpreted in this study extends across the shelf edge, proximal slope, and active toe-thrust belt of the southeastern slope of the Niger delta (Figures 1, 4)(CD Figures 1, 4). Since the late Miocene, the study area has been located in a proximal slope setting, less than 40 km (25 mi) from the shelf-slope break. Large-scale Neogene gravitational collapse of the Gulf of Guinea continental margin began in the middle to late Miocene time, producing a regionally significant toe-of-slope thrust system (Figure 1). Most subsequent deposition has been restricted to intra-slope valleys and "piggyback" basins, between the shelf-edge growth faults and toe-thrust zone.

Megaslides

Damuth (1994) presented a model for the gravity-failure tectonic style of the Gulf of Guinea continental margin slope. Outer-shelf and upper-slope growth faults form as large prisms of sediments collapse basinward with basal detachment above ductile shales. Downslope movement of these prisms terminate in toe-of-slope thrust zones (Figure 4) (CD Figure 4). These "megaslide" prisms are bounded laterally by transtensional "tear" fault zones, often associated with sea-floor craters and mud volcanoes formed by fluids escaping upward along the fault zone (Figure 4) (CD Figure 4).

The depositional fairways of the study area are typically low-relief synforms bordered by slope-parallel, structurally complex zones (Figure 5A) (CD Figure 5A). On 2-D seismic reflection profiles these structurally complex zones are poor data zones commonly interpreted as shale diapirs (Delteil et al., 1974). The 3-D DMO, steep-dip migrated, and coherency-processed data volume help resolve these poorly imaged zones. The reprocessed 3-D seismic data locally show concordant reflections extending through otherwise random data zones, and time slice displays clearly image numerous curvilinear faults within the zone (CD Figure 6). Several channel systems have been traced through these poor data areas, clearly demonstrating that they are poorly migrated data from structurally complex zones and not ductile-shale diapirs (CD Figure 7). Ductile shale with vertical displacement does occur at great depths associated with the growth-fault

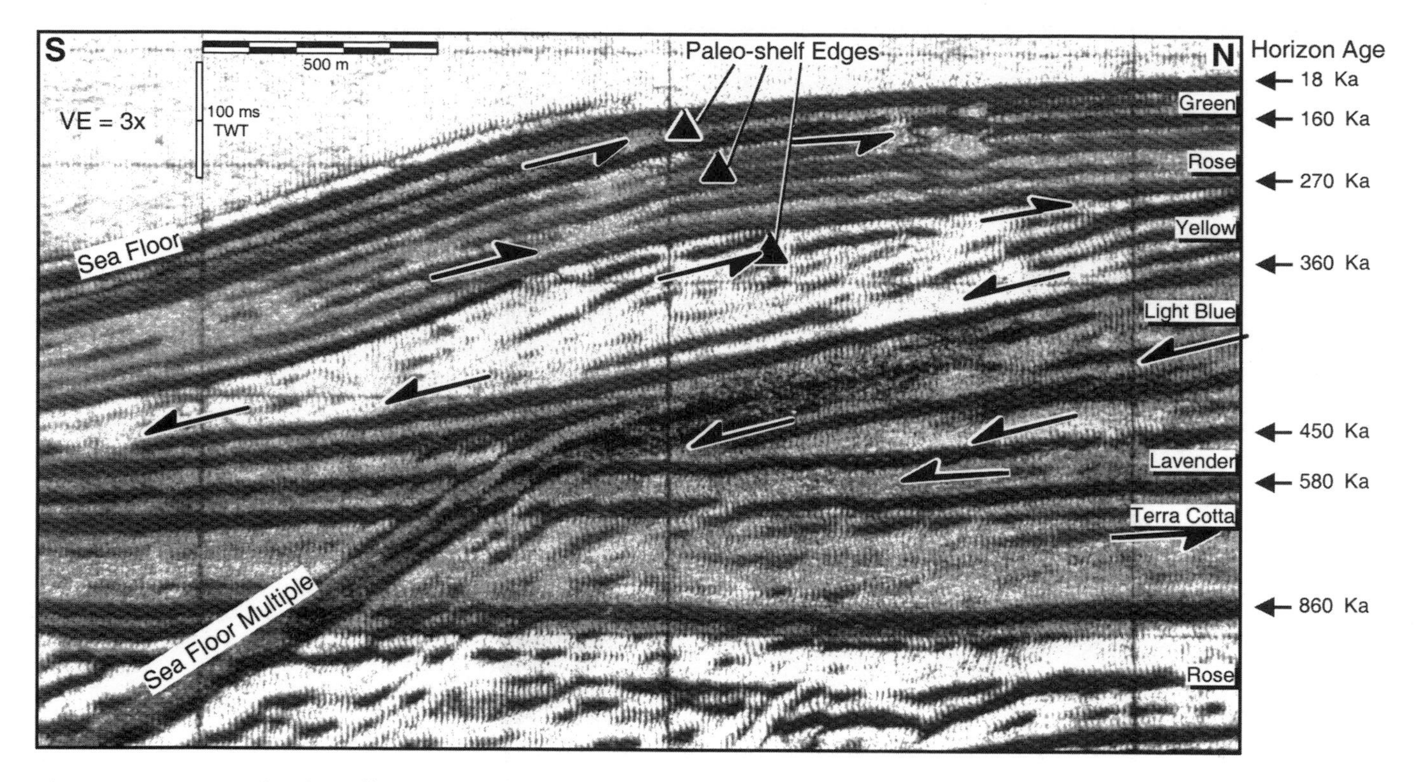

Figure 3—A 2-D seismic reflection profile from the present-day shelf edge showing seismic reflection lapouts (arrows) defining depositional sequence boundaries (see Figure 4 for location). Paleoshelf edge positions (triangles) are identified by mapping the clinoform foreset/topset geometry and internal lapout patterns of each depositional sequence. Color-coded sequence names and the age of each sequence boundary are shown to the right. Ages are based on regional correlation of sequence boundaries into wells and biostratigraphic correlations to a global sea level cycle chart constructed from integration of the Haq et al. (1988) cycle chart and the oxygen isotope record from the deep sea (Trainor and Williams, 1990)(S.A. Bowman, personal communication, 1995). Seismic reflection profile published with permission of TGS Geophysical Company.

detachment zone and toe-thrust faults. Flattening of Figure 5A (CD Figure 5A), at the "datum of major unconformity" is shown as Figure 5B (CD Figure 5B), and illustrates the relatively concordant to slightly convergent reflection patterns across two sediment fairways. The structurally complex zones of numerous curvilinear faults are interpreted as slope-parallel transtensional tear faults along the margins of megaslides with minimal vertical offset.

Slope Valleys

In the study area, the late Pleistocene channel systems near the seafloor (Figure 6) (CD Figure 8) are confined within slope valleys formed above the megaslide prism. These valleys have widths of 7 km (4 mi) or more with only one or two degrees of cross-valley slope. The probability that Gulf of Guinea continental slope Neogene valley systems had little intervalley relief is supported by several channel systems that cross the structurally complex zones from one valley to another (CD Figure 7). Downslope gradient variations of several degrees are indicated by changes in channel form from incision to aggradation across the various structural compartments formed by the growth fault, rollover anticline, piggyback basin, and toe-thrust components of the slope gravity–driven system.

The large, fault-bounded sea-floor megaslide of Figure 4 (CD Figure 4) forms a sea-floor valley within which occurs the intraslope basin gravity-flow system imaged in Figure 6 (CD Figure 8). This valley formed by downslope movement of a stratigraphic volume bounded upslope by down-to-the-south growth faults, and along the margins by curvilinear transtensional tear faults. The resulting megaslide narrows southward, forming a constricted valley that subsequently opens out onto a piggyback basin floor. The latest Pleistocene depositional system was deposited within the same tectonically controlled mega-slide valley during the last sea level lowstand. The late Pleistocene system consists of upper slope channels that coalesce downslope into a single channel complex feeding into a lobate depositional thick (Figure 6A) (CD Figure 8A). The lobate depositional thick is confined within the southern end of the megaslide valley by two down-to-the-north syndepositional faults that blocked further transport to the south. These down-to-the-north

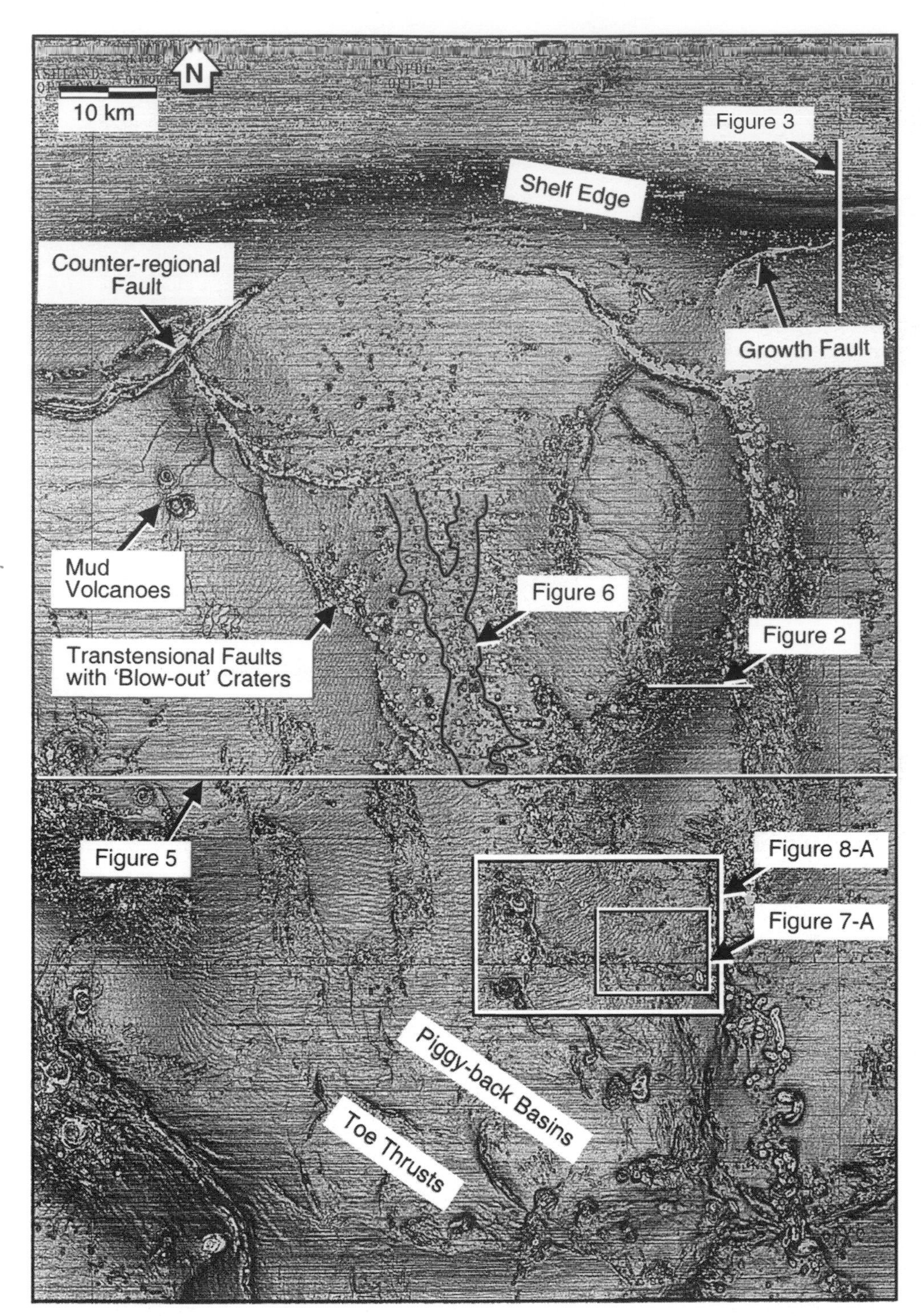

Figure 4—Shaded bathymetric map of the entire study area (Figure 1) displayed to emphasize linear trends. The featureless plain at the north is the shelf. Shaded relief accentuates the shelf edge, counter-regional growth faults, growth faults, transtensional fault zones along the margins of the megaslide, small-scale sea-floor craters, mud volcanoes, piggyback basins, and the toe-thrust zone. The bold outline locates the turbidite system, shown in Figure 6A, within the sea-floor "slide valley."

faults may have formed in response to differential extension where the mega-slide narrows (Figure 4) (CD Figure 4).

DEPOSITIONAL SYSTEM

Regional sequence stratigraphic analysis of the eastern shelf and continental slope of the northern Gulf of Guinea clearly demonstrated that the study area (Figure 1)(CD Figure 1) occurs basinward of depositional shelf edges (Figure 3)(CD Figure 3) throughout the Neogene. This observation encouraged the use of deepwater depositional models for interpretation of the slope facies. To establish a preliminary working hypothesis for deeply buried Gulf of Guinea continental slope gravity-flow systems, the late Pleistocene interval of the study area was examined.

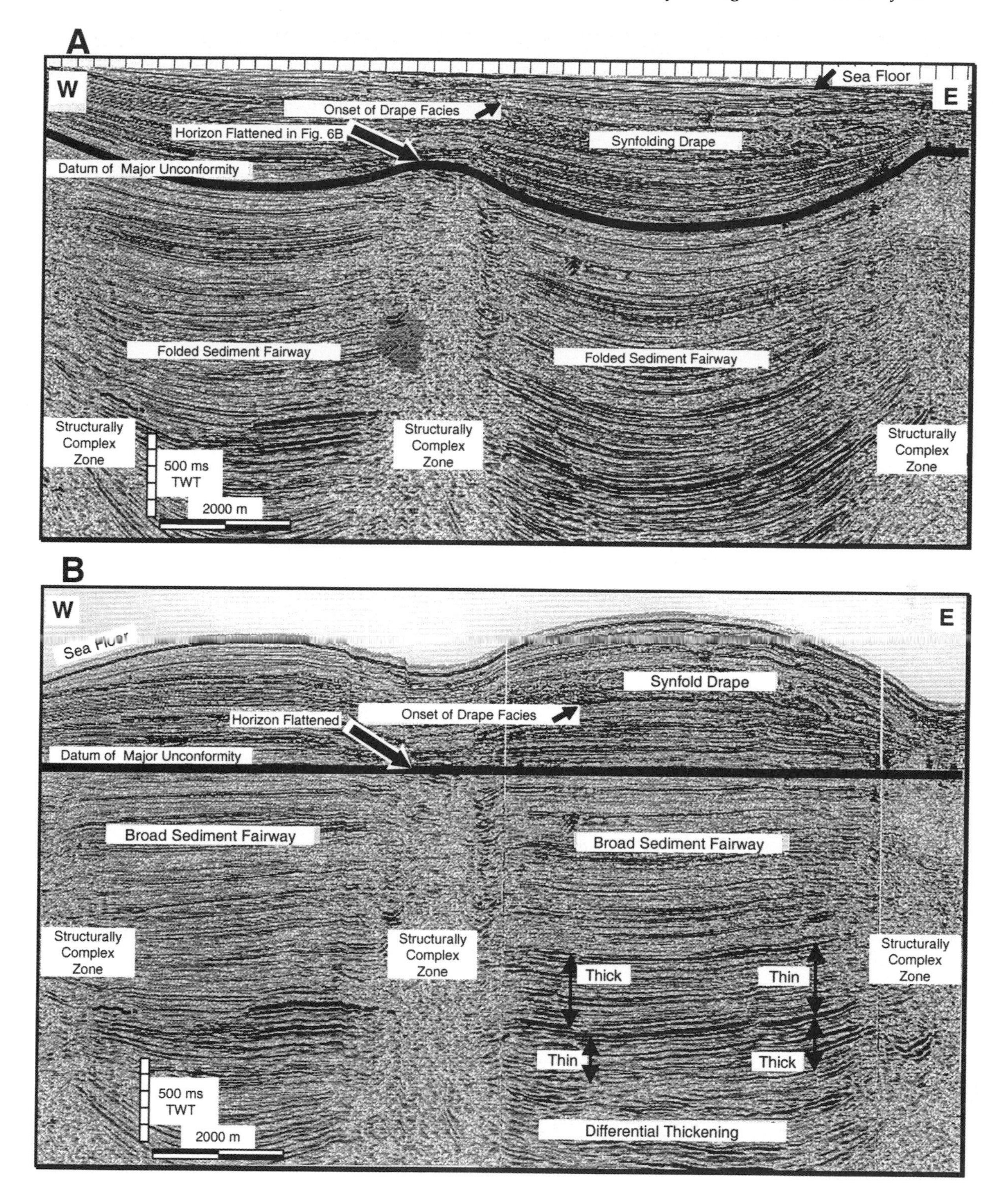

Figure 5—Depositional-strike seismic reflection profiles showing the folded sediment fairways separated by structurally complex zones (see Figure 4 for location): (A) Present-day configuration with the sediments draping and infilling the antiform-synform structure below the major unconformity. This results in a relatively flat present-day sea floor (top of displayed section). (B) The same profile flattened on the "datum of major unconformity" illustrating the relatively concordant to slightly convergent configuration of the sediments below the unconformity. Most of the pre-unconformity sediments appear to have been deposited relatively horizontally, on a sea floor with minimal topography similar to the present-day sea floor. Some differential subsidence resulting in compensation sediment accumulation is indicated by differential thickening of correlative strata across the depositional fairways (thicks vs. thins). The structurally complex zones are interpreted from coherency time-slices as complex fault zones in which concordant strata are poorly imaged on seismic reflection profiles.

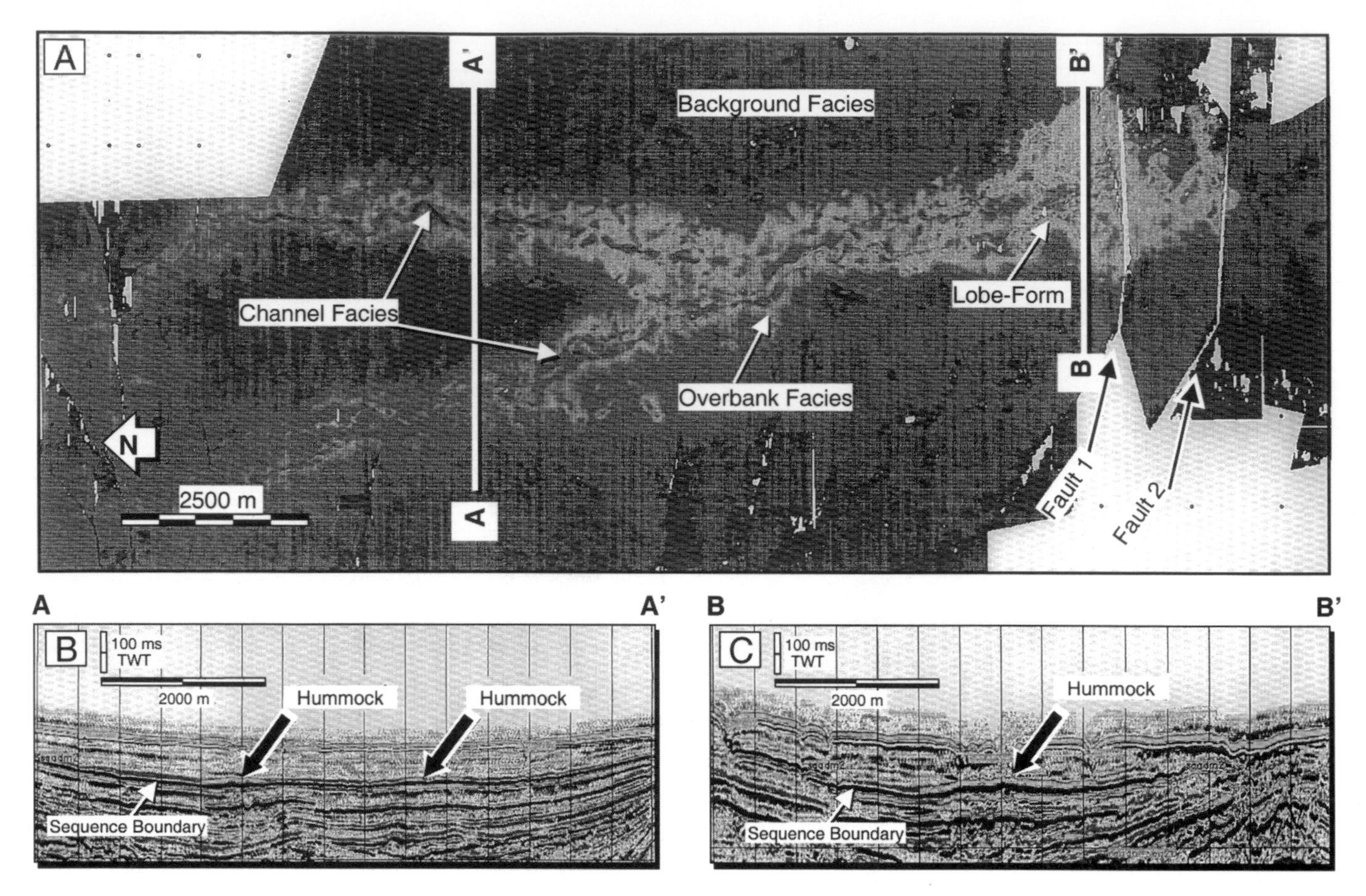

Figure 6—Late Pleistocene Gulf of Guinea continental slope depositional system imaged by RMS (root mean square) amplitude extraction between 15 and 250 ms (TWT) below the sea floor (sequence Green of Figure 3) (see Figure 4 for location of this depositional system): (A) An RMS amplitude extraction showing a curvilinear low-amplitude pattern (darker) flanked by high-amplitude (light) facies interpreted as slope channels flanked by levee and overbank deposits. The channels converge and extend into a lobe where transport of high-amplitude facies sediment is blocked from downslope continuation by down-to-the-north faults, (B) Seismic reflection profile at location *A–A′* showing hummocky facies directly overlying the sequence boundary [zero crossing below the amplitude peak (black)], interpreted as a channel/levee complex, (C) Seismic reflection profile at location *B–B′* showing the broad hummock formed by deposition of a lobe confined by fault-controlled sea-floor relief.

Late Pleistocene Depositional System

Using the 3-D seismic reflection volume, an RMS amplitude display was generated from an interval between 15 and 250 ms (TWT) below the present sea floor (Figure 6A) (CD Figure 8A). This interval images the most recent depositional sequence identified as the Green sequence on Figure 3 (CD Figure 3), and interpreted as latest Pleistocene to Recent (Armentrout et al., 1999). The display shows the elements of the most recent lowstand gravity-flow depositional system. This latest Pleistocene lowstand depositional system consists of three segments, each of which is imaged on Figure 6A (CD Figure 8A). These segments are (in order): (1) upper-slope curvilinear hummocky facies, interpreted as channel elements (Figure 6B) (CD Figure 8B), converging down-slope; (2) single linear to sinuous hummocky facies, interpreted as channel and nested channel elements, grading farther down-slope; (3) slope-basin lens to sheet packages, interpreted as an intraslope basin depositional lobe element (Figure 6C) (CD Figure 8C)(descriptive terms after Pickering et al., 1995). The laterally thinning geometry on either side of the hummocky seismic facies suggests levees formed by overbank deposition from turbidity flow events (Figure 6B) (CD Figure 8B). The lobe element at the distal end of the late Pleistocene turbidity system is approximately 60 km (36 mi) from the age equivalent shelf edge.

Late Pliocene Depositional System

The RMS amplitude extraction of Figure 7A (CD Figure 9A) shows a late Pliocene depositional system between 1.1 and 1.2 s (TWT) (sequence 300 of Armentrout et al., 1999; ~3 to 4 Ma). This depositional system occurs toward the southeastern portion of the study area, where the map view (Figure 7A) (CD Figure 9A) is extracted from a 30-ms (TWT) interval directly above the sequence boundary. The sequence boundary is interpreted below the bidirectional downlapping terminations of the hummock shown in Figure 7C (CD Figure 9C). The mapped interval is between approximately 340 m (1100 ft) and 440 m (1400 ft) below the sea floor; the sea floor

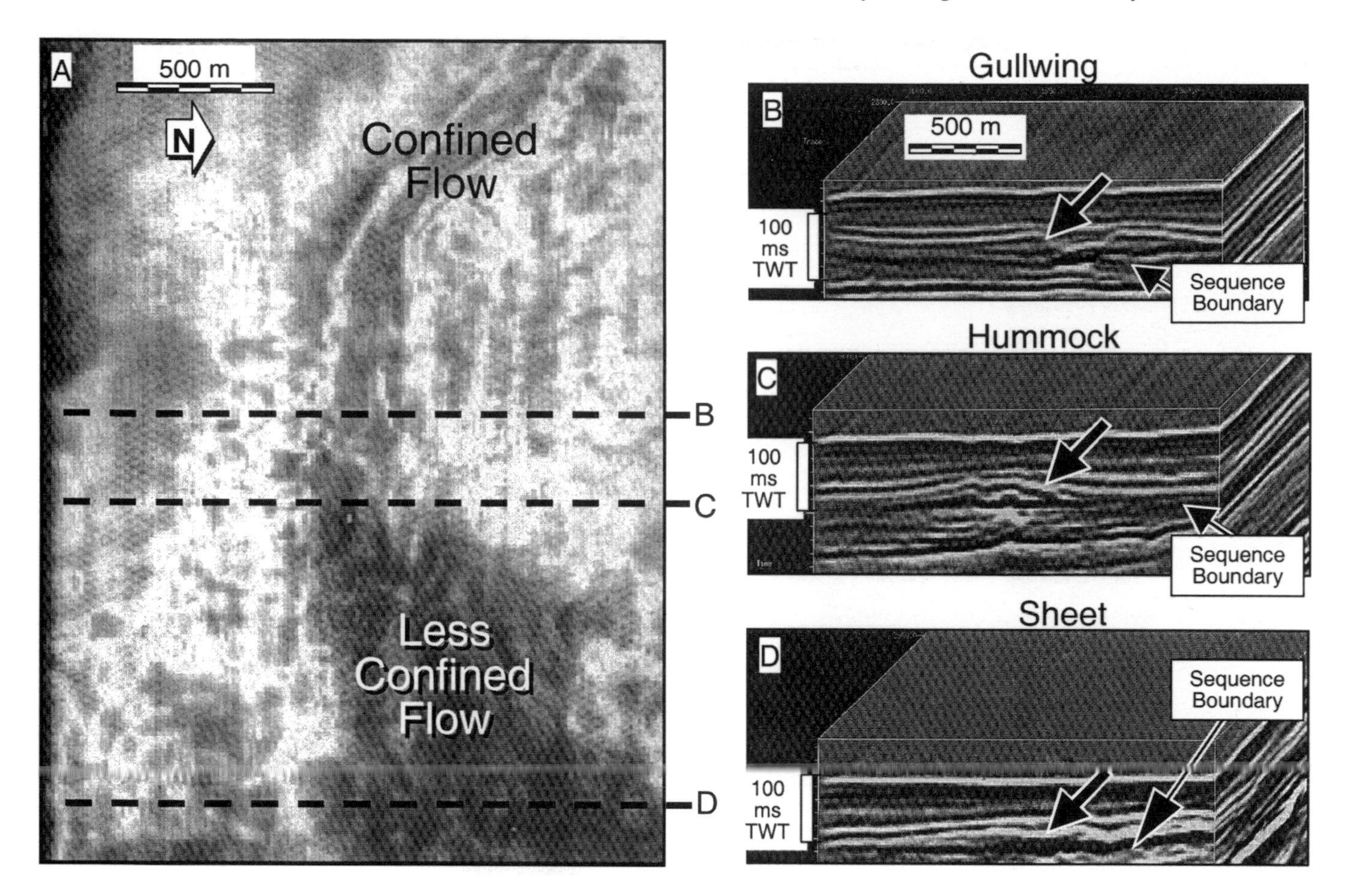

Figure 7—The RMS amplitude extraction of the southeastern terminus of an early Pliocene channel-to-sheet system between 1.1 and 1.2 s (TWT)(see Figure 4 for location). (A) Mapped RMS amplitude pattern showing a curvilinear channel element with high-amplitude margins suggestive of sand-prone facies along a levee system. Reflection profiles are shown for each geometric facies along the turbidite system from: (B) a channel (confined-flow) element to (C) a hummocky transitional facies with small channel forms on the crest, to (D) a sheet (less-confined flow) element. Arrows point at the top of the profile interval imaged on the map display.

averages about 760 m (2500 ft) in water depth across this area.

The mapped pattern shows a curvilinear element grading into a sheetlike high-amplitude facies (black). The channel axis has a low amplitude facies (gray) with high-amplitude margins (white); the channel segment has cross-sectional geometries of symmetrical hummocks, "gullwing" profile (Figure 7B) (CD Figure 9B), and is interpreted as a channel/levee complex. The channel/levee complex is flanked by relatively high amplitude facies (white to light gray) that grade laterally to low amplitude facies (the black area at the upper left margin of Figure 7A) (CD Figure 9A). These channel-flank facies are interpreted as overbank deposits. Transitional between the gullwing channels and the sheet facies are broad hummocks with very-small crestal channel forms (Figure 7C) (CD Figure 9C), interpreted as a depositional lobe with distributary channels. At the eastern end of the channel, a very high-amplitude facies (black to dark gray) spreads outward and has a cross-sectional geometry of a sheet (Figure 7D) (CD Figure 9D). This sheet facies is interpreted as an intraslope basin fan.

This down-system progression of seismic facies suggests a confined-flow channel/levee element through which sediment passed into the less confined flow sheet element forming an intraslope basin floor fan. The distance along the interpreted transport pathway places this Late Pliocene basin floor fan more than 110 km (66 mi) from its correlative shelf-edge.

Latest Miocene Depositional System

The RMS amplitude extraction map of Figure 8A (CD Figure 10A) shows a late Miocene depositional system (sequence 900 of Armentrout et al., 1999; ~5.2–6.5 Ma). This depositional system consists of a channel element that expands into a sheet when entering an intraslope basin. The image is extracted from a 30-ms (TWT) interval directly above the sequence boundary between 2.8 and 3.0 s (TWT), as shown on Figure 8B (CD Figure 10B). This interval is between approximately 1700 and 1900 m (5600–6200 ft) below the sea floor; the sea floor over this area averages about 760 m (2500 ft) in water depth.

The sequence boundary imaged on the seismic profile of Figure 8B (CD Figure 10B) has no underly-

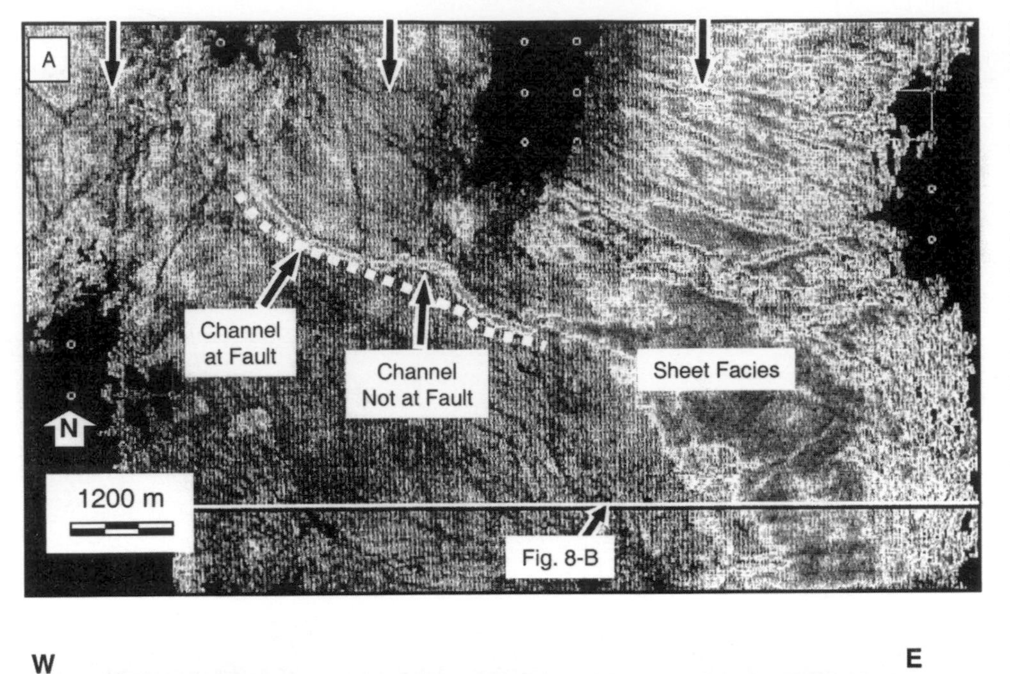

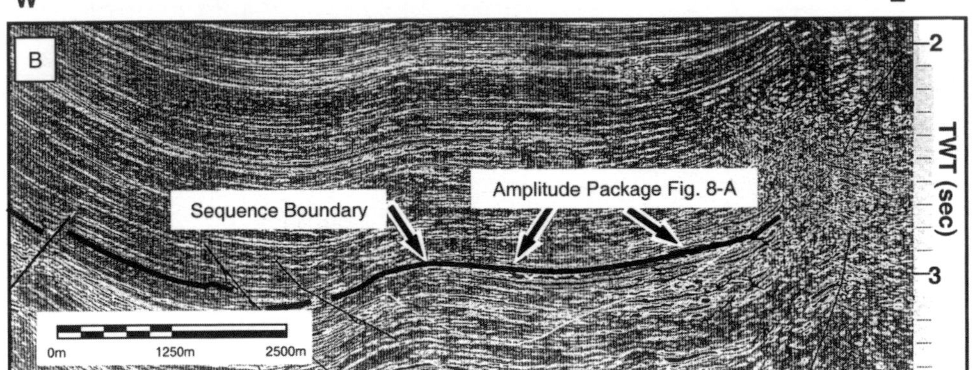

Figure 8—Deeply buried late Miocene turbidite system 2.8 to 3.0 msec (TWT) below the sea floor in the southeast corner of the study area (location on Figure 4): (A) An RMS amplitude map extracted from a 30-msec interval directly above the sequence boundary, illustrating a channel element extending into a sheet element. (B) Seismic profile showing the location of the sequence boundary above which the amplitude was extracted. The channel element crosses two slope valleys through a northwest-trending piggy-back basin overlying a toe-thrust fault complex. The channel character of the linear amplitude pattern is confirmed where the channel-form bends away from the fault.

ing truncation and only very subtle lapout patterns above, but can be identified along the axis of the channel system displayed on Figure 8A (CD Figure 10A). The channel element in Figure 8A (CD Figure 10A) extends from northwest to southeast across two valley systems (south-directed arrows), traversing a piggyback basin along a toe-thrust fault, until it expands into a sheet seismic facies interpreted as an intraslope basin floor fan. Reconstruction of the valley, based on interpretations of differential sediment accumulation patterns on multiple seismic reflection profiles flattened at several successive reflection horizons, suggests that the entry point of the channel system was at a relatively high bathymetric position compared to the floor of the intraslope basin. That relatively high entry point allowed for gravity-flow sediment transport into the sea floor low both to the north (apparent upslope) and south (downslope) of the channel mouth.

The late Miocene distal fan facies are more than 120 km (70 mi) basinward from the age equivalent shelf edge, which occurs north of the study area.

RMS Amplitude Facies

The mapped RMS amplitude distribution of the above three examples illustrates three dominant amplitude facies (Figures 6A, 7A, 8A; see CD figures for color versions) (CD Figures 8A, 9A, and 10A):

(1) Low amplitude facies (black to dark gray) occur over most of the study area and lateral to the curvilinear and fan-shaped higher amplitude facies. This low-amplitude facies is interpreted as muddy hemipelagic sedimentation.
(2) Intermediate amplitude facies (gray), confined to valley axes associated with channel complexes, is interpreted to be overbank deposits of mixed mud, silt, and sand.
(3) Intermediate- to high-amplitude facies (white to light gray), confined to valley axes and associated with channel complexes and distributed over the valley floor basinward of channel complexes, is interpreted to be channel-lag and basin floor fan deposits of silt and sand.

Rock/physics models for these RMS amplitude facies, calibrated to similar seismic facies at the same stratigraphic depth in wells 50 km (30 mi) to the northeast of the study area, provided support for the predrill lithofacies interpretations described above. Subsequent drilling of exploration wells in the study area supports the general lithofacies and depositional system models interpreted here from the 3-D seismic data.

Channel Types

Interpretation of geometric seismic facies, from reflection profiles perpendicular to the depositional axis of the map-view channel elements, illustrates three types of channels:

(1) Erosional; incised into older strata (lowermost channel of Figure 2) (CD Figure 2).
(2) Erosional/depositional; initial incision followed by aggradation (Figure 2) (CD Figure 2).
(3) Depositional; aggradational with no apparent incision into older strata (Figures 6B, 7B, 7C) (CD Figures 8B, 9B, 9C).

The incision along intervals of some channels, and the occurrence of levees and overbank facies, are interpreted to suggest transport of mixed grain-sized sediments by turbulent flow.

Map-view patterns of channel elements range from straight to sinuous (Figures 6A, 8A) (CD Figures 8A, 10A, 11). The channel elements sometimes stack into nested channels (Figure 2) (CD Figure 2). Each channel is typically wider than deep, and is probably comprised of even smaller individual channel-fill deposits below the resolution of the seismic data (CD Figure 12). Late Pleistocene Gulf of Guinea continental slope channel elements can be traced upslope, where they split into progressively smaller channel systems toward the shelf margin. These channels "disappear" toward the top of the slope, where the spatial resolution of the seismic data is less than 3 m (10 ft). Any channel interval that shallows to less than 3 m (10 ft) in depth will not be imaged. The small channels may connect to incised channels at the shelf edge, but this cannot be confirmed with the existing data.

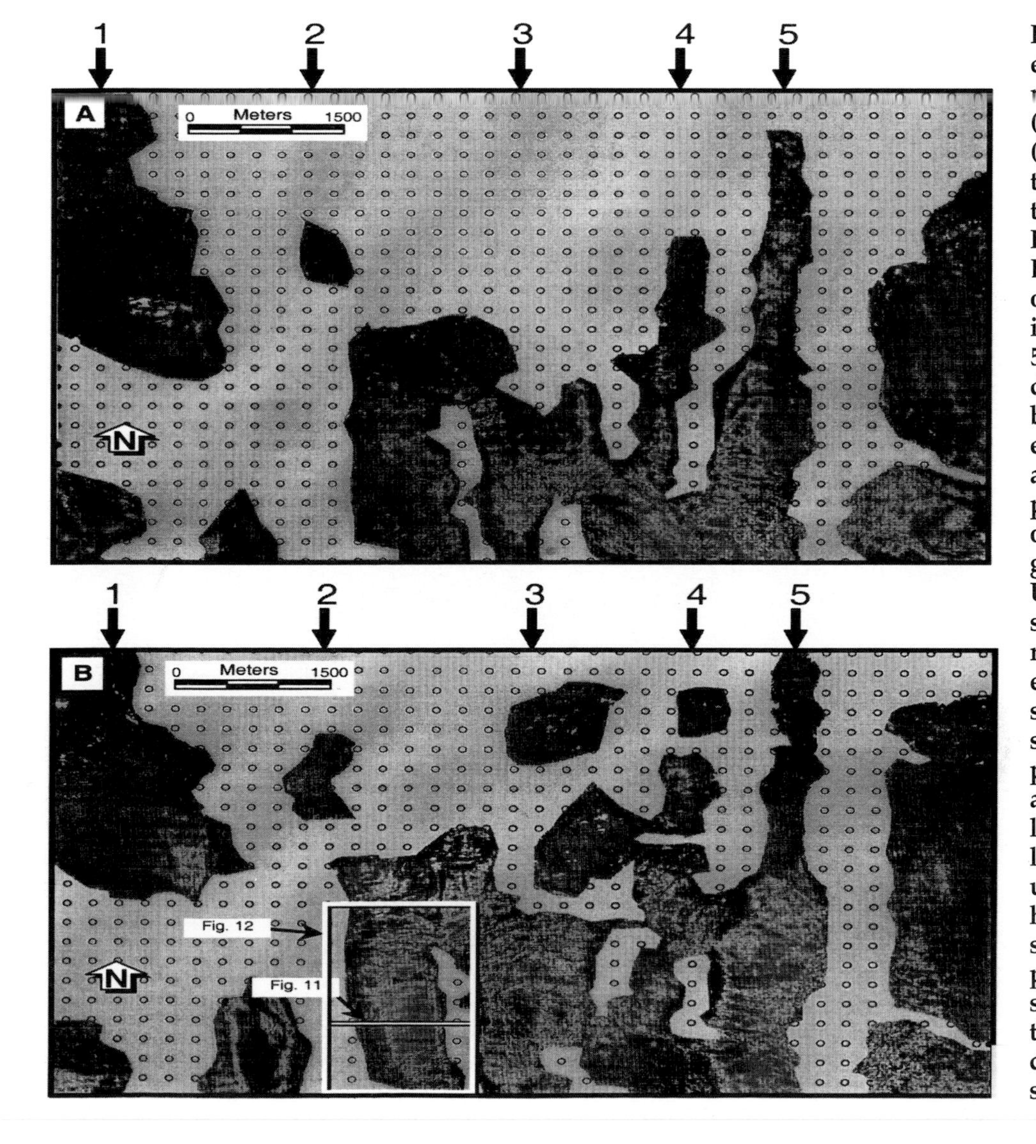

Figure 9—An RMS amplitude extraction maps of two of six mapped depositional systems (*A–F*) for five slope valleys (1–5) extending downslope through the northern part of the study area shown on Figures 1 and 4 (A) Sequence B of Figure 10; (B) sequence C of Figure 10. Each extraction is from a window extending 50 ms directly above the depositional sequence boundary. Rock-physics modeling suggests that the high amplitudes, imaged as light patterns, are sand-prone and occur within the darker background mud-prone facies. Uninterpreted areas are either structurally below the recorded data volume (northern area of Figure 9A), or are so complexly faulted that stratigraphic correlation is not possible (northerly trending areas of Figures 9A, B). The low-amplitude pattern of valley 2 during sequence B (Figure 9A) contrasts with the high-amplitude pattern of sequence C (Figure 9B). The pattern of high-amplitude seismic facies occurring within the five slope valleys over six depositional sequences is shown in Figure 10.

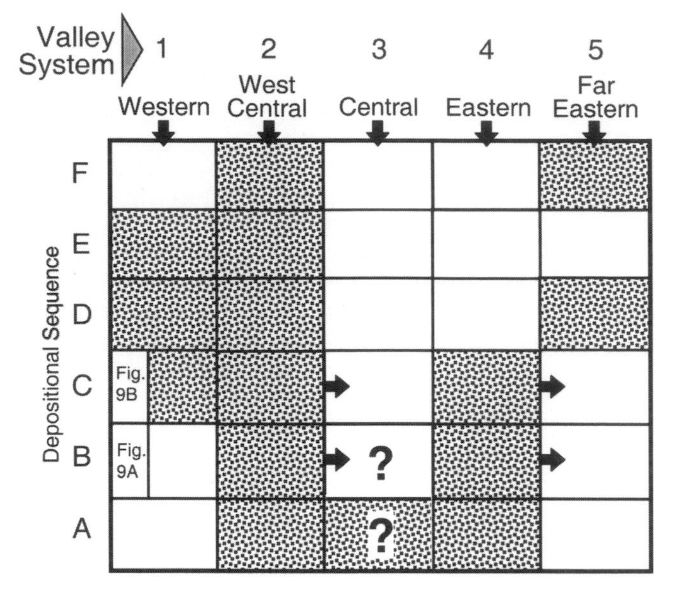

Figure 10—Pattern of sand-prone high amplitude facies (speckled pattern) within five slope valley systems (1–5) during six depositional sequences (*A–F*). The RMS amplitude extraction maps for sequences B and C are shown in Figure 9. Question marks indicate uncertain interpretations. Horizontal arrows suggest sediment input from adjacent valleys where toe-thrust piggyback basins connect valley systems across the slope.

Depositional System Evolution

The Niger delta complex consists of several major sandy river systems, primarily the Niger, Benin, and Cross rivers (Figure 1) (CD Figure 1). These sediment-supply systems shift laterally through time by channel avulsion. The Gulf of Guinea continental slope depositional systems also show a pattern of lateral shifting between slope valleys, probably in response to the lateral shifting of the river systems supplying the sediment and local gravity-driven tectonic modification of the slope valley–floor relief. The slope valley depositional system is interpreted from the occurrence of high-amplitude facies as a proxy for sandy deposition within the turbidite system (white to light gray) (Figure 9) (CD Figure 13).

Comparison of the displays of depositional sequence B (Figure 9-A) (CD Figure 13A) and sequence C (Figure 9B) (CD Figure 13B) illustrates the lateral shifting of interpreted sand-prone sediment input among five slope valley systems. Variations in occurrence of high-amplitude facies for six Pliocene sequences (A–F) are shown in Figure 10 (CD Figure 14). The observed pattern suggests successive sequences have laterally variable sediment supply systems. The absence of high-amplitude facies with the upper-slope segment of a valley system suggests the absence of coarse-grained sediment input from the shelf (see Figure 9B for valley 5) (CD Figure 13B). The appearance of high-amplitude facies within the middle-to lower-slope segments of a

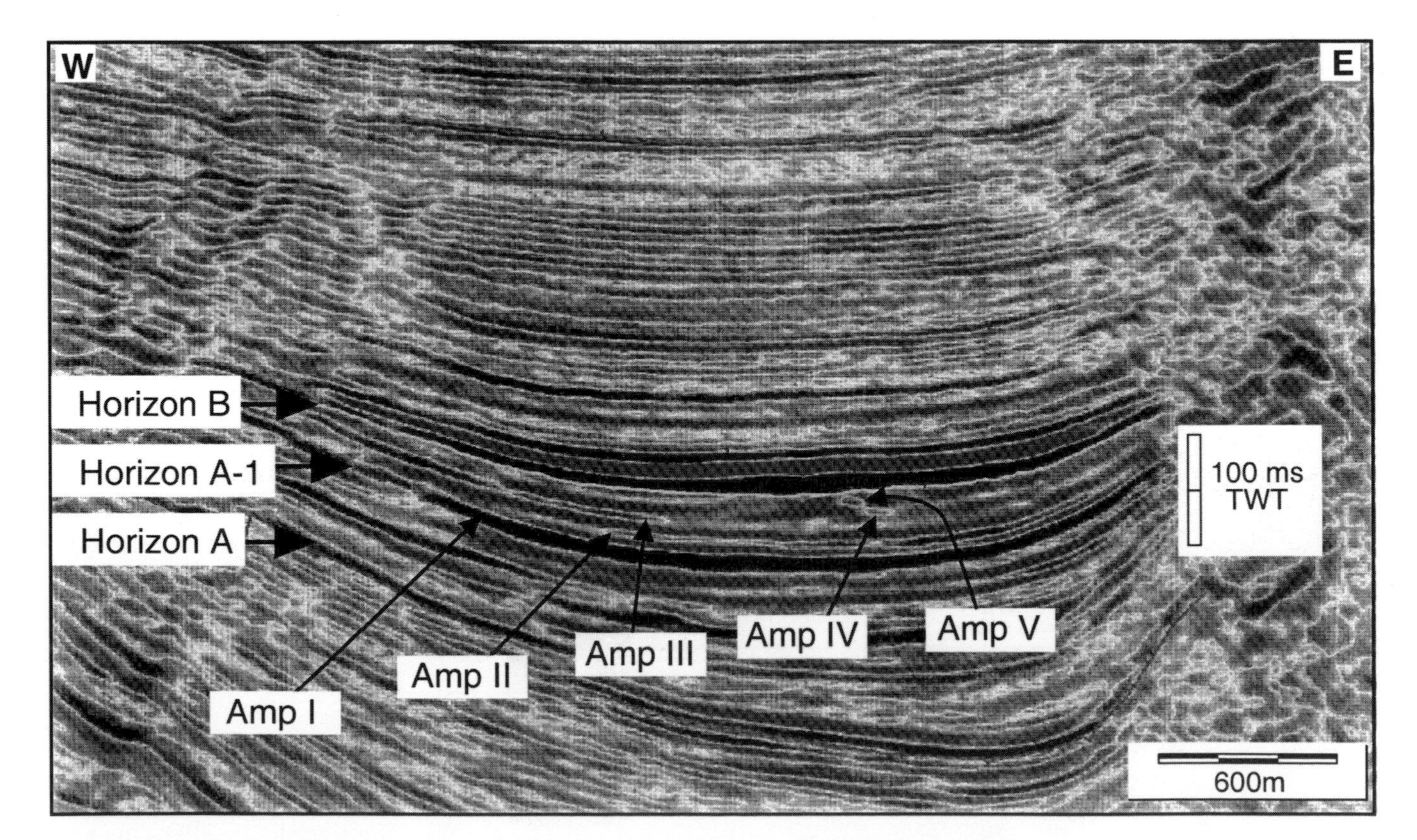

Figure 11—Seismic reflection profile perpendicular to the transport direction down the valley 2 (see Figure 9B for location). Five high-amplitude (black) reflections (I–V) were correlated between sequence boundaries Horizon A-1 and Horizon B, and mapped through the 3-D volume to illustrate the depositional stacking pattern of inferred sand-prone facies (see Figure 12).

valley system suggests either bypass of the upper slope or lateral input from another valley system, such as observed for the late Miocene turbidite system illustrated in Figure 8A (CD Figure 10A). The late Miocene cross-slope transport from one valley system to another is through a toe-thrust piggyback basin, clearly indicating the tectonic control on gravity-flow system transport and deposition.

Gravity-flow depositional events seek coeval seafloor topographic lows. This results in compensation sedimentation, with each successive event seeking lows between previous depositional mounds. Detailed mapping of one depositional sequence within one slope-valley basin illustrates this pattern of sediment accumulation. At the southern end of the West Central Valley (valley 2 of Figures 9B, 10) (CD Figures 13B, 14), the RMS amplitude extraction map of Figure 9B (CD Figure 13B) shows the composite amplitude for the Horizon A to Horizon B interval of Figure 11 (CD Figure 15). Detailed correlation of amplitude peaks on every 20th line identified specific depositional elements between Horizon A-1 and Horizon B (Figure 12) (CD Figure 16). The mapped pattern of amplitudes I through V suggest the following pattern:

Amplitude I: Deposition of a sheet element across the entire slope basin, suggesting relatively uniform accommodation and sediment accumulation.

Amplitude II: Deposition of a sheet element shifting eastward coincident with an isochron thickening (Figure 11), suggesting differential subsidence along the eastern margin. Channel elements at the north suggest basinward extension of the downslope transport system.

Amplitude III: Deposition of a sheet shifting toward the west and north suggesting some compensation sedimentation coincident with tectonically driven subsidence. The subsidence was probably related to initiation of east–west striking normal faults, due to anticlinal folding formed by deep-seated gravitationally driven slope failure.

Amplitudes IV and V: Development of progressively more channel elements in the basin center with sheet elements to the north and south. This is interpreted to reflect continued structural evolution of the east–west-trending anticline, resulting in differential subsidence to the north and south forming additional accommodation. These areas of continued subsidence are accentuated by differential uplift across the basin center, resulting in channelized incision and bypass of most sediment across the relatively rising area with little if any accommodation.

The extensional fault system within the slope basin has compartmentalized the reservoir interval into a series of rotated fault blocks, each containing various components of the channel-to-sheet depositional elements of the gravity-flow depositional system (CD Figure 17).

Observations of gullwinglike morphology along some channel elements suggests overbanking of turbulent flow events. No chaotic seismic facies, associated with mass wasting events, were observed, suggesting that the deposits between Horizon A-1 and Horizon B are dominantly a turbidite depositional system. The scale of the channel elements and the relative thickness of the lobe and sheet facies are uniform throughout the depositional sequence. These patterns suggest that the variations in facies

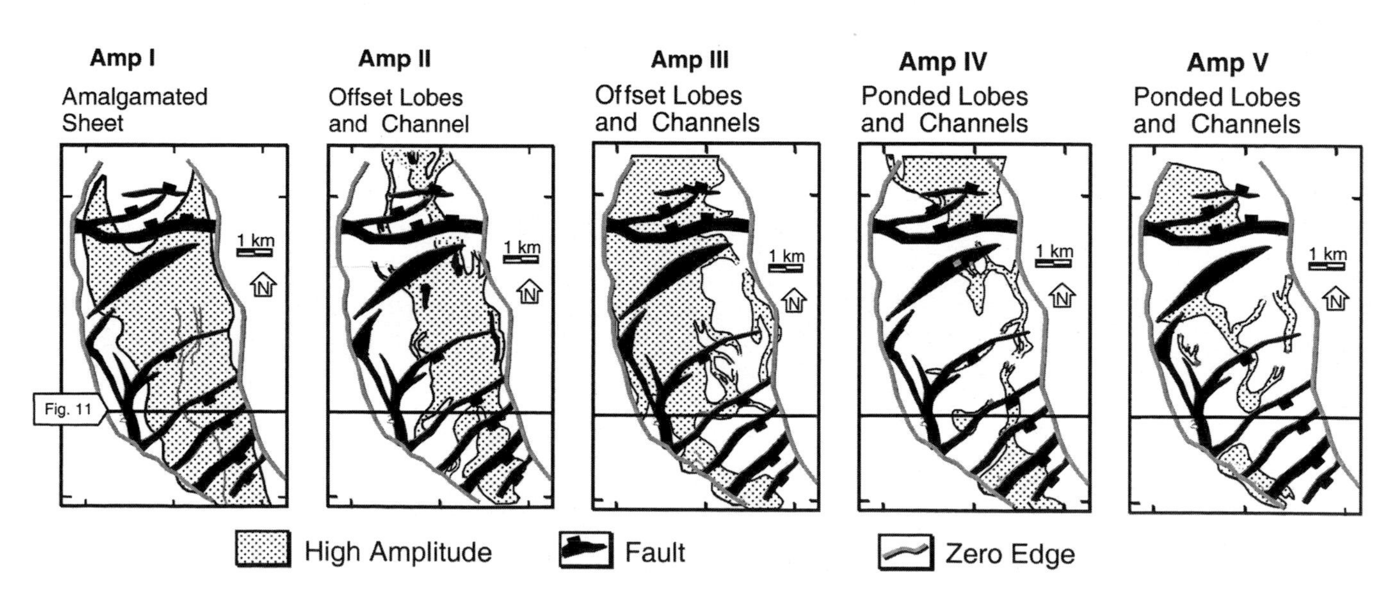

Figure 12—Interpreted maps illustrating the distribution of high-amplitude facies within the depositional sequence between Horizons A-1 and B of Figure 11 (see Figure 9B for map location). The lateral shifting of high-amplitude facies from Amp I to Amp III suggest compensation deposition. The upward progression of dominantly sheet to dominantly channel facies is interpreted to suggest overall decreasing accommodation.

distribution are related to the structural evolution of the basin and consequent accommodation, rather than avulsion of the sediment supply system or decrease in grain size of the transported sediment.

SUMMARY

Application of the sequence stratigraphic methodology facilitates selection of intervals for RMS amplitude extractions that provided map view of seismic facies patterns from deeply buried sequences that are consistent with patterns observed on many late Pleistocene submarine fan systems. Inspection of seismic reflection profiles across the mapped features results in observation of the expected cross-sectional geometry associated with channel, levee, lobe, and sheet elements. The observation of channel incision and overbank deposition suggests that the gravity-flow systems of the study area are dominantly turbidity-flow systems. The mapped length of the channel elements and the location of the depositional lobes and fans are controlled by the tectonically formed valley and intraslope basin geometry.

The systematic application of seismic sequence stratigraphic methodology and deepwater architectural element analysis allows for definition of gravity-flow depositional systems directly above the sequence boundary in each of the three examples presented in this chapter, as well as the other nine sequences mapped in the full study. This provides strong support for the predictive model of the seismic sequence stratigraphic model of Vail (1987) and Posamentier and Vail (1988).

Deepwater, gravity-flow sands are deposited in seafloor lows and are laterally discontinuous due to both the spatial occurrence of sediment supply systems and

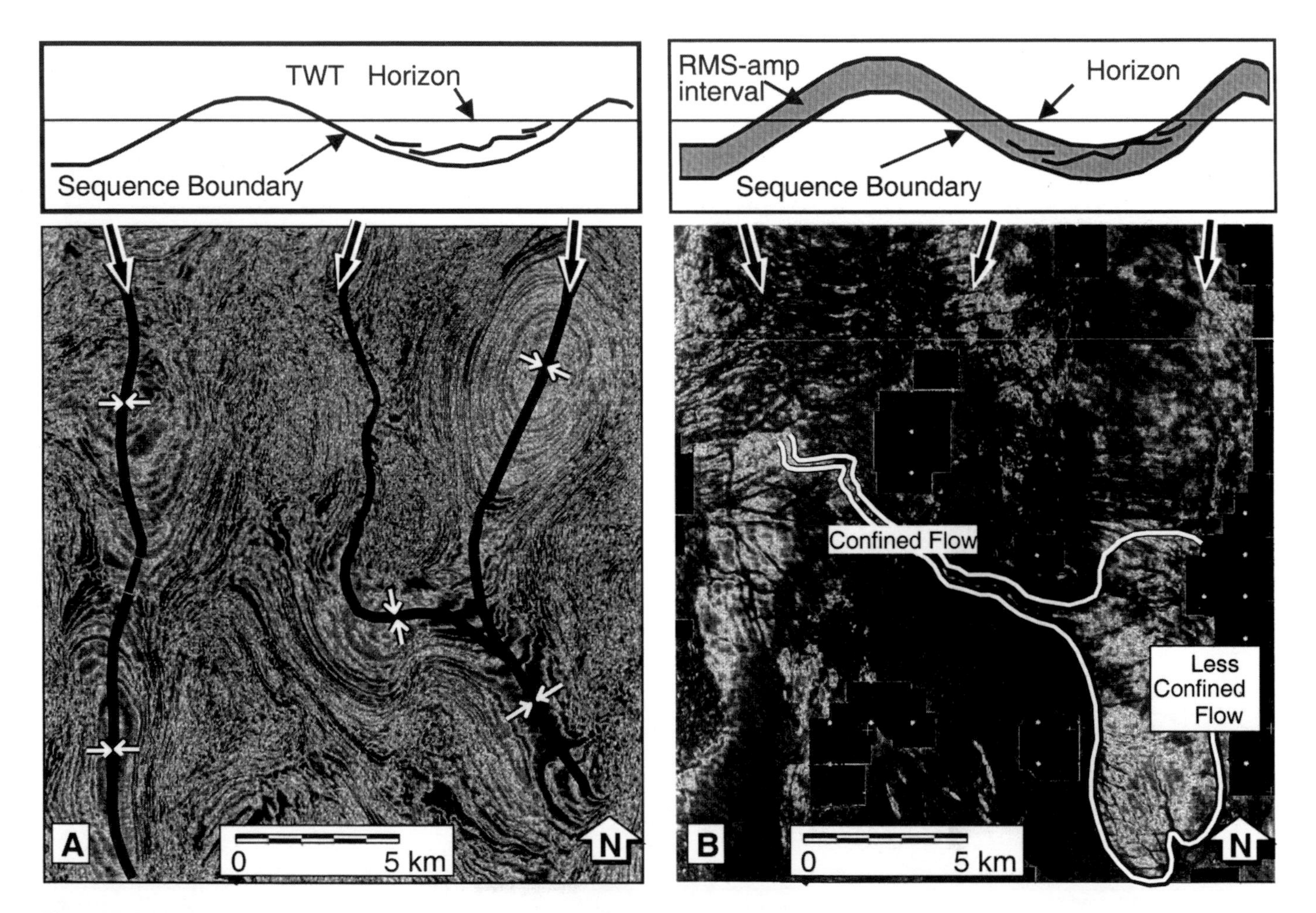

Figure 13—Some 3-D seismic images from the central part of the Gulf of Guinea continental margin slope study area (Figure 1), illustrating two primary types of displays. (A) Seismic horizon slice showing the structural pattern formed by seismic reflections intersecting the two-way-time (TWT) surface. (B) Seismic interval map showing RMS amplitude extracted upward from the depositional sequence boundary through a 50-ms interval. The image shows a confined flow channel element expanding into a less confined flow sheet element. Possible sediment input locations are shown by arrows pointing downslope into synclinal slope valleys. Both images are from approximately 2.9 s (TWT). Figure 13B is the same turbidite system as displayed in Figure 8A. This turbidite system was sourced through a channel complex, across two slope valley trends from west to east, following a course through a piggyback basin associated with a base-of-slope toe-thrust. By overlaying structural and stratigraphic maps for the same stratigraphic interval, potential traps of laterally discontinuous reservoir facies can be identified.

internal variability due to compensation sedimentation. The 3-D seismic data provide a means of mapping much of this lateral variability. Overlaying a present-day structural contour map (Figure 13A)(CD Figure 18A) and a gravity-flow depositional systems map (Figure 13B)(CD Figure 18B) of the same stratigraphic interval results in superpositioning of potential trap and reservoir. This prediction of optimal trap and reservoir juxtaposition, when evaluated within the context of the petroleum system, provides the petroleum explorationist with a clear definition of drilling targets.

ACKNOWLEDGMENTS

The authors gratefully acknowledge the permission of Mobil Producing Nigeria, BP-Amoco, and Mobil Technology Company to publish this paper. Numerous co-workers provided technical input throughout the study: Jeff Faber, project facilitator; George Gail, regional geologist; and Larry Fearn, biostratigrapher provided an extra measure of support. The manuscript has benefited from critical review by Roger Bloch, Arnold Bouma, John Jeffers, Brad Macurda, and Paul Weimer.

REFERENCES CITED

Armentrout, J. M., 1996, High-resolution sequence biostratigraphy: examples from the Gulf of Mexico Plio-Pleistocene, *in* J. A. Howell and J. F. Aitken, eds., High resolution sequence stratigraphy: innovations and applications: Geological Society Special Publication 104, p. 65–86.

Armentrout, J. M., L. B. Fearn, K. Rodgers, S. Root, W. D. Lyle, D. C. Herrick, R. B. Bloch, J. W. Snedden, and B. Nwankwo, 1999, High resolution sequence biostratigraphy of a lowstand prograding deltaic wedge: Oso field (late Miocene), Nigeria, *in* B. Jones and M. Simmons, eds., Biostratigraphy in production and development geology: Geological Society Special Publication, in press.

Bahorich, M., and S. Farmer, 1995, The coherence cube: The Leading Edge, p. 1053–1058.

Damuth, J. D., 1994, Depositional history of the Niger delta slope: AAPG Bulletin, v. 7, p. 320–346.

Delteil, J.-R., P. Valery, L. Montadert, C. Fondeur, P. Patriat, and J. Mascle, 1974, Continental margin in the northern part of the Gulf of Guinea, *in* C. A. Burk and C. L. Drake, eds., The geology of continental margins: New York, Springer-Verlag, p. 297–311.

Doust, H., and E. Omatosola, 1990, Niger Delta, *in* D. B. Edwards and W. Santogrossi, eds., Divergent/passive margin basins: AAPG Memoir 48, p. 201–239.

Haq, B. U., J. Hardenbol, and P. R. Vail, 1988, Mesozoic and Cenozoic chronostratigraphy and cycles of sea-level change, *in* C. K. Wilgus, H. W. Posamentier, C. A. Ross, and C. G. St. C. Kendall, eds., Sea level changes—an integrated approach: SEPM Special Publication 42, p. 71–108.

Loutit, T. S., J. Hardenbol, P. R. Vail, and G. R. Baum, 1988, Condensed sections: the key to age determination and correlation of continental margin sequences, *in* C. K. Wilgus, H. W. Posamentier, C. A. Ross, C. G. St. C. Kendall, eds., Sea level changes—an integrated approach: SEPM Special Publication 42, p. 183–213.

Mitchum, R. M., Jr., P. R. Vail, and J. B. Sangree, 1977a, Seismic stratigraphy and global changes of sea level. Part 6: Stratigraphic interpretation of seismic reflection patterns in depositional sequences, *in* C.E. Payton, ed., Seismic stratigraphy—applications to hydrocarbon exploration: AAPG Memoir 26, p. 117–133.

Mitchum, R. M., Jr., P. R. Vail, and S. Thompson III, 1977b, Seismic stratigraphy and global changes of sea level, part 2: the depositional sequence as a basic unit for stratigraphic analysis, *in* C. E. Payton, ed., Seismic stratigraphy—applications to hydrocarbon exploration: AAPG Memoir 26, p. 53–62.

Mitchum, R. M., Jr., 1985, Seismic stratigraphic expression of submarine fans, *in* O. R. Berg and D. G. Woolverton, eds., Seismic stratigraphy II: an integrated approach to hydrocarbon exploration: AAPG Memoir 39, p. 117–136.

Pickering, K. T., J. D. Clark, R. D. A. Smith, R. N. Hiscott, F. Ricci Lucchi, and N. H. Kenyon, 1995, Architectural element analysis of turbidite systems, and selected topical problems for sand-prone deep-water systems, *in* K.T. Pickering, R. N. Hiscott, N. H. Kenyon, F. Ricci Lucchi, and R. D. A. Smith, eds., 1995, Atlas of deep water environments: architectural style in turbidite systems: London, Chapman & Hall, p. 1–10.

Posamentier, H. W., and P. R. Vail, 1988, Eustatic controls on clastic deposition II—sequence and systems tract models, *in* C. K. Wilgus, H.W. Posamentier, C. A. Ross, and C. G. St. C. Kendall, eds., Sea level changes—an integrated approach: SEPM Special Publication 42, p. 125-154.

Prather, B. E., J. R. Booth, G. S. Steffens, and P. A. Craig, 1998, Classification, lithologic calibration, and stratigraphic succession of seismic facies of intraslope basins, deep-water Gulf of Mexico: AAPG Bulletin, v. 82, no. 5A, p. 701–728 (and figure corrections in AAPG Bulletin, v. 82, no. 12).

Risch, D. L., B. E. Donaldson, and C. K. Taylor, 1996, Deep-water facies analysis using 3D seismic sequence stratigraphy and workstation techniques: an example from Plio-Pleistocene strata, northern Gulf of Mexico, *in* P. Weimer and T.L. Davis, eds., Applications of 3-D seismic data to exploration and production: AAPG Studies in Geology 42 and Society of Exploration Geophysicists Geophysical Developments Series 5, p. 143–148.

Trainor, D. W., and D. F. Williams, 1990, Quantitative analysis and correlation of oxygen isotope records from planktonic and benthic foraminifera and well log records from OCS well G 1267 No. A-1 South Timbalier Block 198, northcentral Gulf of Mexico, *in* J.M. Armentrout and B.F. Perkins, eds., Sequence stratigraphy as an exploration tool: concepts and practices in the Gulf Coast: Gulf Coast Section SEPM Foundation, 11th Annual Research Conference Proceedings, p. 363–377.

Vail, P. R., 1987, Seismic stratigraphy interpretation procedure, *in* A. W. Bally, ed., Atlas of seismic stratigraphy, Volume 1: AAPG Studies in Geology 27, p. 1–10.

Van Wagoner, J. C., R. M. Mitchum, Jr., K. M. Campion, and V. D. Rahmanian, 1990, Siliciclastic sequence stratigraphy in well logs, cores and outcrops: concepts for high resolution correlation of time and facies: AAPG Studies in Geology 27, 55 p.

Weimer, P., et al., 1998, Sequence stratigraphy of Pliocene and Pleistocene turbidite systems, northern Green Canyon and Ewing Bank (offshore Louisiana), northern Gulf of Mexico: AAPG Bulletin, v. 82, no. 5B, p. 918–960.

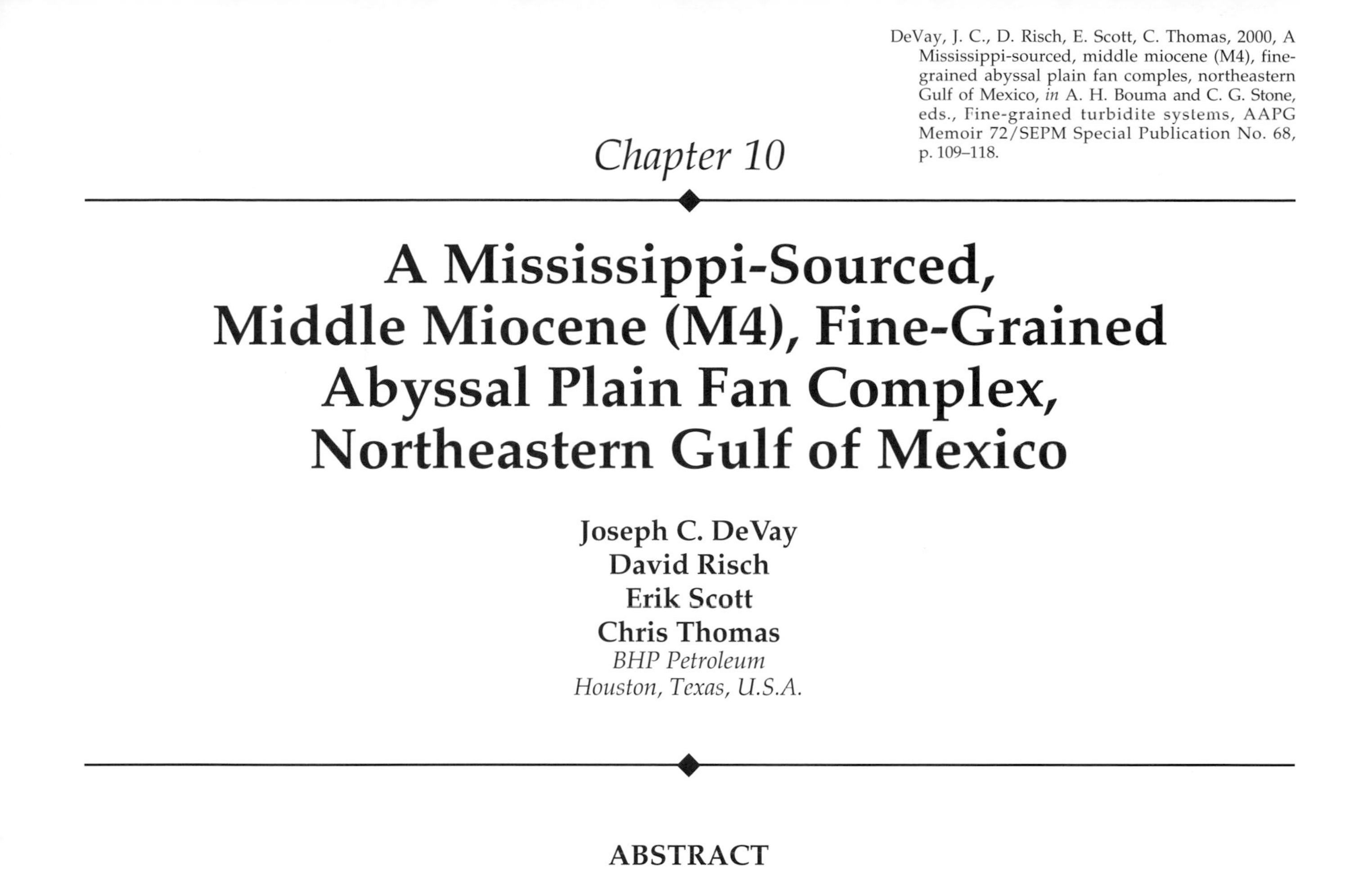

DeVay, J. C., D. Risch, E. Scott, C. Thomas, 2000, A Mississippi-sourced, middle miocene (M4), fine-grained abyssal plain fan comples, northeastern Gulf of Mexico, *in* A. H. Bouma and C. G. Stone, eds., Fine-grained turbidite systems, AAPG Memoir 72/SEPM Special Publication No. 68, p. 109–118.

Chapter 10

A Mississippi-Sourced, Middle Miocene (M4), Fine-Grained Abyssal Plain Fan Complex, Northeastern Gulf of Mexico

Joseph C. DeVay
David Risch
Erik Scott
Chris Thomas
BHP Petroleum
Houston, Texas, U.S.A.

ABSTRACT

Seismic facies mapping and seismic stratigraphic analysis in the northeastern Gulf of Mexico indicate the presence of a middle Miocene fan complex, here termed the M4. This fan complex is interpreted to comprise at least two fourth-order sequences deposited during lowstand by the Mississippi River. Each depositional sequence is well organized in sets of laterally continuous and discontinuous/hummocky seismic reflectors that are interpreted to be sheet-sand lobe, nonleveed channel and distributary channel facies. Sea-floor compensation is shown by east–west migration of the M4 sequences. The average depositional rate across the fan complex was about 0.5 m/1000 yr.

INTRODUCTION

With the recent interest in ultra-deepwater exploration, BHP Petroleum has undertaken a regional seismic stratigraphic interpretation effort, aimed at understanding the distribution and depositional history of abyssal plain fan systems outboard of the present day Sigsbee Escarpment, Northern Gulf of Mexico (Figure 1). Key to understanding sedimentation on the Gulf of Mexico abyssal plain is knowledge of how sediment-laden gravity flows were able to traverse the sediment traps that the complex, updip intraslope salt-withdrawal basins present.

Recently published work in the deepwater Gulf of Mexico over the last decade has concentrated on intraslope salt-withdrawal basin sedimentation and stratigraphy (Bouma and Coleman, 1986; Boyd et al., 1993; Apps et al.,1994; Yeilding and Apps, 1994; Prather et al., 1998) and reservoir characterization of fields such as Mars, Green Canyon 65, and Bullwinkle (Holman and Robertson, 1994; Mahaffie, 1994; McGee et al., 1994), because of the discovery of significant reserves of hydrocarbons in these basins. A significant body of mainly descriptive work has been accumulated detailing seismic facies architecture (Yeilding and Apps, 1994; Prather et al., 1998). It is less clear in the literature how sediment traverses the complex

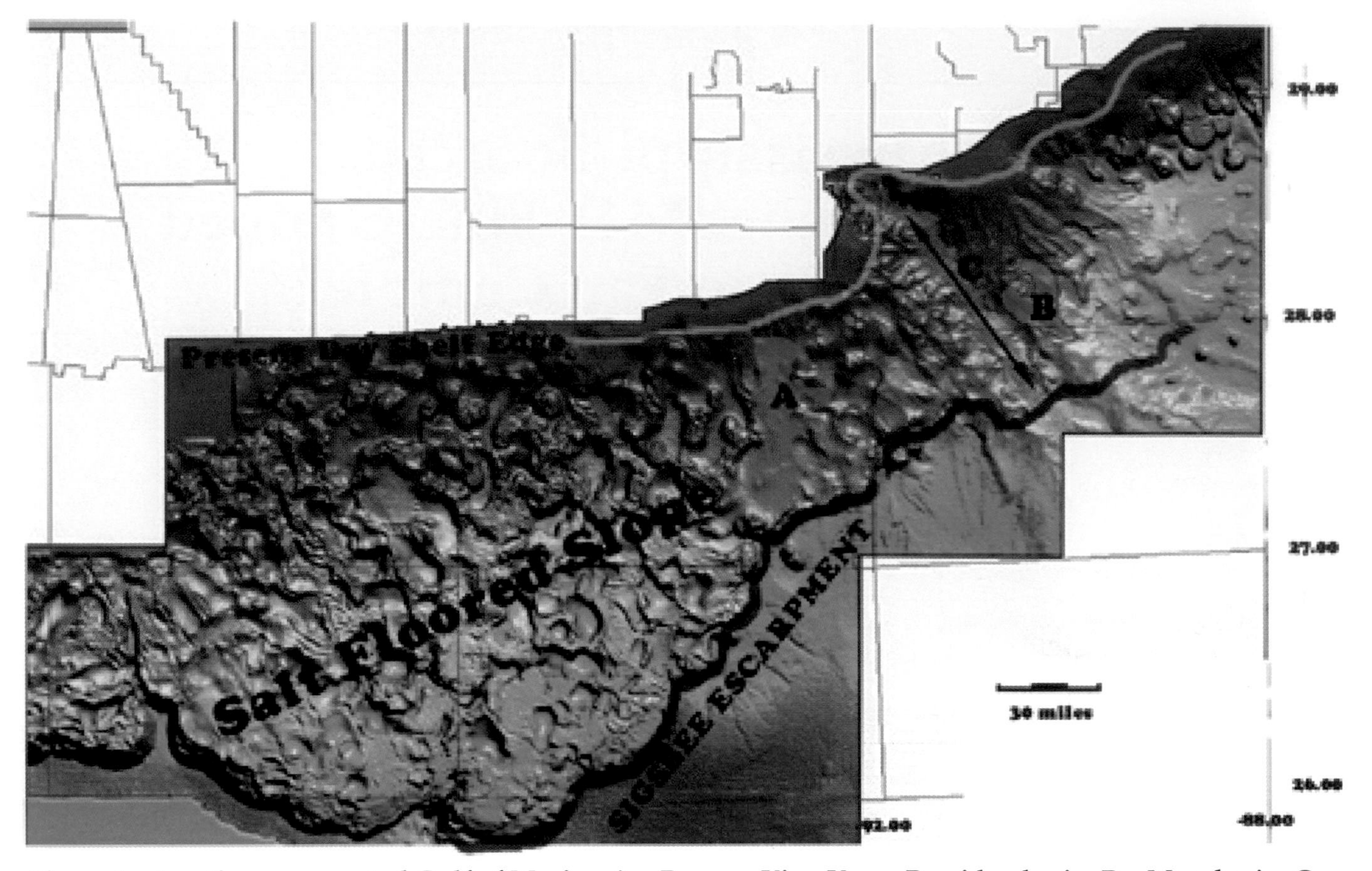

Figure 1—Location map, central Gulf of Mexico. A = Popeye–King Kong–Poseidon basin; B = Mars basin; C = Mississippi fan valley.

intraslope salt-withdrawal basins and has been able to construct abyssal plain fans at various times from the early Miocene to Pleistocene.

GEOLOGICAL OVERVIEW

Prather et al. (1998) provide a detailed summary of intraslope basin stratigraphy and sedimentation that relates seismic facies architecture to basin filling and eventual bypass. Their proposed process of fill and spill works best in relatively low-sediment input scenarios in which the basin fill rate is equal to or less than the rate of salt displacement. In such a process the basin only fills to spill after welding has occurred (Worrall and Snelson, 1989; Prather et al., 1998), with a resultant depositional polarity reversal as the salt-withdrawal basin rotates basinward (Worral and Snelson, 1989). This model (in which the sedimentation rate was high enough to weld-out several basins, but not high enough to reach the abyssal plain) is evidenced by thick Miocene-Pliocene section at the King Kong, Poseidon, and Mars basins (Figure 1), and an unconformity outboard of the Sigsbee Escarpment.

However, during periods of lowstand, when sediment depositional rates exceeded salt displacement rates, intraslope salt-withdrawal basins were filled before salt welding could occur. In this case rather than fill and spill in a systematic way, rapidly accumulated sediment filled, and then bypassed, intraslope salt-withdrawal basins and became deposited directly on the abyssal plain (Figure 1) (Pliocene-Pleistocene Mississippi fan complex).

The sediment-trapping efficiency of intraslope salt-withdrawal basins is directly related to the rate of sediment influx, salt deformation, and supply in response to sediment loading and accommodation space, created by extension within the intraslope salt-withdrawal basin system as a whole. Equal rates of salt deformation (or extension) and sedimentation lead to the development of steeply inclined to vertical sediment/salt interfaces (McGuinness and Hossack, 1993).

The "upbuilding" balance of sediment accumulation and salt deformation rates effectively creates accommodation space in which sediment is captured. Additionally downbuilding or loading out of the salt by accumulating sediment creates accommodation space controlled by the displacement rate of the salt as it is loaded. As a result during periods when sedimentation rates were less than or equal to the rates of salt deformation and slope extension, sufficient accommodation space was created in the intraslope salt-withdrawal basins to trap most of the sediment entering the slope environment.

Conversely, transport to the abyssal plain of sediment by gravity flow requires sedimentation rates sufficient to overwhelm the intraslope salt-withdrawal basin system (the sedimentation rate would have to exceed the rates of salt deformation and slope extension). Buried salt sheets and abyssal plain fan sequences provide direct evidence that the sedimentation rates exceeded intraslope salt deformation and extension rates.

During the Miocene, abyssal plain fan deposits, such as the M4 Fan complex, record the result of sedimentation rates and volumes that exceeded the trapping capacity of the intraslope salt-withdrawal basins. The Pliocene-Pleistocene Mississippi Fan system may provide potential recent analogs for this process (Figure 1).

M4 FAN COMPLEX

The middle Miocene M4 fan complex (Figure 2) was deposited in two major (fourth-order) sedimentation episodes at the onset of the TB3 second-order cycle within the 10.5-Ma, 3.1-third-order sequence (Haq et al., 1988), here named the M4a and the M4b (Figures 3, 4). Within the area of seismic resolution, the M4 lowstand deposited a gross fan system approximately 300 km wide and at least 300 km long on the abyssal plain (the limit of the seismic grid, Figures 3, 4). The abyssal M4 fan system, with an average present gross thickness of approximately 400 m and an area of approximately 90,000 km^2, represents about 36,000 km^3 of compacted fan complex deposits that were transported across the slope to the abyssal plain. If one assumes a 2:1 compaction ratio, the volume of sediment would be about 72,000 km^3 deposited at an average rate of 0.5 m/1000 yr across the fan.

The main sediment pathway across the slope to the M4a subfan was approximately 500 km long and traversed an extended intraslope environment en route to the abyssal plain. The local slope system, normally a bypass facies dominated by slope mud (Nelson and Nilsen, 1984; Normark and Piper, 1991), was comprised of an intraslope salt-withdrawal basin and canyon complex related to the presence of extensive sheets of thick salt.

Consideration of the distance and numerous intraslope basins each flow event had to traverse suggests deposition on the M4 abyssal fan required prolonged sediment gravity flow duration. Gravity flows that have been quantified in terms of duration and speed (Dengler et al., 1984; Piper et al., 1988) suggest velocities in the range of 50 to 300 cm/s. The observed range of velocities and a travel distance of 500 km imply single-flow event duration from around 47 hours at 300 cm/s to nearly 12 days at 50 cm/s.

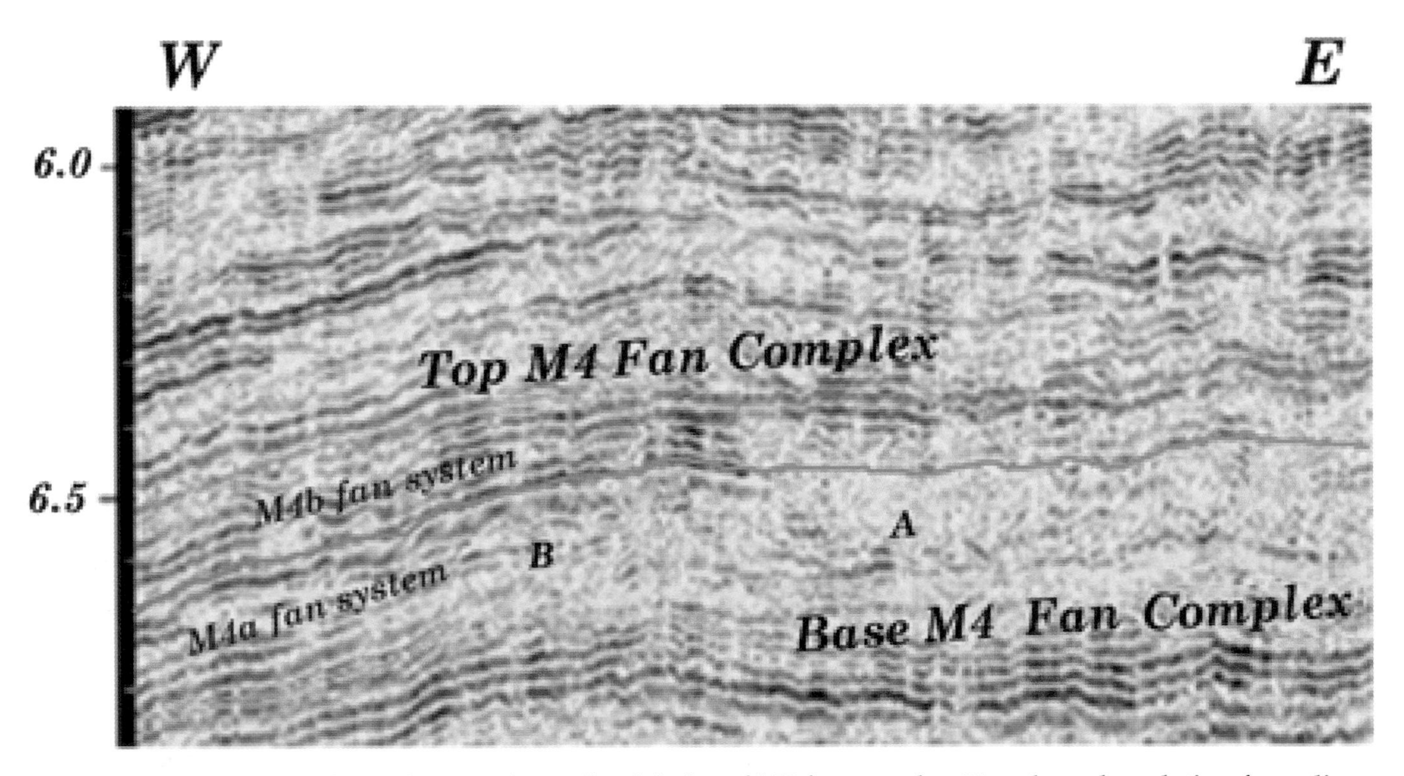

Figure 2—Strike section through main channelized facies of M4 fan complex. Note lateral gradation from discontinuous chaotic reflectors to laterally continuous high amplitude reflector packages. Discontinuous reflectors at *A* are interpreted to be nonleveed aggraded channel facies that are interdigitated with laterally continuous reflectors interpreted to be sheet-sand lobes. Seismic data courtesy of Geco-Prakla.

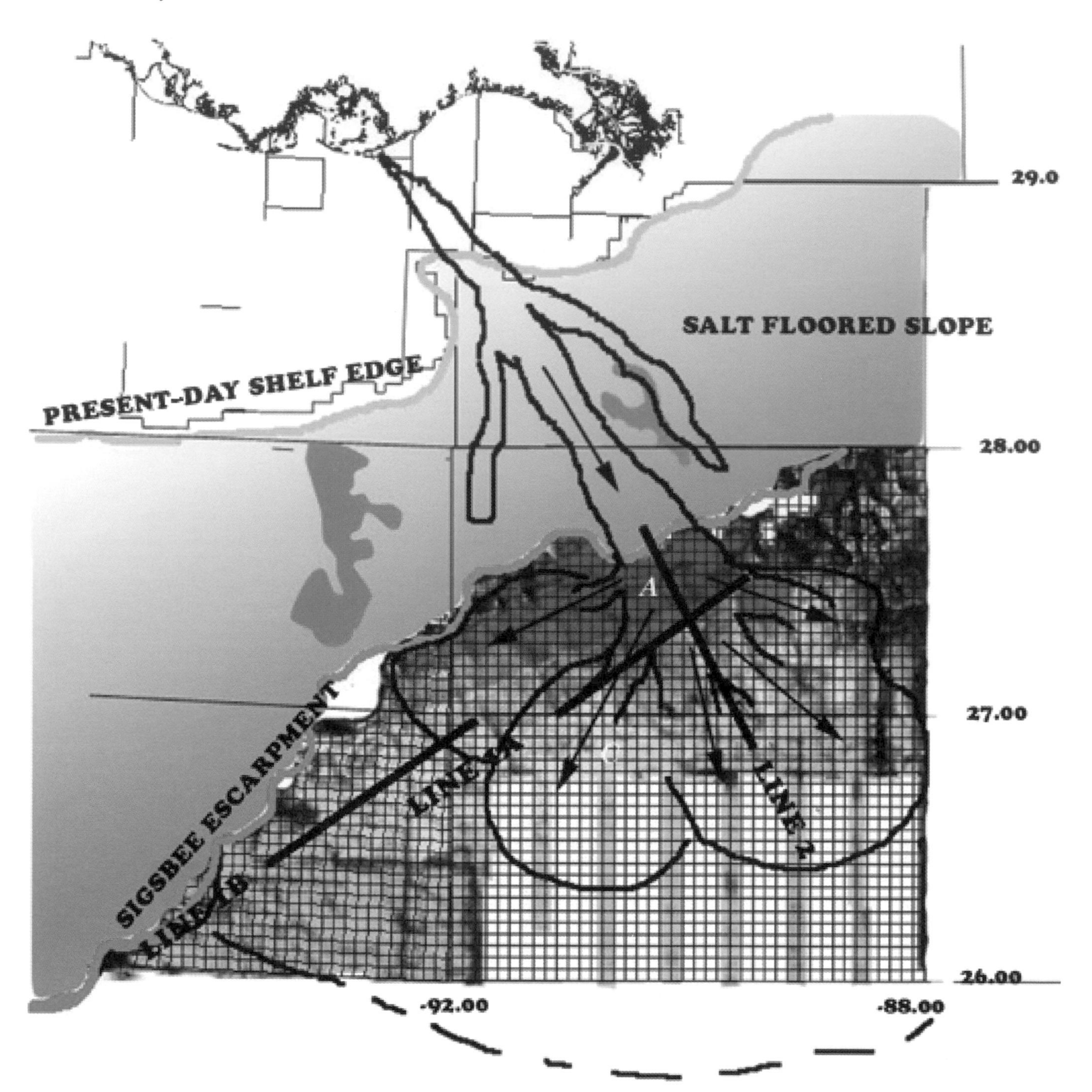

Figure 3—M4a fan system facies map and isopach. Dark blue areas are very thin to condensed, red areas are thick. Distribution of facies is based on seismic facies mapping of discontinuous chaotic, continuous parallel and high-amplitude parallel-divergent reflector packages. Fan subdivisions based on convergent reflector terminations that define fourth-order cycles such as fan lobes. Location A: areal distribution of discontinuous chaotic reflectors interpreted as aggradational, nonleveed channel facies. Location C: areal distribution of parallel, high-amplitude reflectors interpreted as sheet-sand lobe facies. Arrows indicate inferred direction of sediment transport.

The M4a sequence records the initial M4 fan complex deposition followed by the M4b sequence. Each fourth-order cycle has a regular seismic facies architecture that is interpreted to include nonleveed channel, sheeted distributary nonleveed channel, and sheeted lobe depositional facies (Figures 2, 5, 6).

Mapping of these seismic facies shows that each fourth-order cycle contains a discrete aggradational nonleveed channelized facies at the base of slope through which sediment entered the abyssal plain (Figures 3, 4, 6). Laterally the discontinuous aggradational, nonleveed channelized facies

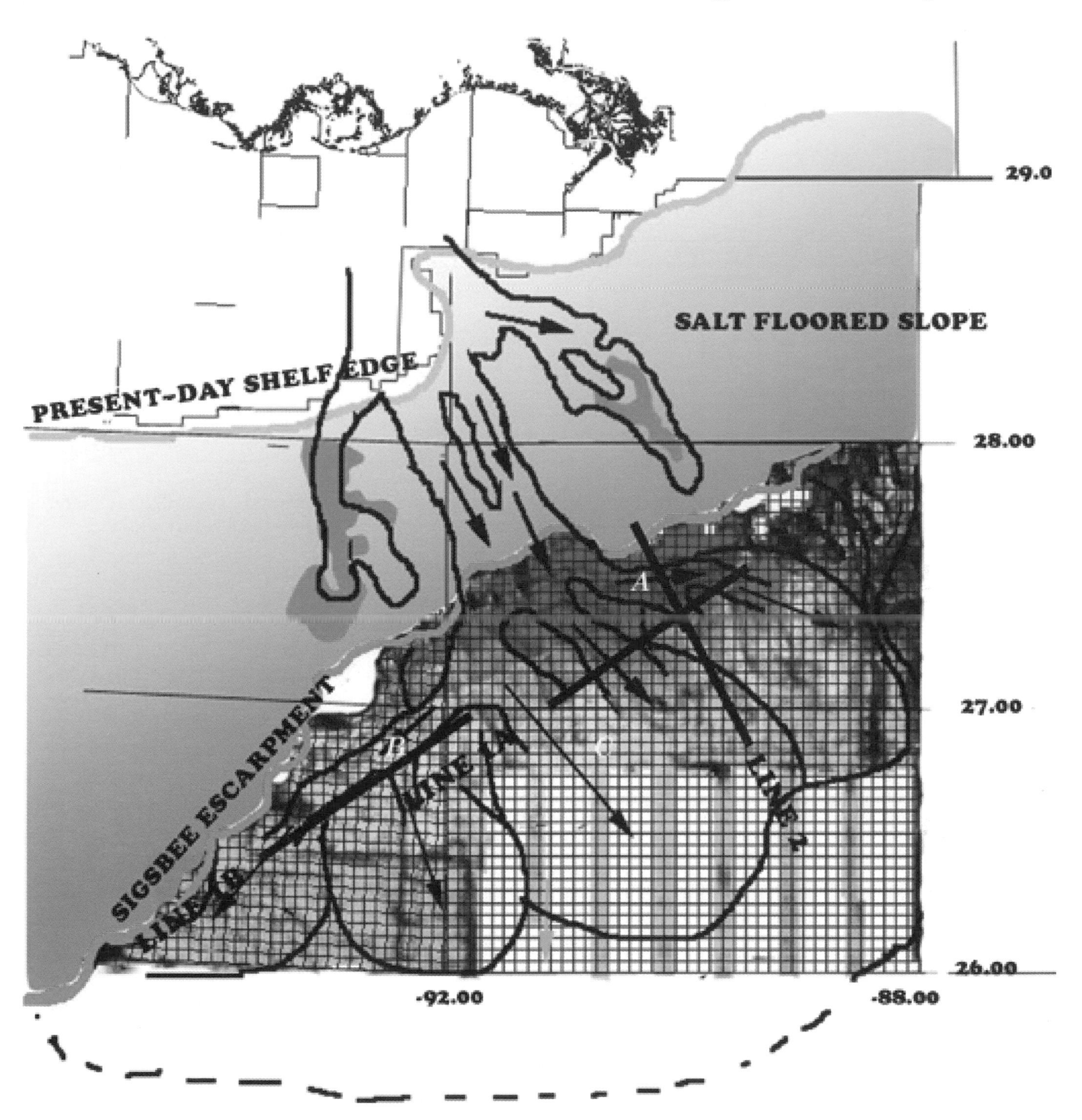

Figure 4—M4b fan system. Dark blue areas are very thin to condensed, red areas are thick. Distribution of facies is based on seismic facies mapping of discontinuous chaotic, continuous parallel and high amplitude parallel-divergent reflector packages. Fan subdivisions based on convergent reflector terminations that define fourth-order sequences such as fan lobes. Location A: areal distribution of discontinuous chaotic reflectors interpreted as aggradational, nonleveed channel facies. Location B: areal distribution of discontinuous chaotic reflectors interpreted as nonleveed distributary channels. Location C: areal distribution of parallel, high-amplitude reflectors interpreted as sheet-sand lobe facies. Arrows indicate inferred direction of sediment transport.

pass into sheeted, nonleveed, distributary channels and interdigitated sheet lobes (Figure 2), which lose reflectivity and continuity with increasing distance from the sediment entry point (Figure 5, M4a section).

The sediments were deposited on the abyssal plain in a fan system that was roughly equivalent to the 18 ka Pleistocene Mississippi fan system in scale and facies.

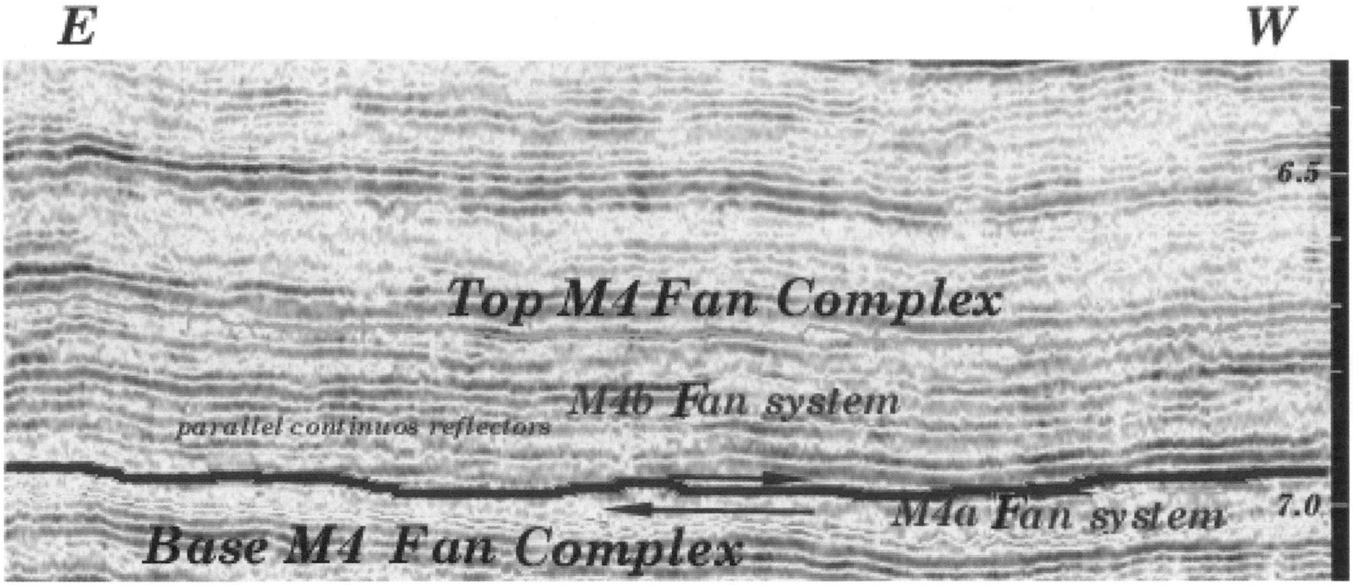

Figure 5—Strike section showing laterally continuous sheet-sand lobe facies of M4b fan system. Note lapout of M4a fan system to west and loss of reflector strength and continuity. Nonleveed distributary channel oblique section is noted at location. Seismic data courtesy of Geco-Prakla.

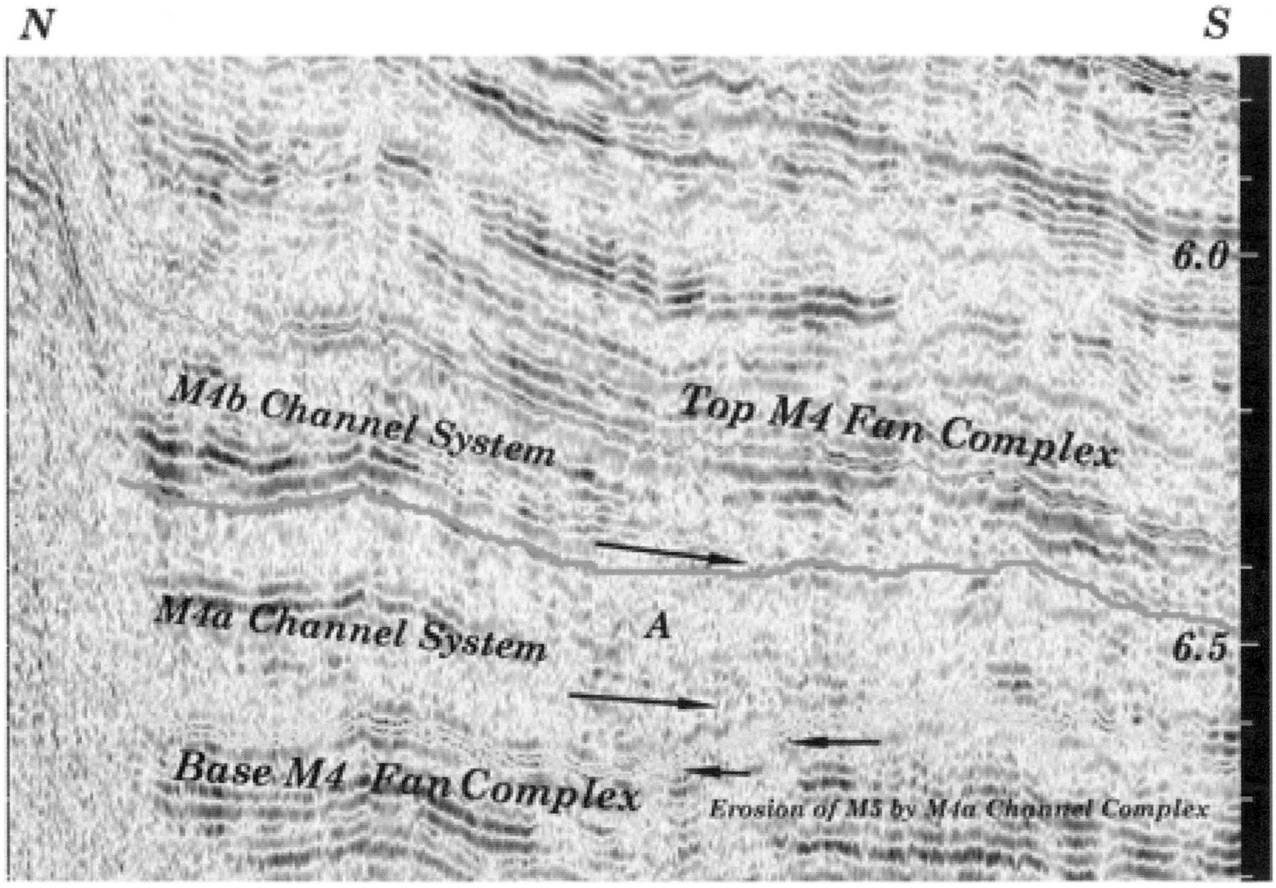

Figure 6—Seismic Line 2 (Figures 3, 4) dip section through M4 Fan Complex. Aggradational nonleveed channel systems of M4a and M4b mapped in Figures 3 and 4. Note erosion at base of slope where M4a channel system "dumps out" onto abyssal plain and strong progradation of system indicated by downlapping reflectors suggesting high depositional rates. Two principal depositional cycles of M4a and M4b are clearly defined by the internal sequence boundary defined by reflector convergence at *A*. Note also lateral interdigitation downdip of chaotic discontinuous and continuous parallel reflector packages. Seismic data courtesy of Geco-Prakla.

M4 FAN FACIES

M4a Channelized Facies

Lateral variation in both dip and strike directions is shown in the seismic sections that transect this area (Figures 2, 5, 6). Seismic reflection data in the direction of depositional dip are shown in Figure 6. At the base of slope a, discontinuous, chaotic, and hummocky facies predominate (Figures 3, 4, 6). These facies connect stratigraphically updip to M4 intraslope salt-withdrawal basins and canyons beneath the present-day canopy of salt.

The discontinuous, hummocky facies are interpreted to be aggradational, nonleveed channel and slump facies. Individual channels appear to be roughly 3–5 km wide and stacked vertically, or nested, within a 60-km-wide area of channel sedimentation (Figures 3, 4, 6). Laterally and downdip the channelized facies interdigitated with high-amplitude, continuous reflectors (Figures 2, 6).

Channelized facies within the M4a fan are concentrated at the base of slope entry point of sediment from the salt-withdrawal basins/extended slope to the abyssal plain (Figures 3, 6). These facies are characterized by strong aggradation and progradation (indicated by steeply dipping downlapping reflectors: Figure 6) of nonleveed distributary channels that pass laterally into sheet lobe facies (Figures 2, 3, 4, 6).

M4a Sheeted Distributary Channel & Lobe Facies

High-amplitude, wide areal extent, continuous reflectors dominate the middle part of the M4a fan system (Figures 2, 3, 5). Internal offlap of individual reflectors and southward-directed downlap indicate progradation of the fan system away from the channelized facies of the central upper fan (Figures 2, 5). Moderate truncation of underlying reflectors is common in the high-amplitude continuous facies. Broad, relatively thin, hummocky, discontinuous facies are also present in areally limited areas of the high-amplitude continuous facies and map out as discrete, "minor," channel systems (Figures 2, 4).

In several intraslope salt-withdrawal basins, the high-amplitude continuous facies is penetrated by wells. The penetrations show that this facies corresponds to thickly bedded sand. Similar facies tested in the Pliocene section at Auger field (Prather et al., 1998), and middle Miocene at Mars (Mahaffie, 1994), show this facies to be sand-rich. The average net-to-gross ratio of similar high-amplitude continuous reflector facies is around 35%, based on seismic facies analyses calibrated to well data in the Green Canyon and Mississippi Canyon protraction areas.

The laterally extensive, high-amplitude continuous-reflector facies is interpreted to be sheeted sand-rich lobe facies. Minor feeder channels are present within this facies as well. Offlap of reflector sets and infill of interlobe areas (Figure 2) show interlobe compensation. The vertical scale of each reflector set is approximately 50 to 70 ms (~70 to 100 m) thick in the middle fan area.

The middle fan high-amplitude continuous-reflector packages pass laterally into progressively dimmer, thinner continuous-reflector packages until the limit of seismic resolution is reached for individual reflector sets and the M4a fan becomes transparent (Figure 5). The dimming out of the reflector packages is interpreted to be the result of loss of acoustic contrast as a result of decreasing sand content. The thickness of the fan, even though largely transparent, is still as thick as outcrop examples of individual fan units such as Tanqua Karoo in South Africa (Bouma and Wickens, 1994) and Annot in southeastern France.

M4b Fan System

The M4b system contains the same facies set as the M4a system but is areally larger. The entry point of sediment to the M4b fan is west of the central channelized facies in the M4a fan. Unlike the M4a fan, the M4b fan has a well-developed distributary channel system (Figure 4) that extends across the middle fan area. Sea-floor compensation is demonstrated by onlap of the M4b onto the M4a fan system (Figures 2, 5) and the westward migration of the main depocenter of the M4b fan system (Figure 4).

The upper central part of the M4b fan is comprised mainly of discontinuous, aggradational channel facies (Figures 2, 4, 6) that pass laterally and stratigraphically downdip into high-amplitude, continuous reflectors (Figures 2, 5, 6). Late-stage erosional channel systems are also common in the M4b, particularly toward the top of the complex in the upper and middle fan area (Figures 4, 5). These late-stage channel systems pass laterally into amplitude reflectors downdip.

The lower part of the M4b fan complex thins with increasing distance from the sediment entry point and becomes acoustically transparent when the individual reflectors go below the limit of seismic resolution. As with the M4a continuous-reflector facies, amplitude decreases with increasing distance from the base of slope sediment entry point.

M4 Fan Complex Depositional Model

The M4 fan complex is interpreted to be a mud-rich fan system similar in character to the Tanqua Karoo fan complex described by Bouma and Wickens (1994), with the exception of the extended slope and intraslope salt basins. It is a three component, aggradational system comprised of (1) upper fan intraslope salt-withdrawal basins and canyons, (2) middle fan nonleveed channel and slump facies, and (3) lower fan sheeted nonleveed distributary channel and lobe facies (Figure 7).

The geometries of the three main acoustical depositional facies create a very good areal view of the distribution of the principal parts of the lower M4 fan (Figures 3, 4). The lateral extents of sheet-sand lobes based on the acoustic facies areal distribution results in extended slope model sheet-sand lobes that are larger than in the Tanqua Karoo model. The process by which extensive laterally monotonous sheets of sand-rich sediment are

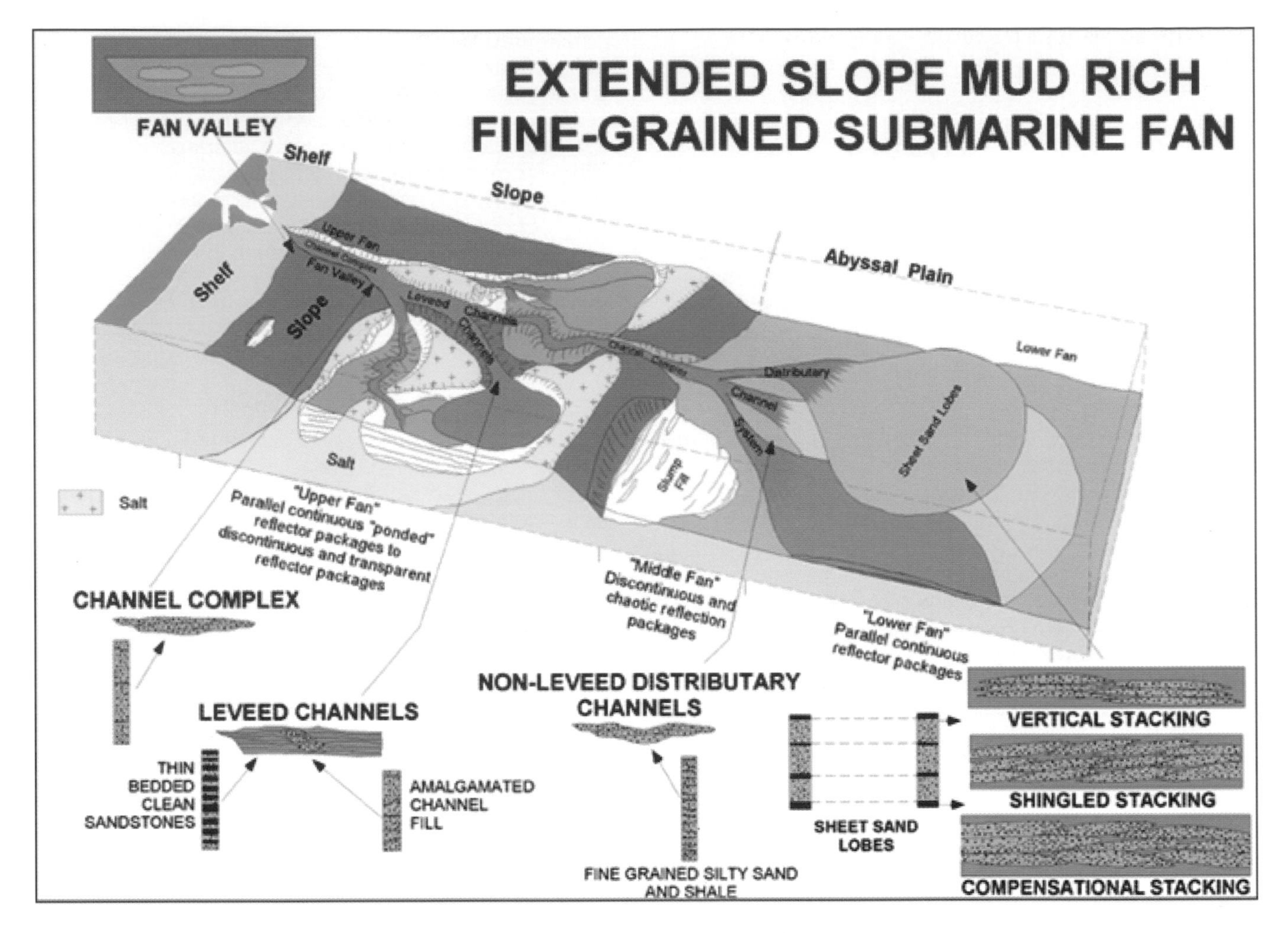

Figure 7—Extended slope mud-rich, submarine fan model (modified from Bouma, 1997).

laid down would require long-duration, continuous sediment-gravity flow support. As already noted long-duration flow is needed for sediment-gravity flows to traverse the slope. Also needed are high sedimentation rates with which to overwhelm the intraslope salt-withdrawal basin complex.

The large volume of sediment transported to the abyssal plain, approximately 4 km^3/100 yr, suggests that the Mississippi River was probably directly feeding sediment into the M4 slope canyon system. Also, failure of the M4 slope, slumping of canyon walls, and resedimentation of sand and mud from intraslope basins probably contributed significantly to the volume of M4 abyssal fan sediment.

CONCLUSIONS

The large influx of M4 sediment resulted in the deposition of an extensive, nonleveed distributary channel and sheet-sand lobe lower-fan complex at the base-of-slope outboard of the Sigsbee Escarpment on the abyssal plain. The sheet-sand lobe facies is sand-rich where penetrated in salt-withdrawal basins and is characterized by continuous, monotonously isopachous sheets several thousand square kilometers in extent. With the exception of topography presented by growing folds along the Atwater foldbelt, the only sea-floor topography that affected sheet-sand lobe deposition was previously deposited M4 lobes.

The M4 abyssal fan is a complete slope/abyssal fan complex in an atypical environment in which the extended salt-floored slope was a distributed depocenter. Rather than a simple bypass facies, the extended slope is an environment of intraslope salt-withdrawal basins that trap the sediment output of the Mississippi River during sea level highstands or waning periods of fan deposition. Only during periods of waxing sedimentation or lowstands does the sediment output of the Mississippi River overwhelm the intraslope salt basins and deposit sediment on the abyssal plain.

During periods of lowstand the sediment accumulation rate may exceed the rate at which the slope salt sheets can deform. As a result the intraslope salt-withdrawal basins are unable to accommodate the rapidly accumulating sediment. The intraslope salt-withdrawal basins fill to spill then become bypassed as sediment gravity flow events carry sediment out to the abyssal plain.

At flow rates comparable to modern examples, the

large distances that had to be traveled by discrete sediment gravity flows from shelf to the abyssal plain indicate prolonged flow periods. The very broad areal extent and uniformity of sheet-sand lobe facies reflector packages suggests large volumes of sediment were transported in sheeted, nonconstrained flows that resulted in amalgamated packages of aggraded sheet-sand lobes.

REFERENCES CITED

Apps, G. M., F. J. Peel, C. J. Travis, and C. A. Yeilding, 1994, Structural controls on Tertiary deep water deposition in the northern Gulf of Mexico, *in* P. Weimer, A. H. Bouma, and B. F. Perkins, eds., Submarine fans and turbidite systems—sequence stratigraphy, reservoir architecture, and production characteristics, Gulf of Mexico and international: Gulf Coast Section SEPM Foundation 15th Annual Research Conference, p. 1–7.

Bloom, A. L., 1983, Sea level and coastal morphology of the United States through the late Wisconsin glacial maximum, *in* H. E. Wright Jr., ed., Late Quaternary environments of the United States: Minneapolis, University of Minnesota, v. 1, p. 215–229.

Bouma, A. H., and J. M. Coleman, 1986, Intraslope basin deposits and potential relation to the continental shelf, northern Gulf of Mexico: Gulf Coast Association of Geological Societies Transactions, v. 36, p. 419–427.

Bouma, A. H., and H. DeV. Wickens, 1994, Tanqua Karoo, ancient analog for fine grained submarine fans, *in* P. Weimer, A. H. Bouma, and B. F. Perkins, eds., Submarine fans and turbidite systems—sequence stratigraphy, reservoir architecture, and production characteristics, Gulf of Mexico and international: Gulf Coast Section SEPM Foundation 15th Annual Research Conference, p. 23–34.

Bouma, A. H., 1997, Comparison of fine grained, mud-rich and coarse grained, sand-rich submarine fans for exploration-development purposes: Gulf Coast Association of Geological Societies Transactions, v. 47, p. 59–64.

Boyd, J. D., M. J. Mayall, and C. A Yeilding, 1993, Depositional models for intraslope basins, variability and controls (abs.): AAPG Annual Meeting Program with Abstracts, p. 79–80.

Dengler, A. T., P. Wilde, E. K. Noda, and W. R. Normark, 1984, Turbidity currents generated by Hurricane Iwa: Geo-Marine Letters, v. 4, p. 5–11.

Haq, B. U., J. Hardenbol, and P. R. Vail, 1988, Mesozoic and Cenozoic chronostratigraphy and cycles of sea level change, *in* Wigus, C. K., et al., eds., Sea-level change: an integrated approach: SEPM Special Publication 42, p. 71–108.

Holman, W. E., and S. S. Robertson, 1994, Field development, depositional model, and production performance of the turbiditic "J" sands at Prospect Bullwinkle, Green Canyon 65 fields, outer shelf Gulf of Mexico, *in* P. Weimer, A. H. Bouma, and B. F. Perkins, eds., Submarine fans and turbidite systems—sequence stratigraphy, reservoir architecture, and production characteristics, Gulf of Mexico and international: Gulf Coast Section SEPM Foundation 15th Annual Research Conference, p. 139–150.

Mahaffie, M. J., 1994, Reservoir classification for turbidite intervals at the Mars discovery, Mississippi Canyon 807, Gulf of Mexico, *in* P. Weimer, A. H. Bouma, and B. F. Perkins, eds., Submarine fans and turbidite systems—sequence stratigraphy, reservoir architecture, and production characteristics: Gulf Coast Section SEPM Foundation 15th Annual Research Conference, p. 233–244.

McGee, D. T., P. W. Bilinski, P. S. Gary, D. S. Pfeiffer, and Sheiman, 1994, Geologic models and reservoir geometries of Auger Field, deepwater Gulf of Mexico, *in* P. Weimer, A. H. Bouma, and B. F. Perkins, eds., Submarine fans and turbidite systems—sequence stratigraphy, reservoir architecture, and production characteristics, Gulf of Mexico and international: Gulf Coast Section SEPM Foundation 15th Annual Research Conference, p. 245–256.

McGuiness, D. B., and J. R. Hossack, 1994, The development of allochthonous salt sheets as controlled by the rates of extension, sedimentation, and salt supply, *in* J. M. Armentrout, R. Bloch, H. C. Olson, and B. F. Perkins, eds., Rates of geologic processes: Gulf Coast SEPM Foundation 14th Annual Research Conference, p. 127–139.

Nelson, C. H., and T. H. Nilsen, 1984, Modern and ancient deep sea fan deposits: SEPM short course 14 notes, p. 197–300.

Normark, W. R., and D. J. W. Piper, 1991, Initiation processes and flow evolution of turbidity currents: implications for the depositional record, *in* R. H. Osborne, ed., From shoreline to abyss: contributions in marine geology in honor of Francis Parker Shepard: SEPM Special Publication 46, p. 207–230.

Piper, D. J. W., A. N. Shore, and J. E. Hughes Clarke, 1988, The 1929 Grand Banks earthquake, slump and turbidity current: Geological Society of America Special Paper 229, p. 77–92.

Prather, B. E, J. R. Booth, G. S. Steffens, and P. A. Craig, 1998, Classification, lithologic calibration, and stratigraphic succession of seismic facies of intraslope basins, deep-water Gulf of Mexico: AAPG Bulletin, v. 82, p. 701–728.

Salvador, A., 1991, The geology of North America, vol. J, The Gulf of Mexico Basin: Boulder, Geological Society of America, 568 p.

Worrall, D. M., and S. Snelson, 1989, Evolution of the northern Gulf of Mexico basin, with emphasis on Cenozoic growth faulting and the role of salt, *in* A. W. Bally and A. R. Palmer, eds., The Geology of North America—an overview: Geological Society of America, p. 97–138.

Yeilding, C. A., and G. M. Apps, 1994, Spatial and temporal variations in the facies associations of depositional sequences on the slope: examples from the Miocene-Pleistocene of the Gulf of Mexico, *in* P. Weimer, A. H. Bouma, and B. F. Perkins, eds., Submarine fans and turbidite systems—sequence stratigraphy, reservoir architecture, and production characteristics, Gulf of Mexico and international: Gulf Coast Section SEPM Foundation 15th Annual Research Conference, p. 425–437.

Coleman Jr., J. L., F. C. Sheppard, III, and T. K. Jones, 2000, Seismic resolution of submarine channel architecture as indicated by outcrop analogs, *in* A. H. Bouma and C. G. Stone, eds., Fine-grained turbidite systems, AAPG Memoir 72/SEPM Special Publication No. 68, p. 119–126.

Chapter 11

Seismic Resolution of Submarine Channel Architecture as Indicated by Outcrop Analogs

J. L. Coleman, Jr.
F. C. Sheppard, III[1]
T. K Jones[2]
BP Amoco
Houston, Texas, U.S.A.

ABSTRACT

Economical production of hydrocarbons from submarine fan reservoirs requires a clear understanding of reservoir architecture and fluid flow properties. Seismic profiling has been used to delineate the extent of hydrocarbon-bearing reservoir architecture and associated flow heterogeneities. Outcrop study constrains the architectural interpretation options in field development plans. Seismic modeling of an outcrop of a submarine fan channel complex was undertaken to illustrate the acquisition and processing requirements and interpretation limits for a generic exploration seismic program. Unless appropriately designed, conventional, exploration-grade seismic data will not have the frequency content and resolution capability to image clearly submarine channel complexity of a detail commonly observed in outcrop.

INTRODUCTION

Understanding the geological details buried beneath the resolution of reflection seismic profiling has been a continuing challenge ever since its advent. The common way to improve this understanding is to model geological outcrops of relevant sections and study the interplay between lithology, stratal surfaces, density, porosity, and fluid content within a petroleum system context. Throughout the deepwater exploration drilling boom of the 1980s and 1990s, 2- and 3-D seismic data were being touted by many as the field-scaled reservoir delineation tool. In order to test the capability of reflection seismic data to meet this challenge, it was decided to attempt to model the outcrop face at a scale, noise, and frequency content similar to conventional industry exploration seismic acquisition parameters.

STUDY AREA AND INTERVAL

A prime outcrop candidate lies on the north shore of the Arkansas River, just west of North Little Rock, Arkansas, in an abandoned rock quarry. Here an almost 1000-m-long, 60-m-high exposure of Carboniferous (lower Pennsylvanian) Jackfork Formation sandstone, conglomerate-breccia, and shale (Figure 1). Stratigraphically, this exposure rests approximately 790 m below the top of the Jackfork Formation in this area. Below the quarry face and exposed in the quarry floor is a thick dark gray shale. This channel complex is at least 9.6 km wide and 16 to 24 km long. It appears to pinch out about 4 km north of the quarry. At least 14 channels are exposed here, some with relief exceeding 3.6 m. Most of the channels have flow indicators oriented west–southwest; however, a few have southeast

[1]Presently with 3DX Technologies, Inc., Houston, Texas, U.S.A.
[2]Presently with Diamond Geophysical Service Corporation, Houston, Texas, U.S.A.

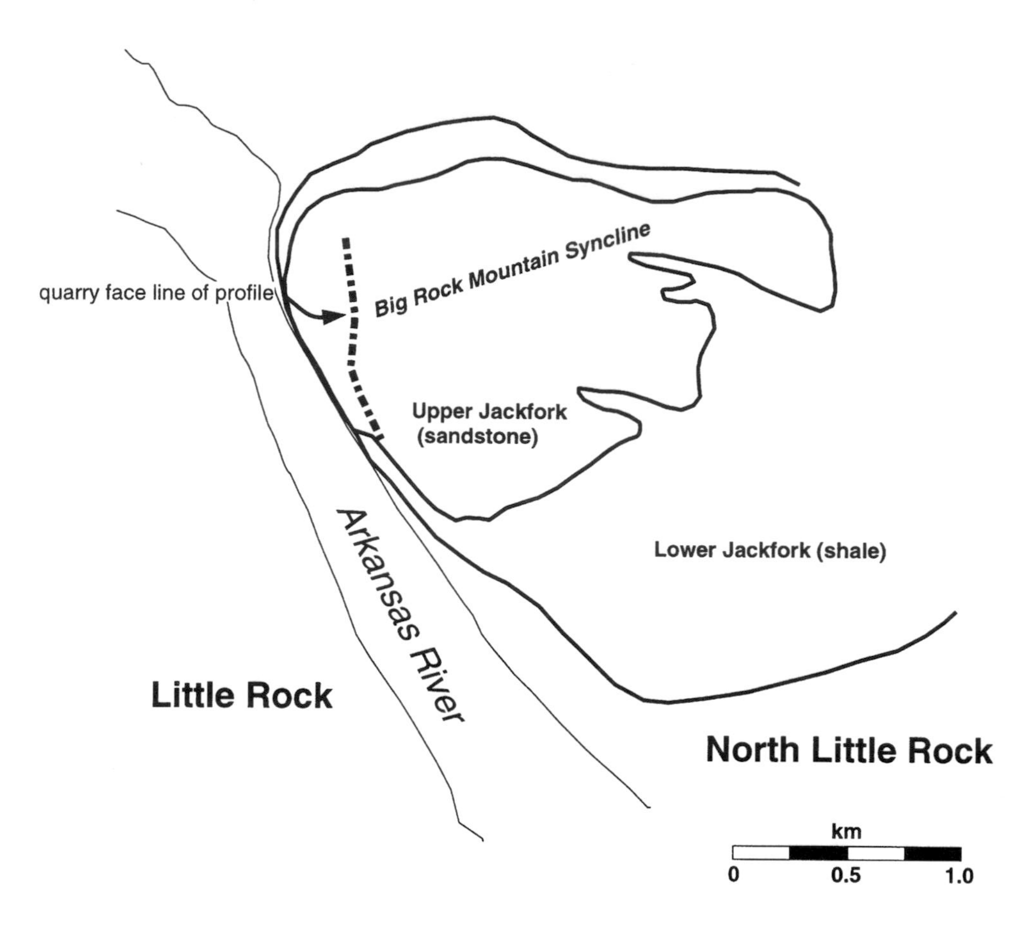

Figure 1—Location map of Big Rock Quarry, North Little Rock, Arkansas (modified from Stone and McFarland, 1982).

markers (Stone and McFarland, 1982). The sand content varies across the quarry face from a net sand-to-gross interval of about 75% on the western side, to near 100% near the center, to approximately 75% on the east (Jordan et al., 1991).

The entire Jackfork section at this quarry represents the episodic filling of a canyon or upper fan channel system, with numerous bypass and abandonment surfaces. Thick black shales overlie this section, which, in turn, are succeeded by younger channel-fill deposits.

The Jackfork Formation of central Arkansas consists of approximately 2000 m (6500 ft) of sandstones and shales originally deposited in deepwater (i.e., beyond the shelf-slope break) by a variety of gravity flow processes (Coleman, this volume). The formation, which has group status in Oklahoma and consists of five mappable formations, is routinely divided into two units in central Arkansas. In the study area, the Jackfork is loosely divisible into an upper sandy unit and a lower shaly interval. The lower shaly unit hosts much of the local and regional deformation in the Little Rock area, so the true stratigraphic thickness of the entire Jackfork is unknown.

PREVIOUS WORK

This quarry, termed the Big Rock Quarry in recent geological literature, has been well studied and discussed over the past 10 years (Jordan et al., 1991, and Cook , 1993, and references therein). The Jackfork Formation has been routinely interpreted to be a deepwater sandstone complex composed primarily of turbidites, muddy and sandy debrites, and shales that were deposited in slope canyon-fill and channel-levee-overbank systems (Coleman et al., 1994; Shanmugam and Moiola, 1995; Slatt et al., 1997). Although the exposure at Big Rock Quarry has several interpretations ranging from canyon-fill to middle fan channel-fill deposits (Stone and McFarland, 1982; Link and Stone, 1986; Jordan et al., 1991; Kuecher, 1992; Cook et al., 1994), it is apparently a nearly orthogonal cut across the downcurrent direction, and displays a nearly complete cross-sectional view of a submarine, downslope transport system.

Kuecher (1992) studied this exposure and developed a preliminary seismic model of the quarry. He concluded that the acoustic contrasts presented by the lithologies juxtaposed at channel and slump boundaries in Big Rock Quarry were insufficient to generate seismic reflections. He further suggested that an acquired seismic profile actually captures the acoustic reflectivity of a sand-prone interval, rather than a true reservoir architecture.

METHODOLOGY

In 1988, JLC took a series of telephotographs from the parking lot of the condominium across the river from Big Rock Quarry to build a photomosaic of the quarry exposure. This mosaic was later merged with outcrop gamma-ray logs collected by Jordan et al. (1991) and core data from Link and Stone (1986) (Figure 2). Formally from Jordan et al.'s study, and informally from

Figure 2—Photomosaic of Big Rock Quarry face (location of profile shown on Figure 1).

anybody else's first visit to the quarry, it is immediately apparent that well log correlation tools and principles would never result in an accurate interpretation of the stratigraphic variability of the exposure face. Cook (1993) and Bouma and Cook (1994) confirmed this position with their detailed photomosaic depiction and interpretation of the main quarry face (Figure 3A). In order to create a simple but reasonably accurate geologic model of the quarry face for this study, a line drawing rendition of the photomosaic was created (Figure 4). No attempt was made to correct the parallax problems created during the photography, however, the outcrop gamma-ray logs were adjusted to compensate for the apparent northward thinning of the quarry face sandstones.

To test the effect of scale on interpretation of outcrop and seismic model complexity, Cook's (1993) interpretation (Figure 3A) was scaled to the approximate aspect ratio of the photomosaic (Figure 2). The high degree of lateral heterogeneity is *apparently* reduced substantially, as the bedding surfaces are flattened to near-horizontal (Figure 3B).

A geologic model of the outcrop face was created that contained the interbedded sandstones, conglomerates, and shales of the quarry face as the middle interval. A homogeneous shale unit, 152 m thick,

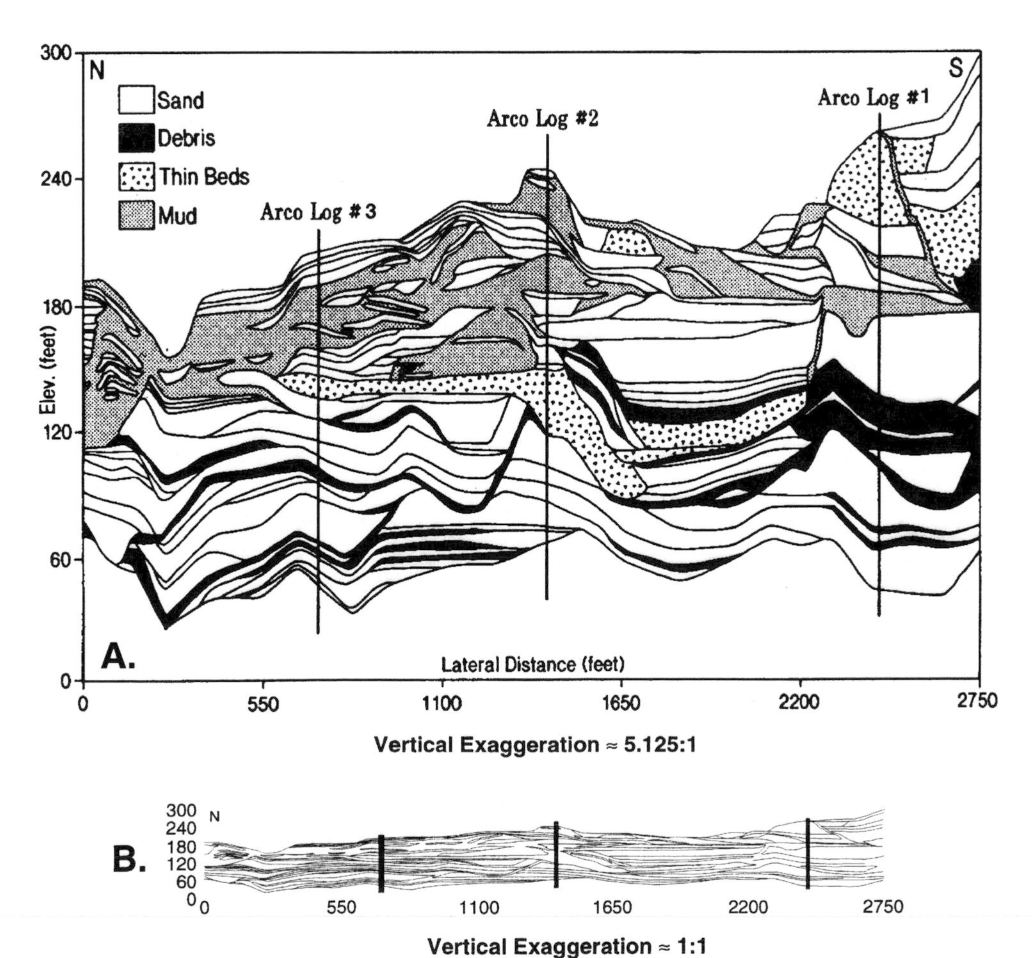

Figure 3—(A) Correlation diagram from quarry photo mosaic (original scalar distortion) (from Bouma and Cook, 1994, used by permission). (B) Correlation diagram from quarry photo mosaic (differentially "squeezed" to approximate photomosaic and seismic aspect ratio and scale) (simplified from Bouma and Cook, 1994).

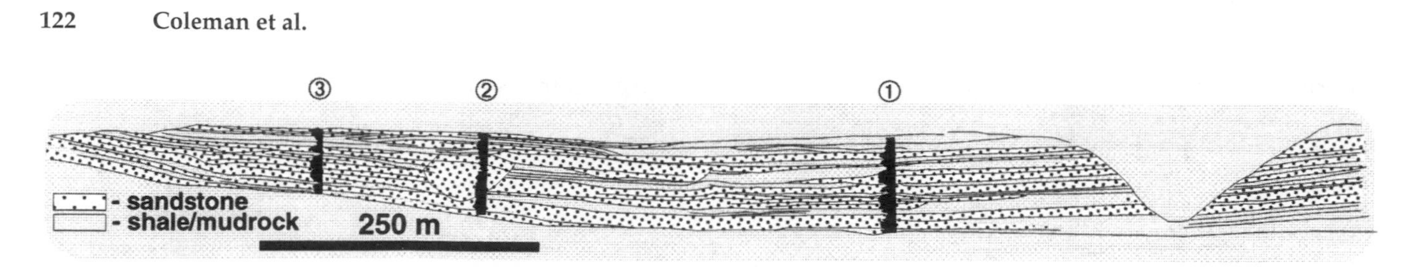

Figure 4—Line drawing from photomosaic of Big Rock Quarry face, with outcrop gamma-ray logs from Jordan et al. (1991).

and representative of the dark gray shale that is exposed at the western end of the floor of the quarry, was placed beneath the quarry face section. Another shale unit, also 152 m thick and similar to shales exposed on the north side of the Big Rock Syncline, was placed on top of the quarry face section. The large, modern-day, erosional cut near the eastern end of the exposure was artificially "filled" with the overlying shale interval. The model was placed at subsurface pseudodepths between 490 and 915 m. Rock properties and interval velocities consistent with Pliocene-Pleistocene deepwater sediments in the northern Gulf of Mexico were assigned to the model lithologies (Figure 5).

From the geologic model, a noise-free "exploding reflector" 2-D synthetic seismic line with upper band limit of 300 Hz was computed. These synthetic seismic data simulate a 2-D common midpoint stack and were then migrated to simulate a typical seismic profile used for interpretation (Figure 6). The resulting

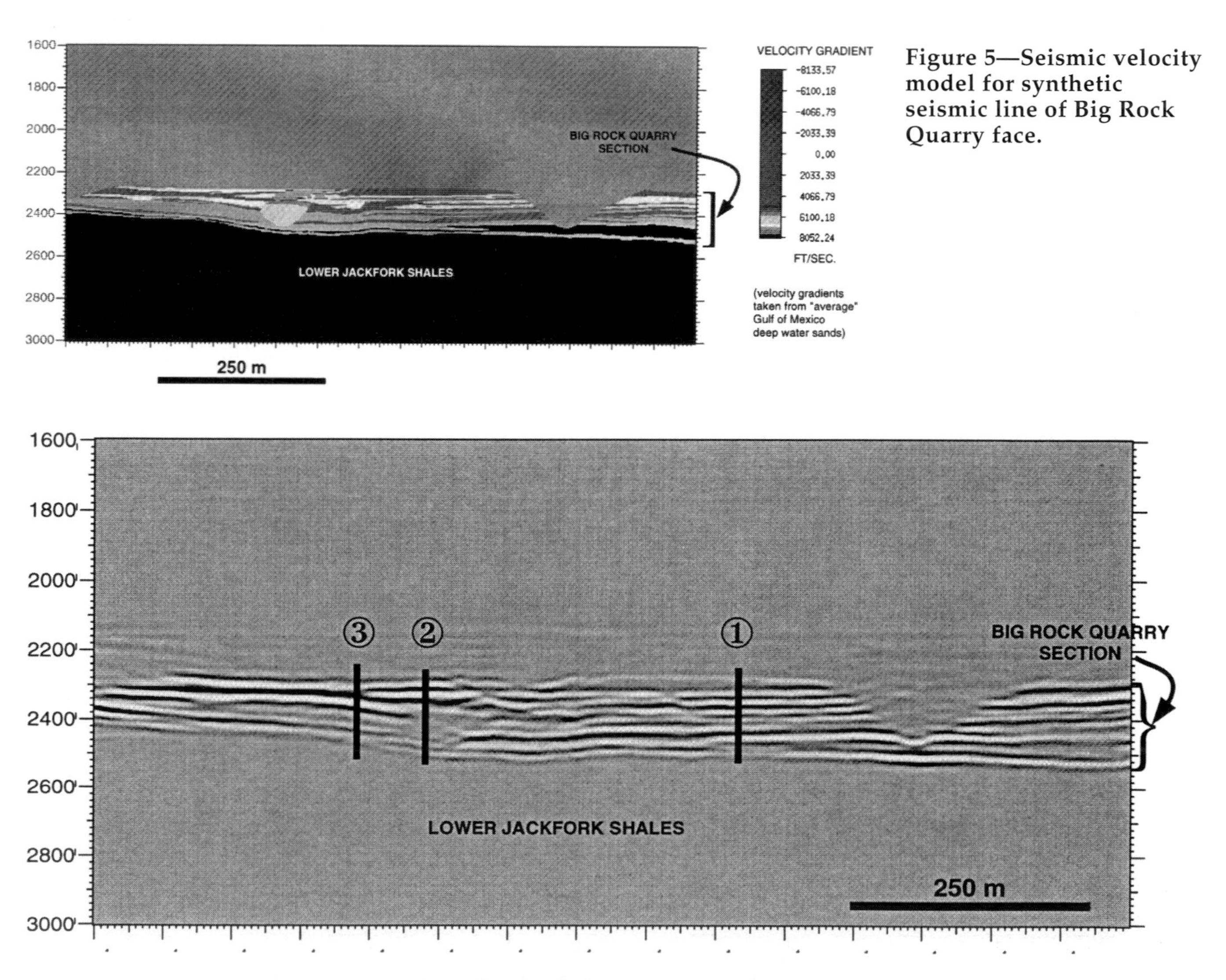

Figure 5—Seismic velocity model for synthetic seismic line of Big Rock Quarry face.

Figure 6—A 300-Hz synthetic seismic line, Big Rock Quarry face, with location of ARCO's outcrop gamma-ray logs from Jordan et al. (1991).

line reasonably depicts the simplified elements of the outcrop photomosaic interpretation (Figure 4). However, on closer examination it is apparent that not all of the detailed stratigraphy of the outcrop was reflected by the wavelets of the synthetic seismic line. From this observation it is apparent that detail, such as that provided by Cook's photomosaic interpretation (Cook, 1993; Figure 3), would not be imagable with the existing seismic model and its processing stream.

The 2-D synthetic seismic data were assigned attributes consistent with conventional 2-D, exploration-grade reflection seismic data with an upper bandlimit of 50 Hz (Figure 7). *No noise was put into the model at this point.* The resulting synthetic seismic data generated using more realistic parameters suggest that the stratigraphic detail of the Big Rock Quarry exposure face would not be decipherable at all with similarly acquired conventional reflection seismic data, yet an overall sandy interval, such as the quarry face model, would stand out from a shale above and below. The thick sand section near the center of the exposure, which is recognizable on the 300-Hz line, is essentially invisible in the 50-Hz version. The variation in net sand-to-overall-gross interval is not apparent at all with the 50-Hz line.

The reality of the 50-Hz line, and the attendant disturbing conclusions, is supported, somewhat, by comparing the modeled 50-Hz data (Figure 7) with seismic data from the Green Canyon area (offshore Louisiana) illustrated by Geitgy (1990, his figure 5B) (Figure 8A). Here the seismic expression of Pliocene-age channel-levee deposits, at a depth of approximately 3560 m in this area, appear very similar to the model Big Rock Quarry seismic signature. The central core of the Green Canyon Block 136 fan channel (which would represent the center of the Big Rock Quarry exposure) is approximately 1 km across on Figure 8A. The entire fan extends 14.5 km in a downcurrent direction and expands to a width of 6.4 km. Unfortunately, the line shown in Figure 8A has no closely associated well penetration to help corroborate the seismic facies interpretation.

The aspect ratio of the modeled Big Rock Quarry 300-Hz seismic line was modified to approximately the seismic aspect ratio of the Green Canyon Block 136 line (Figure 8B). The seismic character of the quarry appears to be entirely consistent with the character depicted in exploration seismic data across the channel core area in Block 136. However, the detail as observable in the quarry face is not immediately evident, possibly because of the scalar distortion caused by converting the quarry seismic line to a conventional vertical time–horizontal distance format. If the entire 9.6 km lateral extent of the Big Rock Syncline sandstone complex were modeled, then perhaps it would more closely resemble the 5.1-km Block 136 channel complex of Figure 8A, with both showing lateral terminations and other external morphological similarities. The internal detail, however, might still be indeterminate.

RECOGNITION AND SOURCE OF LIMITATIONS

It is apparent from this and Kuecher's (1992) study that unambiguous interpretation of detailed stratigraphic content and reservoir architecture by conventional reflection seismic profiles *alone* should be suspect. Kuecher (1992) suggests that the seismic profile captures the acoustic reflectivity of a sand-prone

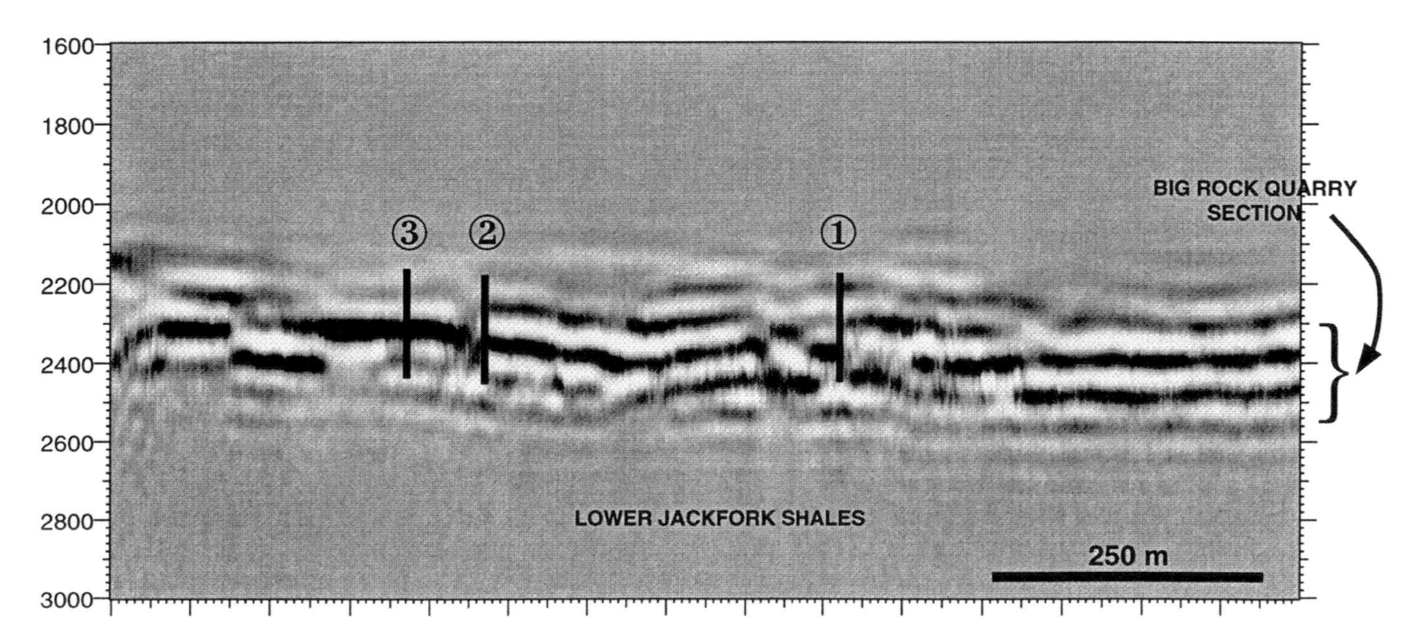

Figure 7—A 50-Hz synthetic seismic line, Big Rock Quarry face, with location of ARCO's outcrop gamma ray-logs from Jordan et al. (1991).

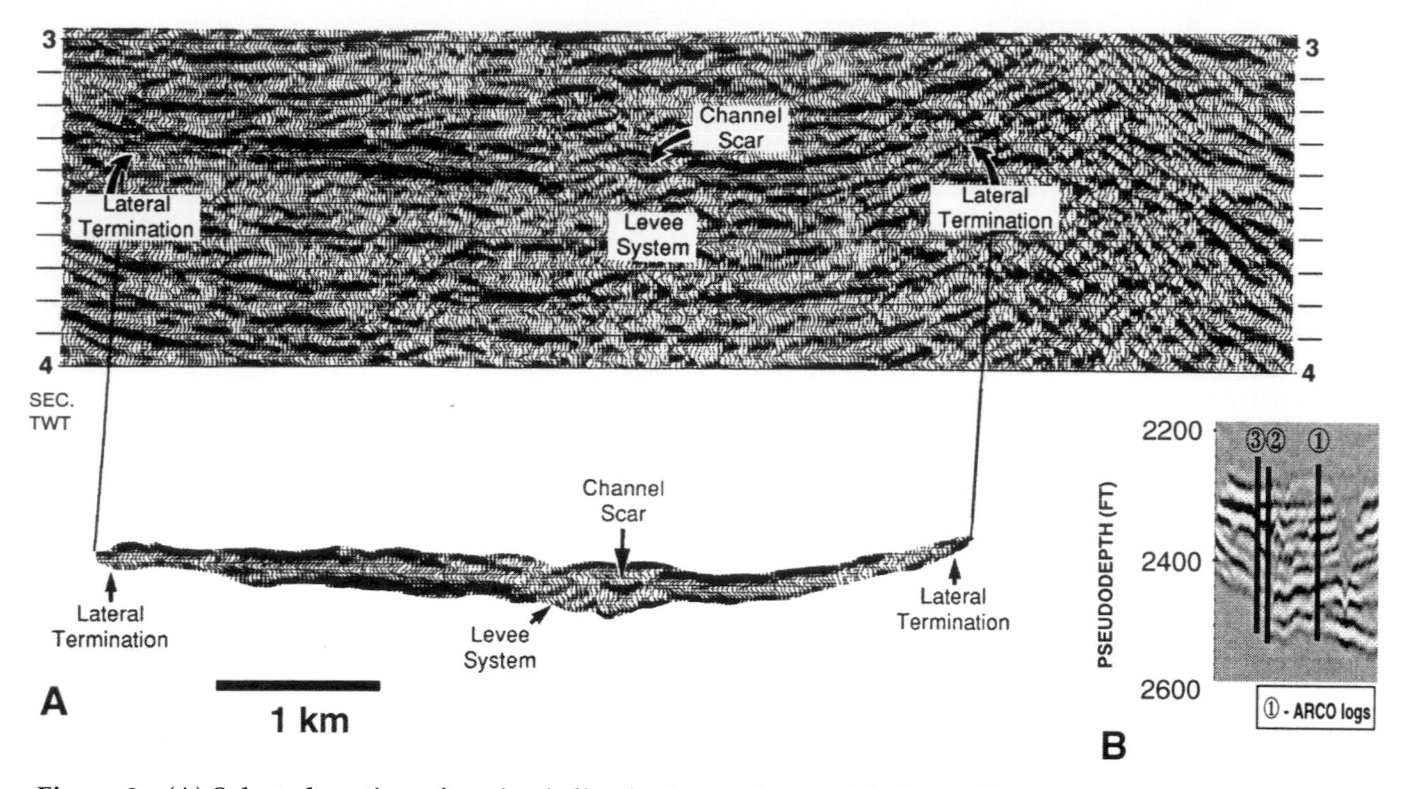

Figure 8—(A) Selected portion of a seismic line in Green Canyon Block 136 illustrating seismic facies character of a profile across a submarine channel/levee complex (modified from Geitgy, 1990, reproduced with permission of the Gulf Coast Sections Society of Economic Paleotologists and Mineralogists Foundation). (B) A 300-Hz synthetic seismic line of Big Rock Quarry face (Figure 6) displayed at approximately the same seismic aspect ratio as in Figure 8A.

interval rather than a true reservoir architecture. The work presented here seems to support that observation and further suggests that only the shape formed by the external boundaries of a sand body might be clearly discernible, leaving the internal architectural elements and potential fluid flow barriers or baffles imprecisely imaged. This observation may be supported by observations made during development of Bullwinkle Field, Gulf of Mexico, where complexity of the reservoir architecture did not become clear until several development wells had been drilled (R. D. Shew, 1997, personal communication, 1997).

In spite of such learnings, many published and proprietary studies show seismic profiles with reasonably detailed architecture, similar in scale to that seen in the Big Rock Quarry outcrop (Weimer and Link, 1991; Weimer et al., 1994). These studies pose the following questions: Since we cannot clearly image the stratigraphic detail of a large outcrop with conventional frequency content seismic data, what level of detail can we actually see on a conventional seismic line? What seismic events give us the sedimentary geometries that are routinely interpreted? How meaningful are seismic facies interpretations from a seismic dataset that is not capable of detecting the size and boundaries of the depositional elements interpreted? Are we seeing a self-similar, or even fractal, organization of deepwater systems, as suggested by Valasek (1992), in that similarly shaped, but subseismic-scale resolution, features build into similarly shaped, seismically resolvable depositional systems? This concept, although intriguing and potentially quite valuable, has yet to be proven in deepwater depositional systems.

Continued work on large-scale outcrops at the field reservoir level of detail, but compiled at the scale of resolution of conventional exploration-grade seismic data, is needed. Preliminary work using the exceptionally detailed seismic data from the Oligocene-Miocene of offshore Angola (West Africa) and detailed outcrop cross sections from the Permian of West Texas suggest that the problem is solvable (Marotta et al., 1998; Carr et al., 1998; Coterill et al., 1998; Gardner and Borer, this volume). Prather et al. (1998) attempt to overcome some of the problems by using 3-D seismic volumes calibrated with well log and core control from 22 wells. In their study, spatial distribution of seismic reflections and associated interpreted depositional geometry are used to develop a seismic facies classification scheme for seismic profile characteristics. The obvious question, is "How far beyond well control are we safe in extrapolating lithofacies assemblages from seismic facies?"

CONCLUSIONS

Seismic modeling of geologic outcrops indicates that internal sand-body architecture and geometry may not be clearly imagable with conventional seismic profiling. Further work in generating seismic models of large outcrops is needed to help determine what is being imaged on modern, high-quality seismic profiles.

ACKNOWLEDGMENTS

The authors thank BP Amoco for permission to publish this report and for financial support for publication. We also thank Arnold Bouma, Lesli Wood, and Gerald Kuecher for their helpful reviews.

REFERENCES CITED

Bouma, A. H., and T. W. Cook, 1994, Architecture of a submarine channel complex, Jackfork Formation, Arkansas: Louisiana Basin Research Institute Bulletin, v. 4, p. 11–22.

Carr, M., M. Gardner, M. Batzle, T. Melick, and C. Woodland, 1998, Portrait of a basin-floor fan for sandy deep-water systems, Permian lower Brushy Canyon Formation, Delaware Mountains, West Texas (abs.): AAPG annual meeting abstracts with program v. 1, no. A113, 1 p.

Coleman, J. L., Jr., 1990, Comparison of depositional elements of an ancient and a "modern" submarine fan complex: early Pennsylvanian Jackfork and late Pleistocene Mississippi fans (abs.): AAPG Bulletin, v. 74, p. 631.

Coleman, J. L., Jr., 1993, Controls on variability of depositional style in Carboniferous submarine fan complexes of the Ouachita Basin of Oklahoma and Arkansas (abs.): AAPG 1993 annual convention program with abstracts, p. 86–87.

Coleman, J. L., Jr., 1994, Factors controlling the deposition and character of submarine fan complexes: an illustration from the Carboniferous Ouachita Basin (abs.): Geological Society of America abstracts with programs, v. 26, no. 1, p. 4.

Coleman, J. L., Jr., G. Van Swearingen, and C. E. Breckon, 1994, The Jackfork Formation of Arkansas: a test of the Walker-Mutti-Vail models for deep-sea fan deposition: Arkansas Geological Commission guidebook prepared for the 1994 South-Central Section meeting, Geological Society of America, March 20–23, 1994, 56 p.

Cook, T. W., 1993, Facies architecture of deep-water channel deposits, Brushy Canyon Formation, West Texas, and Jackfork Group, Arkansas: Unpublished Master's thesis, Louisiana State University, 108 p.

Cook, T. W., A. H. Bouma, M. A. Chapin, and H. Zhu, 1994, Facies architecture and reservoir characterization of a submarine fan channel complex, Jackfork Formation, Arkansas, *in* P. Weimer, A. H. Bouma, and B. Perkins, eds., Submarine fans and turbidite systems: sequence stratigraphy, reservoir architecture and production characteristics, Gulf of Mexico and International: Gulf Coast Section SEPM Foundation 15th Annual Research Conference Proceedings, p. 69–81.

Coterill, K., A. Champagne, J. Coleman, D. Marotta, M. Pasley, G. Tari, L. Binga, and H. Van Dierondonck, 1998, Sinuous morphologies in submarine channels—scale and geometries in seismic and outcrop indicating possible mechanisms for deposition (abs.): AAPG international meeting abstracts with program volume, p. 20.

Geitgy, J. E., 1990, Characterization of a restricted turbidite sand system, Green Canyon area, offshore Louisiana, *in* Perkins, B. F., ed., Sequence stratigraphy as an exploration tool: concepts and practices in the Gulf Coast: Gulf Coast Section SEPM Foundation 15th Annual Research Conference Proceedings, p. 177–191.

Jordan, D. W., D. R. Lowe, R. M. Slatt, A. D'Agostino, M. H. Scheihing, R. H. Gillespie, and C. Stone, 1991, Scale of geological heterogeneity of Pennsylvanian Jackfork Group, Ouachita Mountains, Arkansas: applications to field development and exploration for deep-water sandstones: Dallas Geological Society Field Trip Guidebook no. 3, 142 p.

Kuecher, G. J., 1992, Seismic mapping of turbidite channels and channel complexes based on predictable scale: Basin Research Institute Bulletin, Louisiana State University, v. 2, p. 46–52.

Link, M. H., and C. G. Stone, 1986, Stop 1—Jackfork sandstone at the abandoned Big Rock Quarry, North Little Rock, *in* C. G. Stone and B. R. Haley, eds., Sedimentary and igneous rocks of the Ouachita Mountains of Arkansas: a guidebook with contributed papers, part 2: Arkansas Geological Commission Guidebook 86-2, p. 1–8.

Marotta, D., C. S. Alexander, K. Coterill, K. Hartman, M. Pasley, T. C. Stiteler, G. Tari, L. Binga, and B. Lehner, 1998, The use of 3-D visualization for understanding Tertiary deep-water clastic systems: a West Africa example (abs.): AAPG international meeting abstracts with program volume, p. 746.

Prather, B. E., J. R. Booth, G. S. Steffens, and P. A. Craig, 1998, Classification, lithologic calibration, and stratigraphic succession of seismic facies of intraslope basins, deep-water Gulf of Mexico: AAPG Bulletin, v. 82, p. 701–728.

Shanmugam, G., and R. J. Moiola, 1995, Reinterpretation of depositional processes in a classic flysch sequence (Pennsylvanian Jackfork Group), Ouachita Mountains, Arkansas and Oklahoma: AAPG Bulletin, v. 79, p. 672–695.

Shew, R., 1997, Deepwater core workshops: reservoir characterization (architecture and properties), presented at the Gulf Coast Association of Geological Societies 47th annual convention, New Orleans, October 15–17, 1997.

Slatt, R. M., et al., 1997, Reinterpretation of depositional processes in a classic flysch sequence (Pennsylvanian Jackfork Group), Ouachita Mountains,

Arkansas and Oklahoma: discussion and reply: AAPG Bulletin, v. 81, p. 449–491.
Stone, C. G., and J. D. McFarland, III, 1982, Stop 1—Abandoned Big Rock Quarry in upper part of Jackfork sandstone, *in* C. G. Stone and J. D. McFarland, III, eds., Field guide to the Paleozoic rocks of the Ouachita Mountain and Arkansas Valley provinces, Arkansas: Arkansas Geological Commission guidebook 81-1, p. 7–13.
Valasek, D., 1992, Self-similarity of high-frequency sequences—implications for correlation and reservoir characterization, Gallup sandstone (Turonian), northwestern New Mexico (abs.): AAPG annual meeting program and abstracts, p. 133.
Weimer, P., and M. H. Link, eds., 1991, Seismic facies and sedimentary processes of submarine fans and turbidite systems: New York, Springer-Verlag , 447 p.
Weimer, P., A. H. Bouma, and B. F. Perkins, eds., 1994, Submarine fans and turbidite systems: sequence stratigraphy, reservoir architecture and production characteristics: Gulf of Mexico and international: Gulf Coast Section SEPM Foundation 15th Annual Research Conference Proceedings, 440 p.

Batzle., M., and M. H. Gardner, 2000, Lithology and fluids: seismic models of the Brushy Canyon Formation, west Texas, *in* A. H. Bouma and C. G. Stone, eds., Fine-grained turbidite systems, AAPG Memoir 72/ SEPM Special Publication No. 68, p. 127–142.

Chapter 12

Lithology and Fluids: Seismic Models of the Brushy Canyon Formation, West Texas

Michael Batzle
Department of Geophysics, Colorado School of Mines
Golden, Colorado, U.S.A.

Michael H. Gardner
Department of Geology and Geological Engineering, Colorado School of Mines
Golden, Colorado, U.S.A.

ABSTRACT

Lithology and fluid information can be extracted from seismic data of deepwater clastics if their relative contribution to the signal is understood. Brushy Canyon Formation outcrop seismic models are constructed for the Western Escarpment of the Guadalupe Mountains using properties from outcrop, normal, and overpressured Gulf of Mexico and North Sea basins to test seismic sensitivity to lithology, fluid, and pressure. Large, clean, gas-saturated, and overpressured sandstones have the best resolution. Hydrocarbon saturation does not necessarily enhance seismic response. Lithology and fluid effects can reduce impedance contrast, resulting in low amplitudes (dim spots). Elevated geopressures preserve porosity producing low velocities and high amplitudes (bright spots). Even in low-impedance contrast intervals, offset-dependent amplitudes increase resolution and indicate hydrocarbons.

INTRODUCTION

Three-dimensional seismic surveys are the current primary exploration tool used for deepwater sandstone reservoir delineation. However, there are many uncertainties concerning what is actually being imaged. Because deepwater reservoirs occur in basins with different petroleum systems and depositional systems, it is often difficult to isolate the lithologic and fluid contribution to seismic response from variations related to architectural style. Despite these limitations, geologic models derived from seismic surveys have become increasingly more robust as seismic resolution and 3-D coverage has increases.

Seismic stratigraphy assumes reflectors follow chronostratigraphic surfaces and define genetic rock packages that comprise sequence stratigraphic models (Vail et al., 1977). In contrast, outcrop seismic models of a Permian carbonate shelf margin indicate that seismic reflectors more closely follow facies and facies-controlled property distributions (Stafleu and Sonnenfeld, 1994). These contrasting geologic interpretations of seismic data highlight the need for a better understanding of the controls governing the seismic response of stratal architecture.

Rock and fluid properties from different deepwater reservoir sand bodies produce extremely variable seismic expressions, ranging from both positive to negative reflectors. The variability in reflector geometry, and in the interpretation of what it represents, reflects the multiple geologic variables that contribute to seismic response. The principal variables include (1) pore fluids, (2) reservoir pressure and depth, (3) spatial distribution of lithology and rock properties, and (4) sand-body size and architecture of encasing nonreservoir strata.

These combine to determine seismic expression and reflect both the present reservoir setting and ancient depositional environment (Figure 1). In general, resolution is best when sand bodies are large and bounded by rocks of contrasting lithology (Figure 1). However, there are numerous exceptions to this generalization. Low sand/shale contrast can produce weak reflections and, under some conditions, even hydrocarbon saturation will produce low amplitudes (dim spots). This investigation models pore fluids, reservoir pressure, and lithology distributions that contribute to the seismic character of deepwater sandbodies.

GEOLOGIC SETTING

Outcrops of the Brushy Canyon Formation expose an oblique depositional dip profile of deepwater sandstone and siltstone (Figure 2) exposed along an 80-km outcrop belt (Gardner and Borer, this volume). A seismic model is constructed across a proximal, siltstone-rich slope succession of low sand content (~10%), dissected by clinoform surfaces modified by slumping. These clastic deposits thicken basinward to form a 400-m thick clastic wedge deposited below the physiographic shelf margin.

Depositional patterns in this proximal setting are related to a stratigraphic hierarchy of seven submarine fans that stack to form the Brushy Canyon fan complex (Gardner and Borer, this volume). A conspicuous depositional pattern shown by this fan complex is an upward change from basin-centered to slope-centered sediment thicks (Figure 3). Basin-centered fans are 100 m thick, and slope-centered fans to the genetic top siltstone are 320-m thick at section M on the cross section. This thickness pattern reverses 30 km from the physiographic break, with basin-centered fans 250 m thick and slope-centered fans 120 m thick. Basin-centered fans onlap below the shelf break, and are thin and generally sandstone-poor in slope facies tracts. Slope-centered fans onlap and fill submarine canyons, contain thick siltstone-dominated clinoform packages, with large isolated sandstones in slope facies tracts. This reorganization in sedimentation pattern is defined by the 40-ft siltstone marker.

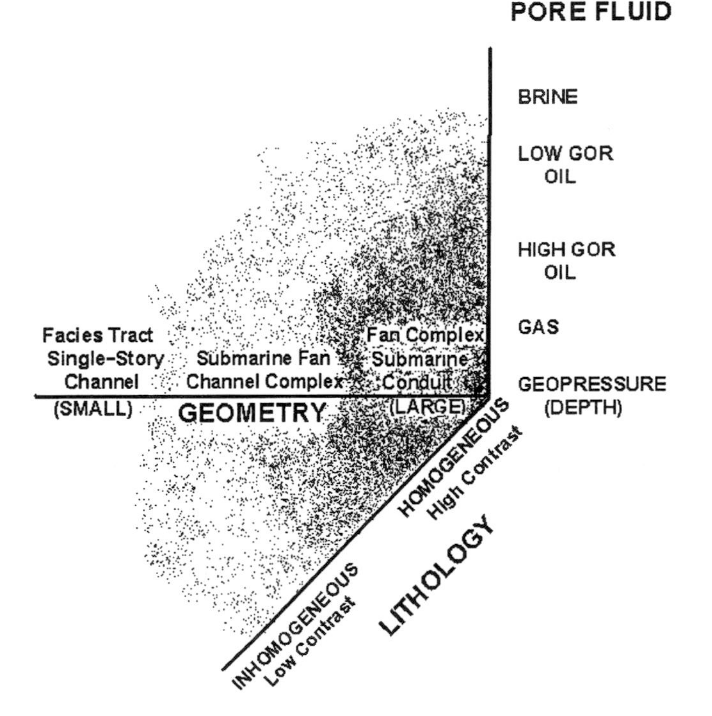

Figure 1—Chart summarizing the relative influence of lithology, fluid, and geometry on the seismc resolution of deepwater sandstone. The stippled region indicates optimum image area. Areas near the origin, in general, have the strongest seismic response. GOR = gas-oil ratio.

ARCHITECTURE

The stratigraphic cross section used in the model is constructed from 24 regional stratigraphic sections tied to digital photomosaics and geologic maps (Figure 3). In outcrops proximal to the paleoshelfbreak, deepwater clastics obliquely onlap an unconformity that floors submarine canyons incised into the underlying Leonardian carbonate shelf margin. Sandstones and conglomerates fill erosional canyons, and slope successions contain slump scar confined channel complexes (Gardner and Borer, this volume).

The lower Brushy Canyon consists of poorly resolved basin-centered fan deposits. Thin and siltstone-rich, fan 1–3 deposits onlap below the shelf margin. Fans 4 and 5 thin and merge by toplap that is indicated by the interbedded sandstone and organic-rich siltstone that form the middle part of the Brushy below the "40-ft" siltstone marker. Above this marker, clinoforms of slope-centered fans separate three multistory channel complexes that rise stratigraphically basinward. Coalesced slump scars confine these kilometer-scale sand bodies encased in siltstone. Siltstones above these sandstones contain laterally continuous, organic-rich siltstones that correlate to back-stepping channel complexes that progressively onlap the margin northward along the outcrop. These organic-rich siltstones commonly drape discordances and clinoform surfaces and have the longest correlation lengths. They record periods of fan abandonment and form stratigraphic cycle boundaries at all scales.

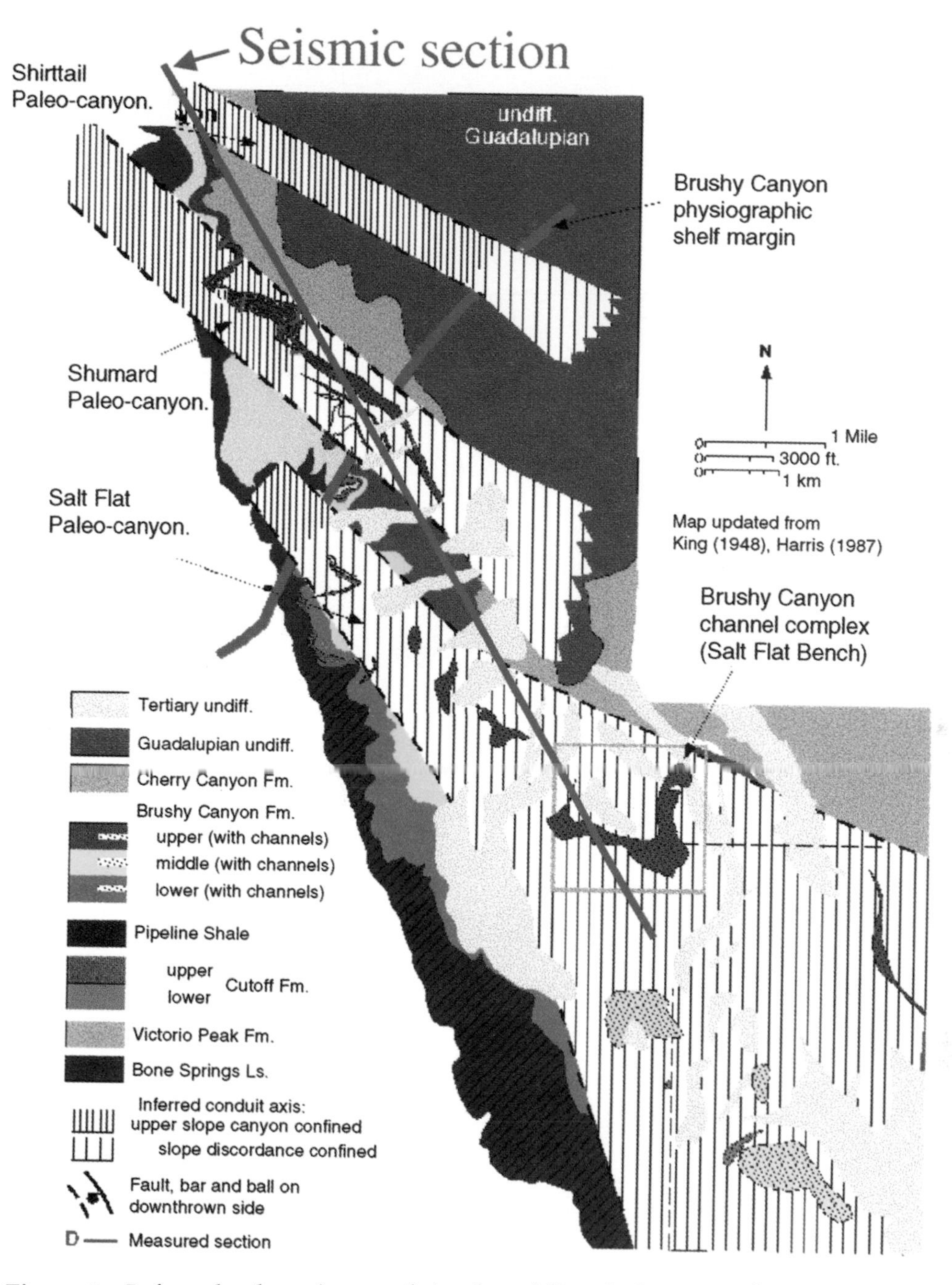

Figure 2—Inferred submarine conduits that obliquely intersect the outcrop belt of the Brushy Canyon Formation along the Western Escarpment of the Guadalupe Mountains. Location of modeled seismic cross section (Figure 4) is indicated by the heavy line.

The digitized seismic model includes (1) stratal surfaces separating rocks of different lithology; (2) formation boundaries and unconformities; (3) thin, laterally continuous, organic-rich siltstone intervals; (4) channel sand bodies; (5) clinoform surfaces; and (6) stratal discordances representing slump scars. This simplified cross section is built to resolve larger seismic-scale features (Figure 4). Rock properties are held constant within each individual sediment body, but are varied spatially across the section by relating lithology to rock properties.

ROCK PROPERTIES

Seismic impedance contrasts may be abrupt to gradual in a vertical sense, and continuous to discontinuous laterally. Velocity and density depend on lithology, porosity, texture, and pore fluid. In general, high-porosity yields low velocity and density.

Rock velocity data from the Brushy Canyon Formation was collected from core plugs along outcrop sections and from well logs at Cabin Lake field.

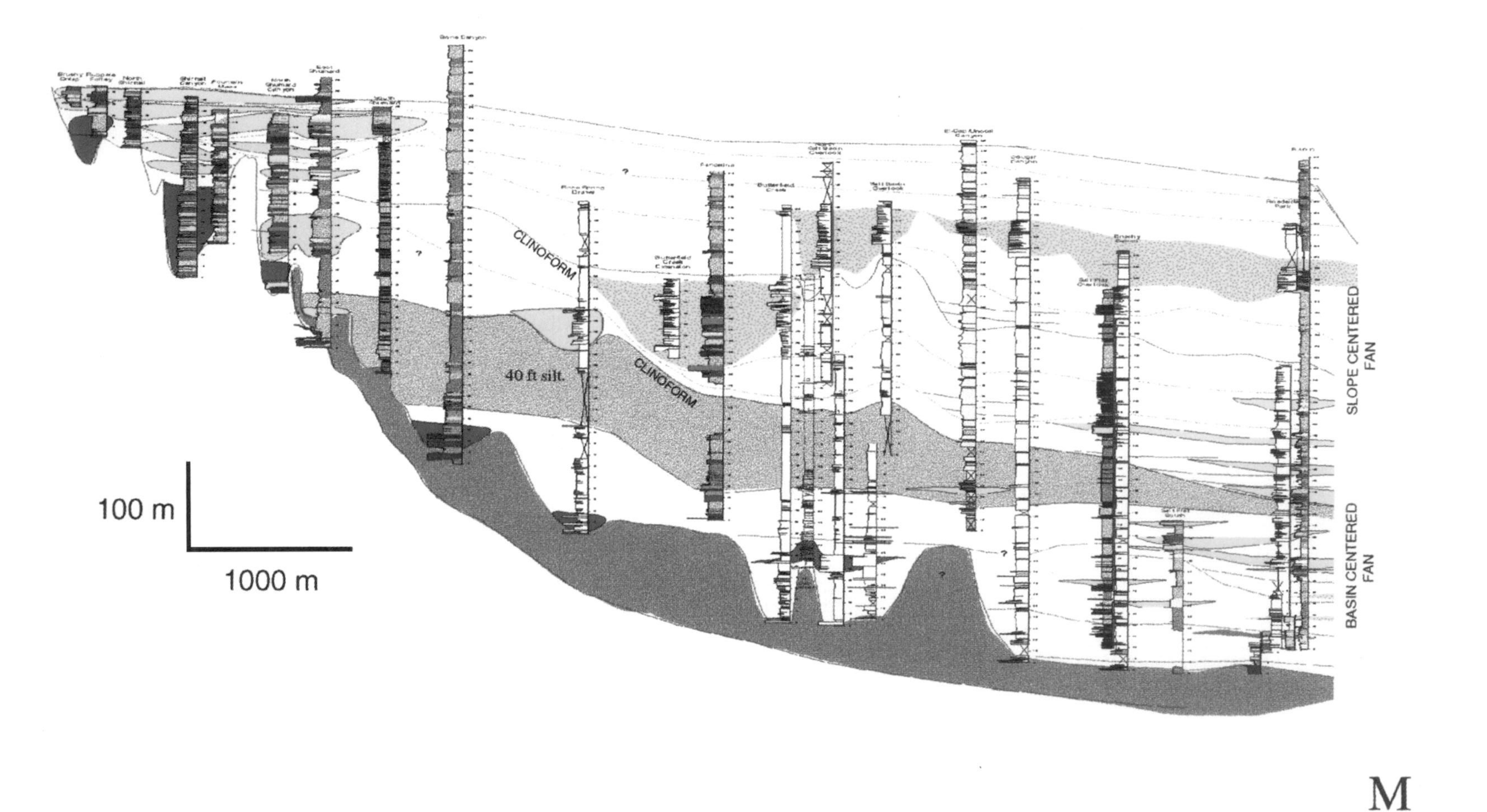

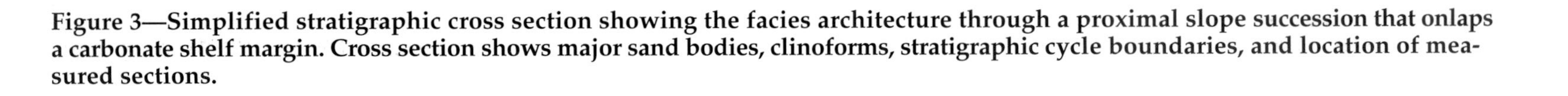

Figure 3—Simplified stratigraphic cross section showing the facies architecture through a proximal slope succession that onlaps a carbonate shelf margin. Cross section shows major sand bodies, clinoforms, stratigraphic cycle boundaries, and location of measured sections.

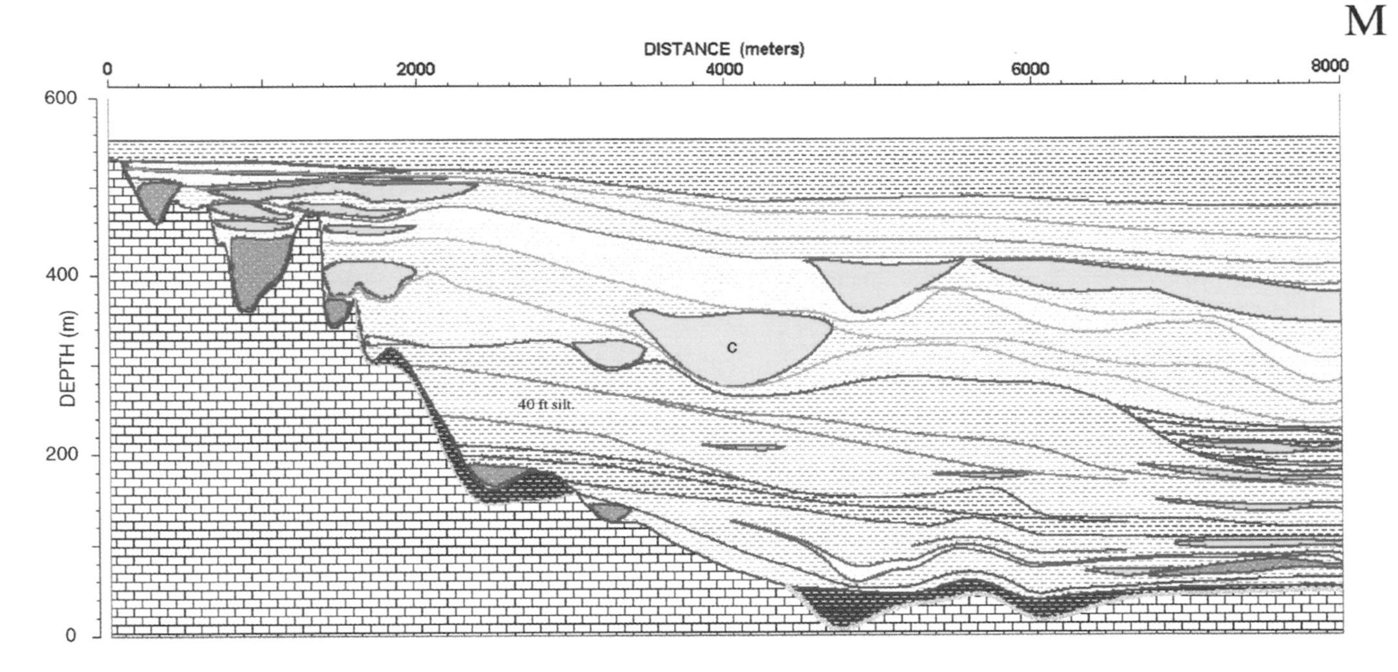

Figure 4—Digitized seismic model based on the geologic section in Figure 3. Sand body marked "C" is the focus of pore-fluid substitution calculations.

Velocity distributions used to construct all seismic models are compared in Figure 5A, B, C. In the three subsurface areas rock properties are derived from well logs.

Principal facies within the cross section include

(1) Carbonate—The composition and pervasive cementation in carbonate rocks promote high velocities and densities. Limestone, dolomite, and megabreccia underlying deepwater clastics have very high velocities (5 to 6 km/s) and densities. Subordinate carbonate concretions in organic-rich siltstones and limestones below the 40-ft siltstone also show high velocities.

(2) Sandstone and conglomerate—Very fine- to medium-grained sandstone in a well-sorted, subangular, subfeldspathic arenite. Structureless (2.3 km/s) and stratified sandstones (4 km/s) have porosities up to 15% and the lowest velocity rock–seismic facies. Subordinate sandstone-matrix–supported conglomerates consist of carbonate clasts and allochems. These debris flows and carbonate clast-rich sandstone are well cemented, dense, and have the highest velocities (~5.7 km/s).

(3) Siltstone—Two siltstone types with varying organic content are present. Organic-poor siltstones are thinly interbedded to interlaminated with sandstones and form up to 200-m thick successions. Depending on sand content, these siltstones show a broad range of low to moderate velocities. Encasing siltstones average about 2.5 km/s and show strong impedance contrasts when in contact with stratified sandstone and conglomerate. However, siltstone is largely indistinguishable from structureless sandstone. Organic-rich siltstones contain thin carbonate concretions, volcanic ash beds altered to claystone, and have total organic carbon values ranging 1–6% and higher velocities and densities. They form thin (<3-m-thick), laterally continuous intervals that extend across the cross section.

Additional rock property data were derived from well logs through the Brushy Canyon Formation at Cabin Lake field in the northeastern Delaware basin (Figure 6). Although burial and diagenesis has produced more pervasive cementation, sonic and density logs show the same rock-seismic facies relationships documented in outcrop. Carbonates underlying the Brushy Canyon have high velocity, density, and impedance, which produces a strong basal reflection. Porous structureless sandstone has the lowest impedance (gamma ray <80 API). Siltstone and organic-rich siltstone have the highest gamma-ray values and show high impedance. Sandstones in contact with organic-rich siltstone are commonly cemented. The cemented sandstone and siltstone contacts produce the most consistent reflectors, with the organic-rich siltstone forming longest correlation length intervals.

SEISMIC MODELS

The stratigraphic cross section was digitized so that the architecture could be populated with different rock and fluid properties to assess their seismic response. Rock velocity and density vary with lithology and pore fluids to define seismic impedance.

These properties do not necessarily correlate to a specific lithology, but their distribution produces impedance contrasts that define sand-body architecture. This architectural framework is held constant, and lithology and fluid properties from three deepwater provinces are substituted and related to seismic expression.

Synthetic sections from the digitized section are generated using the GXII software package. These models use normal incidence ray traces with first-order diffractions. The model is 0.6-km thick and 8-km long (Figure 4) with a shotpoint separation of 80 m. Traces are calculated and gain, frequency filter (wavelet), and noise are applied after ray paths and reflection coefficients are determined. A 40-Hz Ricker wavelet is applied to all models. An automatic gain control (AGC) is applied with 300-ms window. Noise at a –20 dB level is added to sections, band passed from 8 to 50 Hz.

Brushy Canyon Formation

The initial seismic model uses acoustic properties obtained from rock-seismic facies shown in Figure 5A. These values are used to populate each layer or body in the model. Velocities and densities are varied within sediment bodies so that the stratal architecture shown in the cross section has seismic expression. Hence, the absolute amplitude of an individual reflectors should be viewed as a relative signature because a broad range of values can be applied to these sediment bodies.

Stratal architecture depicted on the noise-free cross section is apparent (Figure 7). The strong reflection at the base is related to high-impedance carbonates. The large slope sand bodies have strong positive reflections at their top, but have strong negative reflections at their base because of the

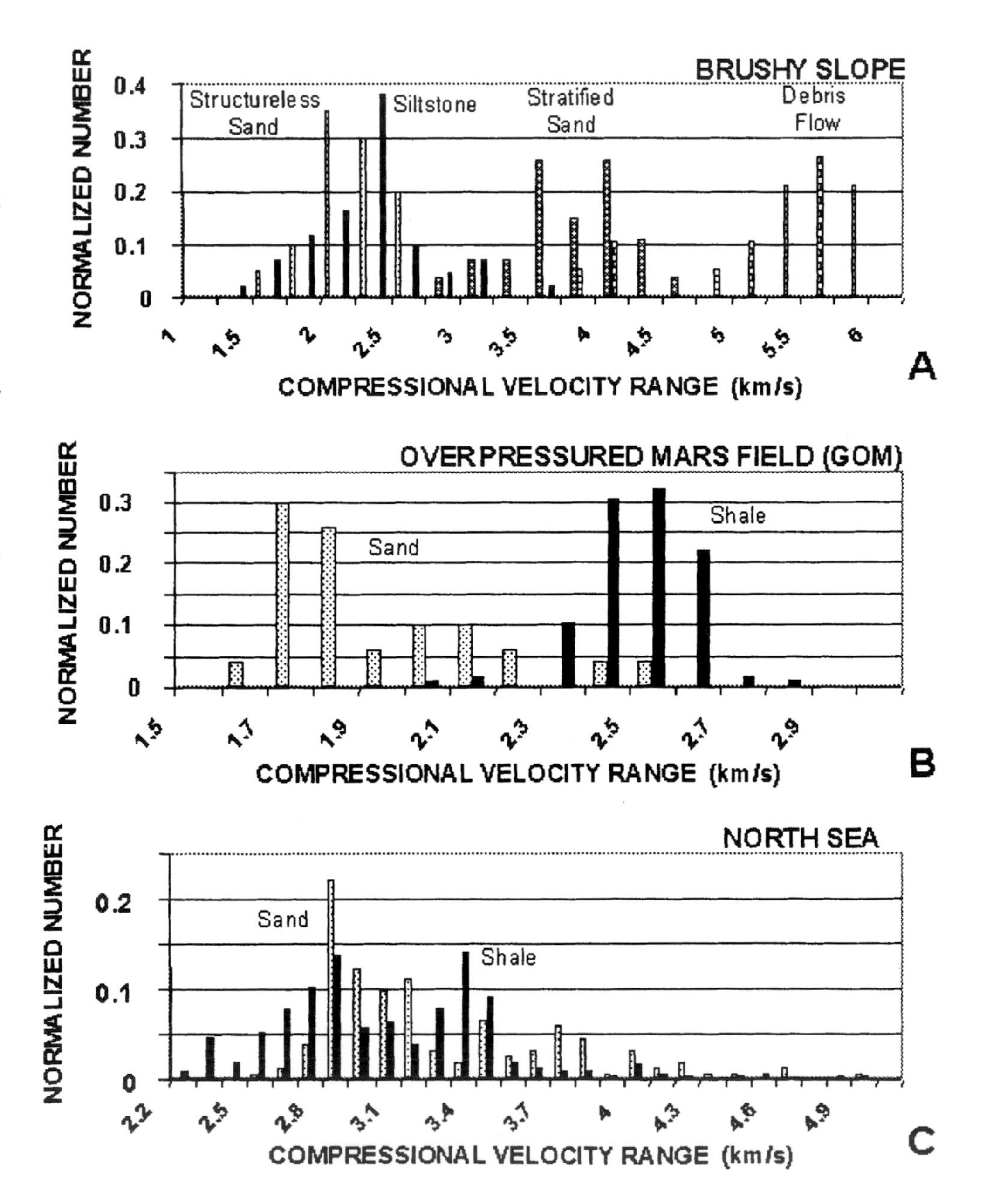

Figure 5—Normalized velocity distributions. (A) Brushy Canyon Formation velocity distributions. Siltstones (black) have velocities significantly lower than stratified sands and conglomerates, but cannot be distinguished from structureless sands. (B) Shale and sand at Mars field. Despite the depth, geopressured sands have low velocity, significantly lower than encasing shales. (C) Normalized velocity distribution for the North Sea reservoir zone. Sand velocities are very similar to shale velocities. High velocity values indicate calcite cementation. GOM = Gulf of Mexico.

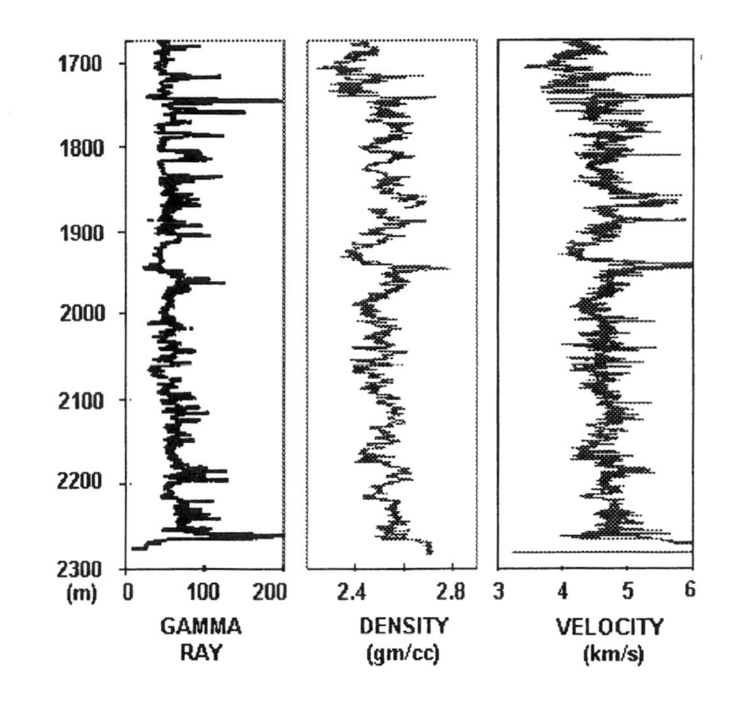

Figure 6—Gamma-ray, density, and sonic logs from the Brushy Canyon Formation in the James A11 well, Cabin Lake field, New Mexico. Thin, high API gamma-ray zones are organic-rich siltstones. Increased cementation is indicated in sandstone in contact with organic-rich siltstones by increased density and velocity.

high-impedance contrast with encasing siltstone facies. The ringing pattern in the lower right corner of the section is produced by the interbedded sandstone and siltstone that record the merging of the upper basin-centered fans. Subtle features within the thick slope-centered succession are resolved in the noise-free model. Clinoform surfaces bounding fans that extend across the outcrop are resolved. Onlap of older deepwater clastics onto the basal unconformity is resolved and provides recognition criteria for resolving basin-centered fans. The organic-rich siltstone capping the Brushy Canyon fan complex forms a continuous

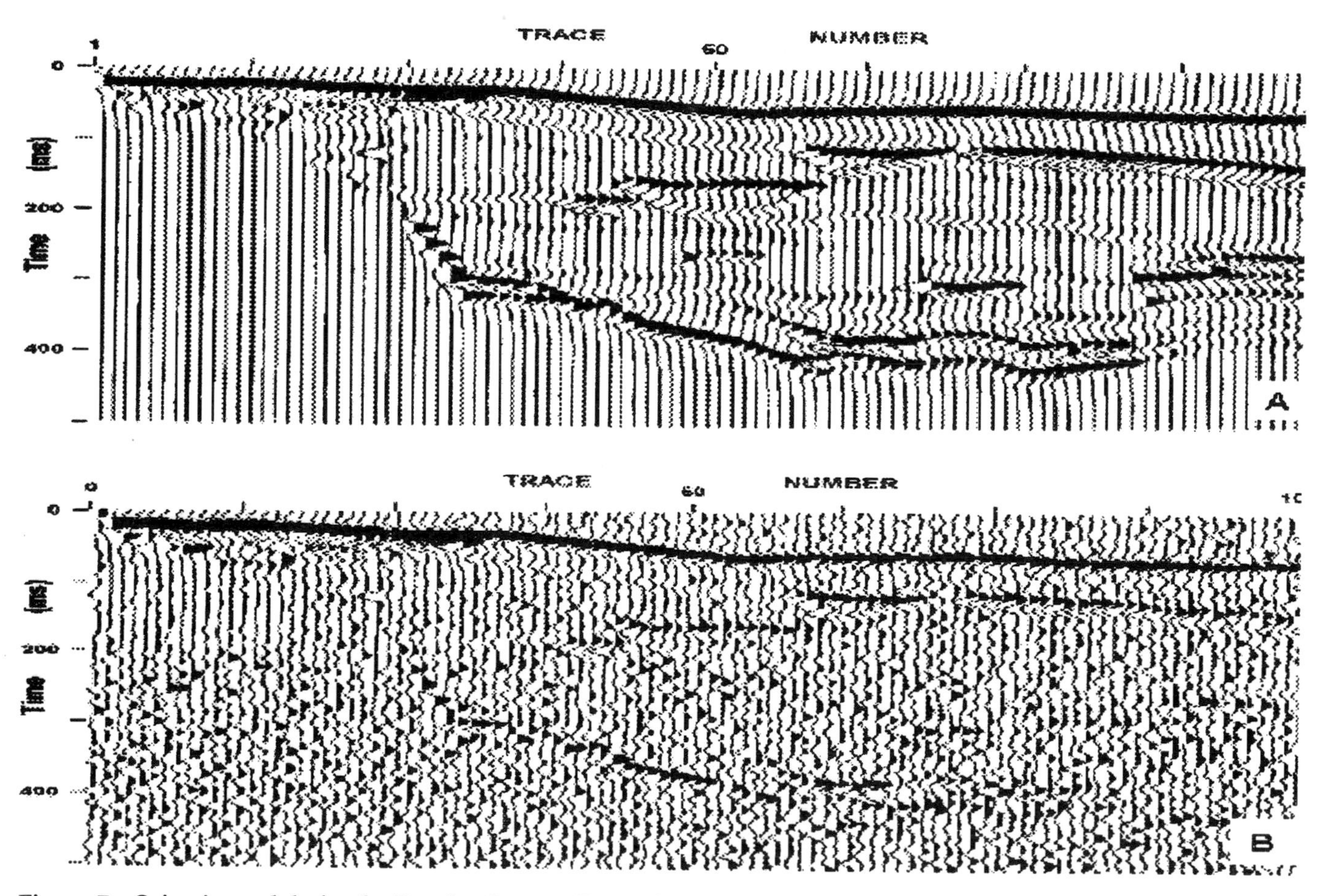

Figure 7—Seismic models for the Brushy Canyon Formation outcrop section (Figures 3, 4) using measured Brushy properties. A zero-phase 40-Hz Ricker wavelet is used with an automatic gain control applied over a 300-ms window. (A) Noise-free, both sand bodies and clinoforms are resolved. (B) Same model with –20 dB noise added. Clinoforms are still visible but difficult to trace. Sand bodies are apparent, but boundaries are difficult to detect.

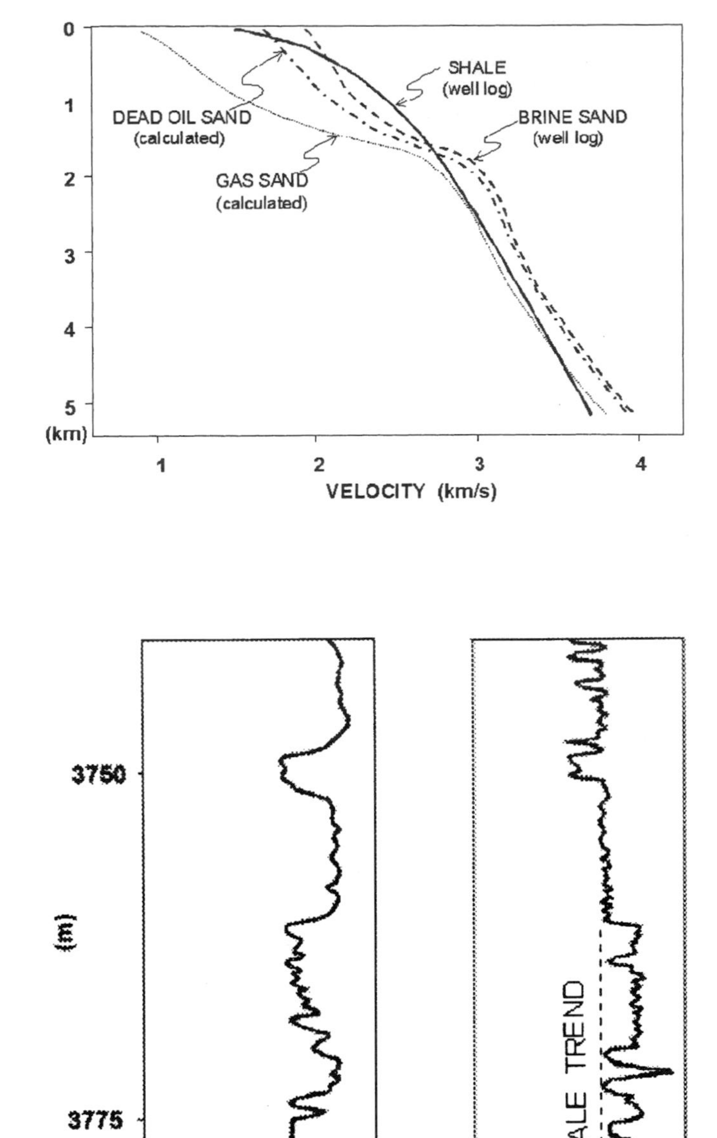

Figure 8—General sand-shale velocity relationships for the Gulf of Mexico under normal pressure conditions (Gardner et al., 1974). At depths below about 1700 m, shales have higher velocities than brine sands. Below this depth, brine-saturated sands are generally faster.

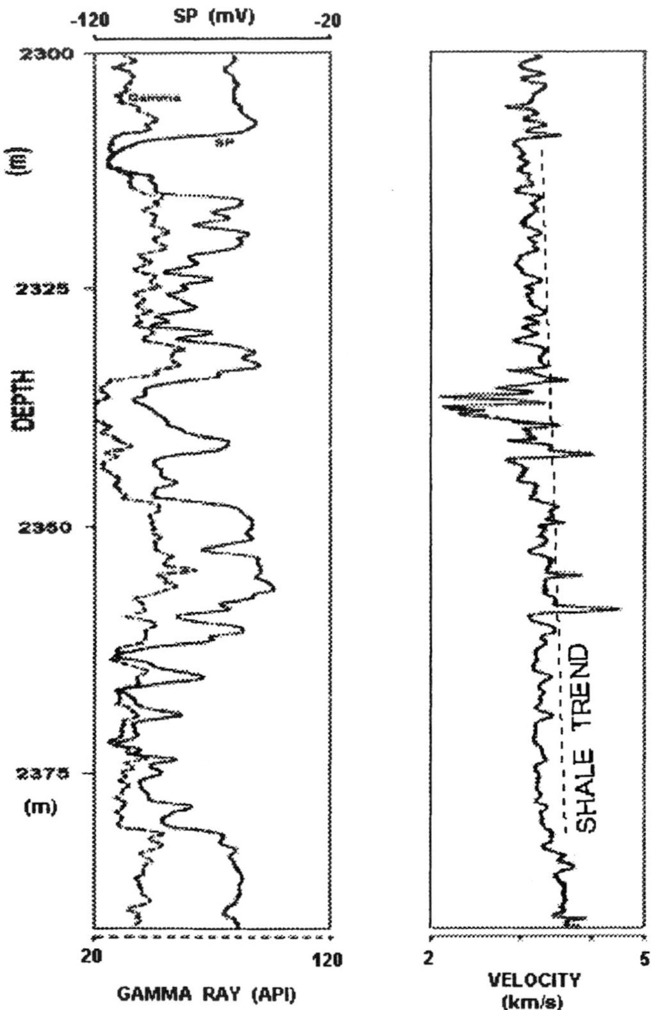

Figure 9—Gamma ray, SP, and sonic logs through a shallow, brine-saturated sand on the continental shelf, offshore Louisiana. These sands have lower velocity than adjacent shales, and hydrocarbon saturation would increase this velocity difference.

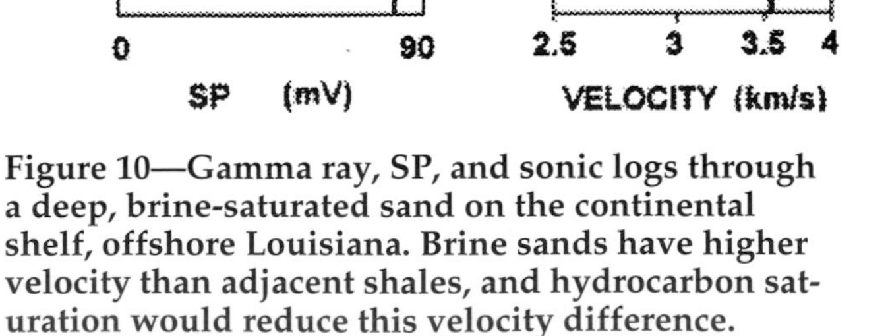

Figure 10—Gamma ray, SP, and sonic logs through a deep, brine-saturated sand on the continental shelf, offshore Louisiana. Brine sands have higher velocity than adjacent shales, and hydrocarbon saturation would reduce this velocity difference.

reflector. Back-stepping channel complexes in slope-centered fans that onlap the shelf margin are poorly resolved. Noise is added to the section to mimic coventional data (Figure 7B).

OUTCROP MODEL CASE STUDIES

Our mapped architecture (Figures 3, 4) forms the base case for substituting rock properties from three other deepwater reservoir settings. Two examples are from the mud-rich Mississippi fan complex. These are compared with sand-rich fan complexes in the Paleocene fill of the North Sea Basin. Two examples representative of deeper burial GOM (Gulf of Mexico) deepwater reservoirs are examined; these settings contain normal and overpressured deepwater reservoirs.

Normal-Pressure Gulf of Mexico Reservoirs

Sandstone and shale properties vary widely in the Gulf of Mexico basin. Reservoir sandstones have velocities that may be higher and lower than encasing shale, which can produce a positive, negative, or no reflection. Figure 8 shows the general trend for normally pressured Miocene sandstone and shale in the GOM (Gardner et al., 1974).

Offshore Louisiana contains a thick Tertiary succession of relatively unconsolidated sediment. In the South Marsh Island area, shelf sandstones are channelized and pond behind subtle topographic highs of deeper salt domes. Brine-saturated sandstones at moderate depths (~500–1500 m) have lower compressional velocities and impedance than shale (Figure 9). Hydrocarbon saturation increases the velocity contrast at shallow depth. With increasing depth, however, this relationship changes and sandstones can have a higher compressional velocity than shale (Figure 10). At greater depths, hydrocarbon saturation decreases sandstone density and velocity. This brings sandstone density and velocity closer to shale and eliminates the reflection.

The seismic model uses deeply buried but normally pressured GOM rock and fluid properties (Figure 10). Sandstone is populated with GOM sand properties, and shale properties are substituted for siltstone and carbonate facies in the model. This produces a much

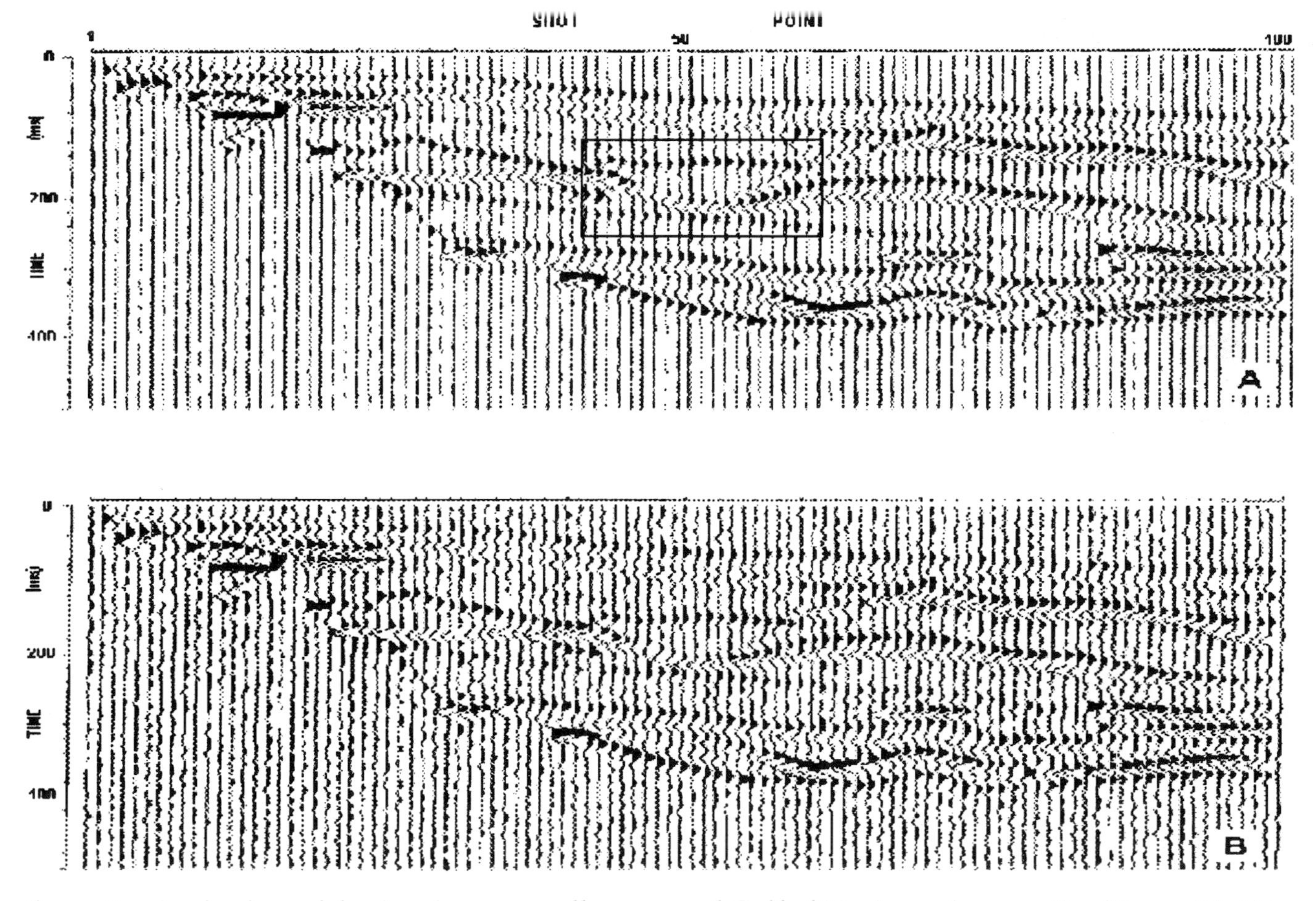

Figure 11—A seismic model using deep, normally pressured Gulf of Mexico rock properties (Figure 10). (A) Noise-free; note that sand "C" (outline) is still clearly visible and slower velocities have stretched the image in two-way transit time. (B) –20 dB noise added.

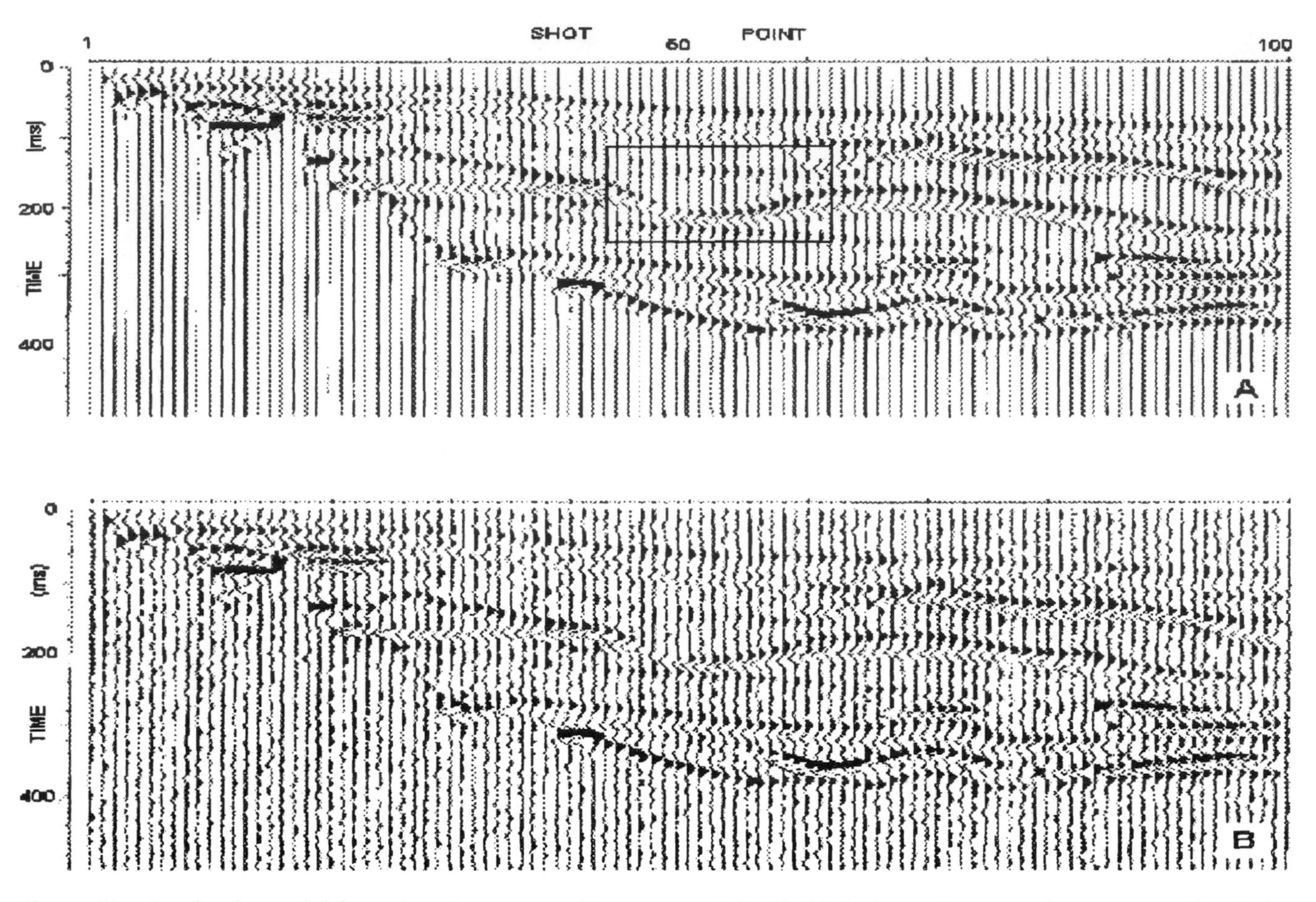

Figure 12—A seismic model for using deep, normally pressured Gulf of Mexico rock properties except with a light oil saturating the outlined sand. (A) Noise-free. Reduced velocities in this sand at normal pressured conditions have reduced the reflection coefficient. A "dim spot" results. (B) –20 dB noise added. With increased noise, this oil channel sand is no longer detectable (compare box with Figure 11). The reflection appears to be a continuation of the siltstone internal clinoform.

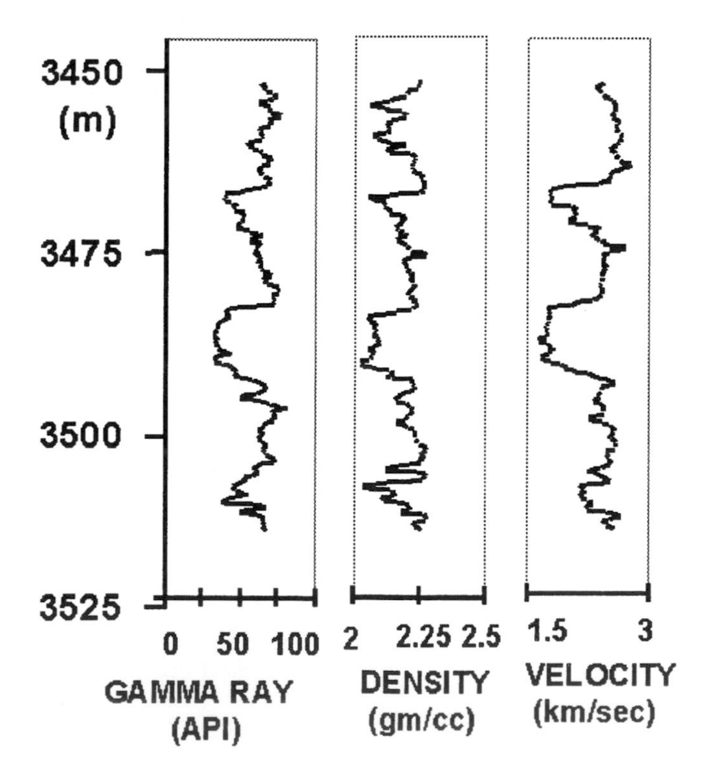

Figure 13—Gamma-ray, density, and sonic logs through the deepwater reservoir sands at Mississippi Canyon block 803, Mars field (after Chapin et al., 1995). These hydrocarbon-saturated sands have very low density and velocity despite the depth due largely to geopressure and lack of diagenesis.

smaller contrast and subdued boundary reflection at the basal unconformity (Figure 11A, B). Channel complex sand-body reflectors, however, resemble the model using outcrop properties. The ringing signal due to topset thinning in the lower right and onlap thinning in the upper left portions of the section are apparent. Reflectors representing the back-stepping channel complex in the upper left of the model are enhanced because of the reduced amplitude of the uppermost reflector. Reflection characters and overall sedimentary geometries are still visible when noise is added (Figure 11B). This enhancement is partly because strong reflectors reduce muting caused by AGC. In addition, lower overall velocities result in a relative time stretch, which has an effect on resolution of increasing the frequency.

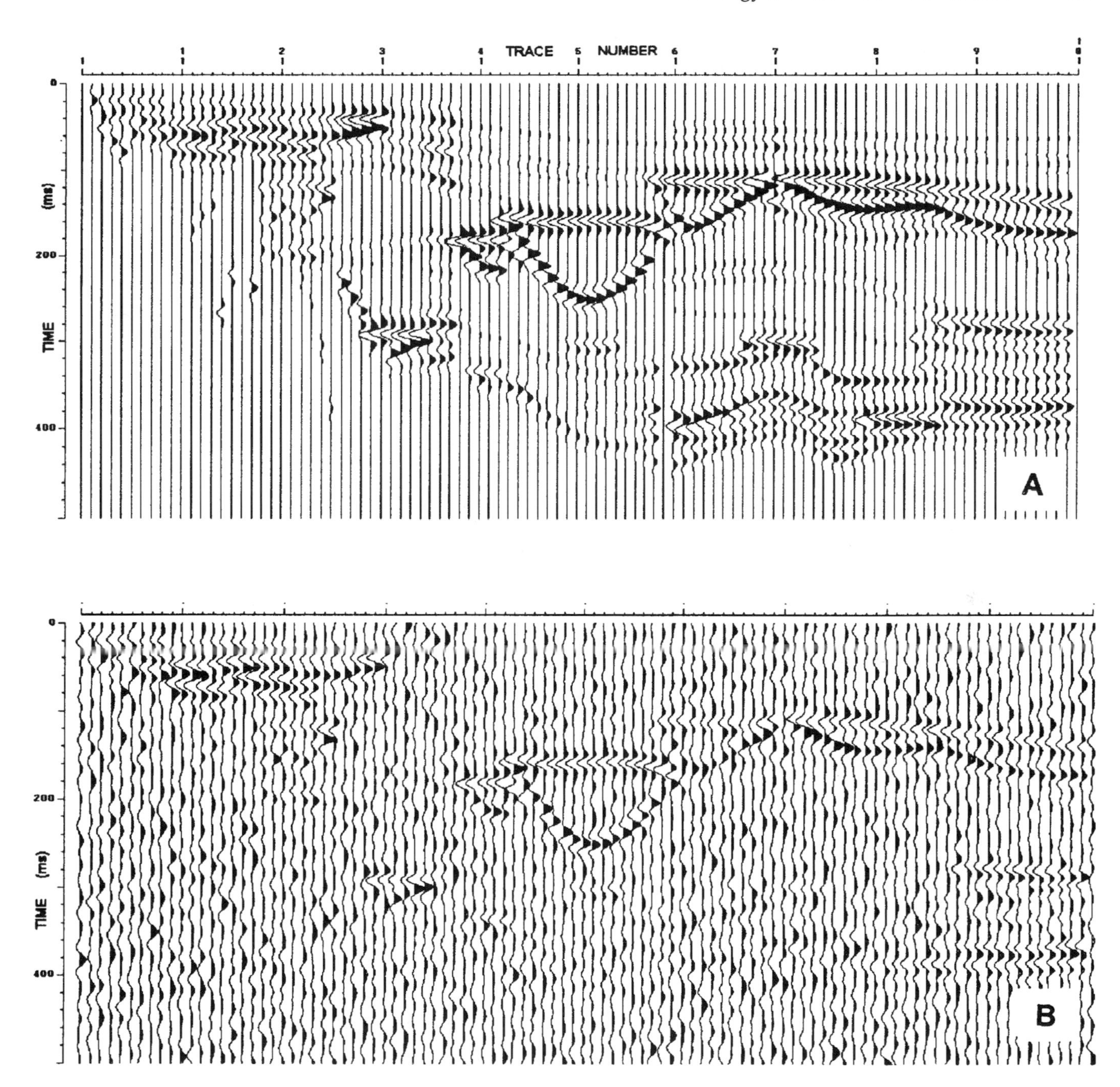

Figure 14—A seismic model using deep but overpressured Gulf of Mexico rock properties from the Mars field (Figure 5B). (A) Noise-free. Low velocities and high-impedance contrasts make the sand bodies obvious. Because of the low sand velocities, the reflection has reversed sign (now negative) and the channel is significantly stretched in time. (B) –20 dB noise added. Despite increased noise, sand is easily detected.

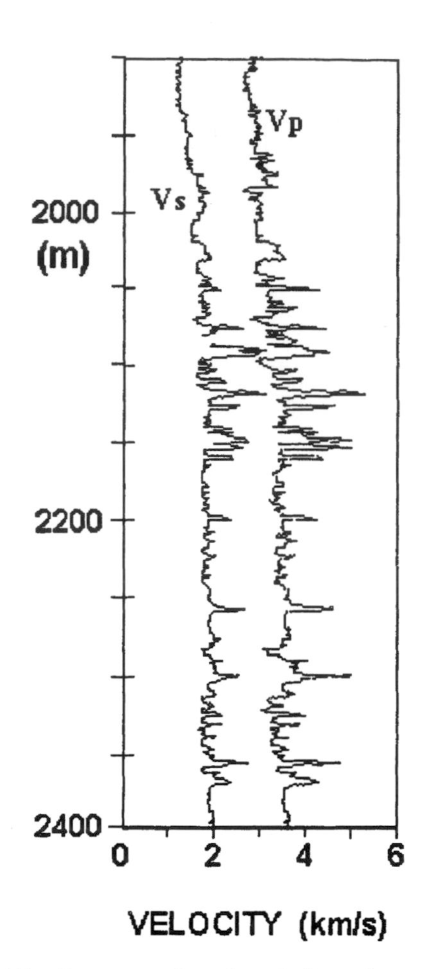

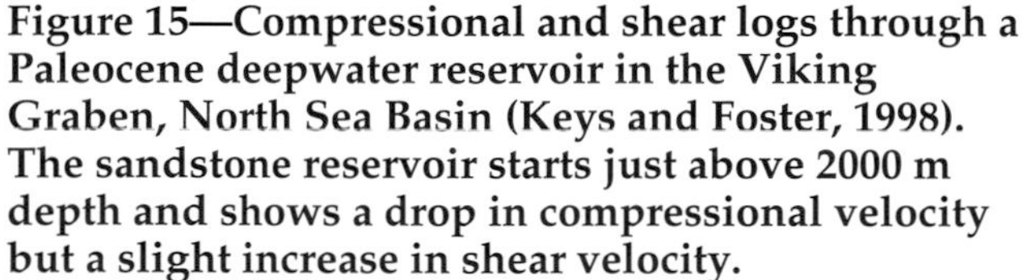

Figure 15—Compressional and shear logs through a Paleocene deepwater reservoir in the Viking Graben, North Sea Basin (Keys and Foster, 1998). The sandstone reservoir starts just above 2000 m depth and shows a drop in compressional velocity but a slight increase in shear velocity.

The influence of hydrocarbons on seismic resolution is assessed through simple fluid substitution. Velocities for typical brine-saturated sandstone are slightly greater than surrounding shales at normal pressures for this depth (Figure 10). Either gas or light, high gas-oil ratio (GOR) oils decrease sand velocity and density almost to the level of encasing shale. Reducing the impedance contrasts produces a "dim spot," or low-amplitude reflector. This is illustrated by sand body "C" (Figure 12A), which was resaturated with hydrocarbons in another model run (Figure 12A, B). The sand-body top reflection is a weak "dim spot" when compared to adjacent brine-saturated sandstone. This hydrocarbon indicator is the opposite of "bright spots" that develop shallower in the section, where low velocities of brine-saturated sandstone are further reduced by hydrocarbons.

Overpressured Gulf of Mexico

To assess the effect of geopressure or high pore pressures, a seismic model was constructed using properties from reservoir sandstones buried to 3.5-km depth at Mars field, in Mississippi Canyon block 803 (figure 13; Chapin et al., 1996). As in the last example, both siltstone and carbonate are replaced with shale velocities (Figure 12).

The combination of overpressure and minimal diagenesis preserves high porosities in sandstones of very low density, velocity, and impedance. In this case, sand densities and velocities are further depressed because of saturation by a high GOR oil. The high-velocity contrast between shale and sandstone, and the overall low velocities are distinctive (almost the same as pure water 1.5 km/s; Figure 13) and out of alignment with burial pressure trends (Figure 8). The overpressured sandstone velocities (1.8 km/s) are substantially lower than those of the encasing shale (2.6 km/s). This produces a large negative contrast between the reservoir sandstone and encasing shale, and produces a strong bright spot even at great depths.

A strong contrast between shale and sandstone produces very strong reflections. Sand body "C" is populated with hydrocarbon-saturated sandstone velocities (Figure 14A, B). Sand-body reflections are strong and easily discernible even in the noise-added seismic section. Because a seismic wave takes longer to pass through these low-velocity rocks, sand bodies are stretched in time and internal channel architecture is resolved. The sand bodies are "bright" at depths that under normal pressures would produce subdued reflections (Figures 11, 12). Brine-saturated sand bodies also have higher velocities and show strong reflections, so that amplitude cannot be used to distinguish brine- from gas-saturated sandstone.

North Sea Basin

The North Viking Graben contains thick, sand-rich, Paleocene deepwater clastics (Keys and Foster, 1996). These sandstones and shales have almost identical velocity distributions (2.6 to 4 km/s; Figure 5C). Compressional (Vp) and shear (Vs) velocity logs show little contrast between sandstone and shale (Figure 15). The strongest reflectors are produced by high-density and low-porosity carbonate-cemented sandstone. These cemented zones and calcite stringers produce the strongest impedance contrasts and inhibit direct sand detection. Hence, the maximum impedance contrast is commonly not at sandstone/shale contacts. These high-velocity, calcite-rich lenses are represented by alternating high- and low-impedance beds in the upper onlap and middle toplap areas of seismic model.

In high-velocity sandstone, resolution is reduced because transit time is contracted through the channel sand body. Reflections of channel sand bodies are also muted because of the lower velocity contrast with encasing shale. Consequently, discrete channel complexes are represented by a single wavelength, and reflection amplitudes are inconsistent and both negative and positive. Hence, sand-body architecture is difficult to distinguish in this setting even when hydrocarbon saturated (Figure 16B).

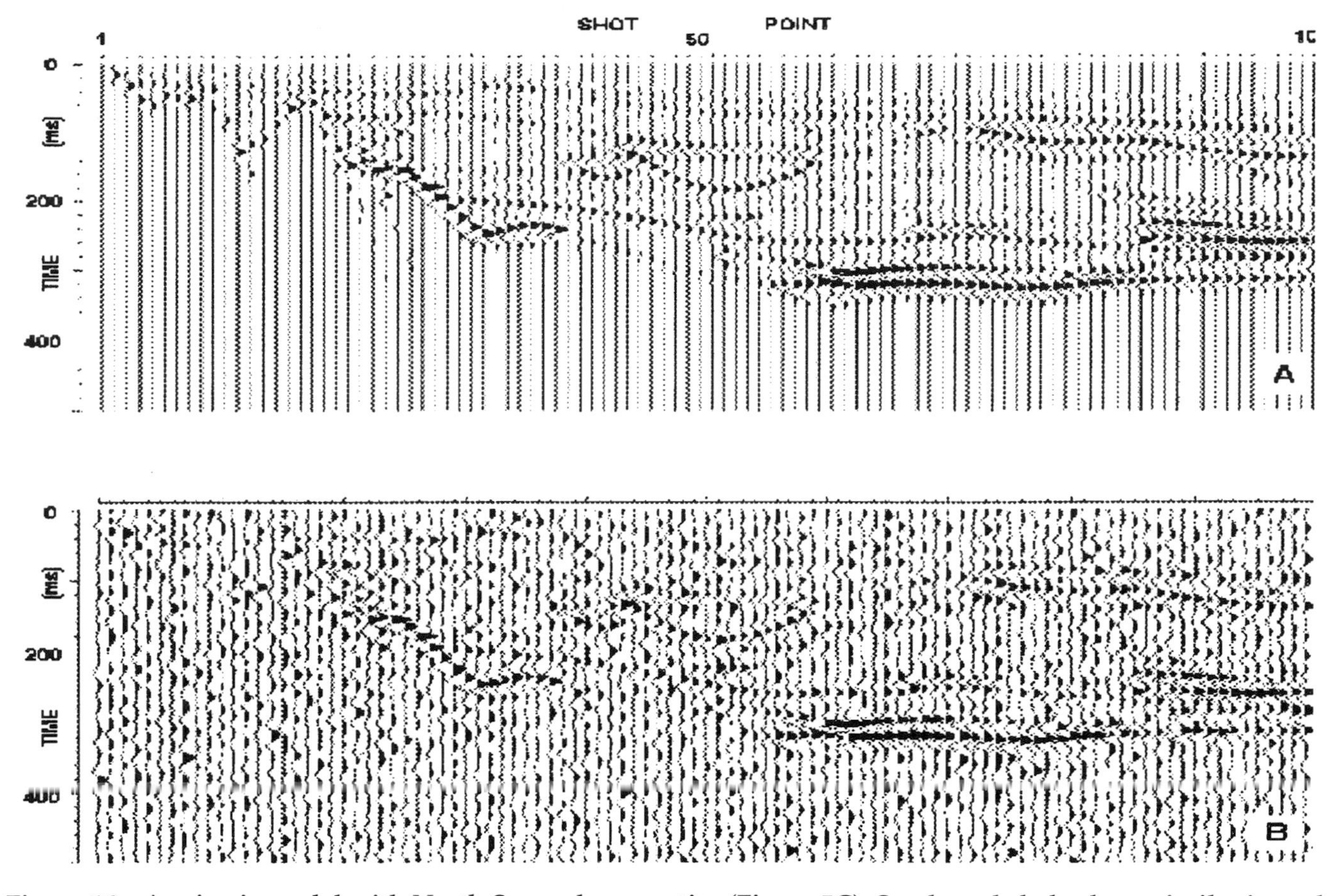

Figure 16—A seismic model with North Sea rock properties (Figure 5C). Sands and shales have similar impedances and sand body reflections are weak. (A) Noise-free; the strongest reflections arise from carbonate-rich beds represented in the lower right of the model. (B) –20 dB noise added. Sand bodies are difficult to distinguish from background shale. Different attributes, such as amplitude vs. offset are needed.

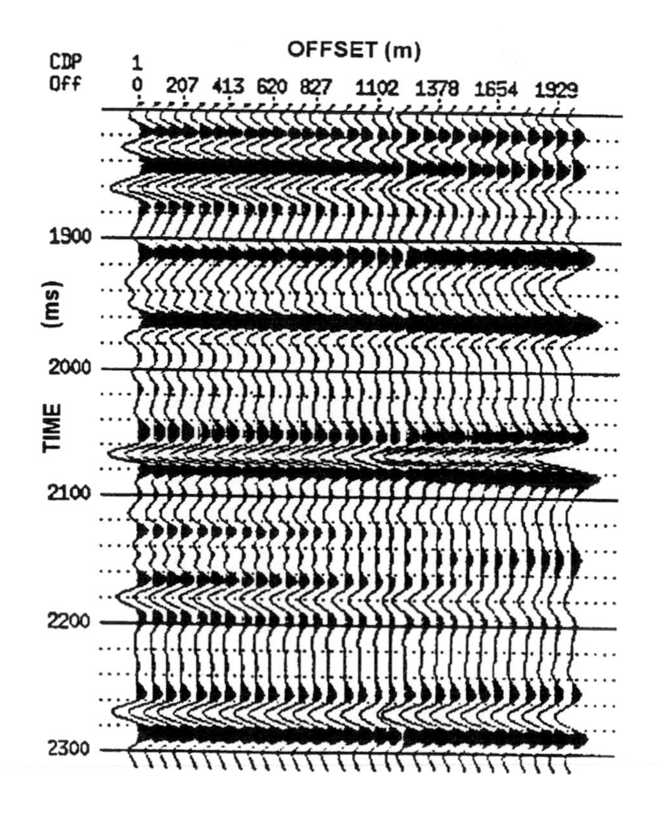

Figure 17—Modeled prestack gather based on the North Sea velocities (Figure 15). Just above 2000 ms, a strong increase in amplitude with offset is seen.

We can expand this analysis to include seismic attributes that detect the presence of hydrocarbons. Amplitude vs. offset (AVO) analysis of a prestack seismic gather looks for changes in compressional reflection amplitude as a function of distance from the source. Reflection amplitude depends on compressional and shear velocities (Vp, Vs) and the incident angle of the seismic wave. Hydrocarbon-bearing sandstone at 1980-m depth show opposite responses on the Vp vs. Vs logs: Vp drops as Vs increases. A decrease in Vp/Vs ratio increases the far-offset (or large-angle) reflection amplitudes. Because hydrocarbons drop Vp but leave Vs unchanged, the Vp/Vs ratio decreases in hydrocarbon-saturated sandstones and produces an increase in AVO. A calculated prestack gather using GOM logs (Figure 10) is shown in Figure 17. The reflector above 2000 ms shows a strong amplitude gain, with offset from left to right. The actual common midpoint gather (CMP) from this location shows a strong AVO response

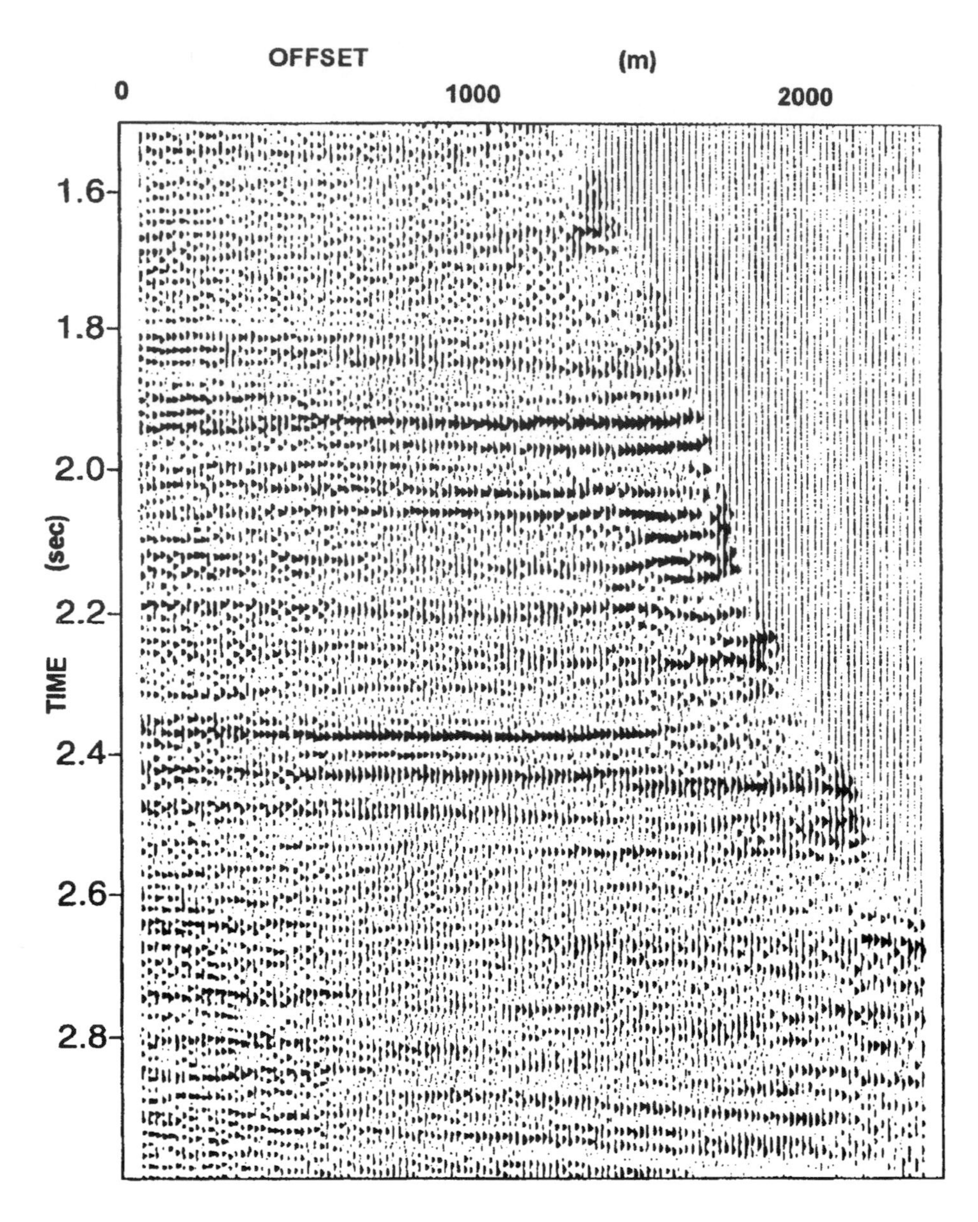

Figure 18—Prestack common midpoint gather (CMP) recorded at the North Sea well location (Keys and Foster, 1998). Although the typical stacked reflection was weak, this hydrocarbon-saturated deepwater sand (>2000 ms) does give a strong amplitude vs. offset signature.

(Figure 18; leftmost traces are strongest). Near-trace/far-trace subtraction techniques resolve gas-saturated sandstone from noise in the section.

DISCUSSION

Comparison of these various seismic models demonstrate the strong control that lithology, pore fluid, depth, and pressure have on seismic signature. These physical factors combine to control the resolution of deepwater sand bodies (Figure 1). The brighter, red-shaded region in Figure 1 indicates the optimum conditions for resolving sandbodies. Figure 1 shows that large, clean, gas-saturated, and overpressured sandstone has the highest resolution. Low sand/shale contrasts produce weak reflections, and hydrocarbon saturation that can produce low amplitudes (dim spots). Velocities and impedances are lower in sandstone saturated with gas, or light, high gas-oil ratio oils. Reservoir pressure has a dramatic effect on seismic resolution. Higher than normal pore pressures often arrest compaction and produce abnormally high-porosity sandstone at great depths.

Scale is another important factor governing seismic resolution. Although often mapped and termed channels, the seismically resolvable channel form feature in all models is a compound sand body consisting of deposits recording multiple depositional events. The size of these slope channel complexes reflects their confinement within a larger depression and does not represent the scale of the depositing flow. Hence, relating facies scale depositional processes to seismic-scale bodies is considered untenable.

Seismic resolution of deepwater architecture is also dependent on the noise level produced by surface heterogeneities. When noise is added to mimic conventional data the effect lowers the resolution of individual sand bodies (Figure 7B). For example, the correlation of isolated sand bodies shown in Figure 7B seems plausible. The feature at shotpoint 60 might erroneously be interpreted as a fault-displacing sandstone about 60 ms to the left. This interpretation is supported by an apparent similar offset in the lower reflector at 360 ms. Such an interpretation would predict a reservoir that is compartmentalized by faults. The alternative and correct interpretation shows sandstone architecture related the stratigraphic stacking pattern of deepwater channel complexes in a prograding slope succession.

CONCLUSIONS

Seismic response for deepwater clastics can vary substantially, ranging from bright positive or negative reflections to nearly invisible. Primary factors include (1) size and geometry, (2) lithology and property contrast, (3) pore fluids, and (4) pore pressure. Signal frequency content, seismic velocity, and internal sand-body structures will modify expression. The impedance contrast with the surrounding lithologies determines the magnitude of the reflection but is not constant or simple for the same lithologies. Hydrocarbon saturation lowers sand impedance, but reflection strength may weaken depending on the properties of the surrounding material. Increased pore pressure generally increases sand body reflections and enhances the hydrocarbon response.

ACKNOWLEDGMENTS

Support was provided by members of the Slope and Basin Consortium. The members of the consortium are listed in Gardner and Borer, this volume. Mark Sonnenfeld helped conduct fieldwork and Brad Sinex assisted in building the first seismic models. Phillips Petroleum provided subsurface data from Cabin Lake field. Fred Armstrong, resource manager at Guadalupe Mountains National Park, was extremely helpful in providing logistics support and park permits to conduct this work.

REFERENCES CITED

Chapin, M. A., G. M. Tiller, and M. J. Mahaffie, 1996, 3-D architecture modeling using high-resolution seismic data and sparse well control: an example from the Mars "Pink" reservoir, Mississippi Canyon area, Gulf of Mexico, *in* P. Weimer and T. L. Davis, eds.: AAPG Studies in Geology 42 and SEG Geophysical Development Series 5, AAPG/SEG Tulsa, p.123–132.

Gardner, G. H. F., L. W. Gardner, and A. R. Gregory, 1974, Formation velocity and density—the diagnostic of stratigraphic traps: Geophysics, v. 39, p. 770–780.

Keys, R., and D. Foster, 1998, Comparison of seismic inversion methods on a single real seismic data set (1994 open workshop results): Society of Exploration Geophysicists, in press.

Stafleu, J., and M. D. Sonnenfeld, 1994, Seismic models of a shelf-margin depositional sequence: upper San Andres Formation, Last Chance Canyon, New Mexico: Journal of Sedimentary Research, v. B64, p. 481–499.

Vail, P. R., R. M. Mitchum, Jr., R. G. Todd, J. M. Widmeir, S. Thompson III, J. B. Sangree, J. N. Bubb, and W. G. Hatlelid, 1977, Seismic stratigraphy and global changes of sea level, *in* C. E. Payton, ed., Seismic stratigraphy—applications to hydrocarbon exploration: AAPG Memoir 26, p. 49–212.

Browne, G. H., R. M. Slatt, and P. R. King, 2000, Contrasting styles of basin-floor fan and slope fan deposition: Mount Messenger Formation, New Zealand, *in* A. H. Bouma and C. G. Stone, eds., Fine-grained turbidite systems, AAPG Memoir 72/ SEPM Special Publication No. 68, p. 143–152.

Chapter 13

Contrasting Styles of Basin-Floor Fan and Slope Fan Deposition: Mount Messenger Formation, New Zealand

Greg H. Browne
Institute of Geological and Nuclear Sciences
Lower Hutt, New Zealand

Roger M. Slatt
Department of Geology and Geological Engineering, Colorado School of Mines
Golden, Colorado, U.S.A.

Peter R. King
Institute of Geological and Nuclear Sciences
Lower Hutt, New Zealand

ABSTRACT

Late Miocene Mount Messenger Formation exposures in north Taranaki, New Zealand, demonstrate contrasting styles of deepwater basin-floor fan and slope fan development. Some of these attributes may have analogs in subsurface thin-bedded, deepwater reservoirs.

Basin-floor fan settings are characterized by thick-bedded sandstone lithofacies (central lobe) and thin-bedded sandstone/siltstone lithofacies (lobe fringe). The thick- and thin-bedded sandstones were deposited by high-density mass flows. Stratigraphically higher slope fan units are invariably thin bedded. They display scouring at various scales and well-developed sedimentary structures that are indicative of deposition by turbidity flows. The slope fan depositional settings include individual and nested channels, and vertically stacked and shingled levee complexes.

INTRODUCTION

Present-day coastal exposures in north Taranaki, New Zealand, display a middle and late Miocene deepwater sedimentary succession more than 2 km thick. These strata dip 2–10° in a southwest direction, almost parallel to the orientation of the cliff transect (Figure 1; CD-ROM Figure 1). The succession is predominantly siliciclastic, although volcaniclastic detritus occurs throughout the section, particularly in the basal 100 m. The middle portion of the overall succession is dominated by siliciclastic sandstones, assigned to the Mount Messenger Formation (King et al., 1993). In the coastal section this formation is about 800 m

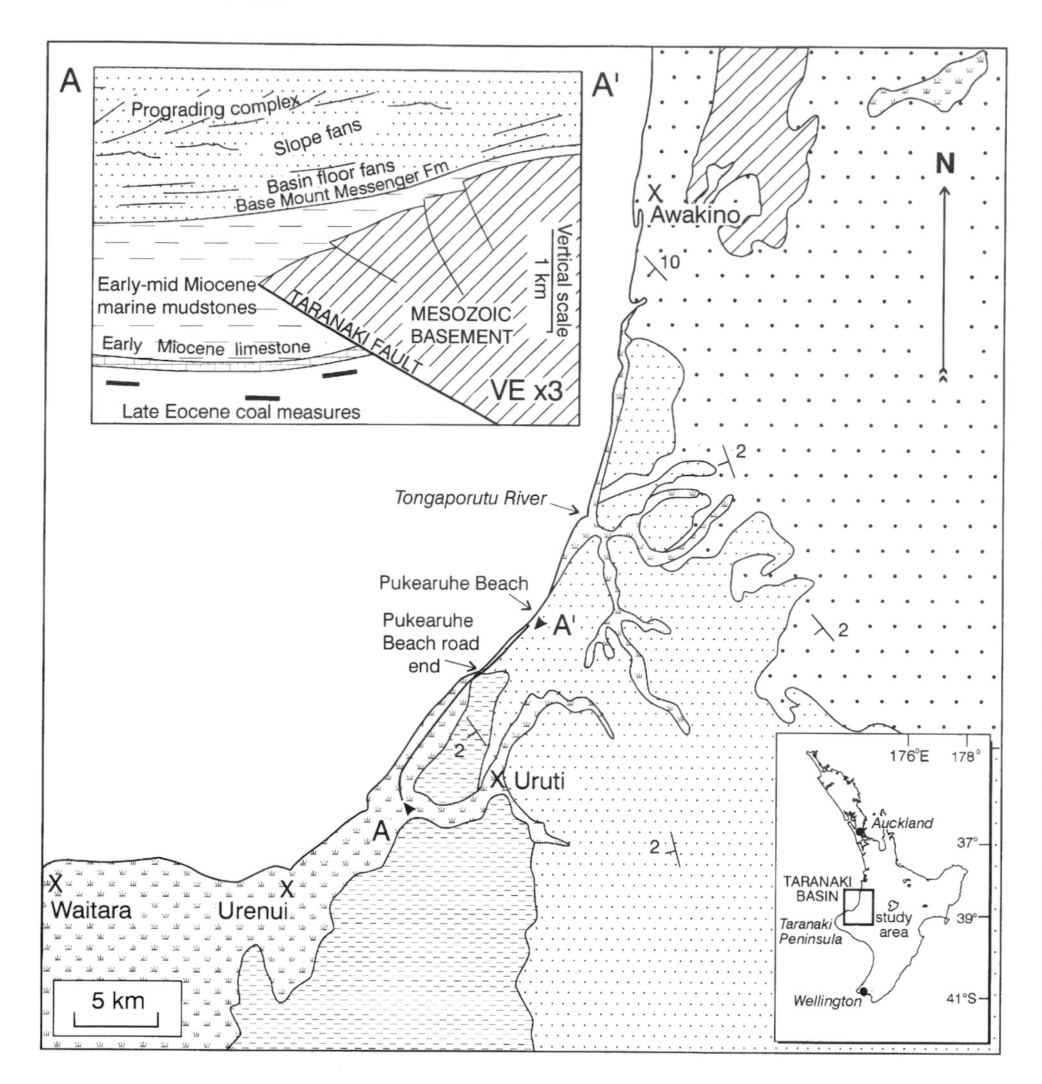

Figure 1—Generalized locality map, showing the distribution of the Mount Messenger Formation (and other units) in the vicinity of the north Taranaki coastal section. The dense stipple pattern on the map is the Mount Messenger Formation, the dashed line pattern is the Urenui Formation. Inset map shows the North Island of New Zealand. Inset cross section A-A′ shows a generalized interpretation of subsurface stratigraphy, based on a seismic reflection profile (PR83-14) located parallel to the coastal section. The third-order late Miocene succession consists of a basin-floor fan to slope fan interval (Mount Messenger Formation) and a prograding complex (middle to upper slope) interval (Urenui Formation).

thick, and has a late Miocene age. This Mount Messenger interval forms part of a third-order progradational lowstand succession, but internally comprises a series of fourth-order depositional cycles (King et al., 1994; Diridoni, 1996; King and Thrasher, 1996). The stacking architecture of these fourth-order cycles is such that different facies components of the third-order aggradational and progradational lowstand system are evident at different stratigraphic levels in the section (Figure 2). Stratigraphically lower fourth-order cycles are dominated by basin-floor fan elements, which pass upsection into cycles with abundant slope fan deposits. Overlying this entire succession is another series of fourth-order cycles that are dominated by prograding complex siltstones and assigned to the Urenui Formation (Figures 1, 2).

This paper presents some of the general stratigraphic and petrophysical attributes of Mount Messenger Formation strata exposed along the north Taranaki coastline, between Tongaporutu River mouth and Pukearuhe Beach road end (Figure 1). This exposure is particularly useful in providing an analog for assessing geometry and performance of subsurface Mount Messenger reservoirs that produce oil and gas in the Ngatoro and Kaimiro fields in northern Taranaki Peninsula (Figure 1, inset map; CD-ROM Figure 2). Mount Messenger outcrop data may also be applicable to fine-grained, thin-bedded, deepwater sandstone reservoirs elsewhere, such as the Ram/Powell and Mahogeny fields in the Gulf of Mexico (Browne and Slatt, 1997; Coleman et al., 1998).

BASIN-FLOOR FAN FACIES ASSOCIATION

Basin-floor fan depositional units, containing lower to middle bathyal foraminiferal microfauna, are well represented in the coastal section from near Awakino in the north to just south of Tongaporutu (Figure 1; King et al., 1993). The overall basin-floor succession comprises a series of fining-upwards cycles, each of fourth-order duration. Three main facies associations are recognized within each cycle, according to the dominant lithology

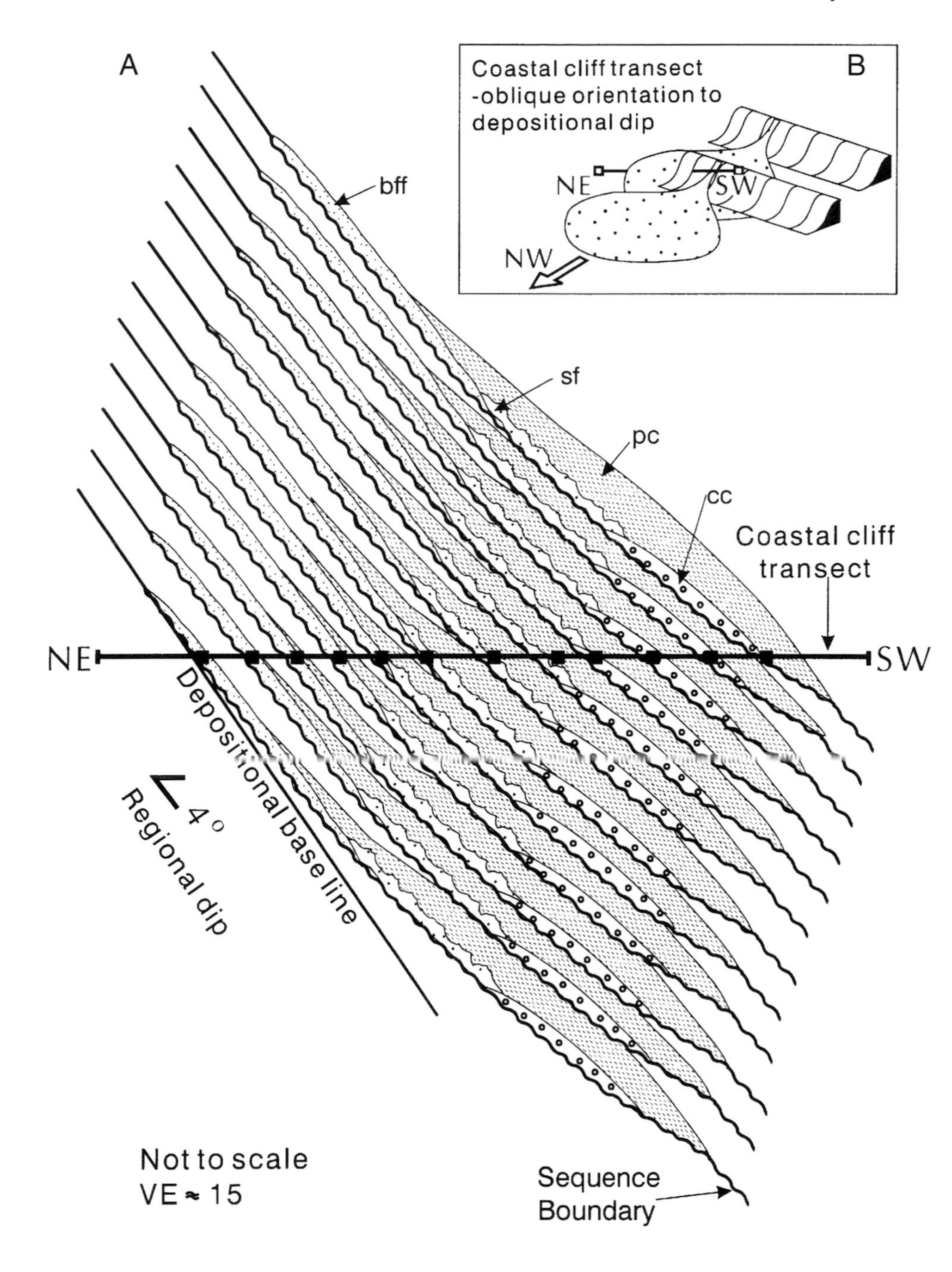

Figure 2—(A) Conceptual model of fourth-order sequences within a third-order aggradational and progradational succession exposed along the north Taranaki coast section. Original depositional dip was to the northwest, and strata have since been uplifted and tilted down to the southwest. Outcrop localities are schematically indicated by the black squares. Oldest cycles (to the northeast) intersected by the cliff transect are dominated by basin-floor fan (*bff*) deposits and immediately overlying cycles by slope fan (*sf*) deposits (all Mount Messenger Formation). Uppermost cycles (to the southwest) are dominated by prograding complex (*pc*) slope siltstones, interspersed with canyon complex (*cc*) conglomerates (all Urenui Formation). (B) Paleodepositional perspective of the oldest fan (in NE) to youngest slope (in SW) cycle exposed along the present-day coastal cliff transect.

and bedding style. In stratigraphically ascending order these are: thick-bedded sandstone, thin-bedded sandstone and siltstone, and siltstone. The lower two sand-rich facies associations are summarized here, but are described in greater detail by King et al. (1993, 1994), Jordan et al. (1994), Browne et al. (1996), Browne and Slatt (1997), and Browne and Slatt (in press).

Thick-Bedded Basin-Floor Fan Units

Thick-bedded sandstones occur at the base of each fourth-order cycle within the stratigraphically lower part of the Mount Messenger section (Figure 2). Individual sandstone beds range in thickness up to about 3 m (Table 1), although the thickest packages often consist of one or more amalgamated beds (Figure 3). The sandstones are overtly massive, moderately well sorted, fine to very fine grained, and display laterally continuous sheetlike geometries (Figure 3). Beds generally have sharp bases and tops, and low-amplitude scours feature at the bases of some beds. Although massive bedding is most typical, some beds show faint parallel to convolute lamination, ripple lamination, loading, dish structures, and flame structures. Intervening siltstone beds are much thinner (generally a centimeter thick), and are often tuffaceous and very bioturbated. Overall, sandstone dominates siltstone by a ratio of approximately 9:1 within the thick-bedded intervals.

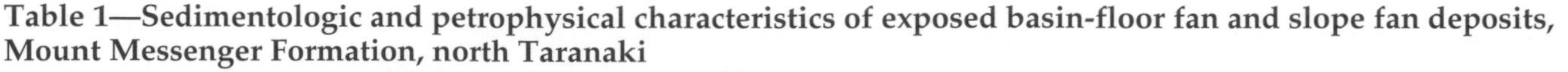

Table 1—Sedimentologic and petrophysical characteristics of exposed basin-floor fan and slope fan deposits, Mount Messenger Formation, north Taranaki

Sedimentary Feature	Basin-Floor Fan: Thick-Bedded Facies	Basin-Floor Fan: Thin-Bedded Facies	Slope Fan
Sandstone bed thickness	<3 m (typically <1.0 m) Often amalgamated.	<0.6 m (typically <0.2 m)	<1.2 m (typically <0.3 m)
Siltstone bed thickness	<0.15 m (typically <0.05 m)	<0.4 (typically <0.1 m)	<1.5 m (typically <0.1 m)
Sandstone:siltstone ratio	9:1	7:3 to 6:4	Variable: 8:2 to 2:8
Sandstone grain size	Fine to very fine	Fine to very fine	Fine to very fine
Sandstone sorting	Moderately well sorted	Moderately well sorted	Moderately well sorted
Sandstone sedimentary structures	Abundant massive sandstone (T_a). Tops of some beds rippled	Abundant massive sandstone (T_a). Common parallel (T_b) and rippled (T_c) beds	Abundant parallel (T_b), ripple and climbing ripple (T_c) beds with minor massive sandstone (T_a)
Permeability range	100–800 md	20–600 md	20–750 md
Porosity range	20–35%	25–35%	25–35%

Figure 3—Thick-bedded basin-floor fan sandstones, with thin siltstone interbeds (light colored), Mount Messenger Formation, Tongaporutu Beach. A 3-m-thick amalgamated bed is present in the middle of the photograph.

Thin-Bedded Basin-Floor Fan Units

Thin-bedded units occur stratigraphically above the thick-bedded units in each basin-floor fan cycle (Figure 4). Individual sandstone beds are up to 60 cm thick (Table 1), but otherwise their texture and overall appearance are similar to the thick-bedded sandstones described above (Figure 3) (CD Figure 3). Some parallel, convolute, ripple, and climbing-ripple lamination is present, particularly at the top of individual beds. Bases and tops of beds are sharp and generally planar, although some beds show low-angle scours at their base. Siltstone interbeds comprise approximately 40% of the total lithofacies thickness.

At first glance, bedding appears to be continuous and of uniform thickness, although topographic compensation-style bedding is sometimes evident (Figure 4). However, detailed measurements and correlations have revealed that individual beds within the thin-bedded association exhibit dramatic thickness variations over relatively short distances (Browne et al. 1996). For example, thickness

Figure 4—Thin-bedded sandstone and siltstone basin-floor fan lithofacies, Mount Messenger Formation, Tongaporutu Beach. Several sandstone beds (dark colored) show depositional thinning or compensation-style geometries. Pale-colored siltstones at the top of the exposure mark the top of this particular fourth-order, fining-upward, basin-floor fan cycle.

variations of 15 sandstone beds and their associated sandstone partings along a 230-m outcrop transect are plotted in Figure 5. A few sandstone beds can be traced along the entire transect, but most pinch out in at least one direction within distances less than the overall transect length (CD-ROM Figure 3). Figure 6 presents the percentage of sandstone beds that can be correlated between various station pairings (amounting to 28 possible combinations between Stations 1 and 8), versus the horizontal distance spanned by each station pair. The plot shows a clear decrease in sandstone bed continuity with distance. Over a horizontal outcrop distance of 100 m, approximately 50% of the sandstone beds have disappeared, whereas over the total outcrop length of more than 200 m, almost all the sandstone beds have disappeared. The outcrop section is oriented roughly normal to the depositional trend, so these data should be considered as across fan-lobe parameters. From a reservoir management perspective, this depositional style has a significant effect on the stacking architecture, geometry, and lateral continuity of thin-bedded sandstones. We have no data on the lateral continuity of sandstone beds in a dip-parallel orientation.

Petrophysical Attributes

Petrophysical data measured from outcrop are limited to the area around Tongaporutu. Sandstone permeabilities, measured with a portable minipermeameter, are in the range 100–800 md in thick-bedded units, and 20–600 md in thin-bedded units (King et al., 1994; Browne et al., 1996). Porosities calculated from laboratory measurements of wet and dry bulk sample weights are consistently in the range 20–35% within both lithofacies associations (Table 1).

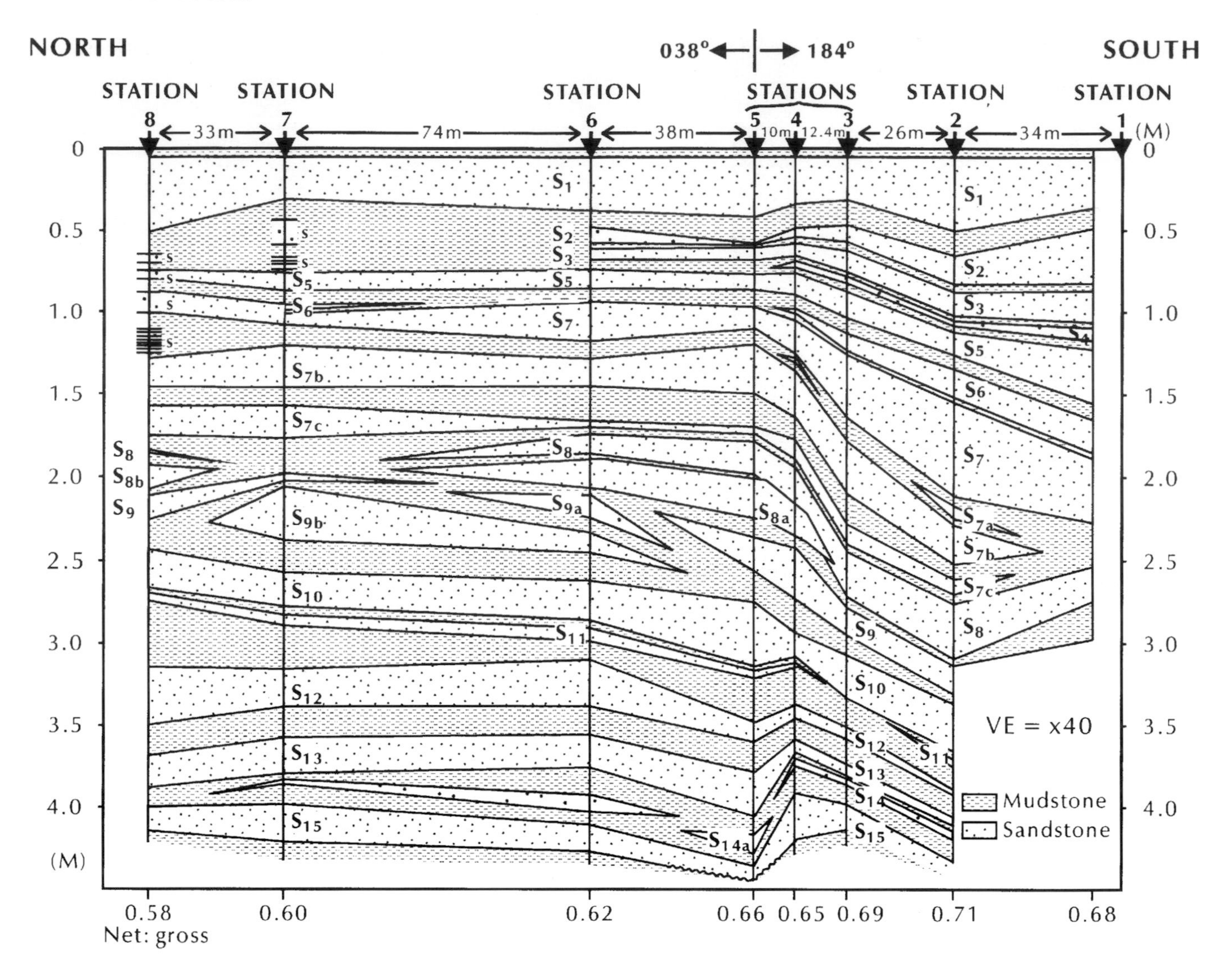

Figure 5—Lateral bed thickness relationships for a 230-m-long coastal-cliff exposure of thin-bedded basin-floor fan sandstones and siltstones, Mount Messenger Formation, Tongaporutu Beach (after Browne et al., 1996). Net sandstone-to-gross-thickness ratios are indicated at respective stations (1–8). Accessible exposure height was a little more than 4 m except in station 1. Sandstone beds were numbered downsection from S_1, correlating from south to north. The correlation datum is a thin mudstone above sandstone bed S_1.

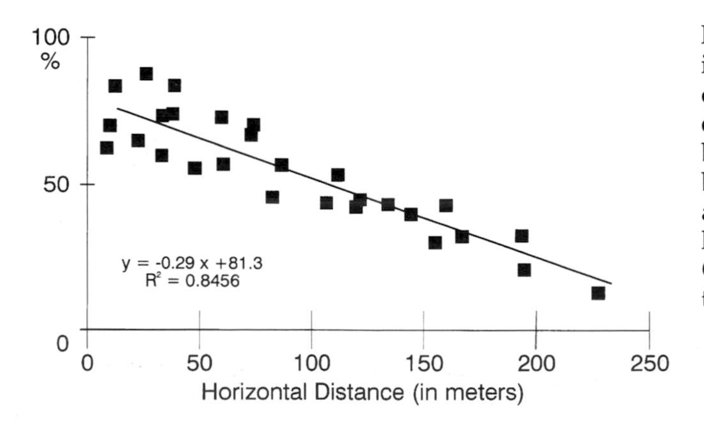

Figure 6—Graphical relationship between bed continuity and distance, for the thin-bedded sandstone beds depicted in Figure 5. The plot indicates the percentage of beds (from a total of 23) that can be correlated between pairs of stations along the cliff transect (i.e., between Station 1 and 2, Station 1 and 3, Station 1 and 4, and so on; 28 possible pair combinations), versus the horizontal distance spanned by each station pairing. Only three beds (~13%) are continuous across the entire transect length.

SLOPE FAN FACIES ASSOCIATION

At least two fourth-order cycles within the Mount Messenger Formation rest stratigraphically above the basin-floor fan cycles and are attributed to deposition on fans at or near the base of the slope (CD-ROM Figure 4). These cycles comprise about 300 m of the total Mount Messenger formational thickness and were deposited in middle to upper bathyal water depths (King et al., 1993). Two depositional associations are recognized within the slope fan succession, as outlined below.

Channel-Fill and Proximal Levee Units

The channel-fill and proximal levee association comprises thinly interbedded sandstones and siltstones contained within channels or interleaved with channels (Figures 7, 8). Depositional morphologies consist of vertically stacked or shingled packages, with discordant bedding orientations. Multiple nested scour and fill bodies are present in places and, less commonly, single-episode scour and fill channels. Complex lateral and vertical bedding relationships are evident where there is a high degree of scouring, and it

Figure 7—Large channel within the slope fan association, upper Mount Messenger Formation, Pukearuhe Beach. The channel was filled initially with siltstone and rare sandstone beds (stratigraphically lower 10 m of fill at right side), and by alternating sandstone and siltstone turbidite beds (center and left). Width of outcrop illustrated is about 25 m. Photograph by L. Homer.

Figure 8—Slope fan channel-fill and proximal levee units, upper Mount Messenger Formation, Pukearuhe Beach. Several nested or shingled channels infilled with sandstone-siltstone turbidite interbeds are illustrated; note the opposing attitude of bedding between different channels. The base of one of these channels is filled with a 2-m-thick matrix-supported conglomerate, containing shell debris and extrabasinal clasts (resistant knobbly ledge at beach level, at lower right by person).

can be difficult to distinguish between channel-fill and adjacent levee deposits. For this reason, we group these facies as one association.

Lithofacies typically consist of sandstone and siltstone turbidite beds with well-developed Bouma divisions ($T_{a\text{–}e}$). Within the sandstone portion of each bed, Bouma T_b and T_c divisions are prevalent, and T_a beds are generally less common. The sandstones generally grade upward into siltstone lithologies (T_d and T_e). *Scolicia* burrows are particularly common in climbing-ripple laminated beds (T_c), where they occur as escape burrows (fugichnia), sometimes with the progenitor echinoid preserved at the top of the burrow. Flame and dish structures are common.

In the largest channel exposed (Figure 7), the basal 10 m of fill material comprises bioturbated siltstone, and thin, centimeter-thick sandstone beds (CD-ROM Figure 5). The dominance of siltstone at the base of this channel appears to be unusual within this outcrop succession, and indicates that the channel was passively backfilled sometime after it was cut. Slightly lower in the succession, a localized debris flow conglomerate is present within an interval of nested channels (Figure 8). This conglomerate contains extrabasinal as well as perigenic clasts, and is interpreted to represent the distal toe deposits of flows debouching from a main feeder canyon at the foot of the slope.

Distal Levee Units

Distal levee units are distinguished from channel-fill and proximal levee units in that scour features are less common and of smaller scale, and siltstone is more abundant in relationship to sandstone (Figure 9). The term "distal" is relative, and is used here to imply deposition some distance (more than a few tens of meters) laterally from the coeval channel system. Sandstone facies are predominantly ripple-laminated,

Figure 9—Detail of distal levee units, slope fan association, upper Mount Messenger Formation, Pukearuhe Beach. This lithofacies comprises mainly bioturbated siltstone, with relatively thin, "wispy" interbeds of sandstone. Scale has 10-cm divisions.

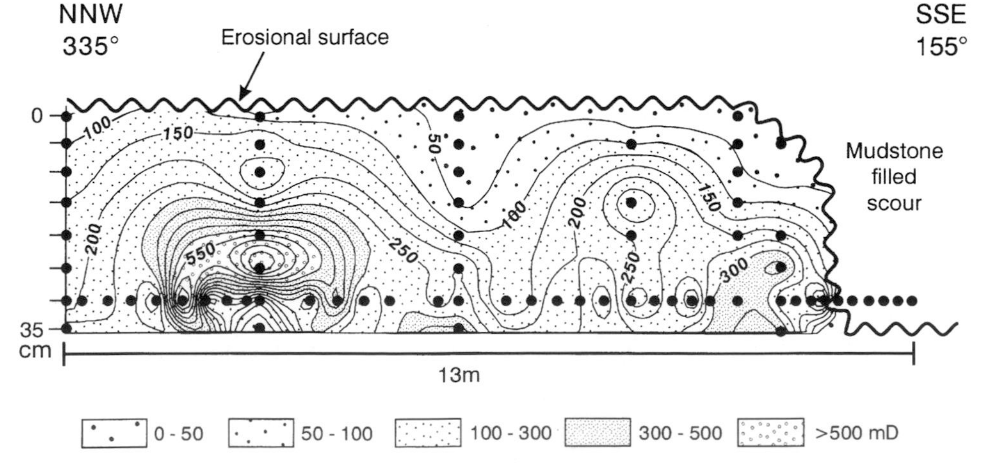

Figure 10—Contoured permeability plot for a 35-cm-thick sandstone bed, comprising Bouma divisions T_b to T_c, within the slope fan (channel-fill and proximal levee) association, upper Mount Messenger Formation, Pukearuhe Beach (note vertical exaggeration). Permeability values (in millidarcies) were obtained from core plugs (depicted by black dots) drilled into the outcrop to minimize the effects of salt (sea spray) crystallization in pore spaces.

and in particular climbing-ripple laminated (T_c); parallel laminated facies (T_b) are also common.

Petrophysical Attributes

Slope fan sandstone samples collected from several outcrop localities have laboratory-measured porosities between 25 and 35% (Table 1). Sandstone permeabilities were measured in detail in a representative, suitably thick (35 cm) and laterally continuous bed from within the channel-fill and proximal levee association (Figure 10; CD-ROM Figure 6). To minimize the effects of salt (sea spray) crystallization within surficial pore spaces, more than 70 core plugs were drilled into this bed over a horizontal distance of 13 m. Minipermeameter measurements were then made on these core plugs. Sandstone permeabilities range from 20 md to around 750 md, and average 200 md. Measured permeabilities were typically 300–750 md in the basal (T_b) portion of this bed, compared with 20-200 md in the overlying climbing-ripple laminated (T_c) portion of the bed. Apart from the vertical permeability trends, there is also considerable lateral variation in permeability within the bed. The lower left portion of the measured transect through the bed displays a concentric permeability distribution related to a flame structure injected into the base of the bed (Figure 10).

DISCUSSION: DEPOSITIONAL STYLE

The Mount Messenger Formation outcrop succession provides an excellent window for examining reservoir attributes of a variety of deepwater lithofacies. Particularly noteworthy are contrasts in bedding geometry, sedimentary structures, and sandstone/siltstone proportions between intervals deposited in basin floor and base-of-slope settings. The entire basin-floor to base-of-slope succession was deposited relatively rapidly, about 500–700 m/m.y. (King et al., 1993). High sedimentation rates are indicated in the slope fan interval in particular, by features such as *Scolicia* escape burrows, climbing ripples, and scour and fill morphologies.

The basin-floor fan bodies appear sheetlike, and in surrounding hill country, bluff-forming thick-bedded sandstone intervals can be traced for several kilometers. Sandstone bed amalgamation is common in thick-bedded intervals, whereas compensation-style geometry is common in thin-bedded intervals. Scours, where present, are chiefly expressed by broad, low-angle erosion surfaces. The marked thickness variations of individual sand beds within the thin-bedded fan intervals reflect small-scale autocyclic changes in sea-floor microrelief and local diversion of flow pathways as the fan lobes were built up.

The physical similarity between sandstones within the thick- and thin-bedded intervals implies that they were deposited by similar flow mechanisms. Their primarily massive appearance, with occasional water-escape features, is consistent with deposition by either high-density, laminar, fluidized flows, or by high-density turbidity currents (King et al., 1993).

The upsection changes from thick-bedded sandstone, to thin-bedded sandstone, to siltstone within each basin-floor cycle could reflect autocyclic lobe switching between central lobe and distal lobe fringe positions, or it could reflect changing flow intensity caused by cyclic changes in relative base level. On balance, we prefer the latter option as the main control on the observed stratal architecture. This conclusion is in part a corollary of our more-regional studies, which indicate a strong influence of relative base level (third order) on the overall late Miocene aggradational and progradational succession in Taranaki Basin.

In the slope fan interval, turbidites are the main sandstone bed type, and typically comprise horizontal (T_b) and climbing-ripple laminated sandstone (T_c). Bases of beds are typically sharp and mildly erosive into underlying strata. Erosional scour surfaces appear to be a major control on lateral bed continuity in the outcrop section, and are often mudlined. Draping of mud at the base of these scour surfaces is potentially a significant permeability barrier in analogous reservoir facies in the subsurface.

The abundance of climbing-ripple laminated sandstones within the slope fan turbidite beds is attributed to the rapid dumping of sediment at the base-of-slope, both within scoured channels and in overbank settings. Conversely, well ordered Bouma divisions have not been recognized in the basin-floor fan interval, as might otherwise be expected from traditional submarine fan models. Most of the basin-floor fan sandstone beds are primarily massive (inferred T_a division). It is conceivable that graded turbidites are present in more distal fan intervals, having been deposited as a result of downdip flow transformation, but such intervals are not exposed in the study area.

The stratal architecture of the slope fan interval is more disorganized than the underlying basin-floor fan interval. The vertical and lateral juxtaposition of nested channels and proximal and distal levee deposits is indicative of shingled fans or sand-rich debris cones developed at or near the base-of-slope. The interfingering nature of these units also indicates that some large-scale compensation-style deposition or fan switching took place. Essentially, after an individual fan or debris cone had built up to a certain height, ensuing flows would have been diverted laterally into topographically lower areas. It seems likely that in the process, flows attained sufficiently high energy to scour the levee flanks of the underlying fan body. As several slope fan units were stacked, the various lithofacies types were superimposed (Slatt et al., 1998, their figure 5).

ACKNOWLEDGMENTS

We gratefully acknowledge the financial assistance of Amoco, ARCO, Conoco, Exxon, Fletcher Challenge Energy Taranaki (Petrocorp), Norsk-Hydro, Schlumberger, and Texaco. Funding was also provided by the New Zealand Foundation for Research Science and Technology, contract C05806. The manuscript was improved as a result of reviews by A. H. Bouma, J. B.Thomas, W. B. Macurda Jr., and B. D. Field. Discussions with colleagues on various aspects of the study, particularly J. Coleman, M. Gardner, A. Melhuish, and R. Spang, were appreciated. Figures were drafted by J. Smith and S. Shaw (GNS), Institute of Geological and Nuclear Sciences contribution 1709.

REFERENCES CITED

Browne, G. H., A. McAlpine, and P. R. King, 1996, An outcrop study of bed thickness, continuity and permeability in reservoir facies of the Mount Messenger Formation, North Taranaki: New Zealand Petroleum Conference Proceedings v. 1, 154–163.

Browne, G. H., and R. M. Slatt, 1997, Thin-bedded slope fan (channel-levee) deposits from New Zealand: an outcrop analog for reservoirs in the Gulf of Mexico: Gulf Coast Association of Geological Societies Transactions, v. 47, p. 75–86.

Coleman, J. L., G. H. Browne, R. M. Slatt, and G. R. Clemenceau, 1998, Comparison of two turbidite intervals: Taranaki Basin (New Zealand) and Gulf of Mexico (GOM) (USA) (abs.): Gulf Coast Association of Geological Societies and the Gulf Coast Section of SEPM Section Meeting abstract, Corpus Christi, Texas, AAPG Bulletin, v. 82, p. 1779.

Diridoni, J. L., 1996, Sequence stratigraphic framework of the Miocene Mount Messenger Formation deepwater clastics, north Taranaki Basin, New Zealand: Unpublished Master's thesis, Colorado School of Mines, Golden, Colorado. 165 p.

Jordan, D. W., D. J. Schultz, and J. A. Cherng, 1994, Facies architecture and reservoir quality of Miocene Mt. Messenger deepwater deposits, Taranaki Peninsula, New Zealand, *in* Weimer, P., A. H. Bouma, and B. Perkins, eds., Submarine fans and turbidite systems: sequence stratigraphy, reservoir architecture and production characteristics, Gulf of Mexico and international: Gulf Coast Section, Society of Economic Paleontologists and Mineralogists Foundation, Houston, p. 151–166.

King, P. R., G. H. Scott, and P. H. Robinson, 1993, Description, correlation and depositional history of Miocene sediments outcropping along north Taranaki coast: Institute of Geological and Nuclear Sciences Monograph 5, 199 p.

King, P. R., G. H. Browne, and R. M. Slatt, 1994, Sequence architecture of exposed late Miocene basin floor fan and channel-levee complexes (Mount Messenger Formation), Taranaki Basin, New Zealand, *in* Weimer, P., A. H. Bouma, and B. Perkins, eds., Submarine fans and turbidite systems: sequence stratigraphy, reservoir architecture and production characteristics, Gulf of Mexico and international: Gulf Coast Section, Society of Economic Paleontologists and Mineralogists Foundation, Houston, p. 177–192.

King, P. R., and G. P. Thrasher, 1996, Cretaceous-Cenozoic geology and petroleum systems of the Taranaki Basin, New Zealand: Institute of Geological and Nuclear Sciences Monograph 13, 243 p.

Slatt, R. M., G. H. Browne, R. J. Davis, G. R. Clemenceau, J. R. Colbert, R. A. Young, N. Anxiona, and R. J. Spang, 1998, Outcrop–behind outcrop characterization of thin-bedded turbidites for improved understanding of analog reservoirs: New Zealand and Gulf of Mexico: Society of Petroleum Engineers Annual Meeting, New Orleans, Paper 5PE49563, p. 845–853.

de V. Wickens, H., and A. H. Bouma, 2000, The Tanqua fan complex, Karoo Basin, South Africa—Outcrop analog for fine-grained, deepwater deposits, *in* A. H. Bouma and C. G. Stone, eds., Fine-grained turbidite systems, AAPG Memoir 72/SEPM Special Publication No. 68, p. 153–164.

Chapter 14

The Tanqua Fan Complex, Karoo Basin, South Africa—Outcrop Analog for Fine-Grained, Deepwater Deposits

H. Deville Wickens
Department of Geology, University of Stellenbosch
Matieland, South Africa

A. H. Bouma
Deptartment of Geology and Geophysics, Louisiana State University
Baton Rouge, Louisiana, U.S.A.

ABSTRACT

The Permian Tanqua fan complex, SW Karoo Basin, South Africa, is undeformed and well exposed, allowing 3-D viewing of outcrops that are laterally continuous over tens of kilometers. The complex consists of six sand-rich turbidite systems, separated by basin shales. The first five form an incrementally prograding, laterally compensating set, whereas the sixth fan, located to the south, downlaps onto Fan 5. Although deposited in a basin flanked by an orogenic belt, the Tanqua fan complex has depositional characteristics similar to fans deposited in passive margin settings. It is a fine-grained, mud-rich bypass system deposited within an unconfined basinal setting. The deposits show architectural and reservoir character changes as they occur from the base-of-slope to their distal termination.

INTRODUCTION

The Tanqua fan complex forms part of the Permian Ecca Group basin fill in the western part of the southwestern Karoo Basin, South Africa (Figure 1). The outcrops cover an area of approximately 650 km^2 and are almost undisturbed tectonically, with a 2–4° dip to the east. These deposits offer a unique combination of superbly exposed large-scale stratigraphy and internal facies architecture that can be used as an analog for fine-grained submarine fans.

The fan complex comprises six individual sand-rich fan systems between 20 and 60 m thick, separated by shale-rich units of similar thicknesses. The sixth fan is located to the south and downlaps onto the fifth fan. The sandstone is very fine to fine grained throughout. Each fan system has a high sandstone-to-shale ratio, whereas the interfan units comprise finely laminated shale and silty shales of hemipelagic and turbiditic origin.

Overlaying the fan complex is a river-dominated deltaic succession (Wickens, 1994; Goldhammer et al., this volume). Both are believed to be the result of an interaction between relative sea level fluctuations and sediment supply within an evolving foreland basin. The tectonic development of the late Paleozoic Cape Fold Belt (CFB) played a fundamental role in the depositional nature of the fan complex.

Although an order of magnitude thinner than seismically mapped complexes in most deepwater settings, the sedimentary characteristics of the Tanqua

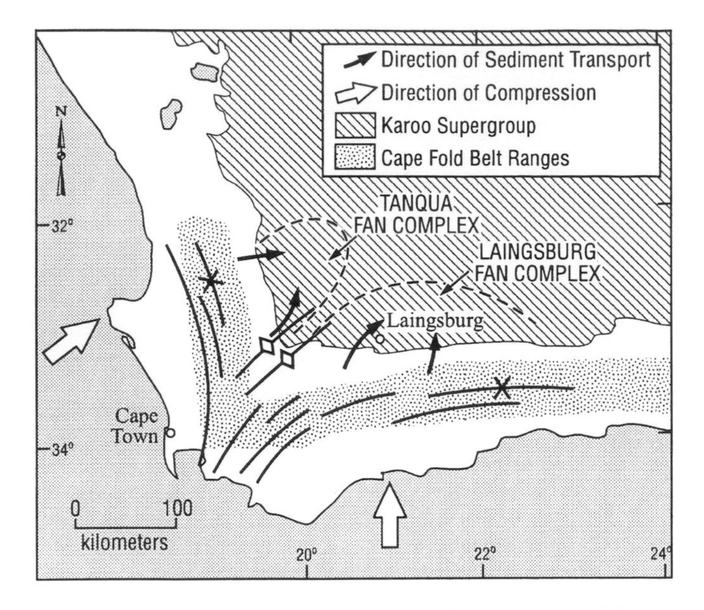

Figure 1—Location and influence of the Cape Fold Belt orogenesis on Ecca subbasin development, and the locations of the Tanqua and Laingsburg subbasins.

fan complex are similar to those of fans deposited in passive margin settings. Furthermore, the progradational nature of the fans, clear evidence for bypassing and depositional compensation for basin-floor topography, suggest an unconfined setting with an overall bathymetry more like that of a passive margin, that is, the downdip part of the fan complex. This most likely relates to the gentle tectonism of the western margin. Contrarily, the coeval Laingsburg fan complex along the southern margin of the Karoo Basin shows similar depositional characteristics but was deposited in a tectonically more active, elongated setting where confinement and basin-floor topography have played a major role in the distribution of the fan systems. The outcrops are now steeply folded.

The "Middle Ecca" stratigraphic succession at Laingsburg (Rogers and du Toit, 1903) was identified by Kuenen (1963) as turbidites. This succession, as well as the Tanqua succession were described as turbidite fan complexes by Wickens (1984). The discovery of Cretaceous hydrocarbon-bearing deepwater reservoirs along the South African south coast in 1988 initiated research on these turbidites. Initial studies were done for SOEKOR (Pty.) Ltd. during 1989/90 (Wickens et al., 1990). SOEKOR permitted the present authors to show the initial results as posters during an AAPG/SEPM annual meeting which resulted in setting up an oil-industry-supported consortium. Three doctoral dissertations, five master's of science degrees, and four honors theses have resulted. A detailed review of the Tanqua fan complex is incorporated in Wickens (1994); other relevant papers include Bouma and Wickens (1991, 1994), Bouma et al. (1995), and Wickens and Bouma (1995).

GEOLOGICAL SETTING

In the southwestern part of the Karoo Basin the Ecca Group comprises an approximately 1300 m thick, seemingly conformable succession of siliciclastic sediments, stratigraphically placed between the Carboniferous glaciogenic Dwyka Group and the Permo-Triassic fluvial Beaufort Group (Figure 2). The Ecca basin-fill succession ranges in age from early to late Permian, that is from ±278 Ma to ±265 Ma. The southwestern Ecca basin is bordered to the west and south by the CFB mountain ranges. It is subdivided into the Tanqua and Laingsburg subbasins; each contains a separate sequence of Ecca Group stratigraphic units above the Collingham Formation (Figures 2, 3).

Tectonically, the Karoo strata along the southern outcrop margin are folded along structures of the CFB, with fold intensity rapidly diminishing toward the north. The Tanqua subbasin deposits, which lie east of the Cedarberg mountain range, have a tectonic dip of 2–4 degrees to the east, except for some gentle E–W folding in the southernmost area and occasional reverse faulting with minor displacement over the central outcrop area. These features clearly illustrate

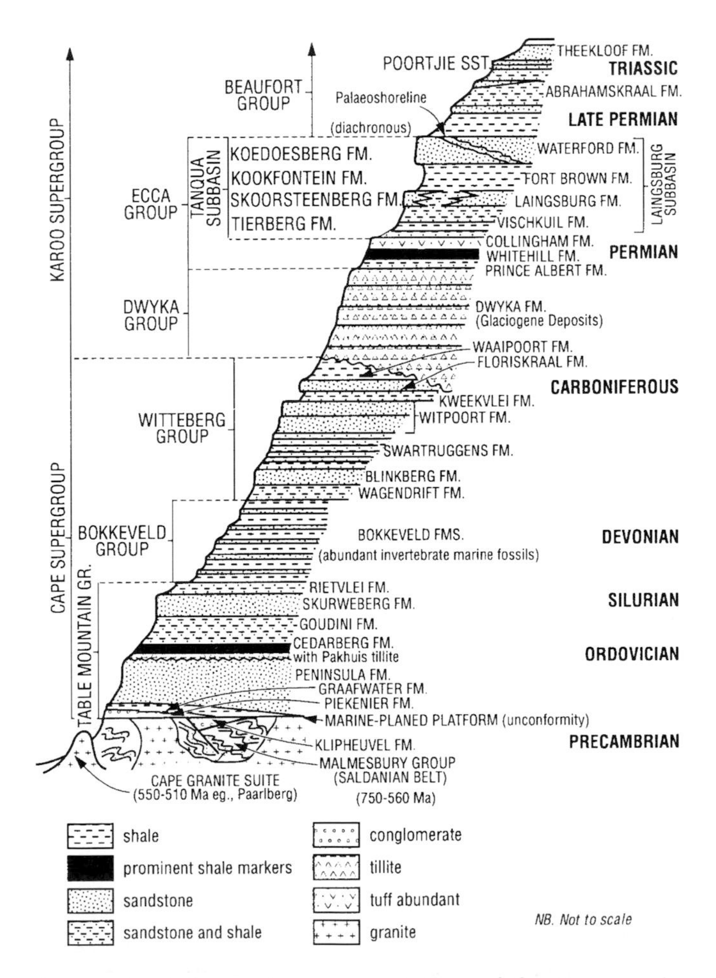

Figure 2—Schematic representation of the Cape and Karoo stratigraphy in the southwestern Karoo Basin.

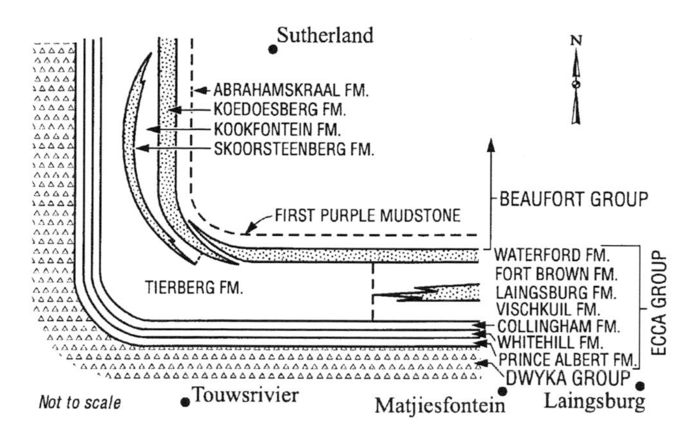

Figure 3—Schematic plan-view representation of Ecca Group stratigraphy in the southwestern Karoo Basin, showing the lack of correlation between the Tanqua and Laingsburg subbasins.

the dominance of northward compression. The spatial and temporal distribution of the submarine fan deposits in both subbasins, as well as the overlapping relationship of the western and southern deltaic systems, are directly related to structural development of the CFB (Wickens et al., 1990; De Beer, 1990).

TECTONIC EVOLUTION OF THE KAROO BASIN

The formation of the Pan-African belts, the extensive Cape and Karoo basins and their correlatives, and the CFB and its correlatives, are all products of the tectonomagmatic activity along the Panthalassan continental margin of Gondwana during the late Paleozoic (De Wit and Ransome, 1992; Veevers et al., 1994; López-Gamundí and Rossello, 1998).

Several episodes of extension and compression characterized southern Gondwana over a period of about 600 Ma since the late Precambrian (De Wit and Ransome, 1992). The late Proterozoic to early Paleozoic extension at about 500 ± 100 Ma, which relates to the breakup of greater proto-Gondwana, led to the formation of an extensive Atlantic-type margin (Cape Basin) along the southern edge of Gondwana. This was followed by the late Paleozoic convergence of 300 ± 100 Ma that relates to the assembly of Pangea and development of "Hercynian"-aged fold-thrust belts such as the CFB. Finally the mid- to late Mesozoic extension of 150 ± 50 Ma involved the breakup of Gondwana with subsequent opening of the southern oceans. A summary of the literature on the pre-Karoo setting can be found in Wickens (1994).

Cape Fold Belt

The early Paleozoic passive continental margin, which formed a large section of the southern edge of Gondwana, evolved into an active collisional margin during the late Paleozoic, which led to the formation of the CFB (Tankard et al., 1982). The CFB developed as a multiple-event orogen spanning some 16 m.y. during the Permian and Triassic. It consists of two branches that formed along previously established structural basement trends in the Saldanian basement rocks (Hälbich et al., 1983) (Figure 1). Major compressional paroxysms at 278, 258, 247, and 230 Ma are indicated by $^{39}Ar/^{40}Ar$ step-heating analysis of cleavage micas from the southern branch (Hälbich et al., 1983).

The northwest-trending Cedarberg Branch of the CFB displays large open folds without axial plane cleavage and is transected by numerous normal faults with approximately the same trend. The east–west-trending Swartberg Branch underwent more intense shortening as indicated by thrusting, overturning, and a high incidence of second-order folds on first-order fold limbs, accompanied by penetrative cleavages in the pelitic cover rocks (De Beer, 1989). Where the Cedarberg and Swartberg branches meet in the Ceres-Worcester area, that is, the syntaxis area, the fold and fault trends are dominantly NE–SW (De Beer, 1989).

Formation of the Karoo Foreland Basin

The Karoo Basin developed as a foreland basin with the fold-thrust belt (CFB) lying along the southern margin of the basin and the subduction zone along the paleo-Pacific margin of Gondwana (Johnson, 1991; Cole, 1992; Wickens, 1994; Veevers et al., 1994; Scott et al., this volume).

Onset of folding related to the first paroxysm (278 Ma) was probably deep-seated and had little effect on the basin configuration (Veevers et al., 1994). However, the time of emergence of the larger syntaxial structures (the Hex River and Baviaanshoek anticlinoria), and therefore initial development of the subbasins may be as early as the 278 Ma-paroxysm. The second paroxysm (258-Ma) resulted in formation of the Gondwana fold belts, including the CFB (Visser and Praekelt, 1996). Rapid downwarping of the basin axis occurred as a result of loading of thrust sheets in the adjacent CFB (Cole, 1992).

MODEL OF DEPOSITION

The postglacial evolution of the Karoo Basin shows two stages of basin fill, namely an initial starved basin stage dominated by fine-grained deposition in an open marine setting (Prince Albert and Whitehill formations), followed by an increase in progradational sedimentation (submarine fan and deltaic deposits). After final disintegration of the Gondwana ice sheets during Sakmarian times (~275 Ma) (McLachlan and Anderson, 1973), large parts of southwestern Gondwana were flooded, followed by postglacial deposition of a dominantly mud-rich succession containing chert horizons and phosphatic nodules (Prince Albert Formation). The overlying black carbonaceous shale of the Whitehill Formation represents deposition of muds under starved anoxic conditions during a period of tectonic quiescence and dormancy of the CFB orogen ["underfilled" stage of Covey (1986)].

The abrupt change from Whitehill conditions to deposition of siliciclastic turbidites and abundant volcanic ash (Collingham Formation) indicates a dramatic change in the tectonomagmatic activity along the Panthalassan continental margin of Gondwana. The volcanic ash most probably derived from volcanoes located in what is now northern Patagonia, where Permian silicic-andesitic volcanic and plutonic rocks crop out. This widespread volcanism seems to correlate with the 278-Ma paroxysm event of the Cape Orogeny. The Collingham Formation is the only unit of the Ecca Group that can be related directly to a major volcanic episode.

Subsidence of the foredeep due to crustal flexure in front of the rising fold thrust belt, probably commenced soon after Whitehill times. Basin plain deposition of clays, interrupted by sporadic influxes of muddy turbidite flows and occasional ashfall events, continued after Collingham time. These deposits constitute the Vischkuil Formation in the Laingsburg subbasin where a gradual increase in the frequency and volume of sand deposition, interpreted as active progradation (Theron, 1967), eventually culminated in the Laingsburg fan complex. These fans may be correlated in different ways with the Tanqua fans (Scott et al., this volume).

In the Tanqua subbasin, basin-plain deposition was interrupted by deposition of five stacked but progradational, arenaceous submarine fan systems (Skoorsteenberg Formation; Figure 4). Fan deposition is believed to have been controlled by fluctuations in relative sea level, probably caused by local tectonic or isostatic factors rather than eustatic changes. Scott (1997) suggests that deposition of the submarine fans in both subbasins took place during a period of tectonic quiescence, that is, between the 278-Ma and 258-Ma tectonic events, and with deposition possibly irregularly switching from one subbasin to the other (Scott et al., this volume).

Turbidite deposition in the two subbasins was succeeded by shelf, prodelta, delta-front, and delta-plain deposition in a gradually shallowing water body. The final phase of Ecca deposition ["overfilled" stage of Covey (1986)] involved progradation of highly constructive deltas from the south, west and northwest into the basin. The deltaic deposits (Koedoesberg and Waterford formations) overlap along the N–E-trending syntaxial structures and both formations thin out against (and across) this inferred basin-floor high. The growth and northeast propagation of the mega-anticlinal syntaxial folds played a vital role in the spatial distribution of the Tanqua and Laingsburg submarine fan systems.

Different rates of subbasin subsidence and subbasin geometry suggest influence on the rate of delta progradation and volume of sediment accumulation. More rapid subsidence of the Laingsburg subbasin probably created more accommodation space, which retarded the Waterford deltas; the more stable western margin may account for the thickness and more rapid progradation of the Koedoesberg deltas.

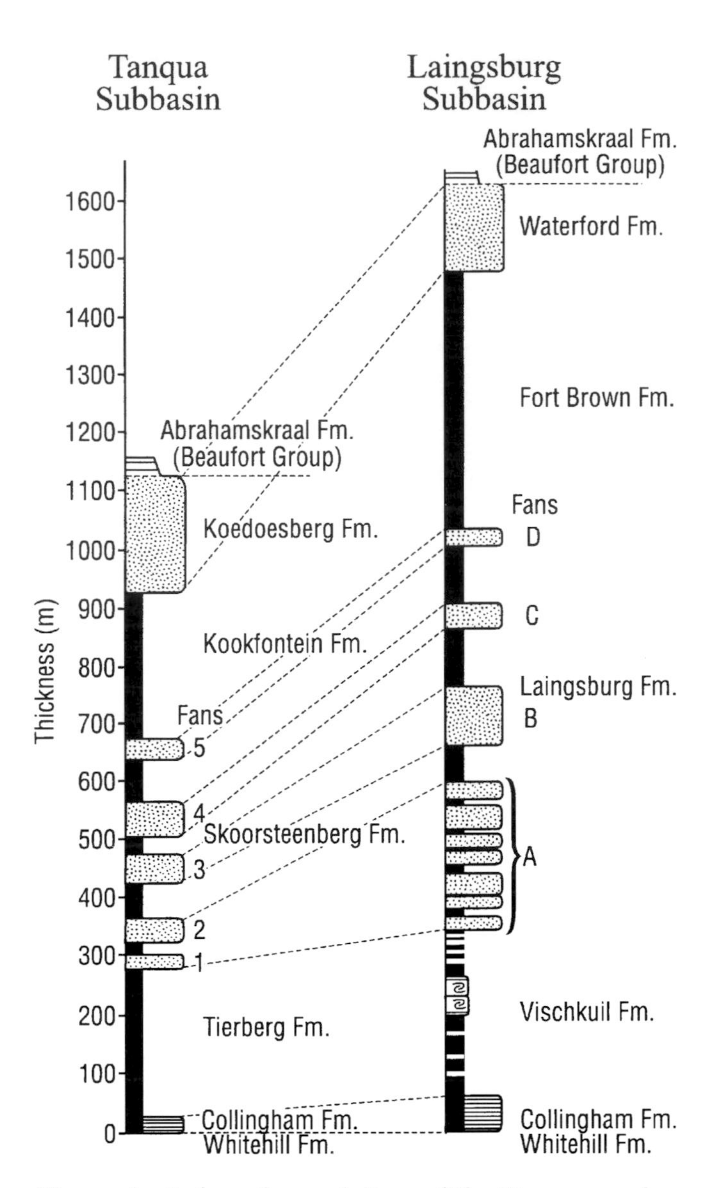

Figure 4—Inferred correlation of the Tanqua and Laingsburg basin-floor fan stratigraphy. Fan 6 is not shown.

GEOLOGY OF THE TANQUA SUBBASIN

The main outcrop area of the Tanqua fan complex (~400 m thick Skoorsteenberg Formation) lies west of the north–south-trending Koedoesberg mountain range. The type locality is Skoorsteenberg ("Chimney Mountain"), a prominent topographic feature rising almost 900 m above the desolate semi-desert plains of the Tanqua Karoo (Figure 5). The formation crops out over approximately 650 km^2 and thins out in a northerly and easterly direction. Thick basin shale units separate the predominantly arenaceous fan systems. The sixth fan ("Hangklip Fan") will not be mentioned here, but is discussed by Neethling (1992), Wickens (1994), and Wach et al. (this volume).

The sandstones are very fine to fine grained, light gray to blue-gray when fresh, poorly sorted, and virtually without any porosity and permeability resulting from diagenetic changes during burial to depths of about 7 km. Five major lithofacies have been recognized: massively bedded sandstone, horizontally and ripple cross-laminated sandstone, parallel-laminated siltstone,

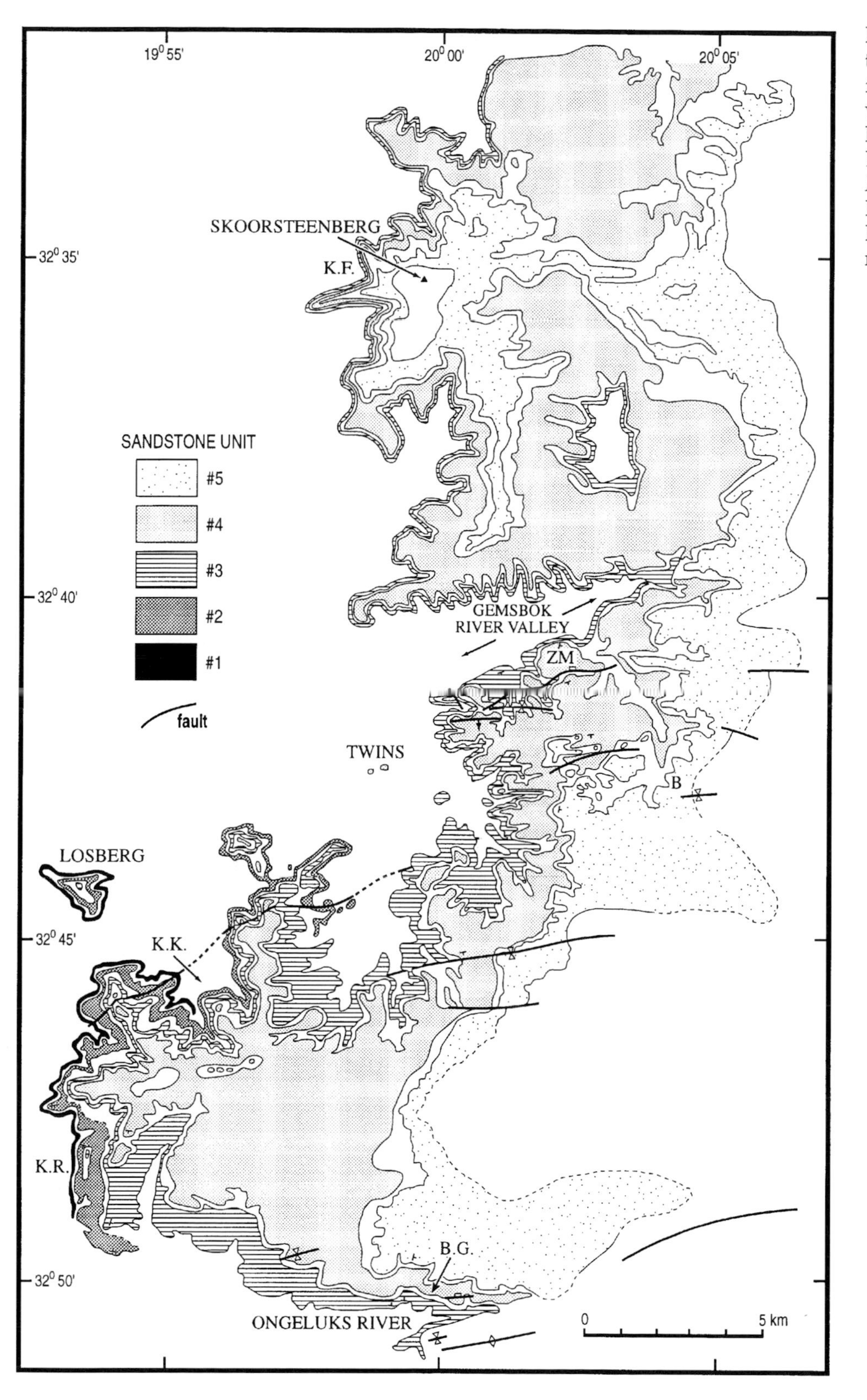

Figure 5—Outcrop distribution of the Tanqua submarine fan systems: Fans 1 through 5, with important geographical locations indicated: *B* = Bloukop; *B.G.* = Bizansgat; *K.K.* = Kanaalkop; *K.L.* = Klipfontein; *K.R.* = Kleine Rietfontein; *Z.M.* = Zoetmeisies Fontein.

parallel-laminated shale, and a micaceous, silty plant-fragment facies. Subordinate facies such as tuff, limestone beds, calcareous concretions, and cherty layers are present but are only of significance in terms of source area tectonics and periods of starvation in clastic sediment supply (primarily as condensed sections).

The massively bedded sandstone facies, characterized by sharp bases and tops, abundant sole marks, ripup shale clasts, and amalgamations are commonly associated with channel-fill or stacked sheet-sand deposits. The ripple cross-laminated and climbing ripple-laminated sandstone and siltstone facies are mostly associated with thin-bedded turbidites in levee-overbank settings, although layers can be medium to thick. These characterize the southwestern part of Fan 3. The micaceous plant fragment facies commonly occurs on top of the sandstone beds, attaining up to 5 cm in thickness.

Mineralogical and geochemical studies by Scott (1997) were conducted to analyze the characteristics of the sediment source terrain, to compare the sandstones from both subbasins, and to attempt to establish a stratigraphic correlation of the individual fans of both subbasins (Scott et al., this volume).

Several locations within a submarine fan comprise thin-bedded sandstones. Basu (1997) and Basu and Bouma (this volume) describe the minor differences between the various locations and the difficulties of applying these to the subsurface and small isolated outcrops of which the location within a fan is not certain.

Reconstruction of the fan systems from paleocurrent and isopach data shows that each successive fan system up to Fan 4 prograded farther into the basin, with an offset stacking pattern to compensate for basin floor topography. This is clearly illustrated by the eastward shift of Fans 2 and 3 compared to the position of Fan 1. Fan 4 is thickest where Fan 3 is thinnest. Fan 4 continued farthest into the basin.

INDIVIDUAL FANS

Fan 1 (Figure 6)

Fan 1 comprises the first sand-rich turbidites deposited in the Tanqua subbasin. It forms the most westerly outcrops of the Skoorsteenberg formation with only the outer marginal part of the fan preserved. It attains a maximum thickness of 24 m (Figure 6). Isopach and paleocurrent data indicate a transport direction to the ENE with the fan axis located to the northwest of the present outcrops.

Outcrops comprise mainly laterally extensive, single or amalgamated, non-channelized sandstone beds, up to 2.6 m thick. They are separated by shale, siltstone, and thin sandstone beds, which often can be interpreted as small-scale thickening-upward units. Lateral changes in bed thickness, sharp top and bottom, ripup clasts, loading features, and sole marks are common (Figure 6). Individual channel fills with a width of up to 100 m and depth of 9 m are locally associated with the sheet sands (Hamman, 1992; M. Martin, personal communication, 1997).

Channel fills at the base of the fan system at Losberg may illustrate an irregular fingering pattern for the outer edge of the different fan units. The small channel fills associated with the single or amalgamated sandstone units elsewhere confirm the above statement. Deposition in the outer fan region is characterized by predominantly massive to parallel-laminated units displaying a sheetlike geometry. The consecutive eastward thinning and eventual pinch-out of the individual sandstone units at Losberg and main outcrop area illustrate basinward progradation of the fan.

Fan 2 (Figure 7)

The isopach trend for this fan shows gradual thinning from approximately 40 m in the southwest to less than 1 m in the northeast, a distance of approximately 12 km. Both direction and location have changed when comparing it to Fan 1 (Figures 6, 7). Three distinct sand-rich units, separated by thin shale units, characterize Fan 2 (Figure 7; Rozman, 1998, this volume). Fan 2 also displays outer fan sheet sand deposits as the more-updip areas have been eroded away. Fan 2 clearly shows progradational characteristics. Each unit prograded farther into the

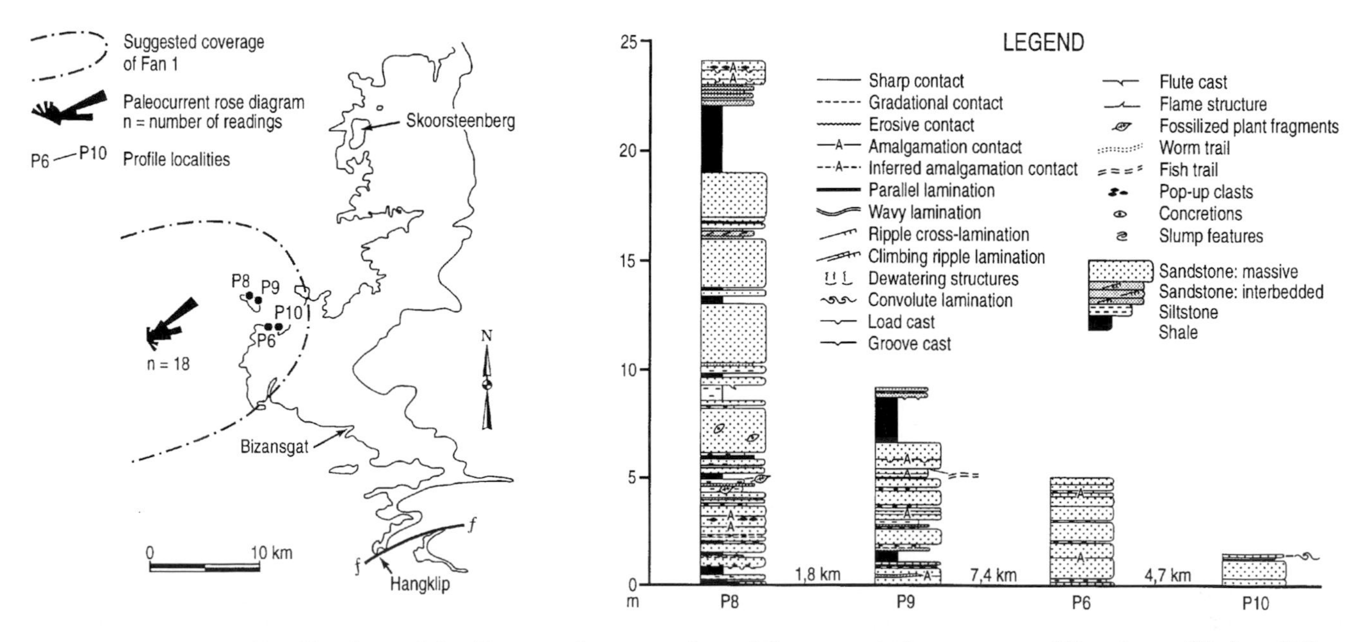

Figure 6—Outcrop distribution of the Tanqua fan complex with suggested coverage of Fan 1, profile localities, paleocurrent directions, and measured vertical sections representing a northwest–southeast cross section through Fan 1. *P8* represents the inferred position for the fan axis.

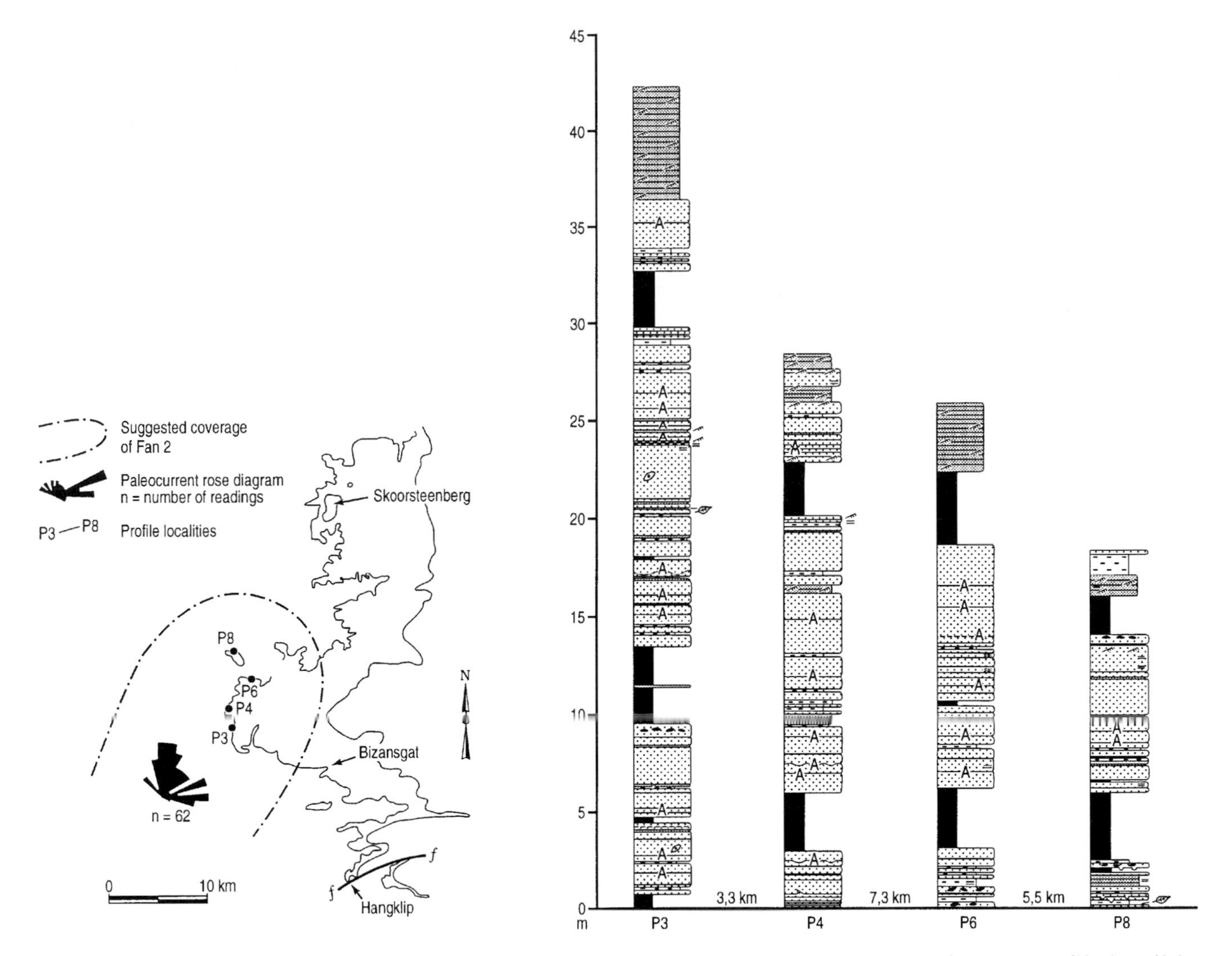

Figure 7—Outcrop distribution of the Tanqua fan complex with suggested coverage of Fan 2, profile localities, paleocurrent directions, and measured vertical sections from south to north illustrating downdip changes in the style of deposition and gradual northward thinning of the fan. For legend, see Figure 6.

basin than the previous one, and was influenced by the topographical expression of the previous fan's depositional axis (Rozman, this volume).

Fan 3 (Figure 8)

Fan 3 outcrops from the Ongeluks River in the south to Skoorsteenberg in the north, a distance of 34 km. It is separated from Fan 2 by a 50- to 80-m-thick succession of shale, siltstone, and very thin sandstones. The thickness of Fan 3 varies from 30 to 50 m in the southern half of the outcrop area and gradually thins northward until it pinches out northeast of Skoorsteenberg. Although more variable in the proximal areas, paleotransport was to the north (Figure 8).

Fan 3 is the only complete basin-floor fan of the Tanqua fan complex. As a result, it shows the greatest variety of facies and architectural elements of all the fans. It consists of a channel-fill complex displayed by the most proximal outcrops at Ongeluks River (Van Antwerpen, 1992), channel-levee deposits and transitional channelized to nonchannelized deposition in the middle area, stacked sheet sand deposits in the northern half, and extensive overbank facies in the southwesterly outcrops (Figure 8). This fan also shows progradational features.

The southwestern part of Fan 3 (~45 m thick) was deposited over the basin-floor high created by Fans 1 and 2. Regionally, this facies is in a position marginal to the fan axis, which lies more to the east and southeast. It consists primarily of an alternation of medium- to thin-bedded ripple cross laminated sandstone and shale and is regarded as levee-overbank facies generated by the axial channels to the southeast. It could also represent deposition from diluted flows marginal to the fan axis, onto and over the topographical high created by Fan 2 (Kirkova, personal communication, 1997).

Outcrops to the north comprise mainly stacked, nonchannelized deposits that gradually thin toward the pinchout area. Broad lenticular sandstone bodies as well as channel fills occur in the pinchout area (J. C. Wyble and S. K. Huisman, personal communication, 1998). Both upper and lower boundaries with the basin shale units are abrupt.

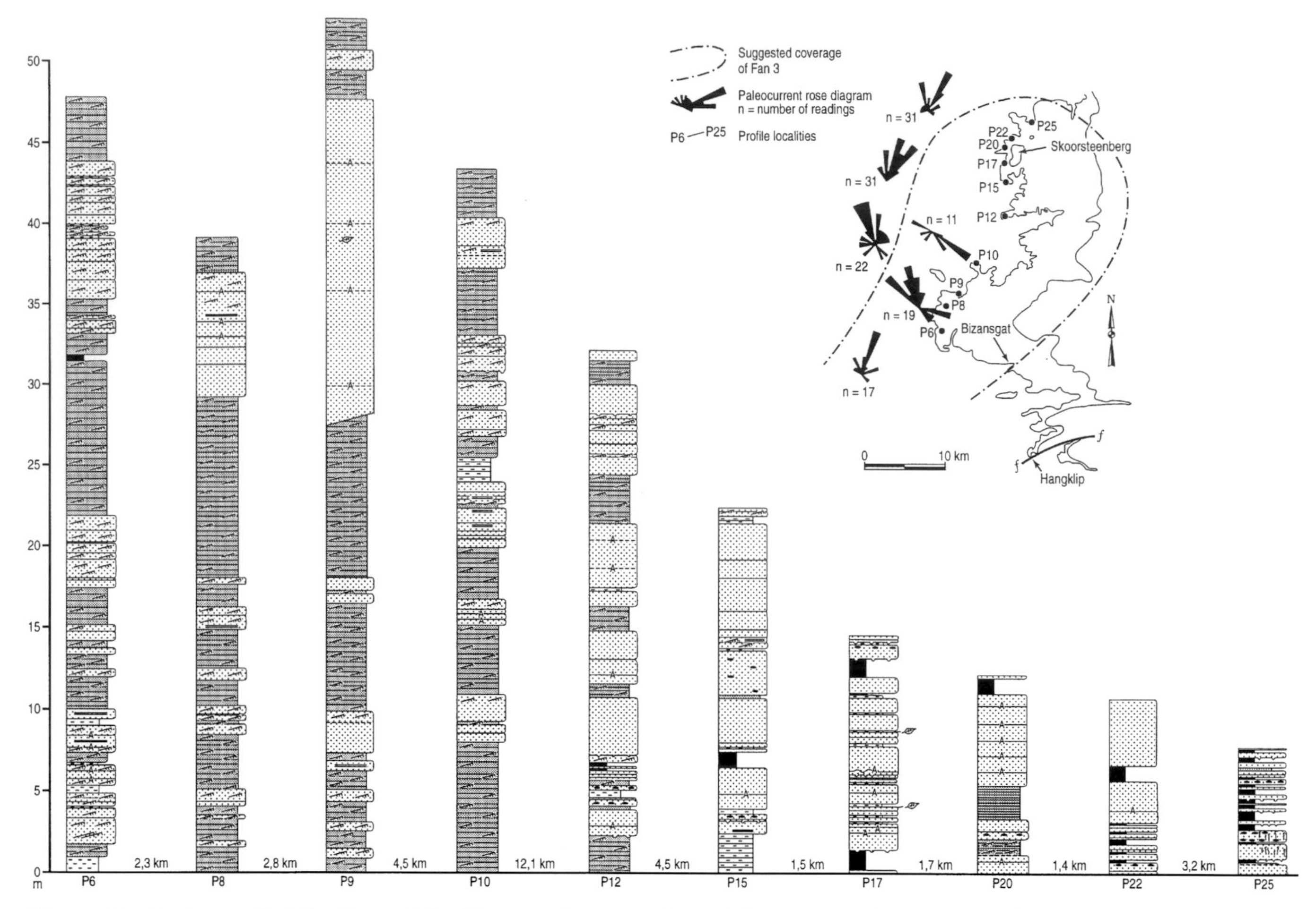

Figure 8—Outcrop distribution of the Tanqua fan complex with suggested coverage of Fan 3, profile localities, paleocurrent directions, and measured vertical sections illustrating the dominance of thin-bedded levee/overbank facies in the southwestern part of the fan, and facies changes and gradual northward thinning over the northern portion. Paleocurrents from profiles P12–P25 are represented in the upper two rose diagrams, whereas the lower rose diagrams represent the southern part of the fan. For legend, see Figure 6.

Fan 4 (Figure 9)

This is regionally the most widespread and thickest fan with sedimentary characteristics, consistent with deposition in a middle to lower fan setting (Figure 9). Laterally extensive and commonly parallel-sided, sheetlike sandstone beds, often arranged in upward-thickening or -thinning packages, characterize the more proximal outcrops. The distal outcrops comprise more thin-bedded sheet sands, in places capped by thick sandstone units (Figure 9). Large single channel fills and channel-fill complexes, comprising predominantly thick, amalgamated units, can be found at or near the top of the succession (D. E. Rehmer and P. R. C. Dudley, personal communication, 1998).

Fan 4 thins to the south, that is, from about 60 m in the Skoorsteenberg area to about 30 m and less in the Kleine Riet Fontein and Bizansgat areas. This may be due to large-scale compensation of basin-floor topography related to the deposition of Fan 3, which thickens to the south. It is also possible that Fan 4 consists of two units: a more mid fan-channelized type in the southern part of the outcrop area and a more outer fan sheet-sand type in the northern area. It is also noted that the shale interval between Fans 3 and 4 is much thicker in the south because it contains more siltstone and thin, silty sandstone beds, which likely reflects the main direction of turbidity flows into the basin.

Fan 5 (Figure 10)

This fan is similar to Fan 4 in areal extent and represents the final phase of submarine fan deposition in the Tanqua subbasin. An arbitrary southern boundary was placed southeast of Bizansgat where Fan 6 (Hangklip fan) downlaps onto Fan 5 (Figure 10). The best exposed outcrops are limited to the vicinity of Skoorsteenberg, where it has a thickness of 35 m.

Thin-bedded, ripple-laminated turbidites associated with channel-fill deposits predominate in the southeastern part of the fan region, whereas laterally extensive, predominantly thick-bedded and amalgamated sandstone units, exhibiting large-scale channel fills and sheet deposits, characterize the succession at Skoorsteenberg. Here, paleocurrents indicate a northeasterly transport direction, whereas the succession north of Bizansgat prograded from the south and southwest. Two phases

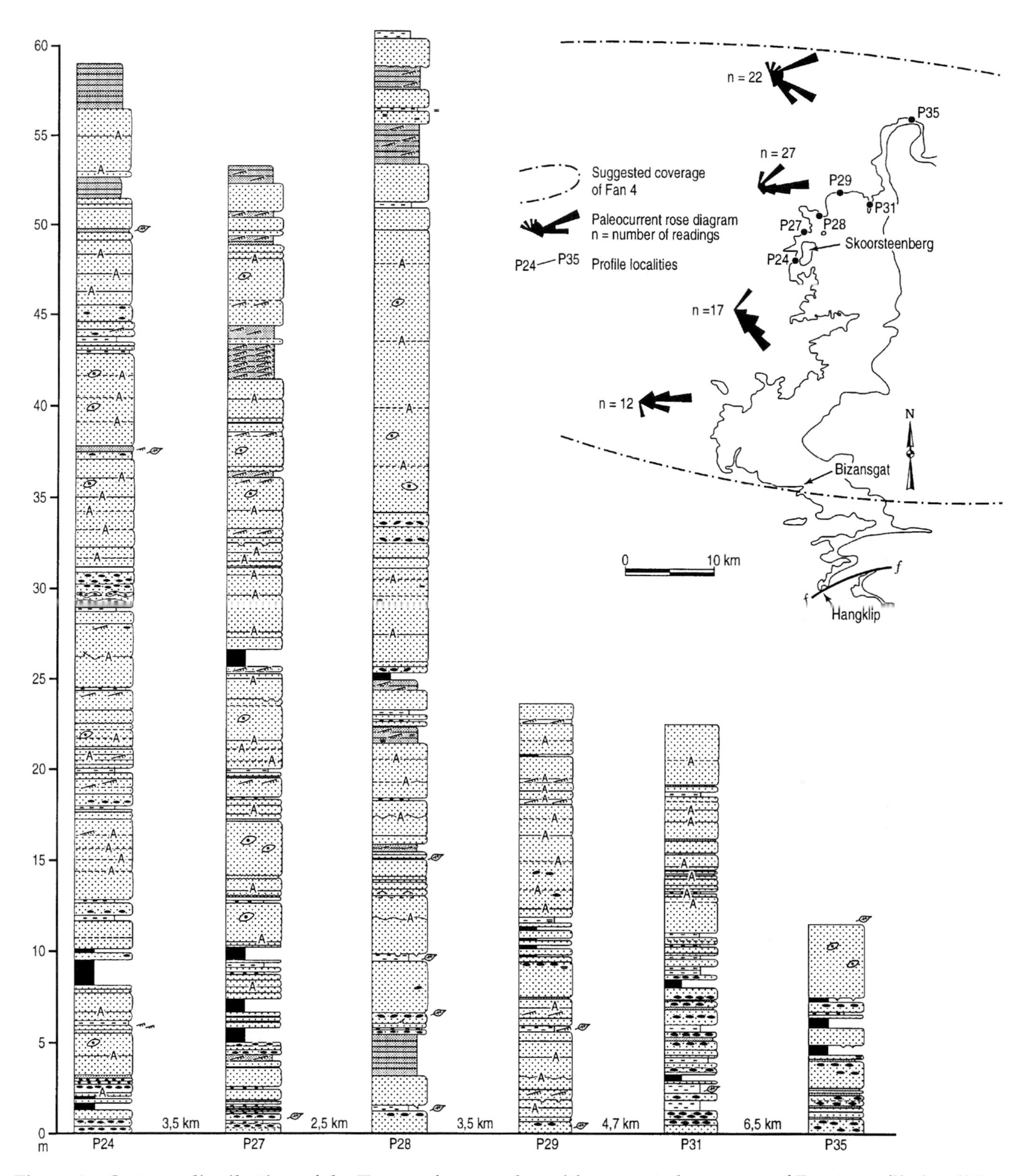

Figure 9—Outcrop distribution of the Tanqua fan complex with suggested coverage of Fan 4, profile localities, paleocurrent directions, and measured vertical sections illustrating the sand-rich nature in the Skoorsteenberg area and the downcurrent change in thickness. For legend, see Figure 6.

of fan deposition therefore seem possible. Kirschner and Bouma (this volume) describe the relationship between the massive channel-fill sandstones and thin-bedded levee-overbank deposits in the southern outcrops.

At Skoorsteenberg the upper contact of this fan is abrupt and overlain by a dark shale containing calcareous concretions. This is succeeded by the first appearance of wave ripple-marked thickening-upward

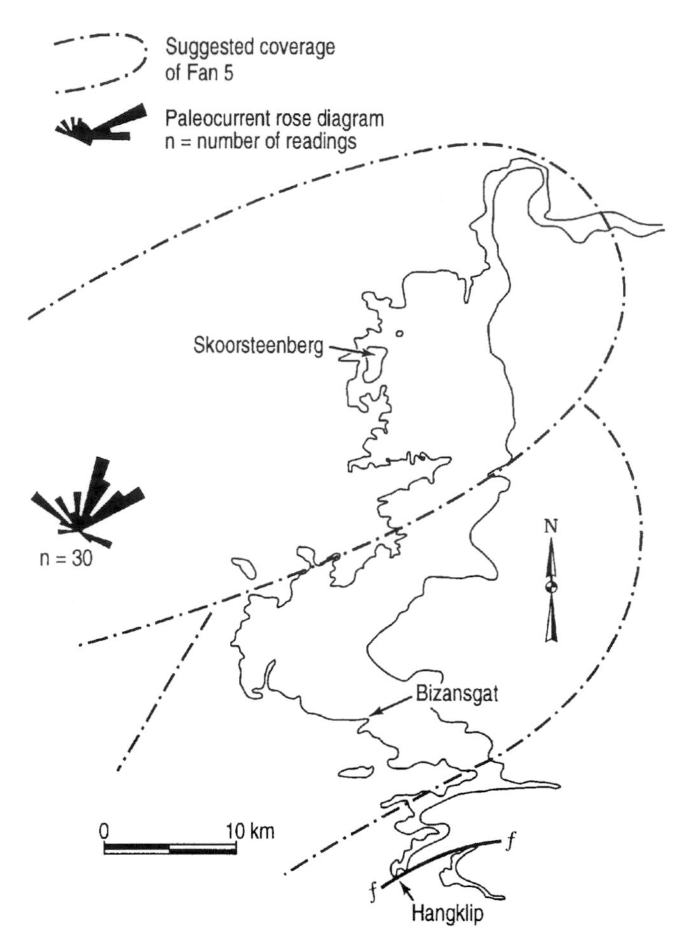

Figure 10—Outcrop distribution of the Tanqua fan complex with suggested coverage for a southern and a northern phase of deposition for Fan 5. Paleocurrent directions are for the northern phase of Fan 5.

cycles and slump units, approximately 60 m above the top of Fan 5 at Skoorsteenberg. These represent distal distributary mouth bar and distributary channel deposits indicating shoaling of the environment with deposition at fair-weather wave base conditions.

SUMMARY

The intracratonic southwestern Karoo Basin developed as an asymmetrical, arcuate foreland basin with a northeast-oriented basin-floor high that divided it into the Tanqua and Laingsburg subbasins. The regional development and structural configuration of the CFB primarily dictated the shape and nature of the Ecca coastline, the development of the subbasins in the foreland domain, and the sedimentation patterns of the submarine fan and deltaic complexes in these subbasins.

The high-quality 3-D exposure of the Tanqua fan complex is due not only to present-day erosion in a semi-desert environment, but also to the very gentle postdepositional deformation. Uplift and erosion following the mid to late Mesozoic breakup of Gondwana, eventually exposed the entire Tanqua basin-fill succession. Access is relatively easy in a relative low relief terrain. The proximal parts of the fans and time-equivalent deltaic succession are removed by erosion.

The following structural and sedimentary features characterize the Tanqua basin fill:

- Virtually no post-depositional tectonic deformation of the turbidite, deltaic, and fluvial successions, except in the extreme southern part.
- A fine- to very-fine grain size for the entire turbidite/deltaic basin-fill succession.
- Lack of sandy turbidites and slumped units above the Whitehill Formation compared to Collingham and Vischkuil deposition in the Laingsburg subbasin.
- Deposition of five discrete basin floor fan successions, separated by thick shale-rich interfan units.
- The total of the five fans form a low sandstone-to-shale ratio (mud-rich) fan/turbidite complex because of the thick basin shale units separating the individual fans. Each fan (fan system), however, has a high sandstone-to-shale ratio.
- Presence of a sixth fan to the south that downlaps onto Fan 5. Fan 6 may be a slope fan.
- Fan systems were progradational and shifted to the east to compensate for basin-floor topography in a seemingly unconfined foreland setting.
- Depositional elements include nested channel fills at the base-of-slope, individual channel fills and levee-overbank in the midfan area, extensive sheetlike depositional lobe deposits in the outer fan area, and possible crevasse channel fills associated with levee-overbank deposits.
- Pinchout areas of fans are characterized by sheet sands displaying lateral variation in thickness, which suggests an uneven fingerlike outer fan margin.
- The fluvially dominated deltaic deposits are abruptly succeeded by high sinuosity meander belt and flood plain deposits.

The above-mentioned attributes are characteristic for most fine-grained turbidite complexes as found in passive margin settings and are atypical of most active margin settings. The sedimentary characteristics of the submarine fan, deltaic, and fluvial deposits indicate a very distant provenance, a wide coastal plain, high-constructive river-dominated deltas, and a wide shelf, which is more typical of passive margin settings, such as the modern Mississippi delta and submarine fan. The reason for mimicking a passive margin setting lies in the distant position of the provenance, the tectonic configuration of the foreland basin margin, and the fact that depositional periods for each fan are much shorter than major tectonic cyclicity. This situation illustrates the danger of using the terms "passive margin fans" and "active margin fans" to characterize the sediments.

As an analog to fine-grained turbidite systems, the Tanqua fan complex is particularly important for adding fundamental geological interpretation to the technologically advanced geophysical datasets on deepwater deposits. Progradational nature of fans, compensation for basin-floor topography, regional facies variation within a fan system, internal reservoir architecture, distribution of heterogeneity at all scales, aspect ratios of architectural elements, and synthetic seismics and well logs are some of the aspects that are currently being studied. Models derived from the Tanqua fans are, therefore, believed to be applicable to many fields in the Gulf of Mexico, off the west coast of central and southern Africa, offshore Brazil, and many basins in the North Sea.

ACKNOWLEDGMENTS

The authors are greatly indebted to the landowners in the study area for their supportive role by allowing access to their properties, the Consortium companies for financial support, and those students who took on the challenges of the remote Ceres and Tanqua Karoo regions.

REFERENCES CITED

Basu, D., 1997, Characterization of thin-bedded turbidites from the Permian Tanqua Karoo submarine fan deposits, South Africa. Unpublished Ph.D. dissertation, Louisiana State University, Baton Rouge, Louisiana, 159 p.

Bouma, A. H., and H. de V. Wickens, 1991, Permian passive margin submarine fan complex, Karoo Basin, South Africa: possible model to Gulf of Mexico: Gulf Coast Association of Geological Societies Transactions, v. 41, p. 30–42.

Bouma, A. H., and H. de V. Wickens, 1994, Tanqua Karoo, ancient analog for fine-grained submarine fans, *in* P. Weimer, A. H. Bouma, and B. F. Perkins, eds., Submarine fans and turbidite systems: sequence stratigraphy, reservoir architecture, and production characteristics: Gulf of Mexico and international. Gulf Coast Section SEPM Foundation 15th Research Conference Proceedings, p. 23–34.

Bouma, A. H., H. de V. Wickens, and J. M. Coleman, 1995, Architectural characteristics of fine-grained submarine fans: a model applicable to the Gulf of Mexico: Gulf Coast Association of Geological Societies Transactions, v. 45, p. 71–75.

Cole, D. I., 1992, Evolution and development of the Karoo Basin, *in* M. J. de Wit and I. G. D. Ransome, eds., Inversion tectonics of the Cape Fold Belt, Karoo and Cretaceous basins of southern Africa: Rotterdam, Balkema, p. 87–99.

Covey, M., 1986, The evolution of foreland basins to a steady state: evidence from the western Taiwan foreland basin: International Association of Sedimentologists Special Publication 8, p. 77–99.

De Beer, C. H., 1989, Structure of the Cape Fold Belt in the Ceres Syntaxis: Unpublished M. Sc. thesis, University of Stellenbosch, South Africa, 134 p.

De Beer, C. H., 1990, Simultaneous folding in the western and southern branches of the Cape Fold Belt: South African Journal of Geology, v. 93, p. 583–591.

De Wit, M. J., and I. G. D. Ransome, 1992, Regional inversion tectonics along the southern margin of Gondwana, *in* M. J. de Wit, and I. G. D. Ransome, eds., Inversion tectonics of the Cape Fold Belt, Karoo and Cretaceous basins of Southern Africa: Rotterdam, Balkema, p. 15–21.

Hälbich, I. W., F. J. Fitch, and J. A. Miller, 1983, Dating the Cape Orogeny, *in* A. P. G. Söhnge, and I. W. Hälbich, eds., Geodynamics of the Cape Fold Belt: Special Publication Geological Society of South Africa, v. 12, p. 165–175.

Hamman, H., 1992, Outer fan lobe deposits of Fan 1, Skoorsteenberg Formation, Ecca Group, South Africa. Unpublished honors project, University of Stellenbosch, South Africa, 15 p.

Johnson, M. R., 1991, Sandstone petrography, provenance and plate tectonic setting in Gondwana context of the southeastern Cape-Karoo Basin: South African Journal of Geology, v. 94, p. 137–154.

Kuenen, P. H., 1963, Turbidites in South Africa. Transactions, Geological Society of South Africa, 66, p. 191.

López-Gamundí, O. R., and E. A. Rossello, 1998, Basin-fill evolution and paleotectonic patterns along the Samfrau geosyncline: the Sauce Grande basin–Ventana foldbelt (Argentina) and Karoo basin—Cape foldbelt (South Africa) revisited. Geologische Rundschau, v. 86, p. 819–834.

McLachlan, I. R., and A. M. Anderson, 1973. A review of the evidence for marine conditions in southern Africa during Dwyka times. Palaeontology Africa, v. 13, p. 37–64.

Neethling, B. C., 1992, Depositional characteristics of submarine Fan 6 in the Tanqua subbasin, Permian Ecca Group, South Africa. Unpublished honors project, University of Stellenbosch, South Africa, 26 p.

Rogers, A. W., and A. L. du Toit, 1903, Geological surveys of parts of the Divisions of Ceres, Sutherland and Calvinia: Annual Report Geological Commission, C.G.H., p. 25–32.

Rozman, D. J., 1998, Characterization of a fine-grained outer submarine fan deposit, Tanqua Karoo Basin, South Africa. Unpublished M. S. thesis, Louisiana State University, 147 p.

Scott, E. D., 1997, Tectonics and sedimentation: evolution, tectonic influences and correlation of the Tanqua and Laingsburg subbasins, southwest Karoo Basin, South Africa: Unpublished Ph. D. thesis, Louisiana State University, 234 p.

Tankard, A. J., M. P. A. Jackson, K. A. Eriksson, D. K. Hobday, D. R. Hunter, and W. E. L. Minter, 1982, Crustal evolution of southern Africa: 3.8 billion years of Earth history: Berlin, Springer-Verlag, 523 p.

Theron, A. C., 1967, The sedimentology of the Koup subgroup near Laingsburg. Unpublished M. Sc. thesis, University of Stellenbosch, South Africa, 22 p.

Van Antwerpen, O., 1992, Ongeluks River channel-fills, Skoorsteenberg Formation. Unpublished honors project, University of Port Elizabeth, South Africa, 69 p.

Veevers, J. J., D. I. Cole, and E. J. Cowan, 1994, Southern Africa: Karoo Basin and Cape Fold Belt, *in* J. J. Veevers, and C. McA. Powell, eds., Permian-Triassic

Pangean basins and foldbelts along the Panthalassan margin of Gondwanaland. Geological Society America Memoir 184, p. 223–279.

Visser, J. N. J., and H. E. Praekelt, 1996, Subduction, mega-shear systems and Late Palaeozoic basin development in the African segment of Gondwana: Geologische Rundschau, v. 85, p. 632–646.

Wickens, H. de V., 1984, Die stratigrafie en sedimentologie van die Groep Ecca wes van Sutherland: Unpublished M. Sc. thesis, University of Port Elizabeth, 86 p.

Wickens, H. de V., G. J. Brink, and W. van Rooyen, 1990, The sedimentology of the Skoorsteenberg Formation and its applicability as a model for submarine fan turbidite deposits in the Bredasdorp Basin: Unpublished report, Soekor, Parow, 47 p.

Wickens, H. de V., 1994, Basin floor fan building turbidites of the southwestern Karoo Basin, Permian Ecca Group, South Africa: Unpublished Ph. D. thesis, University of Port Elizabeth, South Africa, 233 p.

Wickens, H. de V., and A. H. Bouma, 1995, The Tanqua basin floor fans, Permian Ecca Group, Western Karoo Basin, South Africa, *in* K. T. Pickering, R. N. Hiscott, N. H. Kenyon, F. Ricci Lucchi, and R. D. A. Smith, eds., Atlas of deep water environments, architectural style in turbidite systems, London, Chapman & Hall, p. 317–322.

Goldhammer, R. K., H. deV. Wickens, A. H. Bouma, and G. Wach, 2000, Sequence stratigraphic architecture of the Late Permian Tanqua submarine fan complex, Karoo Basin, South Africa, *in* A. H. Bouma and C. G. Stone, eds., Fine-grained turbidite systems, AAPG Memoir 72/SEPM Special Publication No. 68, p. 165–172.

Chapter 15

Sequence Stratigraphic Architecture of the Late Permian Tanqua Submarine Fan Complex, Karoo Basin, South Africa

R. K. Goldhammer
Texaco Exploration
Bellaire, Texas, U.S.A.

H. Deville Wickens
Kuils River, South Africa

A. H. Bouma
Department of Geology and Geophysics, Louisiana State University
Baton Rouge, Louisiana, U.S.A.

G. Wach
E&P Technology Department, Texaco
Houston, Texas, U.S.A.

ABSTRACT

The Late Permian Ecca Group (1300 m thick) in the Tanqua Karoo consists of a basin floor fan complex (400 m thick), overlain by river-dominated deltaic deposits and associated updip fluvial deposits. This succession is subdivided into two "third-order" depositional sequences with several superimposed high-frequency, "fourth-order" depositional sequences. The Tanqua submarine fan complex contains six regionally distinct fan systems (24 to 60 m thick), five of which form a progradational stack, as revealed by their spatial distribution and regional facies variation. The sixth fan is situated to the south and downlaps onto the fifth fan. This long-term (third-order) progradational pattern records a combination of reduced accommodation space and/or increased sediment supply. Each fan system is assigned to the lowstand systems tract of each high-frequency, fourth-order sequence, and the particular attributes of each fan system are a consequence of their respective positions within the lower-frequency third-order sequence.

INTRODUCTION

Deepwater deposits are becoming increasingly more significant as exploration targets for hydrocarbons, as exemplified by recent industry success in Tertiary deepwater clastic systems in offshore Nigeria, Angola, North Sea, and the Gulf of Mexico. There are two critical issues to address regarding the successful location and production of such reservoirs, one pertaining to developing predictive basin-scale exploration models, and the second dealing with the complexities of reservoir heterogeneity at the exploitation scale. Addressing these issues in the subsurface involves a complete integration of datasets that investigate the problem at different, yet overlapping scales, for example, interpretation of regional seismic and/or gravity and magnetics complemented by subregional analysis of well-log and/or core data. In this context, the submarine fan complex of the Tanqua Karoo region in South Africa (Wickens and Bouma, this volume) offers a unique opportunity to investigate the complexities of deepwater deposits from the seismic-scale exploration perspective as well as the reservoir-scale, exploitation vantage point.

The purpose of this paper is to provide a sequence stratigraphic and chronostratigraphic framework for the Ecca basin fill into which the Tanqua fan complex and related updip deltaic and fluvial deposits can be placed. Such a framework provides a context for evaluating the variations in the individual fan systems (e.g., facies architecture of fan systems, geometry, and dimensions of sand bodies) and allows one to link the individual fan systems together in a more predictive manner. The framework proposed is based on regional stratigraphic analysis of different sections from various parts of the basin profile and can serve as a "working model" that will require modification as additional studies are completed.

The intracratonic southwestern Karoo Basin developed as an asymmetrical, retroarc foreland basin related to late Paleozoic–early Mesozoic subduction of the paleo-Pacific plate underneath the Gondwana plate. Its deepest part developed closest to the adjoining branches of the Permo-Triassic Cape Fold Belt (CFB), with local subbasin development (Tanqua and Laingsburg subbasins) attributed to tectonic growth of megaanticlinoria structures of the CFB syntaxis. The latter primarily controlled Ecca deposition in the southwestern Karoo Basin. The Ecca Group in the Tanqua subbasin (1300 m thick) consists of a remarkably well-exposed and undeformed basin-fill succession (Wickens, 1994; Wickens and Bouma, this volume), which is subdivided into seven regionally mappable lithostratigraphic units. These units are from base to top: Prince Albert, Whitehill and Collingham formations (initial post-glacial and starved basin deposition), Tierberg Formation (basin plain), Skoorsteenberg Formation (basin-floor fan complex), Kookfontein Formation (basin plain, prodelta, delta front), and Koedoesberg Formation (delta front and delta plain).

A late Sakmarian-Kungurian age (~275 m.y.) is assigned to the Prince Albert/Whitehill formations (Visser, 1992) and Ufimian-Kazanian (~265 m.y.) (Rubidge, 1991) for the top of the Koedoesberg Formation, using the time scale of Haq and Van Eysinga (1987). This roughly constrains the duration of the entire Ecca Group to be approximately 10 m.y. If onset of submarine fan deposition is taken at 258 m.y. to coincide with the second major tectonic event of CFB orogenesis (Hälbich et al., 1983; Wickens 1994), the duration of submarine fan and ensuing deltaic deposition is approximately 7 m.yr. The duration of the different formations and absolute time calibrations are, of course, unknown.

The Tanqua fan complex (~400 m thick) outcrops over an area of 650 km^2 and comprises six individual submarine fan systems, some attaining 60 m in thickness (Wickens and Bouma, this volume). Owing to present-day outcrop preservations, facies distribution, and the progradational nature of the fans, there is not one specific locality where one can traverse vertically through the total succession. However, by traversing many sections in different parts of the basin, one can link the systems together. This has been our approach in constructing the sequence stratigraphic framework presented below.

SEQUENCE STRATIGRAPHIC AND CHRONOSTRATIGRAPHIC FRAMEWORK

The entire sedimentary fill of the Karoo Basin is identified as a first-order sequence, with its internal stratigraphic architecture closely controlled by orogenic cycles of loading and unloading in the CFB. The tectonic paroxysms dated in the CFB (Hälbich et al., 1983) further provide the basic subdivision of the foreland stratigraphy and is regarded as of second-order cyclicity (Catuneanu et al., 1998). However, the timespan for the total Ecca basin-fill succession equates with third-order duration of sea level cyclicity (Vail et al., 1984) and may be subdivided into two "third-order" (i.e., 1–10-m.y. duration) depositional sequences with several superimposed high-frequency, "fourth-order" (i.e., 0.1–1-m.y. duration) depositional sequences (Figure 1).

We use the term "depositional sequence" in the sense of Vail et al. (1984) and Van Wagoner et al. (1987), who define a sequence as a conformable succession of genetically related strata bounded by interregional unconformities or their correlative conformities (i.e. sequence boundaries). Initially, the term "sequence" was typically linked to "third-order" (i.e., 1–10-m.y. duration) changes in eustatic sea level, but now "sequence" carries no time connotation, and sequences originate via accommodation changes driven by *relative* sea level fluctuation and/or changes in sediment supply (Schlager, 1993). A sequence can be subdivided into component systems tracts defined as a linkage of contemporaneous depositional systems (i.e., shelf, slope, basin), which are a three-dimensional

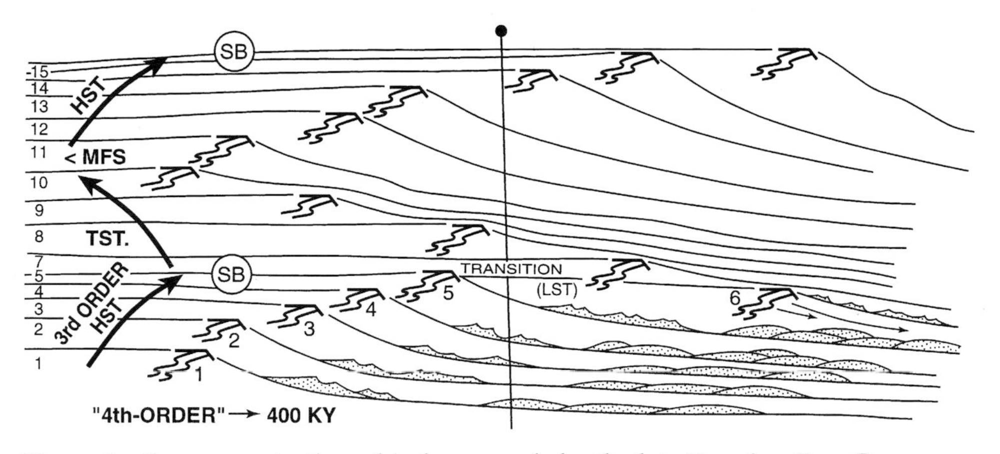

Figure 1—Sequence stratigraphic framework for the late Permian Ecca Group proposed in this paper. Updip is to the left, downdip is to the right. Vertical scale is schematic, as are the relative vertical/lateral proportions. The vertical line in the middle of the diagram approximates the succession preserved near the type section at Skoorsteenberg peak. *HST* = highstand systems tract, *LST* = lowstand systems tract, *MFS* = maximum flooding surface, *SB* = sequence boundary, *TST* = transgressive systems tract. Numbers on left refer to individual "fourth-order" high-frequency sequences.

assemblage of lithofacies (Vail et al., 1984; Van Wagoner et al., 1987). Systems tracts, characterized by predictable large-scale geometry and facies associations, are defined objectively based on position within a sequence, the types of bounding surfaces (type 1 or type 2 sequence boundaries, maximum flooding surface, etc.), and the stacking patterns of higher-order frequency sequences (or parasequences).

After final collapse and retreat of the Dwyka ice margin, probably caused by a relative sea level rise during the late Sakmarian (Visser, 1991), starved basin conditions developed in the southwestern part of the Karoo Basin. Marine conditions prevailed in the basin until at least upper Whitehill time, after which paleontological evidence suggests brackish conditions (Visser, 1992). The dark, fossiliferous shales of the Whitehill are interpreted as a marine-condensed unit with minor superimposed higher-order cyclicity, deposited under highstand conditions (Visser, 1993).

Highstand conditions of third-order duration prevailed until a fall in relative sea level, of fourth-order duration, initiated the first submarine fan to be deposited in the Tanqua subbasin. Subsequent rise in relative sea level caused sudden cessation of sediment supply due to changes in accommodation space, fan abandonment, and a return to highstand suspension-dominated deposition. The one or more dark shale horizons containing calcareous concretions between the fan systems are interpreted as condensed sections. The latter vary in thickness from a few centimeters to about 2 m and generally occur directly above the fan abandonment facies or about halfway between the fan systems. Subsequent fourth-order sea level lowerings with inferred type 1 unconformities, preferentially placed at the bases of the individual fan units, are used to explain the emplacement of these discrete, genetically related stratal packages of turbidites.

Figure 1 illustrates the proposed sequence stratigraphic architecture of the Ecca succession accompanied by Figure 2, which displays the inferred chronostratigraphic relations. In Figure 1, the first third-order sequence is shown as incomplete, lacking its lowstand and transgressive third-order component. This is probably a function of the preserved rock record in the area, and we suspect that the full sequence is probably expressed elsewhere in the basin (i.e., in the subsurface). The upper third-order sequence is complete, with all third-order systems tracts represented. Each third-order sequence consists of higher-frequency "fourth-order" (i.e., 0.1-1.0-m.y. duration) sequences, numbered 1 through 15 in Figure 1. In this hierarchical scheme, the fourth-order sequences are the fundamental "building blocks" of the larger seismic-scale sequences.

The individual fan systems, interpreted here as fourth-order lowstand deposits, each genetically tie to one high-frequency sequence. "Fan 1" is assigned to the lowstand systems tract of fourth-order sequence 2, "Fan 2" to fourth-order sequence 3, and so on (Figure 1). Fan systems 1 through 5 form a progradational stack revealed mainly by their spatial distribution and regional facies variation. This long-term progradational pattern is interpreted here as depicting the third-order highstand systems tract of sequence 1 and records a combination of reduced accommodation space and/or increased sediment supply.

At the type locality of Skoorsteenberg, approximately 900 m of Ecca Group section is exposed (Figure 3). Here the vertical stacking patterns provide an excellent

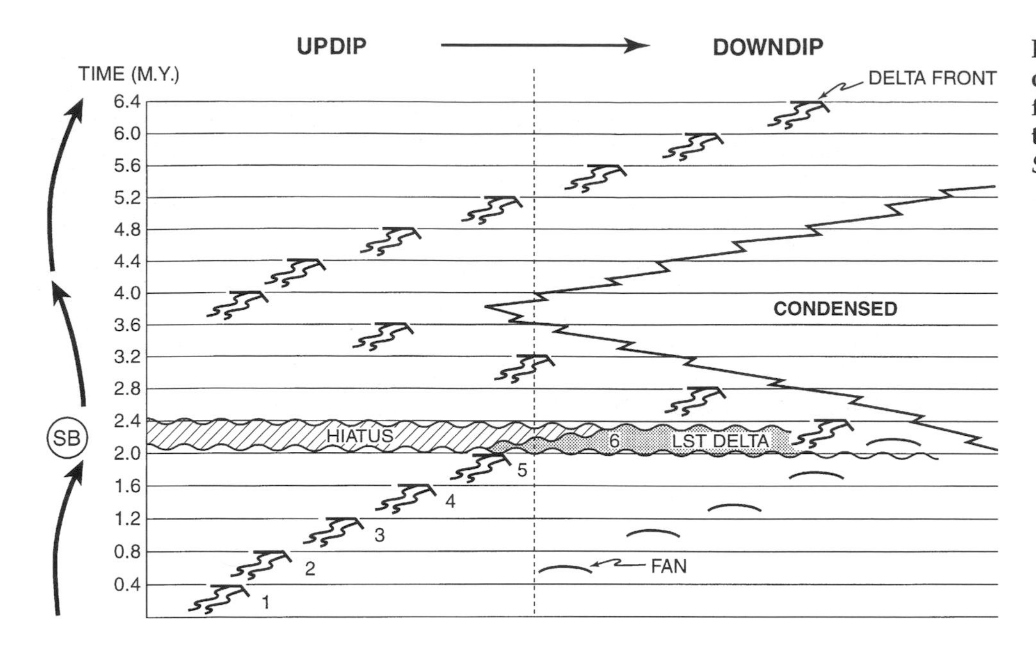

Figure 2—Chronostratigraphy of the sequence stratigraphic framework. Vertical axis is time in millions of years. *SB* = sequence boundary.

example of the sequence architecture proposed in Figure 1. This section exposes Fans 3 through 5, succeeded by prodelta shales and stacked upward-coarsening/thickening deltaic cycles of the Koedoesberg Formation (Figure 3). From the top of Fan 5 one observes systematic changes in facies proportions, sedimentary structures, and so on, indicating an upward shoaling environment that reflects progradation into the basin. Capped by thick sandstone units and dominated by T_a sequences, Fan 3 is interpreted to represent distal fan deposition comprising predominantly sheet sands. Fan 4 comprises predominantly laterally extensive single and amalgamated sheetlike T_{ab} sandstone units associated with large channel fills in places in the upper part of the succession. Fan 4 is interpreted to depict outer fan deposition with stacked and amalgamated sheet sands deposited as shingled lobes. Fan 5 is characterized by large-scale channeling and associated interchannel deposits in the Skoorsteenberg area, with variation from T_{ab} to T_c units. Fan 5 is interpreted as representative of proximal fan deposition marked by channel-overbank complexes and associated lobes. Above Fan 5, prodelta shales (Kookfontein Formation) with concretions signify a major backstep linked with the transgressive systems tract of the upper third-order sequence. These in turn are succeeded by upward-prograding, upward-thickening deltaics and associated fluvial deposits of the Koedoesberg Formation (Figure 3).

Within the context of our sequence-stratigraphic hierarchy, each fan system is assigned to the lowstand systems tract of each high-frequency, fourth-order sequence, and the particular attributes of each fan system are a consequence of their respective positions within the lower-frequency third-order sequence (i.e., third-order highstand, lowstand, etc.). The fourth-order sequence boundaries sit at the base of the major sand packages, which are interpreted as lowstand, and the recessive-weathering shaly intervals are interpreted as the fourth-order transgressive and highstand systems tract of each high frequency sequence. Although these interfan units comprise silty shale in general, especially in the updip position, they contain one or more dark shale horizons associated with carbonaceous concretions or layers. The latter are of early diagenetic origin and indicate periods of starvation in the basin. These horizons are interpreted as condensed sections representing maximum flooding of the 4th order highstand. These shaly intervals would equate updip with high-frequency progradational deltaics (shelf edges in Figure 1).

Fan 6 is not exposed at the Skoorsteenberg peak locality, but outcrops to the south (Wickens and Bouma, this volume; Wach et al., this volume). This fan is interpreted as a slope fan tied to high-frequency sequence 6, which appears to be a basinally restricted sequence of limited aerial extent. It is distinctly different from Fans 1 through 5, lacking sedimentary features indicative of basin–fan floor deposition. Rather, in vertical profile, one observes features indicative of a channelized slope fan, such as scour and fill and channelized slump deposits. In addition, at the Hangklip locality, the basal part of Fan 6 is overlain by deltaic deposits, with wave ripple and hummocky cross-stratification, within a few tens of meters.

Figure 2 illustrates the chronostratigraphic relations of the sequence stratigraphic framework. The time axis is, of course, unconstrained and hypothetical. Given the fact the Ecca Group is of Late Permian age, we speculate that the driving mechanism for the fourth-order sequences is glacioeustasy. In such a scenario, if one links the glacioeustatic forcing to Milankovitch climatic cycles, one might speculate that each high-frequency sequence represents approximately 400,000 yr, driven by Earth's eccentricity cycle, as is the case for the Pliocene-Pleistocene clastic deposits in the U.S. Gulf of

Figure 3—Oblique aerial photograph of the Skoorsteenberg peak section, view to the southeast. Approximately 900 m of section is exposed here. The distance between the two flags is approximately 2 km. The numbers 3, 4, and 5 refer to Fans 3, 4, and 5, respectively.

Mexico. Comparing Figures 1 and 2, a significant third-order condensed interval occurs in both time and space, equating theoretically to a regional diachronous downlap surface. Such an interval would perhaps be recognized in the subsurface with seismic data.

Figure 1 depicts a lowstand delta genetically related to Fan 6. Twenty kilometers to the north of the type locality at Skoorsteenberg, near Katjiesberg, Wickens (1994) has documented in detail a deltaic package that is probably equivalent to this third-order lowstand event. Here, the deltaic succession (locally referred to as the Katjiesberg deltaic member of the Koedesberg Formation) comprises approximately 8 to 10 upward-thickening cycles commonly reflecting delta lobe distributary mouthbar progradation, followed by stacked interdistributary bay-fill and channel-fill deposits. Although there may be arguments for allocyclic control of deposition, that is, inferring higher-order transgressive/regressive sequences, the authors prefer autocyclic processes of delta switching to build the progradational/aggradational succession. G. J. Brink (personal communication, 1992) uses the extensive nature of the delta front cycles, their sharp upper and lower contacts, and the presence of sandstone beds occurring low down in the delta front setting, to argue for higher-order sea level fluctuation. The deltaic succession at Katjiesberg is interpreted to represent third-order lowstand progradation (Wickens, 1994).

ORIGIN OF THE ECCA GROUP SEQUENCES

With regard to causal mechanisms, we cannot differentiate the contributions of eustasy vs. subsidence with regard to the origin of the long-term third-order sequences. Additional stratigraphic work needs to be completed and better dates obtained to enable one to calibrate the duration of the sequences. The role of orogenic loading and unloading in the CFB in driving patterns of subsidence needs to be evaluated. Figure 2 illustrates the chronostratigraphic relations of the sequence stratigraphic framework. The time axis is, of course, unconstrained and hypothetical. Comparing Figures 1 and 2, a significant third-order condensed interval occurs in both time and space, equating theoretically to a regional diachronous downlap surface. Such an interval would perhaps be recognized in the subsurface with seismic data. Given the fact that the Ecca Group is of Late Permian age, we speculate that the driving mechanism for the fourth-order sequences is glacioeustasy. In such a scenario, if one links the glacioeustatic forcing to Milankovitch climatic cycles, one might speculate that each high-frequency sequence represents approximately 400,000 yr, driven by Earth's eccentricity cycle, as is the case for the Pliocene-Pleistocene clastic deposits in the Gulf of Mexico.

Within individual fan systems (i.e., fourth-order lowstand systems tract), however, autocyclic (intrabasinal) processes, such as channel and lobe-type deposition, as well as other random phenomena (local topography) appear to dominate facies architecture, resulting in a less predictable arrangement of reservoir-scale elements (i.e., individual sheet sand bodies, etc). Such autocyclic drivers result in the formation of even higher-frequency cycles ("fifth-order," etc.), which are local in their lateral extent and thus impossible to correlate regionally.

ORDER VS. RANDOMNESS AND IMPLICATIONS FOR RESERVOIR PREDICTION

The sequence stratigraphic hierarchy provides a predictive framework that enhances our understanding of the Ecca Group large-scale basin-fill succession. In addition, the proposed framework places the individual fan systems (i.e., Fan 1 through Fan 6) within a larger context, thus linking them together. Our framework also comments on and has implications regarding order vs. randomness within the system. Simply stated, at what scale is the system more predictable with regard to facies architecture? Obviously, at the third-order scale, the system is fairly deterministic, with a predictable vertical stacking arrangement of facies and overall sand distribution. Armed with such a predictive framework, at any one point in the succession, one can better predict lateral (i.e., updip and downdip) and vertical facies architecture (facies proportions, net sand thickness, etc). The driving mechanism behind the third-order stacking is irrelevant; what is important is the order maintained within the system at this third-order scale.

Moving down in scale, within individual fan systems, however, autocyclic (intrabasinal) processes, such as channel and lobe-type deposition, as well as other random phenomena (local topography) appear to dominate facies architecture, resulting in a less predictable arrangement of reservoir-scale elements (i.e., individual sheet sand bodies, etc). Such autocyclic drivers result in the formation of even higher-frequency cycles ("fifth-order" and higher), which are local in their lateral extent and thus impossible to correlate regionally. Thus, at the higher-frequency scale, there is less predictability within individual fan systems, suggesting that their reservoir-scale architecture is driven more by autocyclic processes which are more random.

CONCLUSIONS

The Late Permian Ecca Group stratigraphic succession records a major phase of basin infilling within the Tanqua Karoo foreland subbasin. It is remarkably well preserved, and undeformed facies succession can be conveniently subdivided into a sequence stratigraphic hierarchy that provides a regional predictive framework. Such a framework links the exploration scale (seismic scale) to the exploitation scale (reservoir scale) and accentuates the utility of these outcrops as analogues for subsurface calibration.

REFERENCES CITED

Catuneanu, O., P. J. Hancox, and B. S. Rubidge, 1998, Reciprocal flexural behaviour and contrasting stratigraphies: a new basin development model for the Karoo retroarc foreland system, South Africa: Basin Research, v. 10, p. 417–439.

Hälbich, I. W., F. J. Fitch, and J. A. Miller, 1983, Dating the Cape orogeny, *in* A. P. G. Söhnge and I. W. Hälbich, eds., Geodynamics of the Cape Fold Belt: Geological Society of South Africa Special Publication, v.12, p. 149–164.

Haq, B. U., and F. W. B. Van Eysinga, 1987, Geological time table, 4th ed., Amsterdam, Elsevier.

Rubidge, B. S., 1991, A new primitive dinocephalian mammal–like reptile from the Permian of southern Africa: Palaeontology, v. 34, p. 547–559.

Schlager, W., 1993, Accommodation and sediment supply—a dual control on stratigraphic sequences: Sedimentary Geology, v. 86, p.111–136.

Vail, P. R., J. Hardenbol, and R. G. Todd, 1984, Jurassic unconformities, chronostratigraphy, and sea-level changes from seismic stratigraphy and biostratigraphy, *in* J.S. Schlee, ed., Interregional unconformities and hydrocarbon accumulation: AAPG Memoir 36, p. 129–144.

Van Wagoner, J. C., R. M. Mitchum Jr., H. W. Posamentier, and P. R. Vail, 1987, The key definitions of stratigraphy, *in* A. W. Bally, ed., Atlas of seismic stratigraphy, volume 1: AAPG Studies in Geology 27, p. 11–14

Visser, J. N. J.,1991, Self-destructive collapse of the Permo-Carboniferous marine ice sheet in the Karoo Basin: evidence from the southern Karoo: South African Journal of Geology, v. 94, p. 255–262.

Visser, J. N. J., 1992, Deposition of the Early to Late Permian Whitehill Formation during a sea-level highstand in a juvenile foreland basin: South African Journal of Geology, v. 95, p. 181–193.

Visser, J. N. J., 1993, Sea-level changes in a back-arc-foreland transition: the Late Carboniferous-Permian Karoo Basin of South Africa: Sedimentary Geology, v. 83, p. 115–131.

Wickens, H. de V. and A. H. Bouma, 1998, The Tanqua Karoo turbidites: facies architecture, heterogeneity distribution and reservoir applications, Unpublished field guide excursion guidebook, Tanqua Consortium, Louisiana State University, 73 p.

Wickens, H. de V., 1994, Basin floor fan building turbidites of the southwestern Karoo Basin, Permian Ecca Group, South Africa. Unpublished Ph.D. dissertation, University of Port Elizabeth, 233 p.

Wach, G. D., T. C. Lukas, R. K. Goldhammer, H. deV. Wickens, and A. H. Bouma, 2000, Submarine fan through slope to deltaic transition basin-fill succession, Tanqua Karoo, South Africa, *in* A. H. Bouma and C. G. Stone, eds., Fine-grained turbidite systems, AAPG Memoir 72/SEPM Special Publication No. 68, p. 173–180.

Chapter 16

Submarine Fan Through Slope to Deltaic Transition Basin-Fill Succession, Tanqua Karoo, South Africa

Grant D. Wach
Theodore C. Lukas
Robert K. Goldhammer
Texaco Upstream Technology
Houston, Texas, U.S.A.

H. Deville Wickens
Consultant
Kuils River, South Africa

A. H. Bouma
Department of Geology and Geophysics, Louisiana State University
Baton Rouge, Louisiana, U.S.A.

ABSTRACT

Superb sections of submarine fan deposits within the Late Permian Ecca Group are exposed within the Tanqua Karoo basin. Five discrete fan systems are capped by shales and deltaic deposits of the Kookfontein and equivalent Koedoesberg formations. Progradation of the deltaic deposits across the basin was in response to a decrease in accommodation space created by relatively high rates of sedimentation within the foreland basin setting. Evidence for the transition from submarine fans to deltaic deposition has been enigmatic, with limited evidence of sediments representative of slope deposition.

The sedimentology and sequence stratigraphy of the Hangklip Fan represents a slope fan and channel complex that shallows up into deltaic deposits, coincident with decresing accomodation space. Erosional slump scars, cutting into laminated shale with chaotic infill of sand intraclasts, point toward slope depositional processes that are not in evidence in the underlying submarine fan deposits. Wave ripples and the trace fossil *Gyrochorte* suggest substantially shallower depositional conditions than either the submarine fan or slope fan deposits, which are devoid of such features.

INTRODUCTION

The Hangklip Fan has been enigmatic, exhibiting some similarities to the underlying submarine fans, but with several glaring contrasts that require additional investigation. The known outcrops of the Hangklip Fan, also referred to as Fan 6, are all located south of the Ongeluks River (Figure 1), in an area where Fans 3, 4, and 5 all appear to thin and pinch out (Bouma and Wickens, 1994; Wickens and Bouma, 1998). Hangklip Fan has a measured thickness of 50 m at the type locality at Hangklip (Figure 2). Lithostratigraphic boundaries have been defined with the lower boundary the Tierberg shale and the upper boundary the deltaics of the Koedoesberg Formation (Wickens and Bouma, this volume).

Past studies recorded by Bouma and Wickens, 1994; Wickens and Bouma, 1998, established the geographic extent of the Hangklip Fan and described the unit as containing "stacked massive, load-formed (slumped?) sandstones, channel fills, and overbank deposits," and observed wave ripples and trace fossils attributed to the *Cruziana* ichnofacies in the upper part of the succession. Earlier interpretations (Bouma and Wickens, 1991, 1994) have suggested these physical and biogenic sedimentary structures may reflect shallower deposition, perhaps

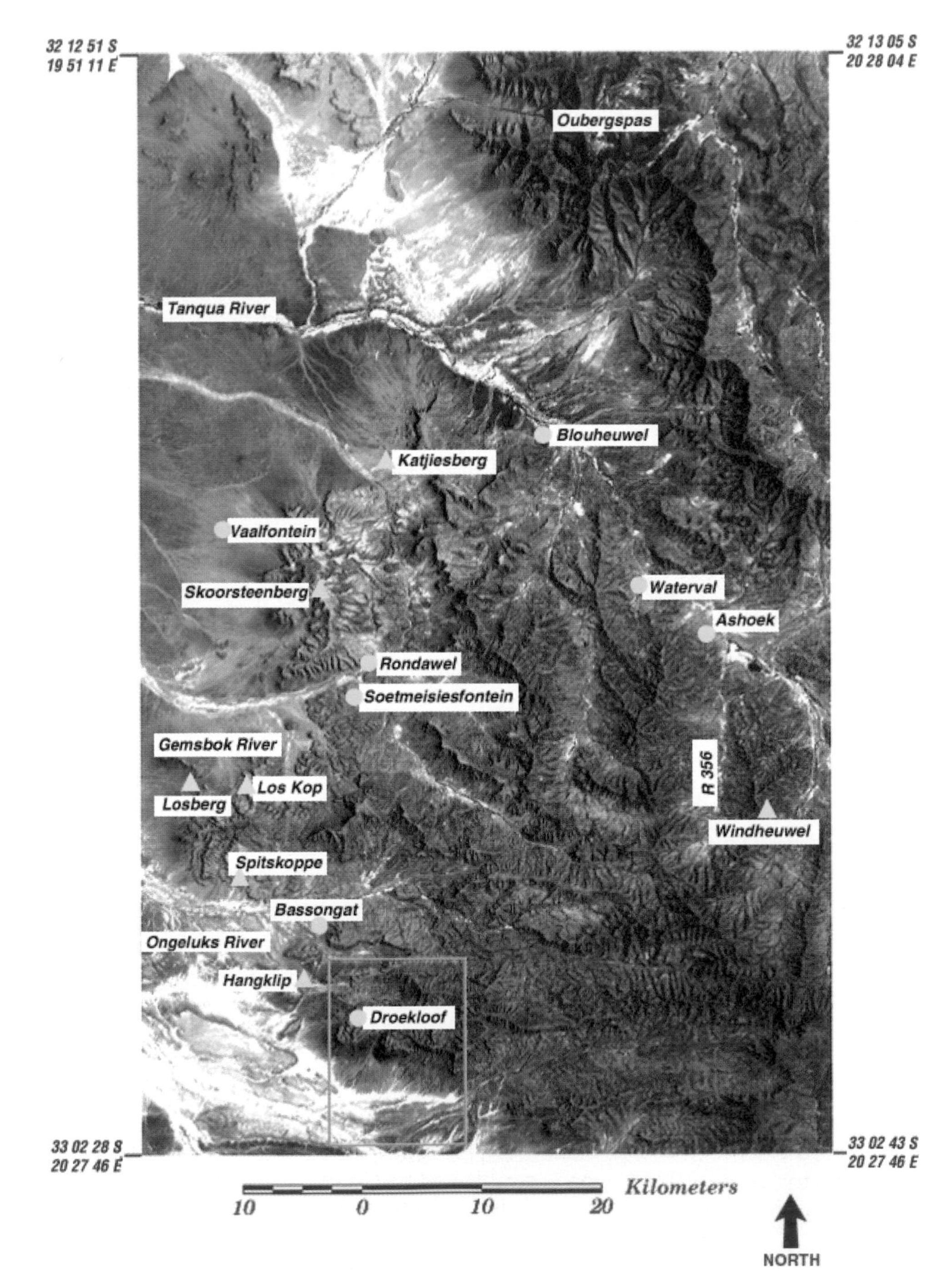

Figure 1—Landsat image of the western Tanqua Karoo showing the extent of the outcrops of fan and deltaic deposits. The study area of the Hangklip Fan is outlined by the box. The Ongeluks River is directly to the north of the study area. The Tanqua River runs east to west in the top half of the photo. The Gemsbok River is in the center. The north–south outcrop between the Tanqua and the Gemsbok rivers includes the type section of the formation at Skoorsteenberg.

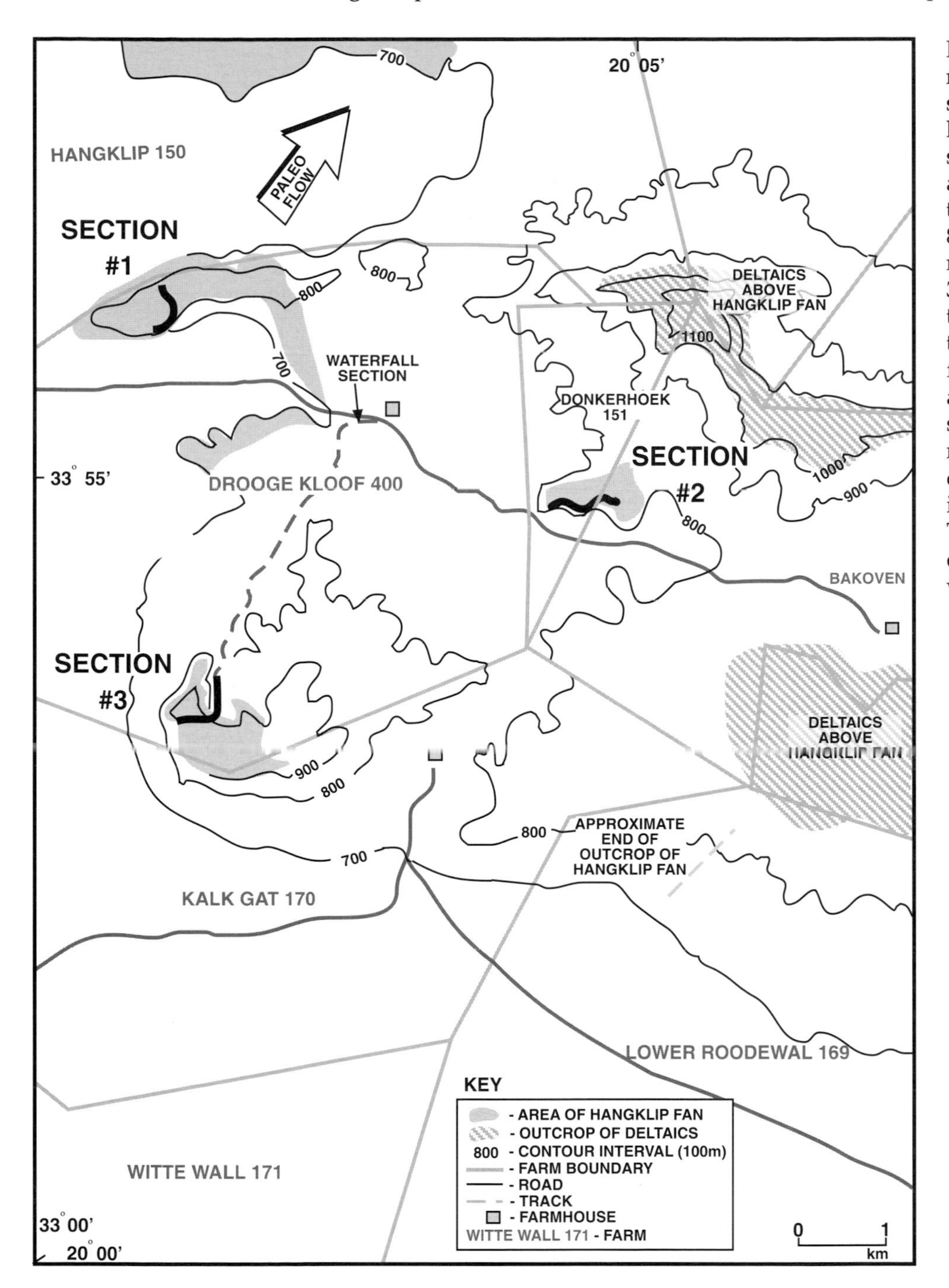

Figure 2—Detailed location map of the Hangklip Fan study area illustrating the location of measured outcrop sections. The type locality is along the southern margin of the Hangklip 150 farm. The 800-m contour at the base of measured sections 1 through 3 correlates with the start of the slope fan succession and the erosional slope scars found at each location. It appears that the erosional slumps in each outcrop are not contemporaneous but are discrete events along this interval of the paleoslope. The direction of paleoflow down the slope to the basin was to the northeast.

along the slope. The reduction in water depth may be attributable to tectonic uplift by the northeast propagation of the syntaxis (Wickens and Bouma, 1998).

Location of Measured Sections

The detailed map of the Hangklip Fan study area (Figure 2) illustrates the locations of the outcrops and sections measured for this study. Gray tones represent the areas of known outcrop exposures, with section 1 on the Droogekloof 400 farm. A small west–east–trending fault appears to cross adjacent to the Waterfall section, which exposes wave-reworked sediments of the upper part of the succession on the downthrown block. Observations, but no measured section, were made of this outcrop exposure. The gray hatched areas are the locations of deltaics of the Koedoesberg Formation. The Hangklip Fan appears to pinch out to the southeast on the Lower Roodewal 169 farm.

Sequence Stratigraphic Framework

The sequence stratigraphic framework of the succession is established by Goldhammer et al. (this volume). The cyclicity of the basin-fill succession was in response to relative changes in sea level throughout the late Permian. With the decrease in basin accommodation space, the depositional systems exhibit characteristics of shallower water conditions, which can be clearly predicted within the sequence stratigraphic framework.

LITHOFACIES AND DEPOSITIONAL ENVIRONMENT DESCRIPTION

Methods

Outcrop section 1 was measured with a tape with visual grain size measurements (1/2ϕ) every 50 cm. The gamma-ray curve (Figure 3) was constructed from the average of 12 scintillometer readings on the surface of the outcrop every 50 cm up the measured section. Sections 2 and 3 were measured using a Brunton level, with visual grain size measurements at every apparent lithology change. Paleocurrent measurements were recorded from available sedimentary structures, particularly wave ripples and scour axes, to determine downslope direction and paleoshoreline. Outcrop photo panoramas and detailed topographic sheets were used to correlate the measured sections. A helicopter overflight aided in this exercise.

Slope and Channelized Slope Fan

The sediments of the slope are comprised of dark-gray laminated siltstone. Measured section 3 (Figure 2) displayed dark grayish-black, organic-rich, siltstones in the lower part of the slope facies section. These organic-rich siltstones were interpreted as higher-order flooding surfaces. The channelized slope fan facies comprises very fine- to fine-grained, well-sorted sandstone beds with scour and fill features along amalgamation surfaces.

Intraslope Slump

The erosional slump and subsequent infill facies comprises a chaotic mix of very fine to fine-grained,

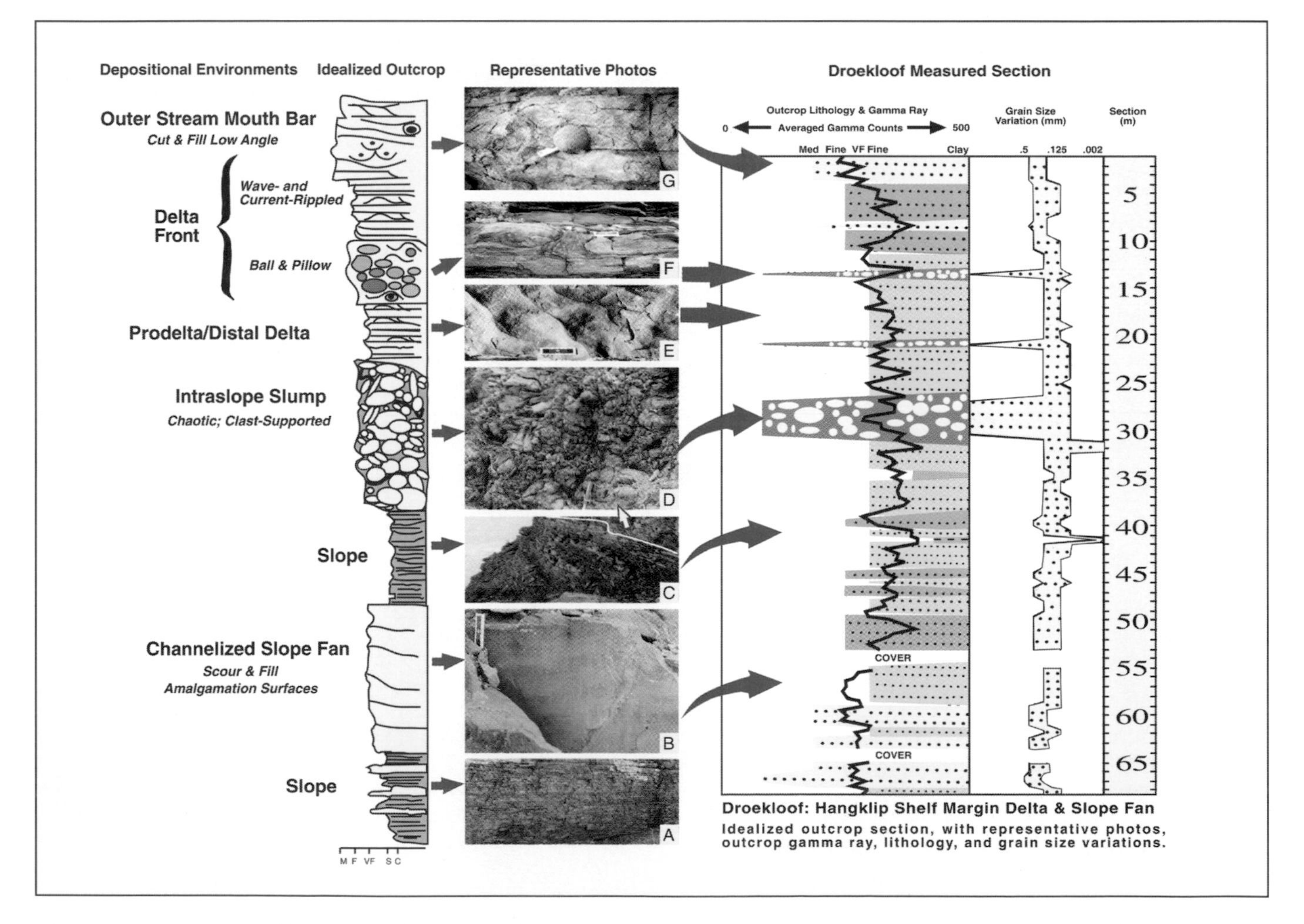

Figure 3—Summary diagram of the channelized slope fan and shelf-margin delta deposits described at the Droogekloof and Hangklip sections of the Hangklip Fan (Fan 6). From left to right: depositional environments and bedding types, an idealized outcrop profile, representative outcrop photos, the actual Droogekloof measured section including gamma-ray curve derived from scintillometer readings, lithologic column, and grain size measurements. Refer to the text for further discussion of the figure and photographs A–G.

rounded, boulder-size intraclasts. The infill is clast-supported, with very little infill of finer-grained sediment filling the voids around the clasts. The clast contacts are deformed, suggesting water-laden sediments undergoing rapid deposition coupled with dewatering adding to the soft-sediment deformation.

Deltaic Facies

Dark-gray laminated siltstones in the prodelta, not unlike those of the slope facies, become very fine-grained sandstones in the thin beds of the distal deltaic deposits. Thin flaser and lenticular beds of very fine sandstone preserve weak, wave-formed ripples.

In the delta front, the lithology is very fine- to medium-grained sandstone, with a marked shift in grain size to medium at the base of the interval, interpreted as outer stream mouth bar. Low-angle cut-and-fill features and trough cross-stratification are present. Wave-formed ripples are apparent with the trace fossil *Gyrochorte* preserved on bedding planes.

DISCUSSION

Idealized Outcrop Section—Shelf Margin and Slope Fan

On the left of Figure 3 are the depositional environments and sedimentary structures associated with the idealized outcrop profile of the slope to deltaic basin-fill succession. The profile was compiled from observations of three measured sections in the area of the Droogekloof and Hangklip farms (Figure 2). The section was drawn omitting repetitive fine-grained slope intervals.

At the base of the section, photo A (Figure 3) illustrates the parallel laminated siltstone and very fine-grained sandstone interbeds of the slope deposits. Photo B illustrates the slope channel and fan deposits with amalgamation surfaces typical of these sediments. The slump scar deposits are shown in photo C. There is significant downcutting and erosion, indicated by the yellow line, of the fine-grained laminated shale of the slope environment. This photo was taken looking northeast along the axis of the scar, downdip toward the basin. Photo D shows in detail the chaotic nature of the fill of the slump deposits within the erosional scour on the slope. Note the clast-clast contacts and soft-sediment deformation, evidence of rapid deposition and deformation initiated by large-scale slides along the surface of the slope. The fill appears to be continuous, with no clear break through the interval, suggesting multiple episodes of infill. The interval is capped with some contorted fine-grained beds of siltstone. Ladder-back wave ripples (photo E), indicative of very shallow conditions, preserve two oblique sets of low-amplitude wave-induced current ripples. Ball and pillow deformation (photo F) represent rapid deposition and subsequent dewatering of sediments, typical of conditions in the delta front. The top photo (G) shows low-angle cut-and-fill and cross-bedded fine- to medium-grained sandstone of the outer stream mouth bar facies capping the succession of the shelf-margin delta. Note the sandstone concretion, not uncommon in sediments of this interval.

Transition from Slope to Shelf-Margin Deltaic Deposits

A schematic section (Figure 4) of the basin-fill succession represented by the transition from the slope to shelf-margin deltaic deposits shows the relative location and development of intraslope slump deposits, as interpreted from the outcrop sections of the Hangklip Fan. Rapid progradation of the shelf-margin deltas, in response to decreased accommodation space within the basin produced an unstable delta front, recorded in part by ball and pillow features observed in the sections (Figure 3). The unstable delta front initiates downslope mass transport and scouring of the slope. Subsequent infill of the erosional scours is chaotic, reflecting the unstable nature of the slope and overlying deltaic deposits on the shelf edge. The same processes can lead to substantial bypass of sediment down the slope to the basin floor, creating the submarine fans that are preserved in the Tanqua Karoo basin. Figure 5 is a schematic view of the depositional system illustrating the transition from the last phase of submarine fan deposition through the intraslope and shelf-margin deltaic deposits of the Hangklip Fan (Fan 6), to the coastal margin deltaics of the Koedoesberg Formation.

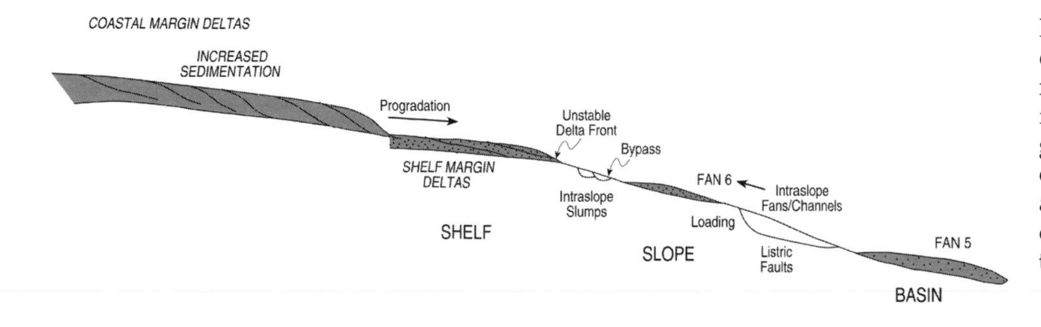

Figure 4—Schematic section of the basin-fill succession represented by the transition from the slope to shelf-margin deltaic deposits. Continuous basinward progradation and sediment infill results in decreased basin accommodation space.

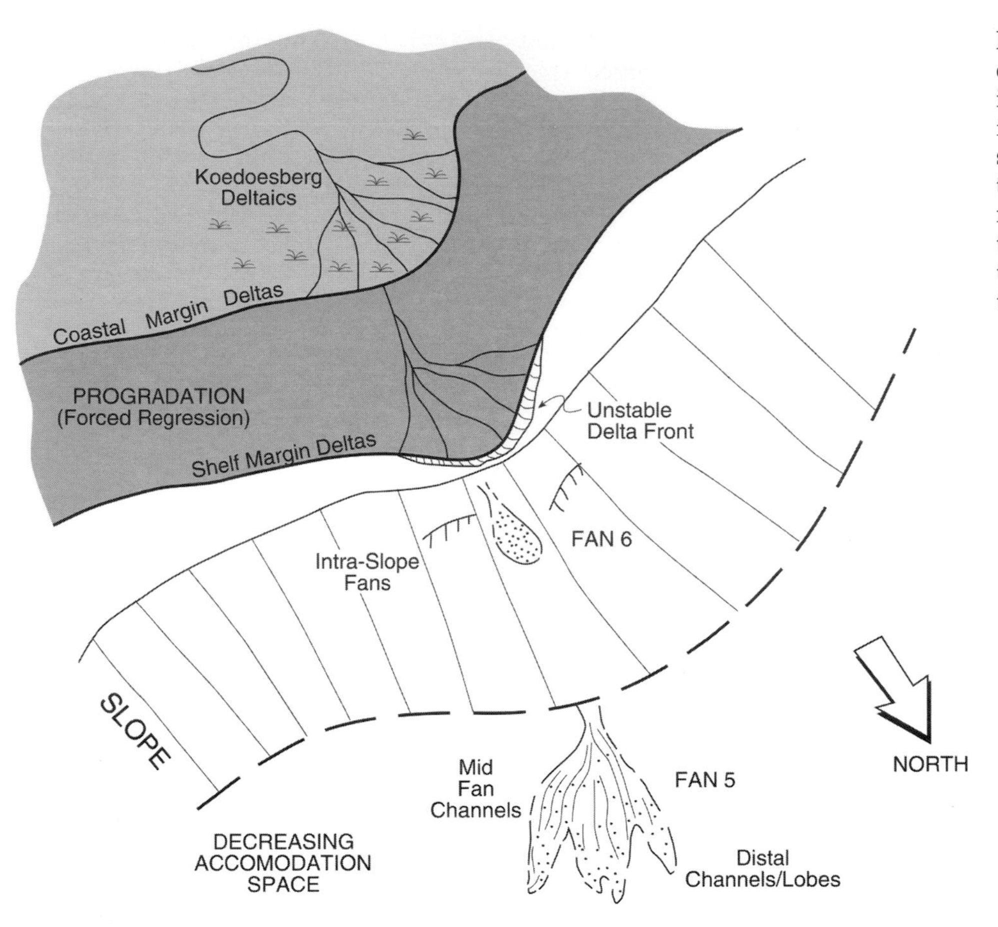

Figure 5—Schematic view of the depositional system illustrating the transition from the last phase of submarine fan deposition to the intraslope and shelf-margin deltaic deposits of the Hangklip Fan (Fan 6), to the coastal margin deltaics of the Koedoesberg Formation.

Paleoflow

Based on the paleocurrents obtained from the sedimentary structures, particularly ripples and the axes of the slump scars, the paleoflow down the slope to the basin is interpreted to be to the northeast, depicted by the arrow in the upper left corner of Figure 2. This direction is consistent with the mapped orientation of the five submarine fan complexes basinward of the slope, and the orientation of the Cape Fold Belt to the west and southwest, which formed the paleoshoreline.

The 800-m contour at the base of the measured sections 1–3 in Figure 2 correlates with the start of the slope fan succession and the erosional slope scars at each of the measured sections. The erosional slump scars occur in a similar position of the measured sections in each outcrop of the stratigraphic succession. They do not appear to be contemporaneous but are discrete events along this interval of the paleoslope.

Architectural Elements

The main architectural element viewed in this study is the scale of the slope slump scars and infill. The erosional surface at the base of the slumps can be traced for 150 m in section 1 to more than 300 m in section 2. The thickness of the slump deposits is 3–10 m. The channelized slope fan deposits comprise beds with amalgamation surfaces that range in thickness from 30 to 150 cm. Each bed may represent a discrete flow event that can be traced for a distance up to 150-250 m, before usually disappearing under cover associated with the finer-grained sediments. The lateral extent of the amalgamated beds is 300–1500 m. The delta front sandstone deposits have a similar extent. The slope deposits extend laterally for distances greater than 2000 m, but the details are often obscured by ground cover formed on these finer-grained sediments. The interlaminated deposits of the prodelta exhibit similar lateral extent.

Processes

Scouring of the slope sediments was likely initiated by traction currents created at the base of turbidity flows moving down the slope. Infill of erosional slump scars could occur during waning current flow, through multiple flow events, or were filled after formation of the slump scars. Evidence for a multiple fill sequence is not readily apparent in the infill of the slumps viewed at outcrop. No surfaces were observed that could be interpreted as depositional hiatuses or evidence of any particular order or grading of the clasts within a slump. Deposition appears to be attributed to one event, perhaps contemporaneous with the initiation of the slope slump scar. Given the chaotic nature of each deposit and the soft-sediment deformation, there is the possibility that a surface could be missed, but this was not apparent. The fill appears to be continuous, with no clear break in the succession. The end of the infill phase was apparently rapid with some contorted fine-grained beds of siltstone capping the succession representing low density turbidity deposition.

Subsurface Comparison and Implications for Reservoir Quality

Slump deposits have been recorded along slope deposits of both passive and active margins. Dixon and Weimer (1998) record large-scale submarine slides visible on seismic that form on low-gradient surfaces associated with the Mississippi Fan, although these features appear to be an order of magnitude larger than the smaller slumps recorded in the Hangklip Fan. Figure 6 is an FMI image illustrating slump deposits from a slope deposition setting in the Gulf of Mexico. The interval shown is similar in scale to those studied in the outcrop of the Hangklip Fan (Fan 6).

Analogous deposits of the slope fans and channels would make fine reservoir targets, with good porosity and vertical and lateral permeability providing reservoir connectivity. The lateral extent of the slope fans and channels would, however, be limited in comparison with the reservoir potential of the downdip submarine fan deposits within the prograding complex. To form reservoirs, sands need to be trapped within topographic lows along the slope, commonly created by slumping or faulting, creating both conduits and traps for the sands. If these traps are not present, the slope is either an area of sediment bypass of sand and silt or there is trapping and aggradation of nonreservoir silt deposits.

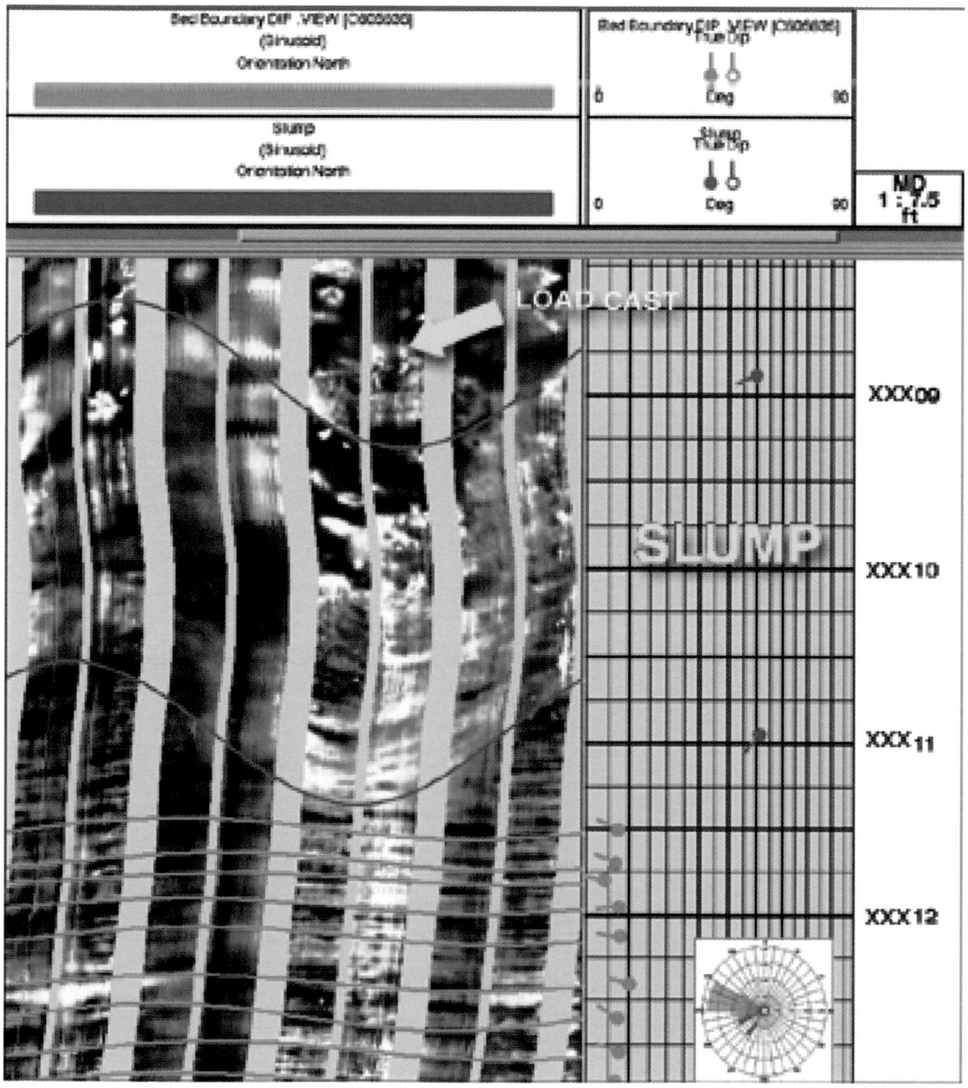

Figure 6—An FMI image illustrating slump deposits from a slope deposition setting in the Gulf of Mexico. The interval shown is similar in scale to those studied in the outcrop of the Hangklip Fan (Fan 6).

CONCLUSIONS

The Hangklip Fan has been enigmatic in the past due to the very nature of the transition it represents from deeper submarine fan deposits through the slope to shallow-water shelf-margin deltaic deposits. The Hangklip Fan represents a slope fan and channel complex and slump deposits of the mid-upper slope, that shallows upward into deltaic deposits, capping the basin-fill succession. Erosional slump scars, cutting into laminated shale with chaotic infill of sand intraclasts, point toward slope depositional processes that are not in evidence in the stratigraphically lower submarine fan deposits. Progradation of these deltaic deposits across the basin was in response to a decrease in accommodation space probably created by relatively high rates of sedimentation, producing increased sediment volumes within the foreland basin setting.

ACKNOWLEDGMENTS

We thank Texaco for permission to publish this work and the Texaco Remote Sensing Laboratory for the Landsat image used in Figure 1. We acknowledge reviewers, particularly James L. Coleman, for suggestions that have improved the manuscript, Linda Lovell for assistance with the illustrations, and Mike Strickler, Greg Isenmann, and John Bretches for obtaining permission to publish Figure 6. Finally, we would be remiss if we did not thank the landowners of the Tanqua Karoo who were kind hosts and gave permission to access their land.

REFERENCES CITED

Bouma, A. H., and deV. Wickens, H., 1991, Permian passive margin submarine fan complex, Karoo Basin, South Africa: possible model to Gulf of Mexico: Gulf Coast Association of Geological Societies Transactions, v. 41, p. 30–42.

Bouma, A. H., and deV. Wickens, H., 1994, Tanqua Karoo, ancient analog for fine–grained submarine fans, *in* P. Weimer, A. H. Bouma, and B. F. Perkins, eds., Submarine fans and turbidite systems: sequence stratigraphy, reservoir architecture and production characteristics, Gulf of Mexico and international: Gulf Coast Section SEPM Foundation, 15th Annual Research Conference Proceedings, p. 23–34.

Dixon, B. T., and P. Weimer, 1998, Sequence stratigraphy and depositional history of the Mississippi Fan (Pleistocene), northeastern deep Gulf of Mexico: AAPG Bulletin, v. 82, p. 1207–1232.

Golhammer, R. K., et al., 2000, Sequence stratigraphic architecture of the Late Permian Tanqua Submarine Fan Complex, Karoo Basin, South Africa, this volume.

Wickens, H. deV., 1994, Basin floor fan building turbidited of the southwestern Karoo Basin, Permian Ecca Group, South Africa. Unpublished Ph.D. dissertation, University of Port Elizabeth, 233 p.

Wickens, H. deV., and A. H. Bouma, 1998, The Tanqua Karoo turbidites: facies architecture, heterogeneity distribution and reservoir applications, unpublished field guide excursion guidebook, Tanqua consortium, Louisiana State University, 73 p.

Slatt, R. M., 2000, Why outcrop characterization of turbidite systems, *in* A. H. Bouma and C. G. Stone, eds., Fine-grained turbidite systems, AAPG Memoir 72/ SEPM Special Publication No. 68, p. 181–186.

Chapter 17

Why Outcrop Characterization of Turbidite Systems

Roger M. Slatt
Department of Geology and Geological Engineering, Colorado School of Mines Golden, Colorado, U.S.A.

ABSTRACT

Building geologic models of turbidite reservoirs for simulation and development planning using only subsurface data suffers from either discontinuous information or information that is displayed at too large a scale of resolution to detect variations in significant geologic properties. Large, continuous outcrops help constrain characterization of reservoirs by providing quantitative lateral and vertical attributes of strata and their bounding surfaces. Besides the traditional tools for examining outcrops, newer techniques include photo imaging, behind-outcrop logging/coring/seismic, gamma-ray/velocity logging, permeability profiling, and ground-penetrating radar. When used in combination, reliable, quantitative characterizations of turbidite outcrops can be developed. Future research should focus on full 3-D quantification of outcrops.

INTRODUCTION

Discipline-integration within the petroleum industry during the past 10 years has changed the way in which geoscientists interact with those in other petroleum disciplines. Considerable geologic integration has been aimed at quantification of geologic properties and variables in order to build subsurface models of structure and stratigraphy. Interaction with petroleum engineers has undergone particularly profound change as geoscientists have been forced from providing reservoir engineers with conceptual geological models to providing data-based, quantitative reservoir models that can address questions such as "How big is the container?" "How will this reservoir style perform?" "How widely must we space our expensive development wells in this geologic setting?" "Should we drill a vertical, slant, or horizontal well?" or "What went wrong?"

Reservoir simulation also has become a common practice for predicting the performance of turbidite (and other) reservoirs and to attempt to optimize reservoir development (Burk et al., 1999). Increasingly, companies are relying on reservoir simulations in their development plans for billion-dollar investments. Geostatistical models are used with varying resolutions of stratigraphic detail. Such models can help provide timely answers for reservoir development projects with short time spans between discovery and production. Reliable simulation results are critical for optimal well placement and successful water-flood projects, and require a good geologic model that captures the heterogeneity of the reservoir.

However, quantifying the heterogeneity of turbidite reservoirs for characterization and simulation is a difficult and complex process, and accounts, in part, for the need for geostatistical models. Subsurface reservoir models built using subsurface data alone suffer from either discontinuous information or information that is displayed at too large a scale of resolution to detect potentially significant variations in reservoir geologic properties.

The common subsurface tools for developing geologic models are well logs, cuttings, cores, 3-D seismic, well test and production information. Well information (including cores and borehole image logs) may provide excellent vertical resolution of data, but correlating between wells can be difficult, leading to considerable uncertainty in the lateral resolution and correlation of structural and stratigraphic features. Three-dimensional seismic horizon slices from 3-D data volumes can provide an excellent image of between-well geometries and lateral attributes, but key structural and stratigraphic features that can influence production, such as bed continuity and bed boundaries, may be below the limit of both lateral and vertical resolution. Well-test and production information provide insight into such reservoir properties as permeability, sand continuity, and faults, but there can be considerable uncertainty in exactly what controls well-test results, particularly when barriers to fluid flow are encountered.

THE VALUE OF OUTCROPS FOR CHARACTERIZATION

Turbidite outcrops help to constrain the characterization of subsurface turbidite reservoirs. Where sufficiently large and continuous, outcrops can provide the necessary quantitative lateral and vertical attributes of turbidite strata and can provide critical information on the nature of bed boundaries and how they might affect reservoir fluid flow. Such outcrops help bridge the gap between well log and seismic scales of resolution.

However, finding suitably large outcrops for characterization in two- and three-dimensions is often difficult, because many outcrops are discontinuous, faulted, or not well exposed. For example, many published results of turbidite bed continuity measurements are from outcrops over which only a few tens to hundreds of meters of bed length could be measured (Chapin et al., 1994; Cossey, 1994; DeVries and Lindholm, 1994; Browne and Slatt, 1997). Enos (1969), Mutti (1992), and Bouma and Wickens (1994) provide data on longer outcrops. Architecture of only a few turbidite outcrops has been documented in sufficient detail for reservoir simulation or modeling (Cook et al., 1994; Gardner et al., 1998; Johnson, 1998; Slatt et al., this volume).

TECHNIQUES FOR OUTCROP CHARACTERIZATION

With the emphasis on constructing quantitative geologic models for application to the subsurface, it becomes necessary, or at least more efficient, to place outcrop information into the more familiar format used by explorationists and reservoir analysts—logs, cross sections, seismic (real and synthetic), and 3-D visualization models. The need for both quantitative data and the conversion of outcrop to "subsurface" data has generated a variety of characterization tools and techniques, as summarized below. Nevertheless, the calibration comes from basic outcrop measurements using the time-honored and traditional rock hammer, hand lens, measuring tape, and Brunton compass.

Outcrop Imaging

Aerial or ground photography to develop outcrop photomosaics for stratigraphic and structural interpretation is particularly useful for bridging the substantial gap between seismic and individual well scales, particularly when photomosaics are placed into electronic files for easy size manipulation and overlay (Dueholm and Olsen, 1993). Photomosaics are easily scanned into electronic files so they can be interpreted on a workstation monitor in much the same manner as a seismic line or cross section is interpreted. Bounding surfaces and faults can be highlighted and facies characteristics can be coded and their proportions measured. However, it is essential that such key features shown on photomosaics are examined and authenticated at the outcrop.

By scaling-up an outcrop image of genetically related bounded facies or beds, and assigning acoustic properties to the intervals, it is possible to develop 2-D synthetic-seismic images for comparison with real seismic data (Batzle and Smith, 1992). Where 3-D outcrops are available, 3-D synthetic seismic images as well as 3-D geologic models for visualization and fluid flow simulation can be developed (Tinker, 1996).

Behind-Outcrop Logging, Coring, and Seismic

Drilling, logging, and coring wells behind an outcrop of good areal and vertical extent provides an excellent means of relating outcrop observations to characteristics seen on typical subsurface wireline logs and cores. The same is true if a high-resolution seismic line can be obtained from behind the outcrop.

Although behind-outcrop coring and logging date back to at least the 1970s, to our knowledge the first attempt at obtaining borehole image logs from behind an outcrop, coupled with core outcrop characterization, was conducted in 1996 in New Zealand (Spang et al., 1997; Slatt et al., 1998). Here, borehole image features were calibrated with sedimentologic and stratigraphic features observed in core (vertical features) and outcrop (lateral features) to delineate well log criteria for interpreting these features laterally away from the wellbore. A nearby high-resolution seismic line also provided a direct comparison with stratification styles and significant bounding surfaces. Similar techniques are now being used in other areas. In a recent study of Lewis Shale (Wyoming) turbidites, Witton (1999) compared outcrop sedimentologic features with borehole image features in order to predict sand continuity away from the borehole.

Outcrop Gamma-Ray and Sonic Velocity Logging

Gamma-ray logging of outcrops, using a hand-held scintillometer or a standard logging truck, has been applied for several years to qualitatively correlate subsurface gamma-ray log response to rock properties and stratigraphic character (Jordan et al., 1993; Slatt et al., 1995). For individuals who routinely work only with subsurface well logs, the outcrop gamma-ray log can provide a reality check because the logs record stratigraphic and sedimentologic properties that are reflected by variations in their natural radioactivity. A hand-held sonic velocity device is also available to obtain velocity profiles of outcrops, which are useful for generating 2-D or 3-D synthetic seismic models (Batzle and Smith, 1992).

Outcrop Permeability Measurements

Outcrop minipermeameters or closely-spaced core plugs are used to develop permeability profiles of rock outcrops (Dreyer et al., 1990; Jordan and Pryor, 1992). Such data, when coupled with outcrop gamma-ray or sonic velocity logs of outcrops, can be used to estimate permeability in equivalent subsurface rocks from which gamma-ray or sonic logs have been obtained. A particularly powerful application of outcrop permeability measurement is to develop 2D (or 3D) outcrop-permeability profiles of a stratigraphic sequence by cross-correlating closely-spaced vertical permeability profiles or making numerous lateral permeability measurements (Kerans et al., 1994). The resultant permeability profile can then be input into a reservoir simulator for 2-D or 3-D fluid flow modeling.

Ground-Penetrating Radar

Ground-penetrating radar (GPR) offers the opportunity to compare outcrop stratigraphic and structural characteristics to pseudoseismic response (Young et al., 1995). The very high resolution afforded by GPR provides an almost one-to-one correspondence with rock geometry, connectivity, and bounding surfaces within the upper 10 or so meters immediately beneath ground surface. Three-dimensional GPR offers promise for developing pseudo-3-D seismic cubes of outcrop features with appropriate 3-D outcrop faces, and the unique ability to characterize the interior structure of a 3-D outcrop volume on a bed-by-bed scale. Recent GPR surveys of turbidite outcrops are presented by Young et al. (1999).

CONCLUSIONS

Outcrops are the backbone of geology. Through the years, most fundamental geologic observations and understandings have evolved through the study of outcrops. Outcrops are particularly important when dealing with subsurface data because they can provide the detailed documentation of lateral and vertical rock attributes that cannot be determined by any other means. In the petroleum industry, outcrops provide a key training component by allowing geologists, geophysicists, and petroleum engineers to see in the field, sometimes in 3-D, what they must interpret in the subsurface with more limited data.

The need for rapid, reliable, quantitative characterizations of turbidite reservoirs is going to increase as more emphasis is placed on reducing both development costs and the time between discovery and development. As demonstrated in this volume, more and more turbidite outcrop studies are being conducted with the goal of quantification of features that affect reservoir performance. Both traditional and newer methods are being applied toward these characterizations.

Of the newer techniques described above, photomosaics allow one to develop a large-scale stratigraphic (and structural) framework. Behind-outcrop high-resolution seismic records can be tied to major stratigraphic features observed at the photomosaic scale. Both outcrop and behind-outcrop logging provides vertical detail of rock features in a format that is readily understandable by people who work with logs on a routine basis. Seismic models can also be made from velocity profiles to correlate with behind-outcrop high-resolution seismic reflection records. Basic outcrop observations and documentation are essential to provide the details of major and minor bounding surfaces mapped on the larger-scale photomosaics, as well as the internal stratigraphy between surfaces. Ground-penetrating radar provides additional small-scale stratigraphic (and structural) detail internal to the outcrop surface. To complete a full reservoir model requires developing permeability profiles of the outcrops. If outcrop permeabilities are not suitable for comparison with a reservoir of interest, then permeability (and porosity) values can be substituted into the reservoir model.

FUTURE DIRECTIONS

Future directions must be aimed at full 3-D quantification of rock properties from the bed to the reservoir scale. Concurrent efforts should be undertaken to improve seismic resolution and imaging of (currently) subseismic scale reservoir properties (Slatt and Weimer, 1999). Also, computing capabilities need to continually evolve to enable input of more quantitatively detailed, and thus more realistic, reservoir building blocks into production simulators.

The full 3-D geological model envisioned through combination of the techniques mentioned above can then either be scaled up for simulation or simulated using the available level of documented detail (Hurst et al., 1999). Different levels of detail can be sequentially added to models to document anticipated changes to reservoir performance during the life of a field. For example, small-scale stratigraphic features might not affect early primary performance but might adversely affect secondary performance (Leonard and Bowman, 1998). Also, detailed models provide the opportunity to test the validity and uncertainties associated with geostatistcal grid-block population methods.

A full 3-D reservoir model can also be used to improve seismic interpretation of subseismic scale features or infer their occurrence within a larger-scale, seismically derived stratigraphic framework. This can be accomplished by electronically overlaying the detailed model onto a seismic reflection record of an analog reservoir at a one-to-one scale. Two- and three-dimensional seismic modeling can add value to this type of effort by modeling the seismic image of various key reservoir architecture components using impedance models designed through outcrop study.

Two additional future directions are worth noting. First, there is renewed effort toward understanding turbidite and related deepwater depositional processes and their products, particularly as applied to predicting bed continuity away from a data point (i.e., wireline log, core, or outcrop). This effort should be continued through experimental flume studies, process-response computer modeling, and outcrop observation until predictive features are identified and verified so they can be interpreted with confidence.

Second, the surface and near-surface environments of modern-day submarine fans have been studied for many years using side-scan sonar, shallow coring, and other techniques. Features on modern fans need to be documented at a resolution that is comparable to 3-D seismic horizon slices for comparative purposes, and to improve interpretation of the latter.

ACKNOWLEDGMENTS

I thank Drs. Leslie Wood, Jack Thomas, and Arnold Bouma for providing critical feedback to an earlier version of this paper. I am indebted to many people with whom I've conducted outcrop and behind-outcrop studies for the past several years, and to the numerous companies that have funded these studies. The recent work of Dr. Michael Gardner and students is particularly worth acknowledgment for demonstrating the approach and value of multitechnique, multiscale, quantitative characterization of outcrops.

REFERENCES CITED

Batzle, M. L., and B. J. Smith, 1992, Hand-held velocity probe for outcrop and core characterization, *in* J. R. Tillerson and W. R. Wawersik, eds., Proceedings of the 33rd U.S. Symposium on Rock Mechanics, Santa Fe, New Mexico, p. 949–958.

Bouma A. H., and H. DeV. Wickens,1994, Tanqua Karoo, ancient analog for fine-grained submarine fans, *in* P. Weimer, A. H. Bouma, and B. F. Perkins, eds., Submarine fans and turbidite systems: sequence stratigraphy, reservoir architecture, and production characteristics: Gulf Coast Section SEPM Foundation 15th Annual Research Conference Proceedings, Houston, p. 23–34.

Browne, G. H., and R. M. Slatt, 1997, Thin-bedded slope fan (channel-levee) deposits from New Zealand: an outcrop analog for reservoirs in the Gulf of Mexico: Gulf Coast Association Geological Societies Transactions, v. 47 p. 75-86.

Burk, M. K., G. L. Brown, and D. R. Petro, 1999, Evolution of the geological model, Lobster Field (Ewing Bank 873) (abs.), AAPG Annual Convention, San Antonio, Texas.

Chapin, M. A., P. Davies, J. L. Gibson, and H. S. Pettingill, 1994, Reservoir architecture of turbidite sheet sandstones in laterally extensive outcrops, Ross Formation, western Ireland, *in* P. Weimer, A. H. Bouma, and B. F. Perkins, eds., Submarine fans and turbidite systems: sequence stratigraphy, reservoir architecture, and production characteristics: Gulf Coast Section SEPM Foundation 15th Annual Research Conference Proceedings, Houston, p. 53–68.

Cook, T. W., A. H. Bouma, M. A. Chapin, and H. Zhu, 1994, Facies architecture and reservoir characterization of a submarine fan channel complex, Jackfork Formation, Arkansas, *in* P. Weimer, A. H. Bouma, and B. F. Perkins, eds., Submarine fans and turbidite systems: sequence stratigraphy, reservoir architecture, and production characteristics: Gulf Coast Section SEPM Foundation 15th Annual Research Conference Proceedings, Houston, p. 69–81.

Cossey, S. P. J., 1994, Reservoir modeling of deepwater clastic sequences: mesoscale architectural elements, aspect ratios, and producibility, *in* P. Weimer, A. H. Bouma, and B. F. Perkins, eds., Submarine fans and turbidite systems: sequence stratigraphy, reservoir architecture, and production characteristics: Gulf Coast Section SEPM Foundation 15th Annual Research Conference Proceedings, Houston, p. 83–93.

DeVries, M. B., and R. M. Lindholm, 1994, Internal architecture of a channel-levee complex, Cerro Toro Formation, southern Chile, *in* P. Weimer, A. H. Bouma, and B. F. Perkins, eds., Submarine fans and turbidite systems: sequence stratigraphy, reservoir architecture, and production characteristics: Gulf Coast Section SEPM Foundation 15th Annual Research Conference Proceedings, p. 105–114.

Dreyer, T., A. Scheie, and O. Walderhaug, 1990, Minipermeameter-based study of permeability trends in channel sandbodies: AAPG Bulletin, v. 74, p. 359–374.

Dueholm, K. S., and T. Olsen, 1993, Reservoir analog studies using multimodel photogrammetry: a new tool of the petroleum industry: AAPG Bulletin, v. 77, p. 2023–2031.

Enos, P., 1969, Anatomy of a flysch: Journal of Sedimentary Petrology, v. 39, p. 680–723.

Gardner M. H., K. Johnson, M. Batzle, M. Sonnenfeld, and B. Sinex, 1998, Geologic building blocks for reservoir characterization—lessons learned from the Permian Brushy Canyon Formation, west Texas, *in* M. H. Gardner, K. Johnson, M. Batzle, M. Sonnenfeld, and B. Sinex, eds., Developing and managing turbidite reservoirs: case histories and experiences: European Association of Geologists and Engineers/American Association of Petroleum Geologists Third Research Symposium, 1998, Almeria, Spain.

Hurst, A., I. Verstralen, B. Cronin, and A. Hartley, 1999, Sand-rich fairways in deep-water clastic reservoirs: genetic units, capturing uncertainty, and a new approach to reservoir modeling: AAPG Bulletin, v. 83, p. 1096–1118.

Johnson, K. R., 1998, Outcrop and reservoir characterization of the Brushy Canyon Formation, Guadalupe Mountains National Park, west Texas, and Cabin Lake field, Eddy County, New Mexico: Unpublished M.S. thesis, Colorado School of Mines, Golden, Colorado, 256 p.

Jordan, D. W., and W. A. Pryor, 1992, Hierarchical levels of heterogeneity in a Mississippi River meander belt and application to reservoir systems: AAPG Bulletin, v. 77, p. 118 120.

Jordan, D. W., R. M. Slatt, R. H. Gillespie, A. E. D'Agostino, and C. G. Stone, 1993, Gamma-ray logging of outcrops by a truck-mounted sonde: AAPG Bulletin, v. 77, p. 118–123.

Kerans, C., F. J. Lucia, and R. K. Senger, 1994, Integrated characterization of carbonate ramp reservoirs using Permian San Andres Formation outcrop analogs: AAPG Bulletin, v. 78, p. 181–216.

Leonard, A., and M. Bowman, 1998, Reservoir management in selected basin floor submarine fan reservoirs from the UKCS, *in* A. Leonard and M. Bowman, eds., Developing and managing turbidite reservoirs: case histories and experiences, European Association of Geologists and Engineers/American Association of Petroleum Geologists Third Research Symposium, 1998, Almeria, Spain.

Mutti, E., 1992, Turbidite sandstones: Agip/Instituto di Geologia, Universitia di Parma, 275 p.

Slatt, R. M., J. M. Borer, B. W. Horn, H. A. Al-Siyabi, and S. R. Pietraszek, 1995, Outcrop gamma-ray logging applied to subsurface petroleum geology: The Mountain Geologist, v. 32, p. 81–94.

Slatt, R. M., G. H. Browne, R. J. Davis, G. R. Clemenceau, J. R. Colbert, R. A. Young, H. Anxionnaz, and R. J. Spang, 1998, Outcrop–behind outcrop characterization of thin-bedded turbidites for improved understanding of analog reservoirs: New Zealand and Gulf of Mexico: Society of Petroleum Engineers Annual Conference, New Orleans, p. 845–853 (SPE49563).

Slatt, R. M., and P. Weimer, 1999, Petroleum geology of turbidite depositional systems: Part II, sub-seismic scale reservoir characteristics: The Leading Edge, p. 562–567.

Spang, R. J., R. M. Slatt, G. H. Browne, N. F. Hurley, E. T. Williams, R. J. Davis, G. R. Kear, and L. S. Foulk, 1997, Fullbore formation micro images logs for evaluating stratigraphic features and key surfaces in thin-bedded turbidite successions: Gulf Coast Association of Geological Societies Transactions, v. 47, p. 643–645.

Tinker, S. W., 1996, Building the 3-D jigsaw puzzle: application of sequence stratigraphy to 3-D reservoir characterization, Permian basin: AAPG Bulletin, v. 80, p. 460–485.

Witton, B. M., 1999, Borehole image analysis of thin-bedded and channelized turbidites and sandy debris flows, Lewis Shale, Wyoming: AAPG Annual Convention, San Antonio.

Young, R. A., Z. Deng, and J. Sun,1995, Interactive processing of GPR data: The Leading Edge, v. 14, p. 275-282.

Young, R. A., B. Peterson, and R. M. Slatt, 1999, 3-D ground-penetrating-radar imaging of turbidite outcrop analogs, in T.F. Hentz, ed., Advanced reservoir characterization for the 21st century: Gulf Coast Section SEPM Foundation 19th Annual Research Conference Proceedings, Houston, in press.

Slatt, R. M., et al., 2000, From geologic characterization to "reservoir simulation" of a turbidite outcrop, Arkansas, U.S.A., *in* A. H. Bouma and C. G. Stone, eds., Fine-Grained Turbidite Systems, AAPG Memoir 72/SEPM Special Publication No. 68, p. 187–194.

Chapter 18

From Geologic Characterization to "Reservoir Simulation" of a Turbidite Outcrop, Arkansas, U.S.A.

R. M. Slatt
H. A. Al-Siyabi
C. W. VanKirk
R. W. Williams
Colorado School of Mines
Golden, Colorado, U.S.A.

ABSTRACT

Detailed geologic mapping of the Pennsylvanian Jackfork Group turbidites in the DeGray Lake area of Arkansas has provided a 3-D geologic model of a mile-long, steeply dipping turbidite succession that is separated into east and west fault blocks by a zone of strike-slip faults. By scaling-up the stratigraphy into four reservoir zones and by choosing two topographic ground surface elevations to represent an unconformity top-seal and an oil-water contact, this outcrop can be considered a pseudoturbidite "reservoir," amenable to production simulation.

"Reservoir" performance simulation was conducted in 3-D, three-phase black oil mode, using vertical- and horizontal-well drilling scenarios for both water drive and depletion drive cases. The results demonstrate the ability to perform simulation on an outcrop to visualize, directly examine, and, hence, better understand the geologic controls on the performance of analog oil and gas reservoirs.

INTRODUCTION

Discipline-integration within the petroleum industry during the past 10 years has changed the way in which geoscientists interact with petroleum engineers. Geoscientists who used to provide reservoir engineers with conceptual geological models have now been forced to provide data-based, quantitative reservoir models which can address questions such as "How big is the container?" "How will this reservoir style perform?" "How widely must we space our expensive development wells in this geologic setting?" "Should we drill a vertical, slant, or horizontal well?", and "What went wrong?"

A common means of addressing such questions is to build geostatistical models (Srivastava, 1994) and then use them for reservoir simulation purposes to predict the performance of the reservoir and to plan for reservoir development. Increasingly, companies are relying on reservoir simulations in their development plans for billion-dollar investments. Reliable simulation results are critical for well placement and water–flood projects.

Outcrops, when sufficiently large, can provide another means of capturing the appropriate level of heterogeneity for placement into simulators (Hurst et al., 1999; Slatt, this volume). Detailed outcrop models of depositional systems may approach the level of heterogeneity of an actual reservoir. Discussions of outcrop-driven reservoir simulations include Cook et al. (1994), Gardner et al. (1998), and Johnson (1998).

We present the results of detailed geologic mapping and correlation of several outcrop exposures and subsurface core borings of the upper Jackfork Group in the DeGray Lake area of Arkansas (Figure 1). This has allowed us to develop a 6000 ft × 1200 ft × 90 ft (1829 m × 366 m x 27 m) 3-D geologic model of a steeply dipping turbidite succession (Figure 2). By assuming that the present-day topographic ground surface represents an unconformity top seal, and by selecting a lower topographic elevation to represent an oil- (or gas-) water contact, it is possible to treat this outcrop as a pseudo-turbidite "reservoir." Standing on the ground surface at this lower elevation is analogous to standing on the fluid contact and being within the reservoir (Figure 2).

"Reservoir" performance was evaluated using different "drilling" scenarios, and geologic controls on performance were identified. It is our contention that visualizing drilling scenarios and their results at an outcrop analog to a reservoir can lead to improved reservoir management through a better understanding of geologic controls on reservoir performance.

GEOLOGIC CHARACTERIZATION AND UPSCALING

Several stratigraphic sections were measured and described on a bed-by-bed basis in the DeGray Lake Spillway and adjacent area. This provided a framework for reliably correlating the 1000 ft (305 m) of turbidite strata exposed across the 1-mile (1.8 kilometer) area that forms the model (Figure 1). Significant variability in the strata was observed, from thin- to thick-bedded, amalgamated to layered sandstones, muddy debris flows, laminated shales, and some conglomerates. On the basis of a variety of sedimentologic and stratigraphic criteria (Al-Siyabi, 1998), these strata have been scaled-up into four broad stratigraphic zones, labeled A, B, C, and D (Figures 1–3) for simulation purposes. Geologic properties for these four zones are listed in Table 1; general characteristics are summarized below.

Zone A

The total thickness of zone A is 246 ft (75 m), with 70% sandstone (net:gross). The most common sandstone types are massive, scour-based sandstone and flat-based sandstone with fluid escape structures; less common sandstones include sandstone with floating mudstone clasts, rippled sandstone, and parallel-laminated sandstone. More than two-thirds (72%) of the beds exhibit sandstone-on-sandstone contacts. The beds are arranged into 10 definable stratigraphic successions,

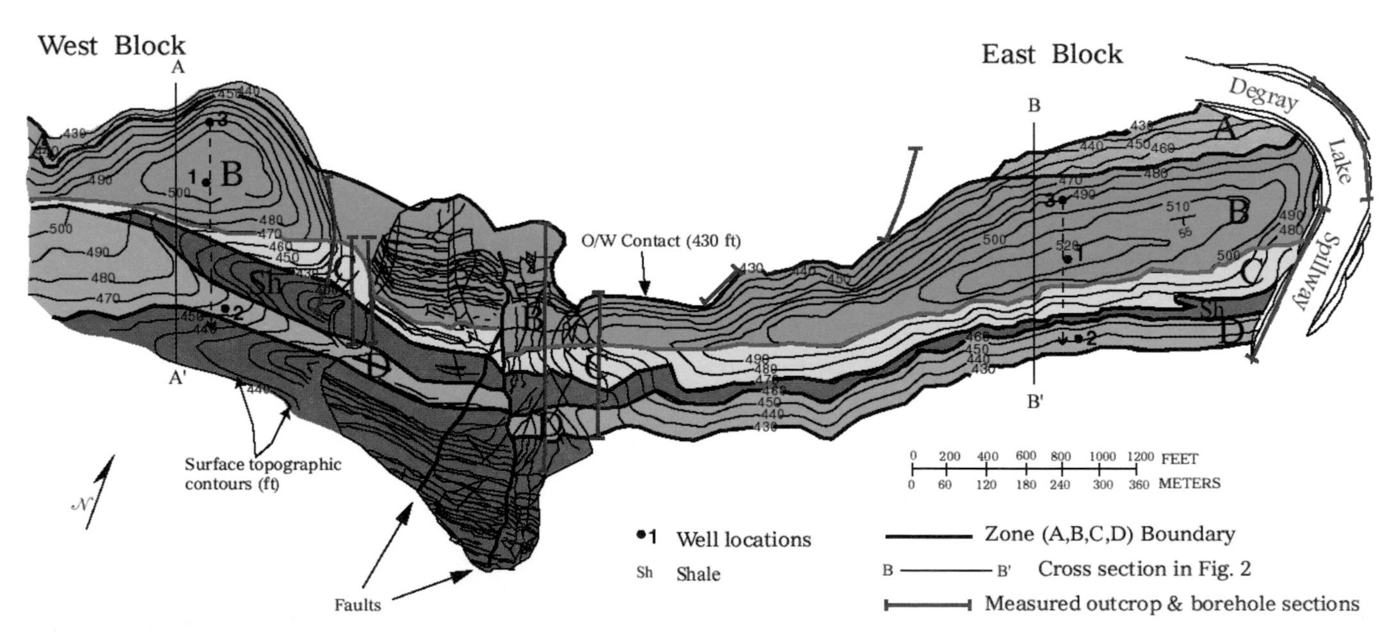

Figure 1—Location of DeGray Lake area, Arkansas. Contours are surface topographic elevations. The contours are used here to represent the "top of structure"; contour interval is 10 ft (3 m). The "oil-water contact" is at a topographic contour elevation of 430 ft (131 m). The reservoir is bounded on the west by the edge of the figure and on the east by DeGray Lake Spillway; both sides are considered in the model to be fault-bounded. Strike of the beds corresponds with the elongate trend of the surface contour lines and the beds dip 55° to the south. Location of a strike-slip fault zone in the area of the present-day DeGray Lake Dam is shown; this fault zone separates the reservoir into West and East Blocks. Reservoir Zones A–D are shown. Cross sections A–A' and B–B' are shown along with the locations of simulated wells 1–3 (Figure 2).

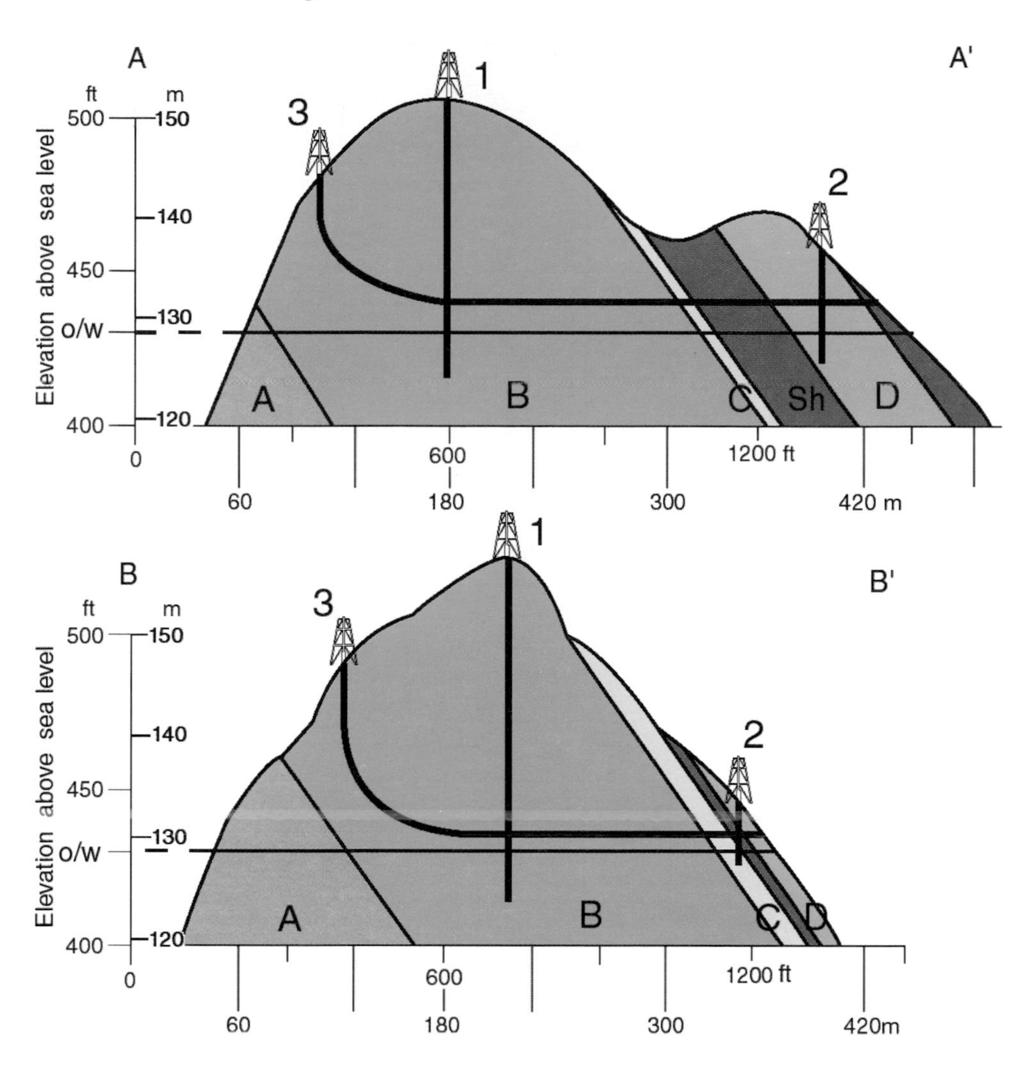

Figure 2—Schematic cross sections (note the different horizontal and vertical scales) showing the stratigraphy and positions of wells in the West and East Blocks. The "oil-water contact" is placed at 430 ft (131 m) topographic elevation. Locations of the cross sections are shown in Figure 1.

Table 1. Geologic Zonation

Zone	Net:Gross	True Gross Thickness in feet (m) *	Por. (%)	Perm. (md)	Sw (%)
A	0.70	118 (36)	29	1000	20
B	0.40	204 (62)	26	300	20
C	0.95	60 (18)	29	1000	20
D	0.96	43 (13)	30	2000	20

*True thickness is above the oil-water contact.

each stratigraphically separated by shale. True thickness above the "oil-water contact" averages 110 ft (34 m) (Table 1).

Zone B

The total thickness of zone B is 364 ft (111 m), with 40% sandstone (net:gross). The most common sandstone types are massive, scour-based sandstone and flat-based, rippled sandstone; less common sandstones include parallel-laminated sandstones, massive sandstone with mudstone clasts, and convolute-bedded sandstone. About half (53%) of the beds exhibit sandstone-on-sandstone contacts. The beds are arranged into 13 definable stratigraphic successions, each stratigraphically separated by shale. True thickness above the "oil-water contact" averages 204 ft (62 m) (Table 1).

Zone C

The total thickness of zone C is 115 ft (35 m), with 95% sandstone (net:gross). The most common sandstone

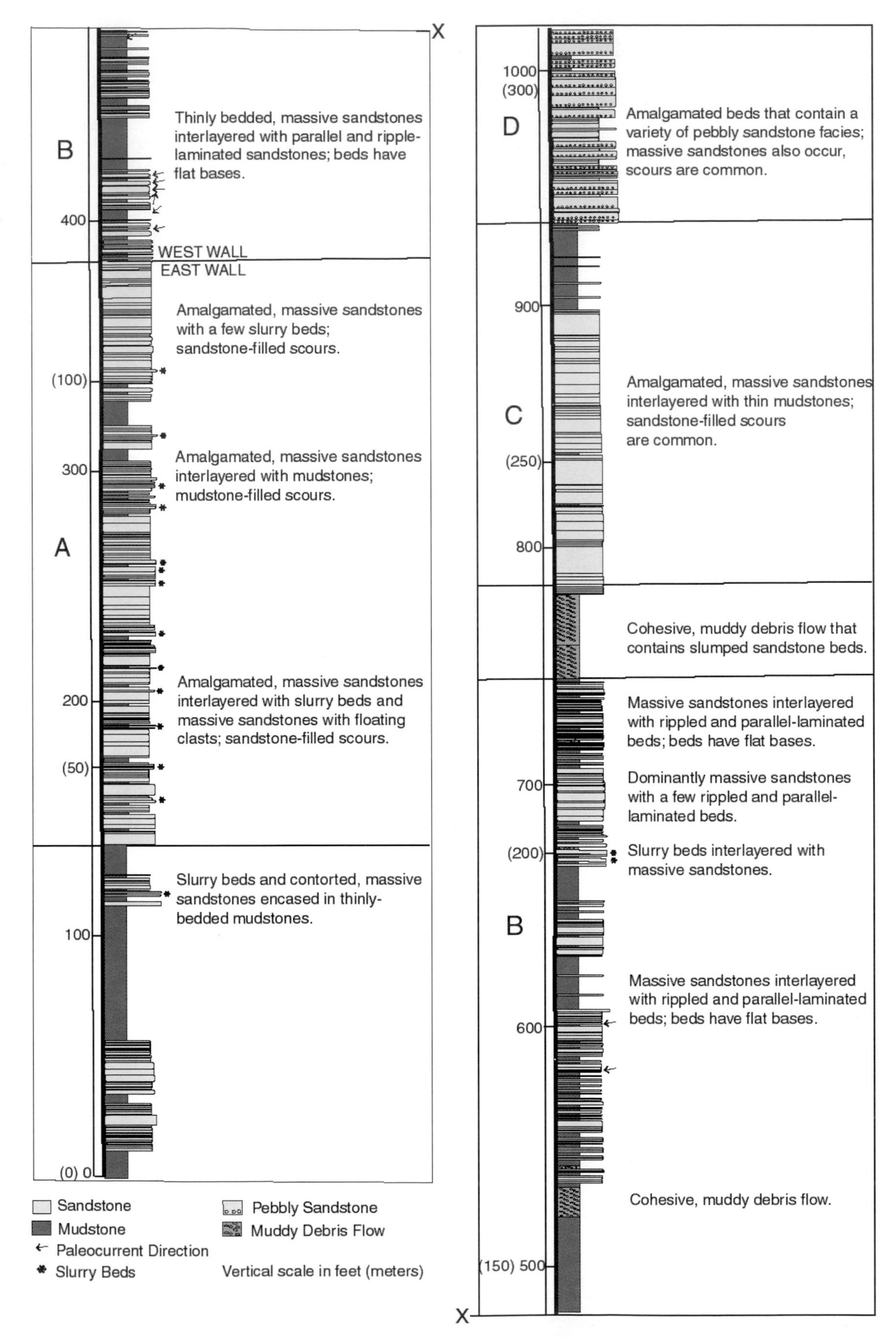

Figure 3—Detailed measured section at DeGray Lake Spillway. Zones A, B, C, and D are shown. There is another 138 ft (42 m) of stratigraphic section beneath the base of Zone A. The base of Zones A and C are considered depositional sequence boundaries (unconformities). The thick shale at 55–110 ft (17–34 m) is a transgressive marine shale. *X* is at the same depth in both columns. See Al-Siyabi (1998) for details.

type is massive, scour-based sandstone; less common sandstones include massive sandstone with mudstone clasts and flat-based, rippled sandstone. More than two-thirds (88%) of the beds exhibit sandstone-on-sandstone contacts. The beds are arranged into six definable stratigraphic successions, each stratigraphically separated by shale. True thickness above the "oil-water contact" averages 60 ft (18 m) (Table 1).

Zone D

The total thickness of zone D is 82 ft (25 m), with 96% sandstone (net:gross). This zone contains the coarsest grained beds. The most common sandstone types are normally graded, pebbly sandstone and massive, scour-based sandstone; less common sandstones include massive sandstone with floating pebbles and inversely graded, pebbly sandstone. Nearly all (90%) the beds exhibit sandstone-on-sandstone contacts. The beds are arranged into 10 definable stratigraphic successions. True thickness above the "oil-water contact" averages 43 ft (13 m) (Table 1).

Because of the closely spaced measured stratigraphic sections and the presence of thick shales and distinctive strata at zone boundaries, Zones B, C, and D can be correlated with a high degree of confidence across the 1-mile area from the DeGray Lake Spillway to the Intake (Figures 1, 2). Zone A is not exposed west of the spillway.

These rocks are neither porous nor oil-bearing; therefore, pseudovalues of porosity, permeability, and water saturation were provided for each zone (Table 1). Although sandstones in all four zones are well lithified by muddy matrix and cement, these values were selected as if the two constituents were not present in the rocks.

The entire stratigraphic succession dips 55° to the south and trends approximately east–west (Figures 1, 2). The interval is dissected by north–south–trending, right lateral, strike-slip faults that have displaced strata approximately 150 ft (46 m) and juxtaposed sandstones against mudstones (Figure 1). This fault zone separates the succession into an East Block and a West Block (Figures 1, 2).

The contours in Figures 1 and 2 are present-day ground surface topographic elevations above sea level. The crest of the "structure" is at a ground or topographic elevation of 520 ft (158 m) in the East Block and 500 ft (152 m) in the West Block (Figures 1, 2). The pseudo "oil-water contact" was placed at a topographic elevation of 430 ft (131 m), which corresponds approximately with the elevation of the DeGray Lake water level.

Thus, the resultant "reservoir" consists of four steeply dipping stratigraphic intervals, mutually separated by thick shales (fluid flow barriers) and laterally divided into two fault blocks. This model results in a hypothetical "oil column" of approximately 90 ft (27 m) in the East Block and about 70 ft (21 m) in the West Block.

"RESERVOIR" SIMULATION INPUT PARAMETERS

Fluid flow simulation modeling was conducted using GeoQuest's ECLIPSE™ program. Both *water drive* and *depletion drive* simulations were conducted under the following conditions: (1) The fault zone that divides the interval into West and East Blocks (Figures 1, 2) was assumed to be sealing, with zero transmissibility; (2) the shales that separate each zone have zero transmissibility, so that the zones are mutually isolated; (3) the oil-water contact was placed at a topographic elevation of 430 ft (131 m) (Figures 1, 2), although the modeled depth was placed at 12,000 ft (3658 m); (4) the dimensions of the entire model were set at 31 cells in the east–west direction, 22 cells in the north–south direction, and four layers (Zones A–D) in the vertical direction; (5) the size of individual cells within the "oil zone" are 100 ft × 200 ft (30 m × 61 m), although for the aquifer case, cells on the south flank of the structure are larger than the oil zone cells; (6) ECLIPSE™ was set in a 3-D, three-phase, black oil mode. A 3-D image of the grid block model set at zero days production is shown in Figure 4A. Figure 4C shows the four layers in cross-sectional view with the oil/water contact at a depth of 12,000 ft (3658 m). Reservoir parameters used in the simulation are provided in Table 2.

"RESERVOIR" VOLUMETRICS

Given the above input data and the standard OOIP (original oil in place) equation, the simulation computed OOIP to be 6.347 MMSTB (million). Oil is distributed in the manner listed in Table 3. The greater volume of OOIP in the East Block than in the West Block is due to its larger areal extent (Figure 1). Although Zone B is the most areally extensive and thickest zone (Figures 1, 2), it does not contain as much OOIP as Zone C because of the higher net:gross and porosity in the latter (Table 1).

DRILLING SCENARIOS

Three drilling scenarios were simulated for both *water drive* and *depletion drive* cases, with one well in each of the West and East fault blocks (Figures 1, 2). These scenarios are: (1) vertical wells through the crest of the structure (highest ground elevation in Figures 1 and 2 in both fault blocks); in both blocks the wells were confined to Zone B; (2) vertical wells on the south flank of the structure, so that the wells were confined to the most permeable Zone D; and (3) south-directed horizontal wells originating on the north flank of the structure in both blocks; the horizontal wells penetrated Zones B, C, and D in each block.

Results of the simulations are summarized in Table 4. Water drive simulations produced more oil than depletion drive simulations. In both cases, the

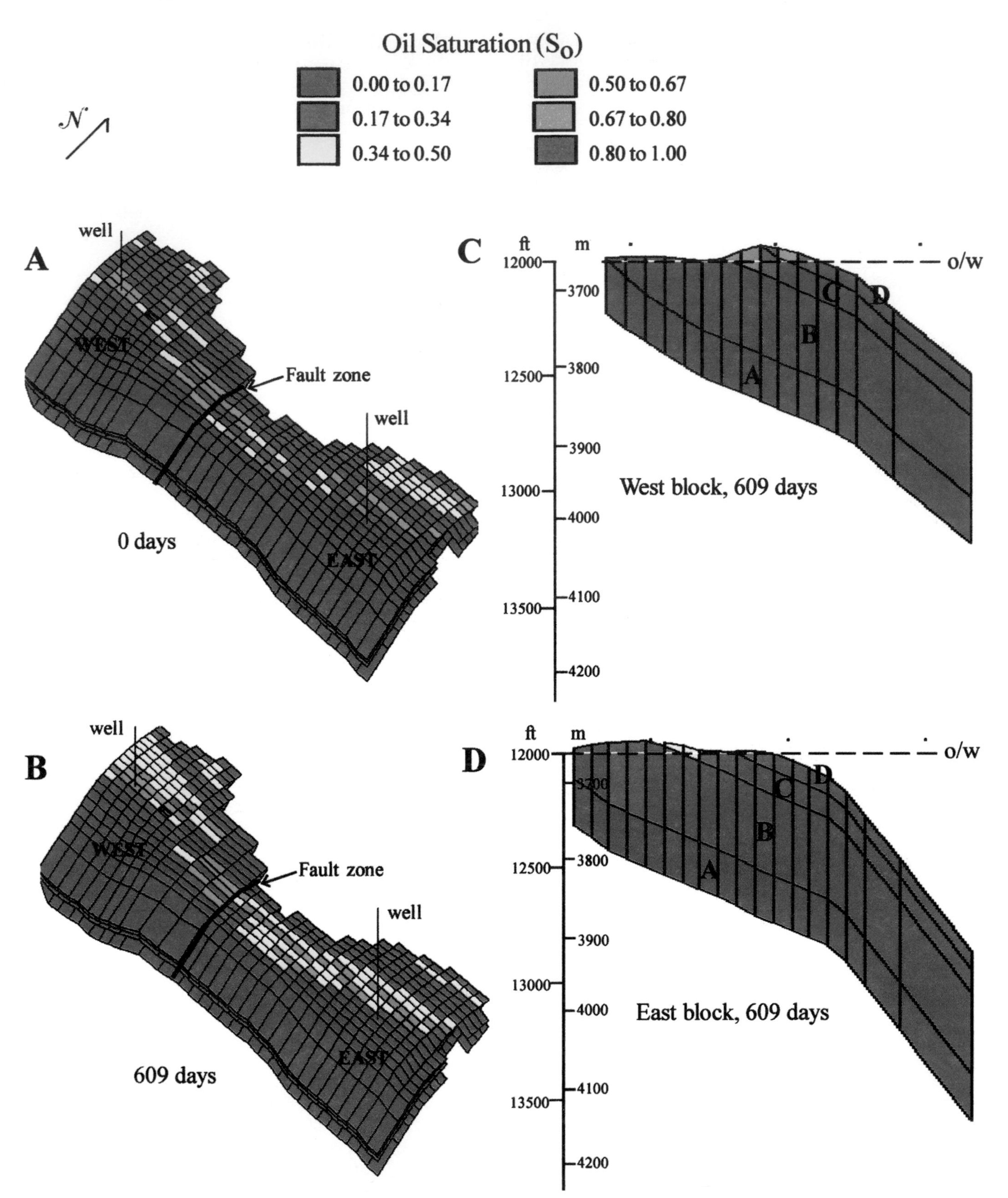

Figure 4—A and B show the grid blocks in 3-D space with oil saturations (S_o) at 0 and 609 days of simulated production. S_o values have been significantly reduced after simulated production. Note the anomalously high S_o just to the west of the fault zone (explained in text). C and D are cross-sectional views of the West and East fault blocks showing S_o after 0 and 609 days of simulated production. Note the anomalously high S_o in the west block (explained in text). The oil-water contact is at 12,000 ft (3658 m) depth.

Table 2. Reservoir Parameters Used in Simulation.

Depth		12,000 ft (3658 m)
Temperature		250° ft
Initial Pressure		8000 psia
Bubble Point Pressure		4000 psia
Oil Gravity		35° API
Gas Specific Gravity		0.75
Simple black-oil PVT data		
	Boi = 1.47	RB/STB at Pi
	Bob = 1.53	RB/STB at Pb
	Rsoi = 851	SCF/STB at Pi
	Uoi = 0.763	CP at Pi
	Uob = 0.414	CP at Pb
	Cs =	2.84 × 10-6 V/V psi
	Cf =	8 × 10-6 V/V psi
Relative Permeability data were generated from correlations, with		
	S_{wi} = 20%	(irreducible water)
	S_{or} = 25%	(residual oil)
	S_{gc} = 3%	(critical gas)

Table 3. Original Oil in Place (OOIP) (MMbbl).

Zone	West Block		East Block	Total
A	—	0.121	—	0.121
B	0.615		1.471	2.086
C	0.822		1.755	2.577
D	1.023		0.540	1.563
Total	2.460		3.766	6.347

least amount of oil was produced from the wells drilled on the crest of the structure, even though these wells drained the most areally extensive and thickest Zone B (Figures 1, 2). Although Zone D is the thinnest zone, more oil was produced from it because of its greater net:gross and permeability (Table 1).

Figure 4 shows an example of the simulation run after 609 days of production using scenario 3 (a horizontal well in each fault block; Figure 2) of the water drive case. Even though the input S_w is 0.20 for each layer (Table 1), the variable initial (zero days) S_o in different cells in the model run result from averaging S_o values within cells that cross the oil-water contact.

Interestingly, the simulation result presented in Figure 4 shows some trapped "attic oil" in the West Block after the 609-day production run (Figure 4C). The oil was trapped within a small, structurally high area along the east side of the West Block, adjacent to the sealing fault zone. The south-directed horizontal well (Figures 1, 2) in this block was unable to reach this trapped oil, resulting in anomalously high oil saturation after the 609-day production run (Figure 4B).

DISCUSSION

The results presented are geologically simplistic because only a four-layer model was used, when, in

Table 4. Simulated Production (MMbbl).

Drilling Scenario	West Block	East Block	Total
Water Drive			
Vertical well on crest structure (Zone B)	0.043	0.219	0.262
Vertical well in high permeability (Zone B)	0.553	0.127	0.680
Horizontal well (Zones B, C, D)	0.780	0.794	1.574
Depletion Drive			
Vertical well on crest structure (Zone B)	0.028	0.115	0.143
Vertical well in high permeability (Zone D)	0.119	0.069	0.188
Horizontal well (zones B, C, D)	0.277	0.475	0.752

fact, internal stratification is considerably more complex (Figure 3). Nevertheless, when discussed at the outcrop, while standing at the elevation of the "oil-water contact" (thus within the reservoir), a number of key reservoir performance points can be illustrated. These points include:

(1) Drilling the crest of a structure does not always produce the largest volume of oil in a reservoir; rather, other geologic factors such as net:gross, porosity, and permeability control volumetrics and production.
(2) Vertical wells targeted to penetrate the strata with the best facies-controlled reservoir quality can be more productive than wells targeted for the thickest strata.
(3) Horizontal wells are more effective than vertical wells in reservoirs with structurally isolated, steeply dipping strata, provided the wells penetrate all of the productive intervals.

ACKNOWLEDGMENTS

We extend our thanks to GeoQuest Inc. for providing ECLIPSE™ software to the Colorado School of Mines, Petroleum Development Corporation of Oman (PDO) for financially supporting Al-Siyabi, and Conoco Inc. for providing financial support used for this research. Laura Sweezey completed the graphics in Figures 1–4.

REFERENCES CITED

Al-Siyabi, H. A., 1998, Sedimentology and stratigraphy of the early Pennsylvanian upper Jackfork interval in the Caddo Valley Quadrangle, Clark and Hot Spring Counties, Arkansas: Unpublished Ph.D. dissertation, Colorado School of Mines, Golden, 272 p.

Cook, T.W., A. H. Bouma, M. A. Chapin, and H. Zhu, 1994, Facies architecture and reservoir characterization of a submarine fan channel complex, Jackfork Formation, Arkansas, *in* P. Weimer, A. H. Bouma, and B. F. Perkins, eds., Submarine fans and turbidite systems: sequence stratigraphy, reservoir architecture and production characteristics: Gulf Coast Section SEPM Foundation 15th Annual Research Conference Proceedings, p. 69–81.

Hurst, A., I. Verstralen, B. Cronin, and A. Hartley, 1999, Sand-rich fairways in deep water clastic reservoirs: genetic units, capturing uncertainty, and a new approach to reservoir modeling: AAPG Bulletin, v. 83, p. 1096–1118.

Gardner, M. H., et al., 1998, Geologic building blocks for reservoir characterization–lessons learned from the Permian Brushy Canyon Formation, West Texas, *in* EAGE/AAPG Third Research symposium, Developing and Managing Turbidite Reservoirs–Case Histories and Experiences, Almeria, Spain, Oct 1998, A010.

Johnson, K. R., 1998, Outcrop and reservoir characterization of the Brushy Canyon Formation, Guadalupe Mountains National Park, West Texas, and Cabin Lake Field, Eddy County, New Mexico: Unpublished M.S. thesis, Colorado School of Mines, Golden, 256 p.

Srivastava, R. M., 1994, An overview of stochastic methods for reservoir characterization, *in* J. M. Yarus and R. L. Chambers, eds., Stochastic modeling and geostatistics: AAPG Computer Applications in Geology, No. 3, p. 3–16.

Gardener, M. H. and J. M. Borer, 2000, Submarine channel architecture along a slope to basin profile, Brushy Canyon Formation, west Texas, *in* A. H. Bouma and C. G. Stone, eds., Fine-grained turbidite systems, AAPG Memoir 72/SEPM Special Publication No. 68, p. 195–214.

Chapter 19

Submarine Channel Architecture Along a Slope to Basin Profile, Brushy Canyon Formation, West Texas

Michael H. Gardner
James M. Borer
Department of Geology and Geological Engineering, Colorado School of Mines
Golden, Colorado, U.S.A.

ABSTRACT

Slope and basin-floor channel sand bodies in the Permian Brushy Canyon Formation comprise a depositional profile, along which changes in the facies architecture of a fourfold channelform hierarchy are compared. Channel complexes form sand bodies with serrated margins consisting of stacked channels that increase in offset basinward. Channels and complexes record "cut," "fill," and "spill" phases of bypass and deposition, with channel and overbank deposition offset in time.

Upper slope siltstones encase the largest channelform sand bodies confined to intraslope depressions. Sediment bypass gives way to deposition downprofile, producing multistory, multilateral, and eventually distributary channel patterns. As complexes widen, "build" phase deposits that precede channelization, and spill-phase overbank deposits, thicken downprofile to equilibrate sandstone volumes inside and outside channels.

INTRODUCTION

Submarine channels are the principal sediment pathway linking the shelf to the basin. Their sediment-fill and bounding surfaces provide insight into fan growth and gravity-flow processes that produce channel form sediment bodies. Despite their prominent role in submarine fan depositional processes, the architecture of submarine channels is poorly understood. Important issues include the (1) controls on channel size and shape, (2) scalar hierarchy of channel sand bodies (Figures 1, 2), and (3) sedimentological criteria that distinguish depositional processes and predict position along a slope to basin profile.

Build-Cut-Fill-Spill Model

The "cut-fill-and-spill" model relates facies patterns in submarine channel and overbank deposits to their position on a slope to basin profile (Gardner et al., 1998). An important premise is that submarine channels generally backfill. Therefore, a fixed point on the depositional profile will record a transition from erosion and bypass, to confined aggradation, to focused, unconfined deposition. These cut, fill, and spill stages of deposition occur at multiple temporal and spatial scales.

The "build-cut-fill-spill" model incorporates the important phase of deposition that may precede channelization. In the upper slope, the "build

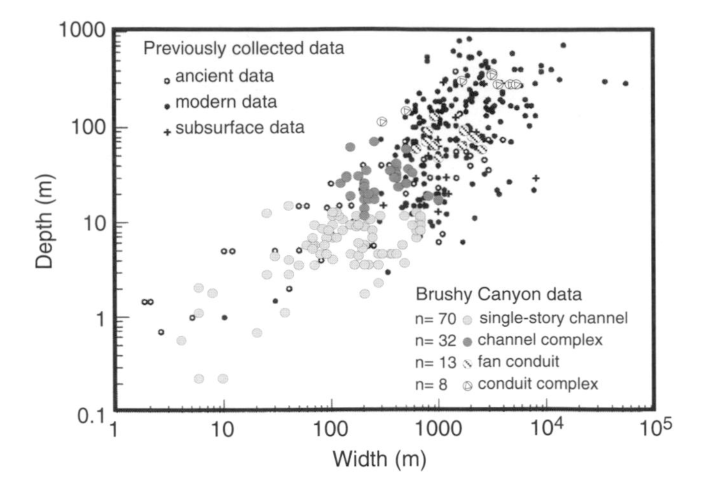

Figure 1—Aspect ratios (width:depth) of Brushy Canyon Formation channelform sand bodies compared with modern and other ancient submarine channels. Note that the larger channels are primarily from modern data reflecting a resolution bias, with outcrops of ancient channels typically not large enough to resolve channelform features remotely imaged using modern data. Modified from Clark and Pickering (1997). The Brushy Canyon channelform hierarchy is shown as patterned dots.

phase" is recorded as an erosional surface of sediment bypass. Erosion and sediment bypass transition to pure bypass and ultimately deposition basinward. As channels extend farther basinward, the physical and temporal separation between the channel fill and the underlying strata decreases.

The preservation of build-cut-fill-spill phases of deposition varies according to position on the depositional profile and/or position in a depositional cycle that records migration of the profile. The percentage of build and spill phase deposits increase downprofile to increase the sandstone percentage within basin-floor successions. Slope and upper-basin floor settings have steep gradients that promote sediment bypass. This produces cut-fill-spill motifs with little or no build-phase deposits. Sandstone percentage is low overall, but locally is high in intraslope depressions confining composite channelform sand bodies. Spill-phase deposits are poorly developed because the topographic depressions are large and difficult to completely backfill. The temporal phases of channel deposition change basinward along the basin-floor profile from complex build-cut-fill-spill, to build-fill-spill, to simple build-spill patterns.

The build-cut-fill-spill model for submarine channel development has important implications for sand bypass and facies prediction. Each depositional phase records different sedimentologic processes and energy trends that directly control the type, distribution, and correlation length of architectural elements and facies. This paper uses four detailed outcrop architecture studies to document proximal-to-distal changes in submarine channel architecture related to variable preservation of the build-cut-fill-spill phases of deposition. Four important attributes of submarine channels are examined: (1) the effect topographic confinement has on channel architecture, (2) the hierarchy of channelform sand bodies, (3) the mechanisms promoting flow confinement, and (4) the timing of channel and overbank deposition.

GEOLOGIC SETTING

Outcrops of the Brushy Canyon Formation in the Guadalupe and Delaware mountains are exposed by Tertiary displacement along basin-and-range faults that define the uplifted western margin of the Delaware Basin, the western subbasin of the Permian basin (King, 1948; Goetz, 1985; Hill, 1995; Figure 3). During middle Permian time, west Texas was the site of a bowl-shaped, epicontinental sea with a restricted southern opening to the ocean through a relict foredeep (the Hovey Channel) (King, 1942; Ross, 1986; Yang and Dorobek, 1995; Hill, 1995; Figure 3). Permian carbonate platforms nucleated on basement highs rim the basin margin and further built shelf-to-basin relief.

The distally steepened carbonate ramp (600 m relief and up to 10° dip) underlying the Brushy Canyon Formation formed a physiographic break that controlled subsequent clastic slope and basin depositional patterns (King, 1948; Pray, 1988; Rossen and Sarg, 1988; Kerans et al., 1992; Kerans and Fitchen, 1994; Zelt and Rossen, 1995). Along the western Delaware Basin margin, this southwest-to-northeast–trending ramp formed an embayment encircling the Brushy Canyon outcrop belt. This ramp margin provides a common reference point for positioning channel complexes located from 7 to 32 km basinward of its terminus. The "outcrop fan complex" is one of three fan complexes that form a bajada-like submarine apron around the northern Delaware Basin (Figure 3).

The Brushy Canyon outcrop is oriented obliquely to Permian sediment transport, with paleoflow indicators shifting from 120° to 85° southward across the outcrop belt (345° trend). Consequently, proximal-to-distal channel morphologies are a composite reconstruction from eight submarine conduits that obliquely intersect the outcrop belt. Change in channel complex architecture is assumed to primarily record depositional processes related to position on a slope and basin depositional profile (Figure 4).

STRATIGRAPHIC FRAMEWORK

The Brushy Canyon Formation in outcrop is part of a submarine fan complex that corresponds with one third-order composite sequence of about 2 m.y.

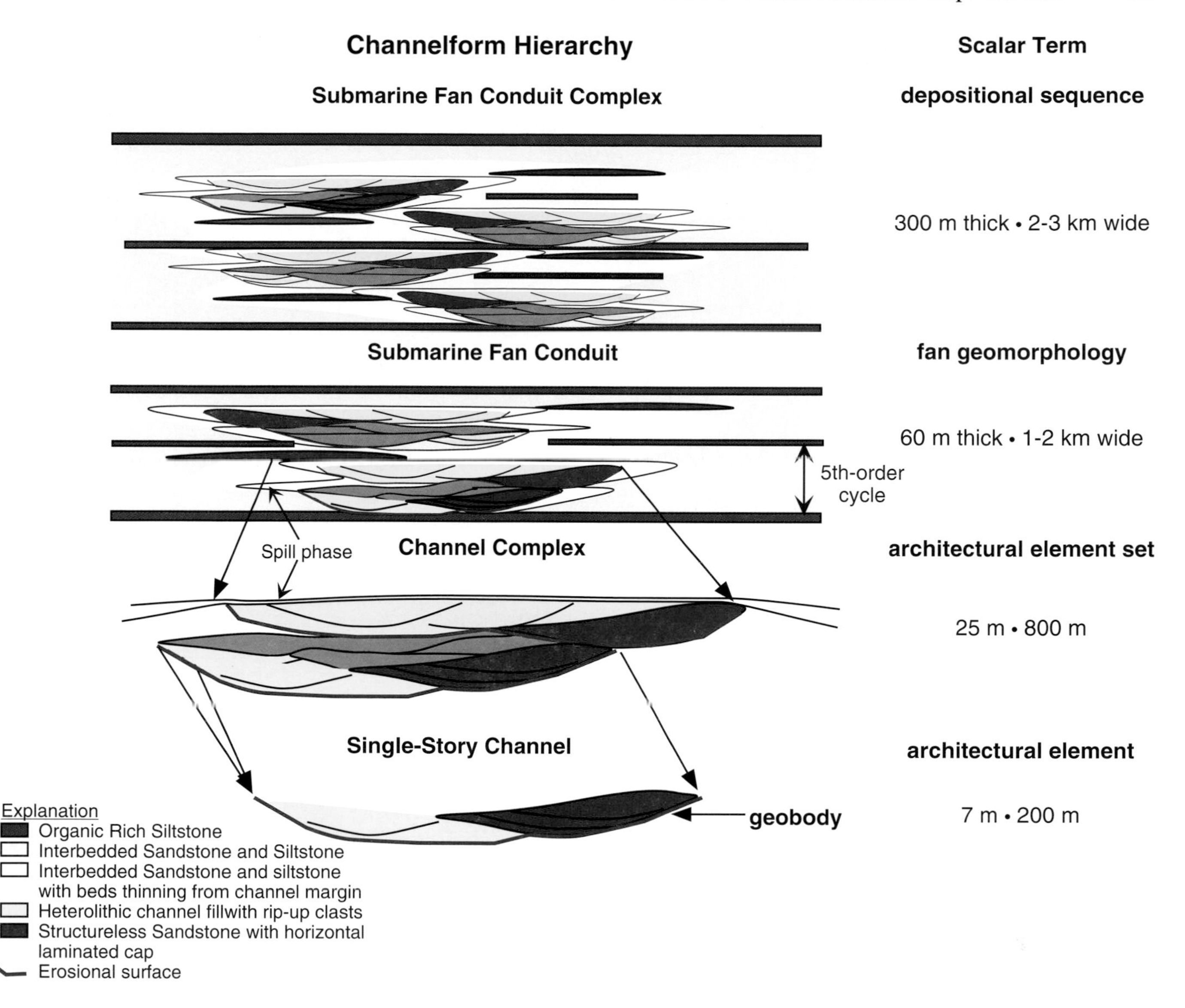

Figure 2—Hierarchy of submarine channelform sediment bodies recognized in the Brushy Canyon Formation in outcrop. In order of decreasing scale: (A) conduit complexes represent fan conduits that remained active through deposition of the fan complex; (B) fan conduits contain more than one channel complex and form kilometer-wide sandstone fairways hundreds of meters thick within a fan; (C) channel complexes are up to 1-km-wide and 40-m-thick multistory and multilateral sandstone bodies with serrated margins; (D) discrete channel fills are up to 7 m thick and hundreds of meters wide and may contain multiple erosive-based sediment bodies.

duration (Vail et al., 1977; Kerans et al., 1992; Figure 5). This fan complex includes the lower part of the Cherry Canyon Formation and is exposed as a 400-m-thick succession bracketed by correlatable siltstone intervals. Each of the eight siltstone-bounded slope and basin cycles up to 90 m thick contain deposits that can be correlated across conduits and show systematic facies changes that correspond to slope and basin positions along a fan profile. Fans are offset across the siltstone intervals that form fourth-order cycle boundaries. These fourth-order cycles in turn contain up to four fifth-order cycles, which can occur as shingled clinoform packages (20 km long and 60 m thick). The thickest part, or clinothem, of a fifth-order cycle is the depocenter along that segment of the fan profile.

This stratigraphic framework permits a comparison of architectural changes in upper and lower slope, base of slope, and basin-floor channel complexes. Architectural element analysis establishes a hierarchy of sediment bodies and bounding surfaces comprising these channelform sand bodies (Figure 2).

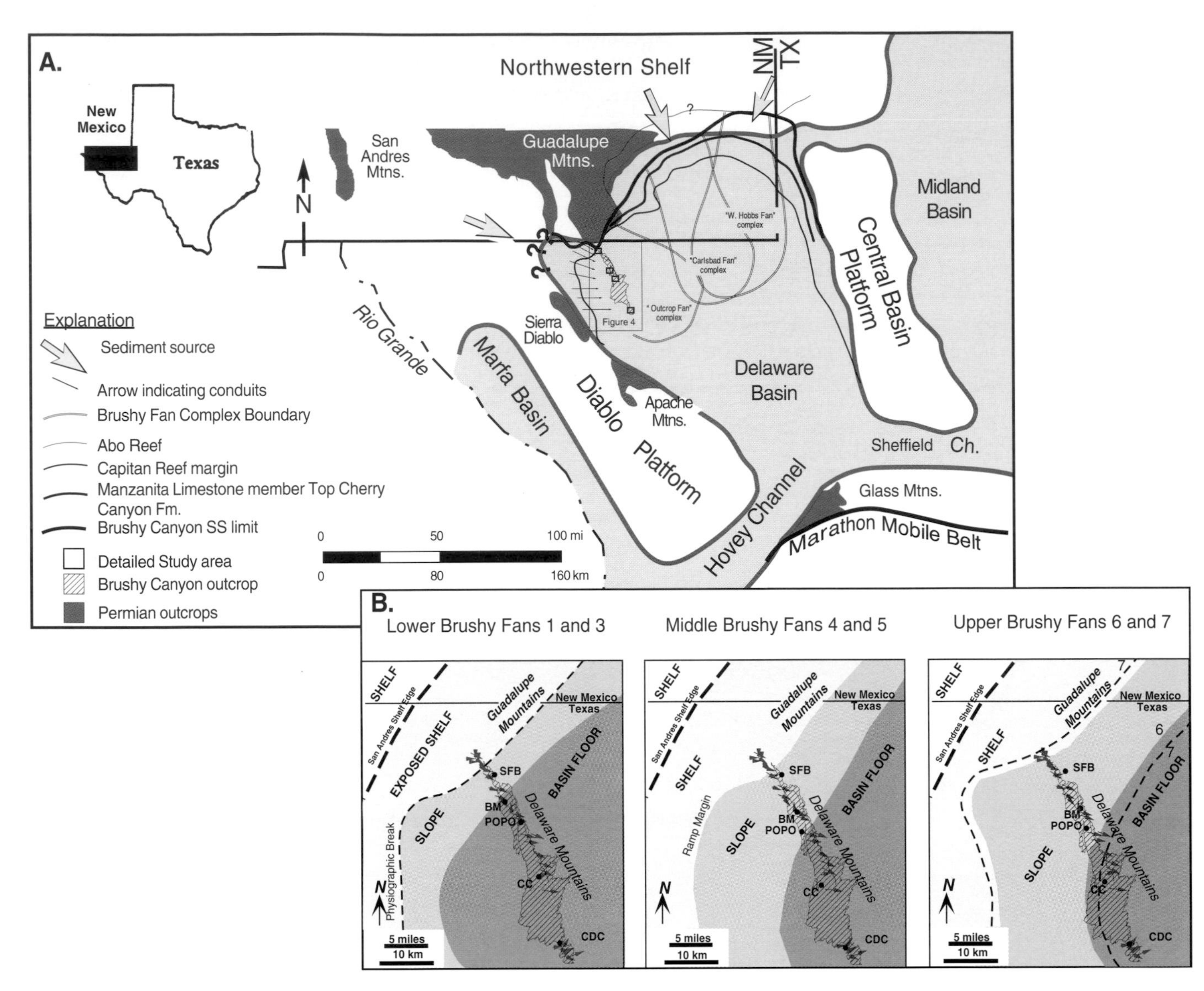

Figure 3—(A) Regional paleogeography of Delaware Basin area (after King, 1948, and Oriel et al., 1967), showing Abo and terminal Capitan shelf-margin trends. Late Leonardian shelf-margin trends (pre-Brushy) probably show similar, although muted, tectonic influence as Abo trend. Mega-embayments in Leonardian carbonate margins are believed to control sand input points, feeding three submarine fan complexes. Small arrows outline Brushy Canyon fan conduits that trend S60°E-trending along Guadalupe Mountains but shift eastward across the outcrop belt. (B) Paleogeographic reconstructions of Brushy Canyon slope and basin facies tracts. Fans 1–3 represent older basin-floor dominated sediment thicks, Fans 4 and 5 represent channelized basin-floor fans, and Fans 6 and 7 occur above the 40-ft siltstone marker and record slope expansion, producing a slope-centered thickness pattern in outcrop. Dark pattern is outcrop belt in Delaware and Guadalupe mountains, with arrows showing position of fan conduits.

A fourfold hierarchy in order of increasing size includes (1) single-story channel fills forming architectural elements, (2) multistory and multilateral channel complexes consisting of two to 26 channels, (3) fan conduits consisting of more than one channel complex, and (4) conduit complexes representing sediment pathways active during fan complex deposition (Figure 2). We have documented a sedimentological hierarchy of structures, sediment bodies, and cross-cutting relationships among bounding surfaces from the four channel complexes, discussed below in a proximal-to-distal order.

UPPER SLOPE CHANNEL COMPLEX, SOUTHERN GUADALUPE MOUNTAINS

A prominent sandstone mesa known as the Salt Flat Bench (SFB) caps the Brushy Canyon Formation at the southern end of the Guadalupe Mountains (Figure 4).

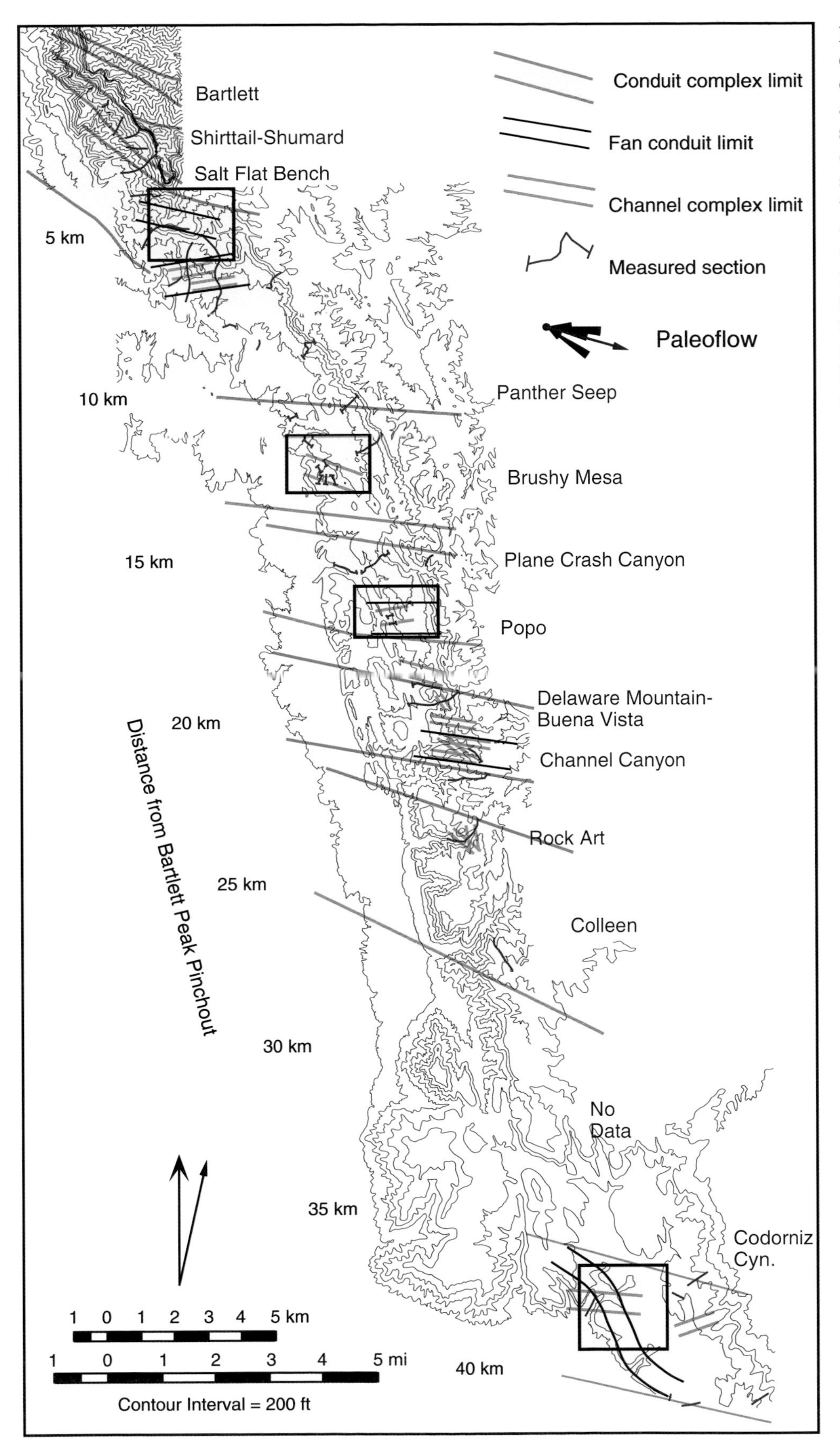

Figure 4—Topographic map of Brushy Canyon Formation outcrop belt in Guadalupe and Delaware mountains showing locality of the four deepwater channel complexes studied. Shaded areas show the position, limits, and orientation of inferred fan conduit complexes (averaged for all fans) that record along-strike variations in sediment supply along the oblique depositional dip outcrop belt; black lines show channel complexes.

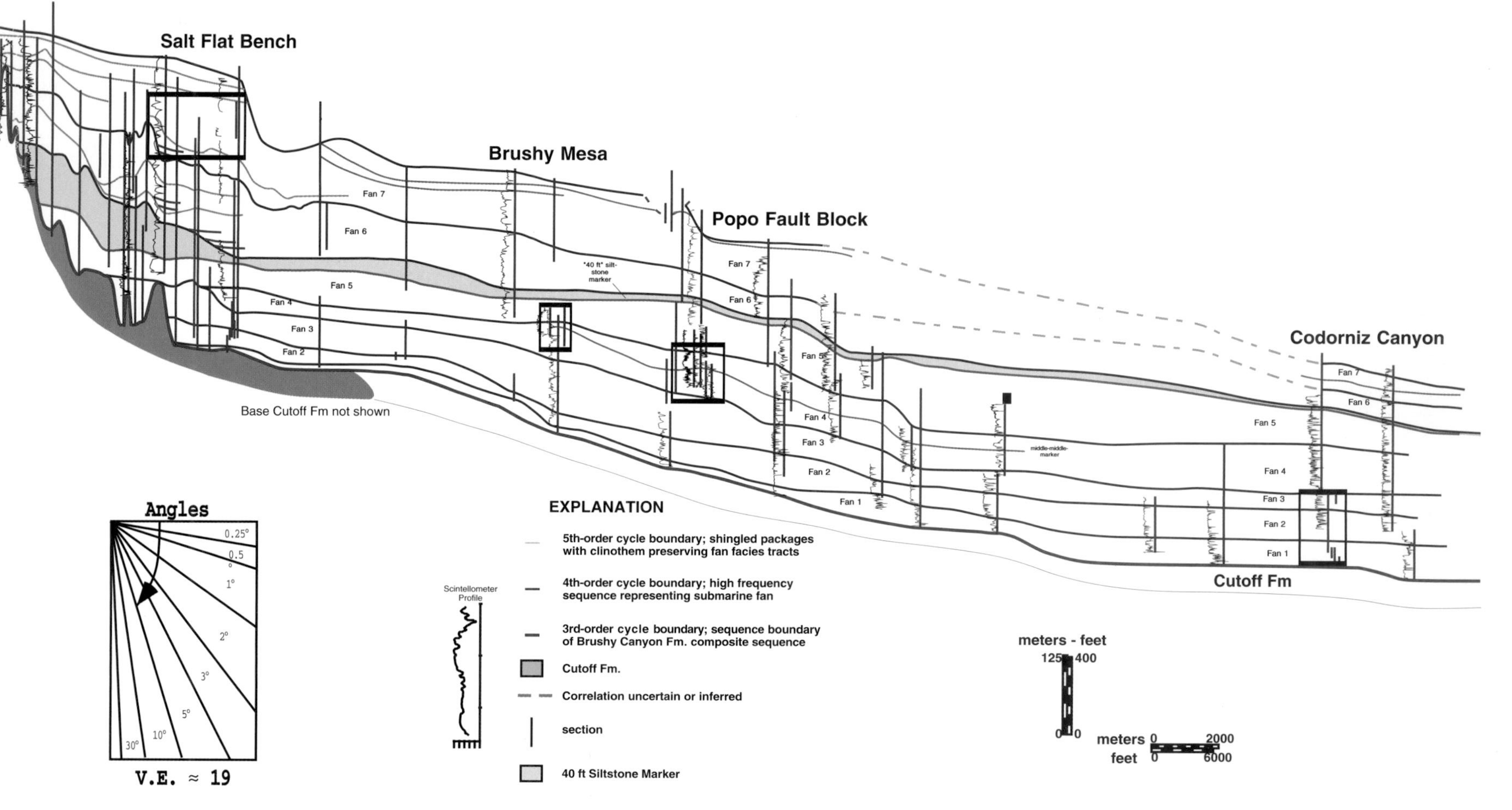

Figure 5—Oblique depositional-dip section of the Brushy Canyon Formation along the Guadalupe and Delaware mountains, west Texas. These basin-restricted deposits of a third-order composite sequence include the basal Pipeline Shale Member and the genetic top siltstone marker within the lower part of the Cherry Canyon Formation. The seven submarine fan cycles are shown by fourth-order cycle boundaries. The 40-ft siltstone marker is shown as a grayed area and records a change in thickness pattern within the fan complex from older basin-centered to younger slope-centered thicks. The right margin of the cross section corresponds with Delaware Mountain outcrops south of Bitterwell Mountain. Inset blocks show study areas of channel complexes discussed in the text.

The SFB occurs in the most basinward of three regularly spaced fan conduits that trend S60°E along the Western Escarpment of the Guadalupe Mountains. Fans 6 and 7 form the majority of the fan complex in this area (Figure 6). The SFB was deposited during the initial transgression of Fan 7. Siltstones below the SFB are organic-poor, contain numerous stratal discordances, and correlate down-profile to large sand-filled channel complexes. Siltstones above the SFB are laterally continuous and progressively increase in organic richness to the "genetic-top" siltstone marker (Sageman et al., 1998). The siltstone-dominated interval above the SFB correlates upprofile to shelfward-stepping, deepwater sandstones that record the final abandonment of the Brushy Canyon fan complex.

The upper slope setting is characterized by large isolated sandstone bodies encased in thick siltstone-rich successions, such as the 40-m-thick and kilometer-scale SFB sand body (Figure 7; Batzle and Gardner, this volume). The SFB is interpreted as an intraslope sand body positioned on the margin of a fan conduit and ponded within an isolated spoon-shaped depression. The SFB outcrop is U-shaped in plan view and represents only about one-half of the depositional sand body geometry. Its outcrops can be divided into proximal and distal strike and dip walls (Gardner and Sonnenfeld, 1996). No single facies is laterally continuous along the entire 2.7-km outcrop length. Wavy-lenticular silty sandstone is the most common facies vertically separating the sandstone beds. It is also the only facies that shows a correlation between bed length and thickness (Johnson, 1998). The sand body contains seven truncated architectural element sets consisting of 26 smaller channelform, wedge-shaped overbank, lobeform, and tabular siltstone architectural elements.

Sandstone content across the outcrop is about 89% (Johnson, 1998; Table 1) but dramatically decreases laterally along strike, where finer-grained turbidite overbank deposits dominate. The high proportions of turbidites in the proximal (western) strike outcrop (610 m long) produce a bimodal sandstone bed thickness distribution, reflecting channelform and overbank elements. Turbidite bed lengths are 28% longer and sandstone conglomerate proportions are the highest (29% vs. 2%) along the distal (eastern) strike outcrop (>1 km long), yet their overall abundance decreases. There is no observed correlation between bed length and facies.

The basal surface of the SFB sand body consists of a series of erosional discordances that form a master surface interpreted to represent coalesced slump scars (Figure 6). In the shallowest part of the depression, this surface is concordant and draped by organic-rich siltstone. Intraslope depressions restrict channel migration, promoting multiple cut-and-fills and confined depositional patterns. The proximal strike outcrop illustrates confined deposition by the channel and overbank deposits that terminate against the basal surface (Figure 7). The distal strike wall illustrates how multiple episodes of erosion and deposition, within a confinement, controls facies architecture (Figure 7). Younger sandstones that thicken toward the depression axis truncate older clast-rich sandstones forming the sand-body base along the eastern distal outcrop. These stratigraphically higher, but older, "perched" deposits are preserved remnants eroded by multiple cut-and-fills in the depression axis. These cut-and-fill surfaces represent an additional surface type that only occurs in association with stratigraphic confinements (Gardner et al., 1995).

Architectural elements that are younger in the proximal strike outcrop than in the distal strike outcrop suggest depositional backfilling of the depression. Additionally, a systematic upward increase in sandstone bed length reflects increased preservation of younger architectural elements within the broader upper part of the (master) confinement. Bed patterns in overbank deposits are also consistent with depositional backfilling of the depression. Turbidite bed lengths are greater in the distal strike outcrop, but their proportion is higher in the proximal outcrop. These observations suggest that deposits recording unconfined flow at distal sites correlate with deposits showing increased aggradation at proximal sites.

LOWER SLOPE CHANNEL COMPLEX, NORTHERN DELAWARE MOUNTAINS

Lower slope deposits in the lower and middle Brushy Canyon Formation record a significant increase in sandstone percentage at Brushy Mesa (BM) relative to more proximal outcrops. Fan 4 dominates BM outcrops, which are fed by a different conduit than the SFB. This conduit's intersection is 13 km basinward from its coeval ramp margin. This setting is basinward relative to the SFB. At BM, Fan 4 forms a 40-m-thick succession, exposing two sand bodies that represent southeast-trending multistory channel complexes (Figure 8). In contrast to the confined architecture at SFB, these isolated siltstone-encased channel complexes show steep serrated margins (Figure 9). One 30-m-thick and 200-m-wide sand body margin shows siltstone interfingering with seven vertically stacked channel sandstone margins that step up and shift laterally along the sand body base. Siltstones are compactionally deformed with bedding rotated at the channel margins and dipping into the channel axis. In these discrete sandstone channels (1–3 m thick), bed length and thickness progressively increase upward as discrete beds amalgamate to form thicker bed sets (Figure 9). Discrete meter-thick sandstone bed sets in the lower half record the amalgamation of multiple high-density gravity flows. Sandstone bed sets show evidence of amalgamation as well as fluidized tractive structures and soft-sediment deformation. Some channel axis deposits have coarse-grained sandstone bases containing centimeter-size siltstone ripup clasts and horizontal laminations with siltstone interbeds (Table 1). Sediment bypass is interpreted from these "left-behind" deposits because they indicate a condensed chronology of many depositional events.

Fine-grained deposits flanking the channelform sand body appear to record active deposition within a lower

Table 1. Summary of Aspect Ratios for Four Brushy Canyon Formation Channel Complexes.

Locality (Facies Tract)	Sand Body Aspect Ratio (W/D measured in meters)		Channel Complex Aspect Ratio (W/D measured in meters)			Channel Stories Aspect Ratio (W/D measured in meters)			Net-to-Gross Sandstone Ratio Outside channel Inside channel
Salt Flat Bench	2000 m × 40 m	50	7.5	††1100 × 69	13	7.6.0	1970 × 9.8	201	89%
(Upper Slope)			7.4	900 × 44.8	38	7.5.5	980 × 13	75	(slump
			7.4	1970 × 52.4	20	7.5.4	980 × 16	61	confined
			7.2	2200 × 168	16	7.5.3	1100 × 13	85	facies)
						7.5.2	1000 × 20	50	
						7.5.1	1000 × 7	143	
						7.4.3	900 × 22	41	
						7.4.2	900 × 9.8	95	
						7.4.1	900 × 13	69	
						7.3.2	1970 × 16.4	120	
						7.3.1	1970 × 36	55	
						7.2.6	780 × 30	26	
						7.2.5	980 × 30	33	
						7.2.4	1400 × 30	47	
						7.2.3	2200 × 26	85	
						7.2.2	2200 × 22	100	
						7.2.1	1570 × 30	52	
						7.1.3	1470 × 16.4	90	
						7.1.2	1200 × 13.2	91	
						7.1.1	656 × 9.8	67	
						7.0.5	1244 × 2.13	584	
Brushy Mesa			4.3?	168 × 22	8	4.3.8	†400 × 7.4	54	43%
(Lower Slope)						4.3.7	†168 × 5	34	83%
						4.3.6	†152 × 4.5	34	
						4.3.5	†146 × 4.1	36	
						4.3.4	†150 × 2.5	60	
						4.3.3	†114 × 4.8	24	
						4.3.2	†100 × 4.2	24	
						4.3.1	†30 × 2.4	13	
Popo Fault Block	600 m × 35 m =	17	4.4	430 × 11.8	30	4.4.2	360 × 8	45	61%
(Base of Slope)			4.3	388 × 22.5	17	4.4.1	380 × 5.7	67	90%
			4.2	375 × 12.4	36	4.3.7	143 × 5.8	25	
						4.3.6	120 × 6.5	18	
						4.3.5	55 × 6.8	8	
						4.3.4	84 × 5.8	14	
						4.3.3	160 × 6.8	8	
						4.3.2	287 × 10.2	28	
						4.3.1	320 × 6	53	
						4.2.2	130 × 7.3	18	
						4.2.1	200 × 2.7	74	
Codorniz Canyon			2.2	137 × 20	7	2.2.4	†16 × 2	8	88%
(Basin Floor)			2.1	134 × 13	10	2.2.3	†40 × 3	13	99%
						2.2.2	†22 × 2.5	9	
						2.2.1	†29 × 4.5	3	
						2.1.5	†85 × 5	16	
						2.1.4	†10 × 1	10	
						2.1.3	†110 × 7	16	
						2.1.2	†75 × 2	38	
						2.1.1	†40 × 3	13	

† Half measurement of channel dimension extrapolated for aspect measurement.
†† At Salt Flat Bench channel complex = architectural element set.
††† At Salt Flat Bench channel stories = architectural elements.

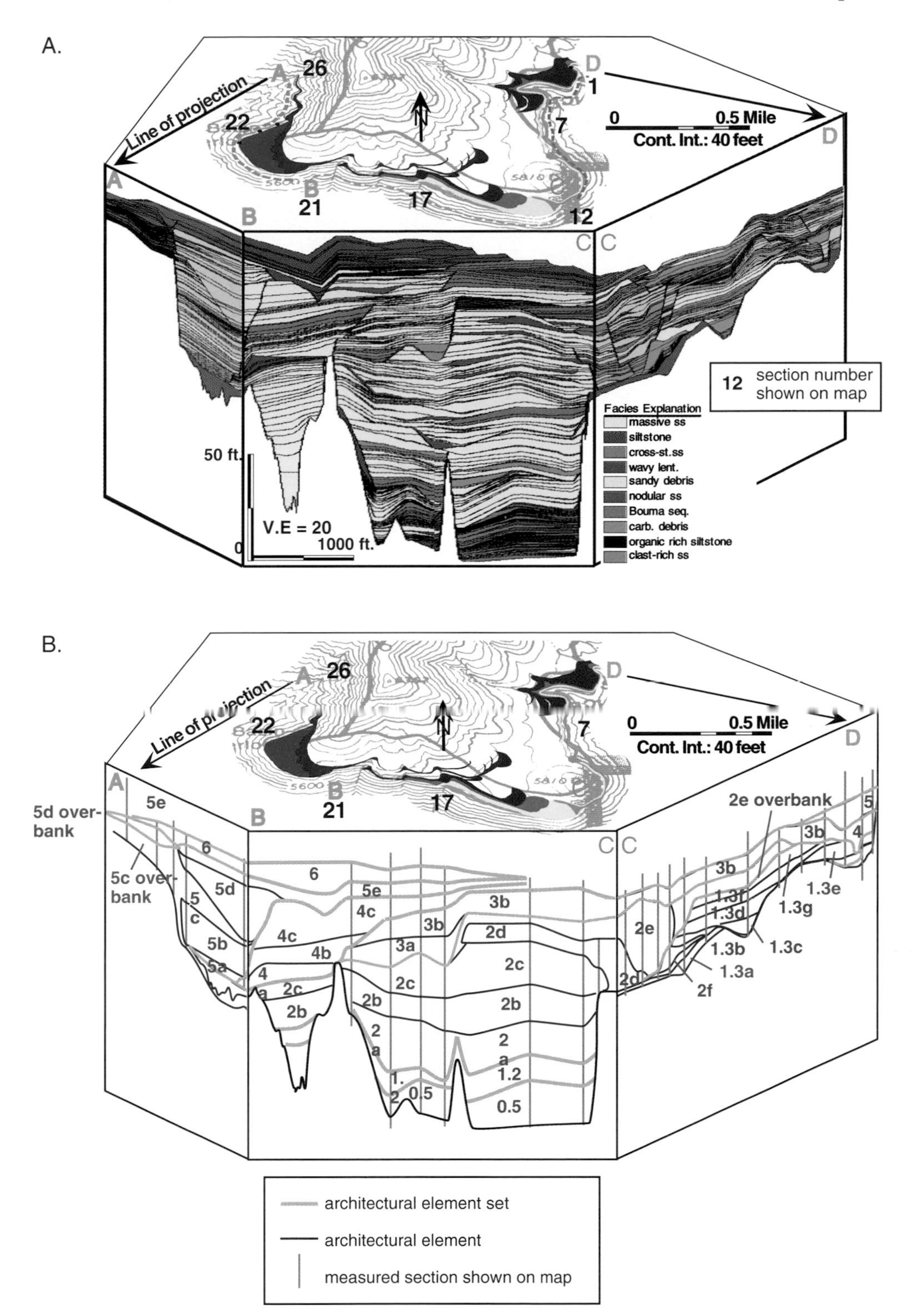

Figure 7—(A) Fence diagram showing facies of upper Brushy Canyon Formation at the Salt Flat Bench. The base of each diagram is the master bounding surface forming a topographic confinement. Note the irregular nature of the basal surface interpreted to represent coalesced slump scars. The panels show the facies architecture along proximal (western) and distal (eastern) strike walls, and the southern dip wall of the sandstone body. No sandstone bed extends across the sandstone body, with the bed length controlled by position within architectural elements shown in B. (B) Fence diagram showing architectural elements and architectural element sets representing truncated channel complexes. Note the truncated deposits forming the sandstone body base along the eastern strike wall. These deposits provide evidence of multiple episodes of cut-and-fill within this topographic depression. The base map shows progressively younger architectural elements toward the north-reflecting depositional backfilling within the topographic confinement.

Figure 8—Views of Fan 4 submarine channel complex and facies exposed at Brushy Mesa. Note the serrated margin that results from stacking of multiple meter-scale channel fills to form the 200-m-wide, 30-m-thick channel complex. Outcrop scintillometer profiles shown.

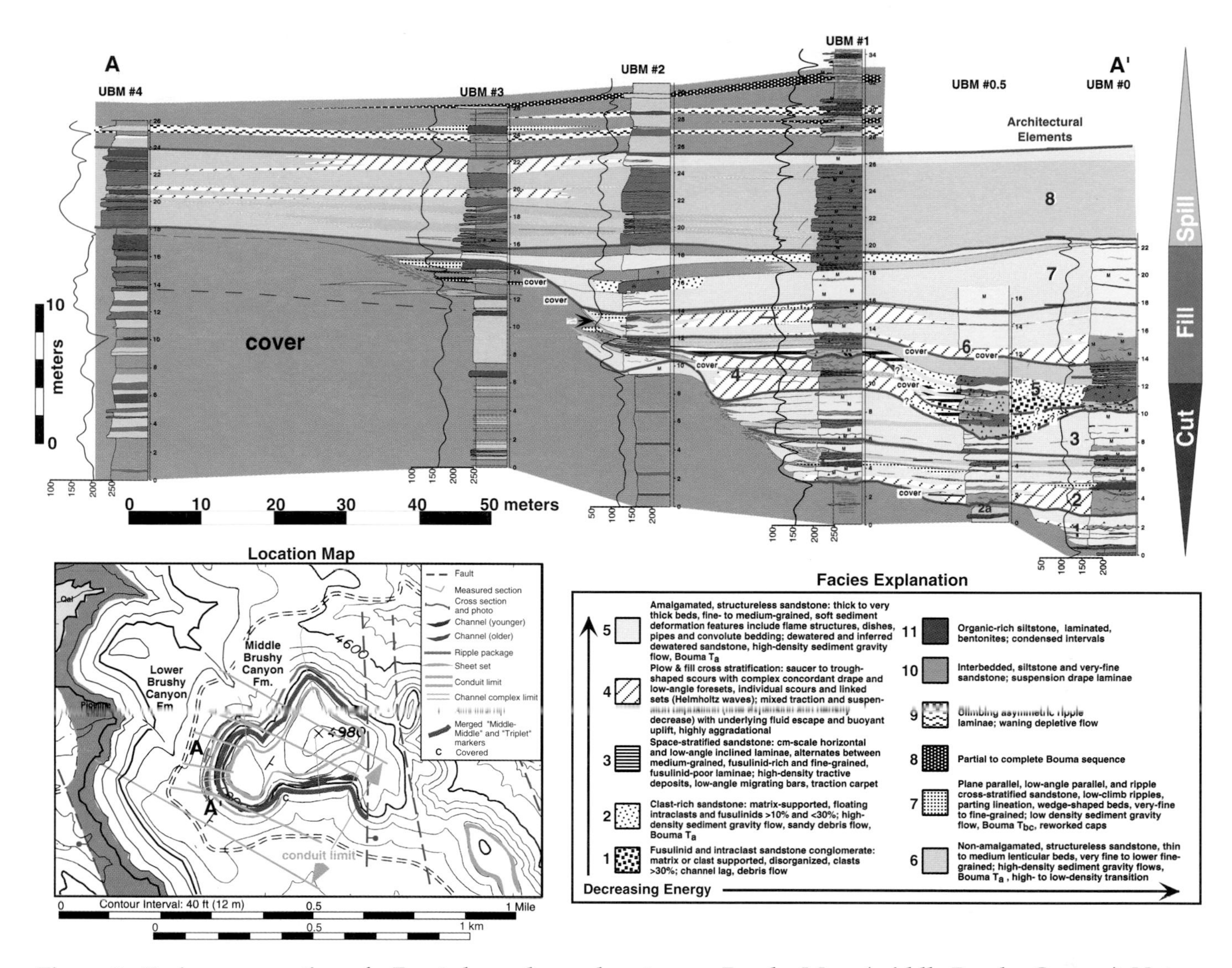

Figure 9—Facies cross section of a Fan 4 channel complex at upper Brushy Mesa (middle Brushy Canyon). Note the high degree of lateral facies change within the channel complex, which is typical of Fan 4 channel fills. Facies correlation length increases and facies diversity decreases upsection through the cut, fill, and spill phases. The channel is multistoried, with a highly serrated margin consisting of many identifiable 1- to 3-m cuts.

slope fan conduit. Interlaminated siltstone and sandstone with rippled sandstone interbeds are common Brushy Canyon slope facies, regardless of proximity to channels. The rippled sandstone interbeds, however, extend from the top and record spillover of discrete channel fills forming the complex (Figure 9). The serrated sand-body margin and coeval ripples in flanking deposits demonstrate the absence of a master surface. In the absence of a master confining surface, a high degree of lateral offset might be expected from one channel to the next. The BM (Brushy Mesa) complex is multistory, not multilateral; therefore, the limited channel offsets must reflect the focusing of sediment from an upper slope confinement. This upprofile confinement repeatedly directed high-density gravity flows that created small channels. The sand-poor overbanks helped confine the subsequent channel fills, which ultimately stacked to form this multistory body.

The sandstone beds capping this channelform sand body show a larger-scale version of this cut-fill-spill depositional pattern. These wedge-shaped bodies consist of amalgamated to nonamalgamated, lenticular and sheetlike sandstones that extend laterally away from the sand body axis to give the complex an overall funnel-shaped geometry. These medium-bedded sandstones thin laterally to interbedded, 1- to 2-m-thick, upward bed-thinning packages of climbing ripple cosets and formsets that decrease upward in frequency and thickness.

BASE-OF-SLOPE CHANNEL COMPLEX, CENTRAL DELAWARE MOUNTAINS

The Popo fault block occurs in yet another fan conduit, 1.5 km wide, intersecting a more basinal position than BM. It lies 5.2 km south of BM, and 25 km from its coeval ramp margin (Figure 10). The base-of-slope to proximal basin-floor deposits of Fan 4 (50 m thick) show an increased sandstone volume (avg. 76%

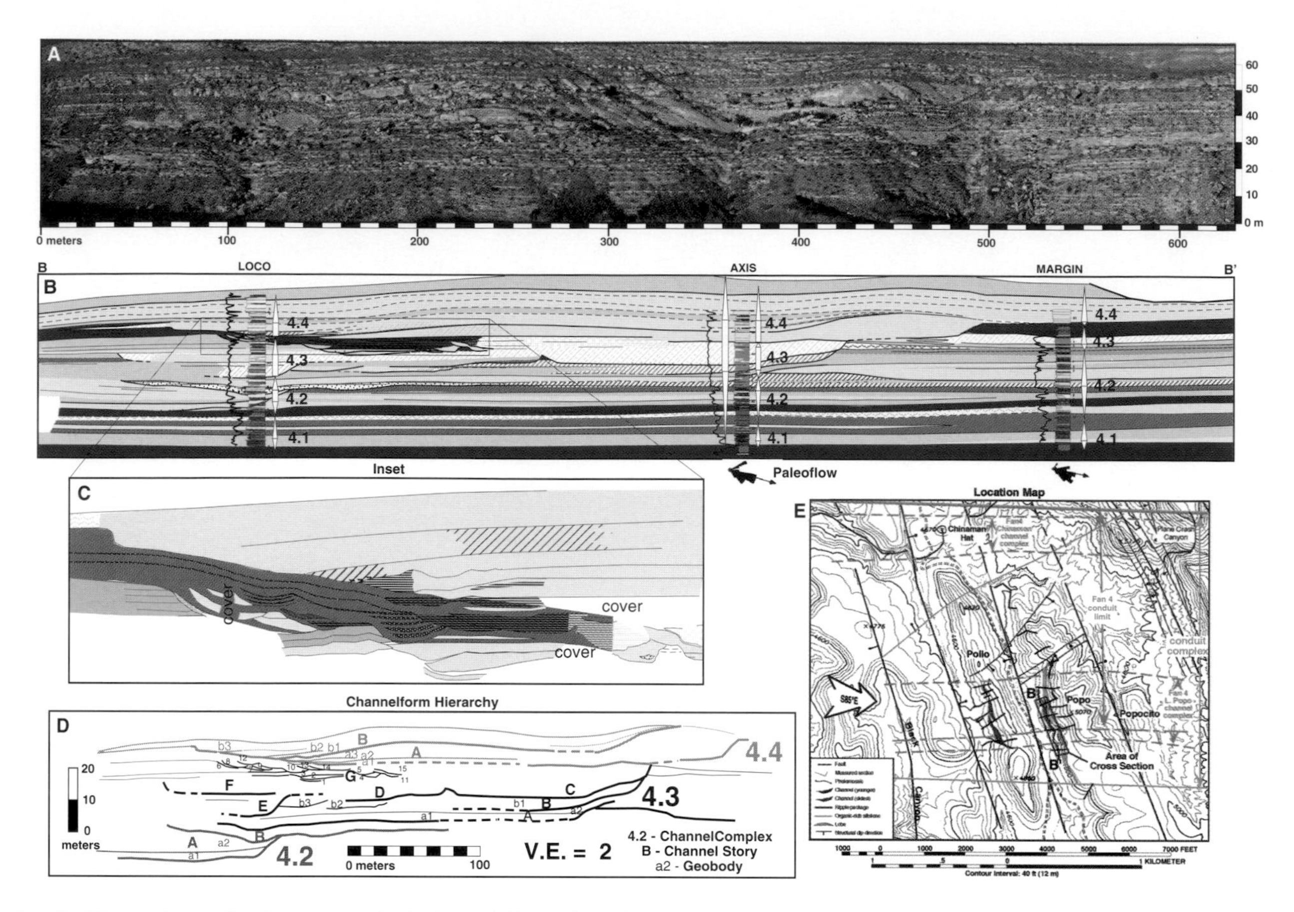

Figure 10—Strike-oriented photomosaic (A) and line-sketch interpretation (B) of Fan 4 submarine channels exposed along the Popo fault block. Outcrop scintillometer profiles shown next to measured sections (see Figure 8 for facies explanation). The outcrop is 25 km basinward of the inferred physiographic margin and exhibits nested offset channel architecture at the base-of-slope to proximal basin floor. The middle Brushy Canyon is the most channelized interval in outcrop. The Popo–Plane Crash area was major sediment conduit where channels repeatedly reinitiated after fan abandonment to form a conduit complex. Inset (C) shows evidence for continued channelization and sediment bypass across the fifth-order abandonment phase. During Fan 4 deposition, the conduit contained two main areas of nested channel complexes, the 600-m-wide Popo sand body and a smaller (lower Chinaman Hat) sand body north of the photo. The Popo area has four fifth-order cycles, three nested channel complexes, and at least 12 single-story channels that stack to form a distinct serrated or stepped southern margin and a less distinct northern margin (D). There is an upward trend from amalgamated high-frequency cuts with abundant sediment-bypass indicators (4.2 and early 4.3), to vertically-stacked (multi-story) cuts (late 4.3), to large offset-stacked, but still highly interconnected, multi-lateral channel stories (4.4). A well-developed, fourth-order spill-phase occurs at the top of cycle 4.4 and bridges the Chinamans Hat and lower Popo complexes. Base map (E) shows Popo area dataset and location of B-B′.

sandstone) relative to BM and the SFB. Progradation internal to Fan 4 is inferred from four offlapping fifth-order cycles. The lower three cycles contain multistory and multilateral channel complexes. These complexes stack vertically, but are also offset 300 m laterally. They collectively form a 35-m-thick and 600-m-wide sand body encased by thin- to medium-bedded sandstone and siltstone. Sand content is 90% within and 61% outside the sand body. The offset stacking of the three channel complexes forms a serrated or stepped sand body margin that erodes siltstone intervals of fifth-order cycle boundaries (Figure 10).

Each channel complex changes upward from highly truncated multistory fills, to heterolithic multilateral fills, to vertically offset multistory sand bodies that show serrated margins. Channel fills comprising complexes shift in both directions across the conduit but generally stack to form only one well-defined margin. The basal multistory and multilateral fills contain frequent erosional surfaces alternating with interbedded siltstone, sandstone with horizontal to inclined laminations and centimeter-thick fusulinid bands, and lag deposits with siltstone ripup clasts. The vertically stacked parts of each complex are the most sandstone-rich. These fills change upward from amalgamated sandstone with Helmholtz waves and aggradational "plow-and-fill" stratification to dewatered structureless sandstone (Figure 8). Where not dewatered this succession is capped by horizontal stratification.

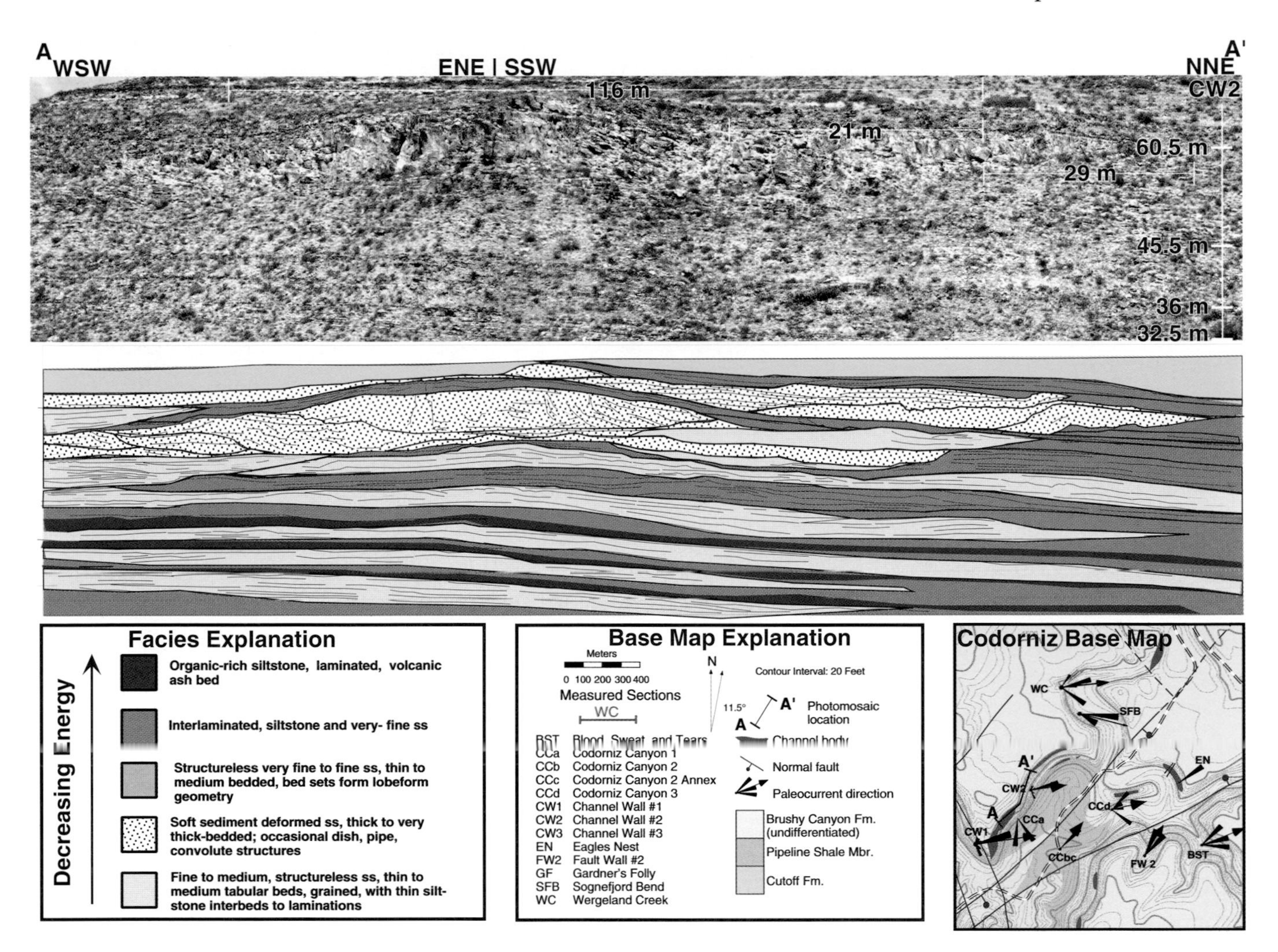

Figure 11—Strike-oriented photomosaic (A) and line-sketch interpretation (B) of Fan 2 submarine channels exposed along Channel Wall, Codorniz Canyon. This wall exposes closely spaced channel complexes encased within a thick succession of tabular sandstone sheets. Channel fills consist of dewatered and soft sediment deformed sandstone forming bi-convex sandbodies that stacked to form a topographic high flanked by younger compensating channelform sandbodies. Thick and sandy build and spill-phase deposits equilibrate the sandstone proportion inside and outside channels.

Channel-flanking deposits consist of interbedded sandstone and siltstone organized as upward-thickening sandstones capped by more continuous siltstone intervals of fifth-order cycles. The sandstones are laterally continuous and locally amalgamated, but generally are interbedded with thin, organic-poor siltstones. Internally, they are structureless, show cryptic stratification, or form thin 300-m-long beds composed of climbing ripples. The sandstone-rich intervals in channel-flanking successions occur near the top of adjacent channel complexes. These intervals contain rippled beds that thin laterally away from the channel and show sediment transport both parallel and transverse to the channel. These spill-phase deposits are eroded by channel rejuvenation during the next fifth-order cycle. Like BM, the multiple episodes of channelization reflect flows persistently directed to this conduit from a confined site upprofile that remained active through deposition of the fan complex.

BASIN-FLOOR CHANNEL COMPLEXES, SOUTHERN DELAWARE MOUNTAINS

Codorniz Canyon provides three-dimensional exposures of channel complexes positioned about 32 km basinward of the ramp margin in the most southerly fan conduit (Figure 4). Although Brushy Canyon sediment transport indicators show a more easterly trend (84°), this area is interpreted to occupy the position farthest from the sediment source along the outcrop belt. Depositional patterns that indicate distal basin-floor position include (1) the thickness and high sand content (65%) of Fan 1 deposits (43 m), reflecting the culmination of a basinward-thickening depositional pattern that is best developed here; (2) the thin Fan 3 interval is a continuation of a basinward-thinning depositional pattern; (3) low conglomerate proportion is consistent with textural sorting trends; and (4) relative to Fan 2 deposits at Colleen

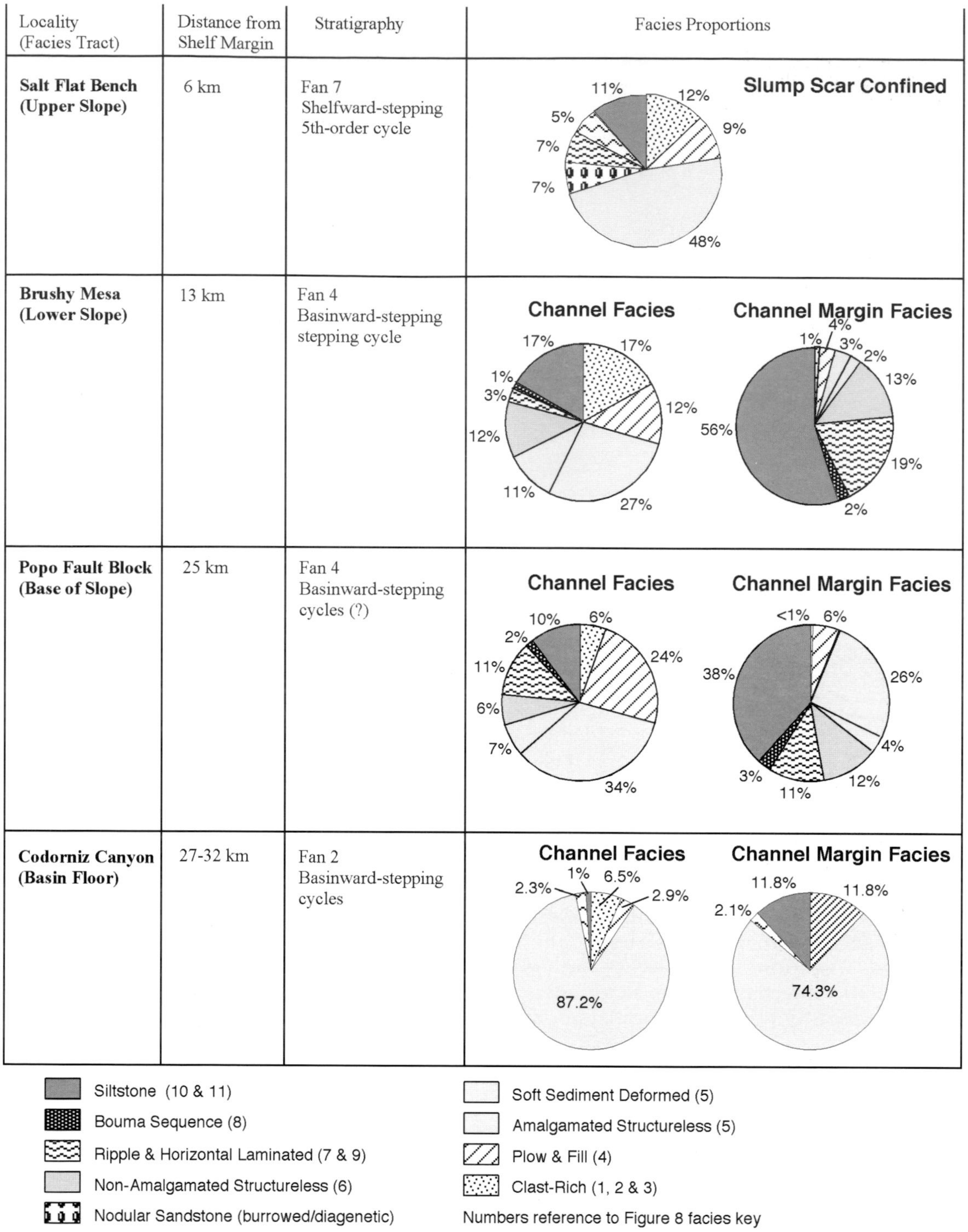
Locality (Facies Tract)
Distance from Shelf Margin
Stratigraphy
Facies Proportions
Salt Flat Bench (Upper Slope)
6 km
Fan 7 Shelfward-stepping 5th-order cycle
Slump Scar Confined
11%
12%
9%
5%
7%
7%
48%
Brushy Mesa (Lower Slope)
13 km
Fan 4 Basinward-stepping stepping cycle
Channel Facies
17%
17%
1%
3%
12%
12%
11%
27%
Channel Margin Facies
4%
1%
3%
2%
13%
56%
19%
2%
Popo Fault Block (Base of Slope)
25 km
Fan 4 Basinward-stepping cycles (?)
Channel Facies
10%
6%
2%
24%
11%
6%
7%
34%
Channel Margin Facies
<1%
6%
38%
26%
4%
3%
11%
12%
Codorniz Canyon (Basin Floor)
27-32 km
Fan 2 Basinward-stepping cycles
Channel Facies
1%
6.5%
2.3%
2.9%
87.2%
Channel Margin Facies
11.8%
11.8%
2.1%
74.3%
Siltstone (10 & 11)
Bouma Sequence (8)
Ripple & Horizontal Laminated (7 & 9)
Non-Amalgamated Structureless (6)
Nodular Sandstone (burrowed/diagenetic)
Soft Sediment Deformed (5)
Amalgamated Structureless (5)
Plow & Fill (4)
Clast-Rich (1, 2 & 3)
Numbers reference to Figure 8 facies key

Figure 12—Summary of architecture of four Brushy Canyon Formation channel complexes.

Figure 13—Diagram summarizing changes in Brushy Canyon submarine channel architecture along a composite slope to basin profile. Along this composite reconstruction, there is a 68% increase in sandstone deposited outside and an 18% increase inside channelform bodies. This increase in sandstone volume outside channel complexes corresponds with a downprofile decrease in channel size. Channel complexes and fan conduits, however, widen downprofile, reflecting increasing offset of both channels and channel complexes, until bed topography from older build-phase deposits begin to control channel pattern and channel complex size decreases.

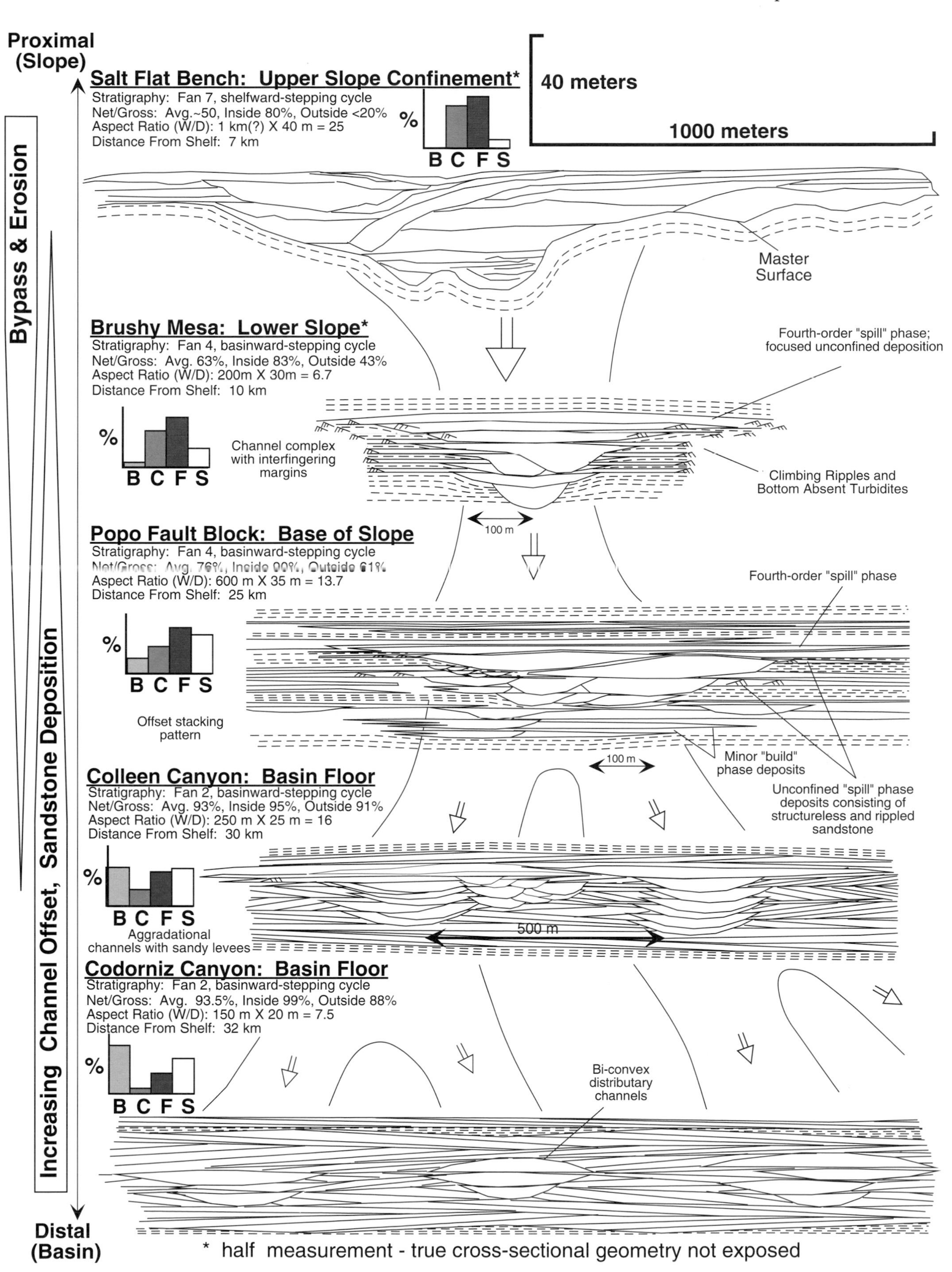

Proximal
(Slope)
Distal
(Basin)
Bypass & Erosion
Increasing Channel Offset, Sandstone Deposition
40 meters
1000 meters
Salt Flat Bench: Upper Slope Confinement*
Stratigraphy: Fan 7, shelfward-stepping cycle
Net/Gross: Avg.~50, Inside 80%, Outside <20%
Aspect Ratio (W/D): 1 km(?) X 40 m = 25
Distance From Shelf: 7 km
%
B C F S
Master Surface
Brushy Mesa: Lower Slope*
Stratigraphy: Fan 4, basinward-stepping cycle
Net/Gross: Avg. 63%, Inside 83%, Outside 43%
Aspect Ratio (W/D): 200m X 30m = 6.7
Distance From Shelf: 10 km
%
B C F S
Channel complex with interfingering margins
Fourth-order "spill" phase; focused unconfined deposition
Climbing Ripples and Bottom Absent Turbidites
100 m
Popo Fault Block: Base of Slope
Stratigraphy: Fan 4, basinward-stepping cycle
Net/Gross: Avg. 76%, Inside 90%, Outside 61%
Aspect Ratio (W/D): 600 m X 35 m = 13.7
Distance From Shelf: 25 km
%
B C F S
Offset stacking pattern
Fourth-order "spill" phase
100 m
Minor "build" phase deposits
Unconfined "spill" phase deposits consisting of structureless and rippled sandstone
Colleen Canyon: Basin Floor
Stratigraphy: Fan 2, basinward-stepping cycle
Net/Gross: Avg. 93%, Inside 95%, Outside 91%
Aspect Ratio (W/D): 250 m X 25 m = 16
Distance From Shelf: 30 km
%
B C F S
Aggradational channels with sandy levees
500 m
Codorniz Canyon: Basin Floor
Stratigraphy: Fan 2, basinward-stepping cycle
Net/Gross: Avg. 93.5%, Inside 99%, Outside 88%
Aspect Ratio (W/D): 150 m X 20 m = 7.5
Distance From Shelf: 32 km
%
B C F S
Bi-convex distributary channels
* half measurement - true cross-sectional geometry not exposed

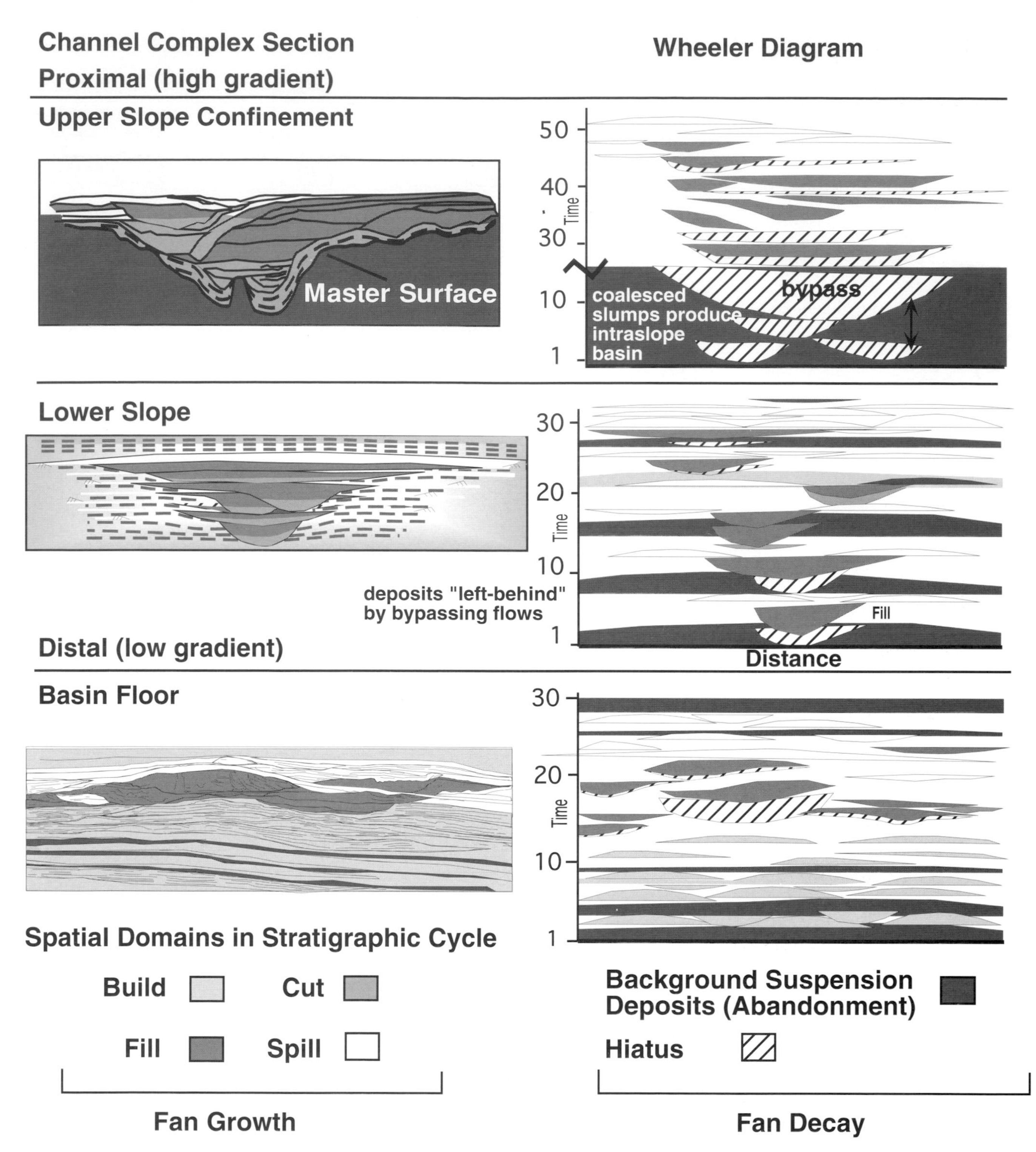

Figure 14—Facies architecture and companion Wheeler diagram summarizing temporal and spatial build-cut-fill-spill phases of submarine channel development along a slope to basin profile. (A) Upper slope channel complex showing the significant time gap between formation of a master bounding surface and depositional filling of the topographic confinement. (B) Lower slope channel architecture and companion Wheeler diagram emphasizes the multiple cut-fill-spill events that stack to form a channel complex. (C) Basin-floor channel architecture showing the high proportion of build-phase deposits that encase small compensating channel complexes.

Canyon, the channel complexes are smaller but show thicker sandstone bedding and slightly increased sand content, demonstrating that the smaller complex size at Codorniz Canyon is not related to decreased sediment supply (Carr and Gardner, this volume). Siltstone intervals that bound fans are laterally continuous across the area, but siltstones that bound fifth-order cycles internal to fans may be eroded.

The lower three fans of the Brushy Canyon fan complex form a 95-m-thick succession with an average sand content of 80% across the 1.7 km^2 area (Figure 11). Fan 2 deposits are the thickest (47 m) and have the highest sand content (94%). Fan 2 consists of four fifth-order cycles containing thick-bedded tabular sandstone bedsets encasing channelform sand bodies (Table 1). Sandstone content is 99% inside and 88% outside channel complexes. Each sand body contains two to 10 channel stories that stack vertically, collectively forming nested complexes 20 m thick and 200 m wide (Figure 11). These multistory complexes are characterized by: (1) close spacing, (2) a high proportion of bi-convex channelform sand bodies, (3) little facies variation in channel fills, and (4) amalgamated bed sets in channel axes that become nonamalgamated beds at interfingering channel margins. The dominant facies include ungraded, structureless sandstone with floating siltstone clasts and local soft-sediment deformation and dewatering features. Channel bases show only minor erosion and in many cases appear to drape underlying bed topography. Load amalgamation is more common than amalgamation due to erosive truncation.

Fan 2 channel complexes at Codorniz Canyon are encased by laterally continuous, thick-bedded sandstones interbedded with thin siltstones that form tabular bedsets (Figure 11). The geometry of a nonamalgamated sandstone bed is lobate or lobeform. These lobeform bodies are interpreted to record compensating deposition by low-gradient, unconfined, high-density gravity flows deposited on the basin floor. Significantly, bed topography of preceding lobeform deposits created preferred gravity-flow pathways that controlled sites of channelization and produced a closely spaced distributary pattern. These comparatively small channel complexes are equivalent in size to base-of-slope channel fills, hence "build" phase deposits contribute significantly to the observed distal increase in Fan 2 sandstone volume. Build-phase bed topography exerts a strong control on channel complex size and shape. Channel-flanking successions in between bracketing "build" and "spill" phase deposits are thicker bedded, contain frequent yet small-scale erosional cuts, complex soft-sediment deformation features, and rippled sandstones (Figure 12).

DISCUSSION

Recognizing the sediment body hierarchy for submarine channelform sand bodies provides insight into architectural changes that occur along a slope and basin profile. The wide scatter in compiled submarine channel aspect-ratios highlights the limited utility of quantitative measurements, unless they are collected within a stratigraphic framework that reflects a hierarchy of architectural elements. For example, a prevailing view holds that the aspect ratio of channelform bodies increases down a fan (Barnes and Normark, 1983/84; Clark and Pickering, 1997; Figure 1). Brushy Canyon data also show a downprofile decrease in channel width, but channel complexes and fan conduits increase in size to the point where channel complexes become distributary and then decrease in size. (Figure 12; Figure 13). These channelforms get bigger for several reasons, despite the decreased size of individual channels. First, the restriction of stratigraphic confinements to upper slope settings produces a systematic downprofile increase in the offset of channel bodies forming channel complexes and channel complexes in fan conduits. Second, upprofile confinement focuses gravity flows to the same site to generate build-cut-fill-spill patterns that construct the hierarchy of channels and channel complexes. These cluster to create fan conduits of enhanced sandstone volume. Third, there is a downprofile increase in sandstone content (build-and-spill phase) of overbank deposits, which promotes lateral offset and widening of the erosional channel complex. Fourth, the lithology contrast between deposits inside and outside channel complexes decreases down the profile. This reflects the increased proportions of build and spill-phase deposits flanking and encasing channel complexes. This depositional pattern reduces the ability to resolve smaller-scale channel bodies, making larger-scale fan conduits the only resolvable channelform sand body.

Master bounding surfaces that cluster, amalgamate, and confine channel fills to construct large channelform bodies most likely occur in proximal slope and canyon settings. Here, a long history of slope adjustment, combined with sediment bypass and erosion, helps develop compound erosional surfaces. Recognizing master bounding surfaces in slope systems is aided by applying stratigraphic criteria also used to distinguish nonmarine valley fills from their fluvial counterparts. Although not analogous in process to the formation of nonmarine valleys, intraslope depressions share a common stratigraphy, reflecting the significant period of time required to develop a master bounding surface and the resulting confined deposition. In the Brushy Canyon, slope confinements are produced by lateral coalesced slump scars (Gardner and Sonnenfeld, 1996). Master bounding surfaces define topographic depressions and "containers" that also direct and focus gravity flows basinward. Furthermore, preexisting degradational topography affects subsequent gravity flows by conserving energy that would otherwise be lost through erosion, but is instead translated downslope to support and maintain basinward-directed flows. This upprofile focusing mechanism has a direct impact on the pattern of channel and flanking overbank deposition.

Submarine channel deposits show a complex internal stratigraphy expressed by repeated episodes of erosion and deposition (Piper, 1970; Walker, 1975; Mutti and Normark, 1987; Clark and Pickering, 1997). This pattern is expressed in Brushy Canyon channel-overbank deposits as an organized record of build-cut-fill-spill deposition that occurs at multiple scales and stacks to form a hierarchy of channelform bodies (Figure 14). In general, a build-cut-fill-spill cycle is initiated by focusing gravity flows through a topographic low. The proportion of deposits that precede channelization increases down the profile to form bed topography during the "build" phase of deposition, which has an increasing influence on channelization basinward. Maintenance and/or erosional expansion of the channel in the "cut" phase occurs during bypass and limited deposition of remnant deposits. The amount of time separating cut-and-fill phases systematically decreases downprofile. Temporal discontinuities are greatest in upprofile positions where master bounding surfaces separate wholly younger strata above from older strata below. Amalgamated vertical sandstone bed successions record depositional backfilling and lateral offset of beds in the "fill" phase of channel deposition.

Deposits that record aggradation and passive backfilling dominate Brushy Canyon channel fills. This contrasts with both the "classic" upward-fining profile of an active channel fill and with the fine-grained passive fill of channels abandoned through updip avulsion. Continued deposition of unconfined gravity flows at the filled channel site produces a "spill" phase of deposition and is a precursor to shifting subsequent flows to a new site. These build-cut-fill-spill phases represent spatial and temporal domains in submarine channel development. They do not represent a particular sediment body type, although skewing of sediment body and facies proportions occurs within these temporal phases. They occur at multiple scales and explain depositional patterns in a hierarchy of channels, complexes, and fan conduits. The stacking of multiple cut-fill-spill channels yields thick successions of channel-overbank successions, but channel and overbank deposition are offset in time.

Although unique and site-specific depositional processes help define the position of a submarine channel on a slope-to-basin profile, the preservation is controlled by stratigraphic position. For example, a proximal slope position for the SFB sand body is indicated by the high proportion of siltstone, conglomerate and slump deposits, slide blocks, and slump scars. This facies architecture occurs only on the upper slope segment of the Brushy Canyon profile, where slope confinements restrict lateral movement of channels. Although older fans contain slope deposits, the fact that these confined slope sand bodies are best developed in younger slope deposits of Fan 7 reflects changing from net bypass to net deposition during the stratigraphic evolution of the fan complex. The repeated rejuvenation of channelization along the basin-floor profile to maintain a fan conduit emphasizes the important control of upper slope and canyon confinements on directing gravity flows basinward. This requires that these proximal confinements remain underfilled until the latest stages of fan complex evolution.

CONCLUSIONS

Sediment gravity flows produce a hierarchy of submarine channel deposits. If data on channel shape and size are going to be used to predict trends and patterns in submarine channel development, and/or for quantitative reservoir modeling, then the hierarchy of bodies that compose a channelform body must be resolved. Architectural element analysis of the Brushy Canyon Formation reveals a fourfold hierarchy of channelform sand bodies. Brushy Canyon channel complexes change downprofile. Large upper slope channelform sand bodies have multiple cut-and-fills, where both channel and overbank deposits are contained within a stratigraphic confinement. Channel complexes widen downprofile because of increasing offset of component channel bodies, and they are encased within higher proportions of overbank deposits that contribute to channel offset and increase the fan sandstone volume.

Stratigraphic changes in facies architecture reflect the "build," "cut," "fill," "spill," spatial, and temporal domains of submarine channel development. Facies and lithology proportion, sediment-body type, and the connectivity and clustering of sediment bodies change within these spatial domains based on the channel complex position along the slope-to-basin profile.

ACKNOWLEDGMENTS

Financial support for this research was provided by members of the Slope and Basin Consortium administered by Michael H. Gardner. Consortium sponsors include ARCO Exploration & Production Technology, Amoco Production Company, British Petroleum, Conoco Inc., Elf Exploration Production, Exxon Production Research Company, Marathon Oil Company, Mobil Exploration and Producing Technical Center, ORYX Energy Company, Phillips Petroleum Company, Shell Offshore Inc., Statoil, Texaco Exploration and Production, Inc., and Unocal Corporation.

Bente Blikeng, Marieke Dechesne, Edith Fugelli, Frode Hadler-Jacobsen, Brian Horn, John Law, Laird Little, Jesse Melick, Jeremy Merrill, Ted Playton, Mark Sonnenfeld, Roger Wagerle, and Conrad Woodland helped collect field data. Mary Carr contributed as project manager. Shell Offshore Inc. supported Kyle Johnson's DEPSIM SFB model. Linda Martin, Leann Wagerle, and Ted Playton helped draft figures. Fred Armstrong, resource manager, Guadalupe Mountains National Park, Tony Kunitz and Mike and Anne Capron, Six Bar Ranch, and Dr. Fred Speck and Jim Dackus, Sibley Ranch, provided access and logistics support. We thank Arnold Bouma, Roger Slatt, Charles Stone, and Mark Sonnenfeld for careful review of an early version of this manuscript.

REFERENCES CITED

Barnes, N. E., and W. R. Normark, 1983/84, Diagnostic parameters for comparing modern submarine fans and ancient turbidite systems, COMFAN: Geo-Marine Letters, v. 3, p. 2–4.

Clark, J. D., and K. T. Pickering, 1997, Submarine channels, processes and architecture: London, Vallis Press, 231 p.

Gardner, M. H., B. J. Willis, and W. Dharmasamadhi, 1995, Outcrop-based characterization of low accommodation/sediment supply fluvial-deltaic sandbodies, *in* The 3d JNOC-TRC International Symposium: reservoir characterization: integration of geology, geophysics and reservoir engineering: Special Publication of the Technology Research Center, JNOC, No. 5, p. 17–54.

Gardner, M. H., and M. D. Sonnenfeld, 1996, Stratigraphic changes in facies architecture of the Permian Brushy Canyon Formation in Guadalupe Mountains National Park, west Texas: Permian Basin Section—SEPM publication 96-38, p. 17–40.

Gardner, M. H., J. M. Borer, and M. Dechesne, 1998, Cut, fill and spill: a new look at the overbank paradigm for sandy deep water systems (abs.): AAPG National Convention Program, Salt Lake City, Utah.

Goetz, L. K., 1985, Salt Basin Graben: a basin and range right-lateral transtensional fault zone—some speculations, *in* W. Dickerson and W. Muhlberger, eds., Structure and tectonics of Trans-Pecos Texas: West Texas Geological Society, Field Conference Publication 85-81, p. 165–167.

Hill, C. A., 1995, Geology of the Delaware Basin, Guadalupe, Apache, and Glass mountains, New Mexico and west Texas: Permian Basin Section-SEPM publication 96–39, 480 p.

Johnson, K. R., 1998, Outcrop and reservoir characterization of the Brushy Mountain Formation, Guadalupe Mountains National Park, West Texas, and Cabin Lake field, Eddy County, New Mexico: Unpublished M.S. thesis, Colorado School of Mines, 256 p.

Kerans, C., W. M. Fitchen, M. H. Gardner, M. D. Sonnenfeld, S. W. Tinker, and B. R. Wardlaw, 1992, Styles of sequence development within uppermost Leonardian through Guadalupian strata of the Guadalupe Mountains, Texas and New Mexico, *in* D. H. Murk and B. C. Curran, eds., Permian Basin exploration and production strategies—Applications of sequence stratigraphic and reservoir characterization concepts: West Texas Geological Society Symposium, Publication 92-91, p. 117.

Kerans, C., and W. M. Fitchen, 1994, Sequence hierarchy and facies architecture of a carbonate-ramp system: San Andres Formation of Algerita Escarpment and Western Guadalupe Mountains, West Texas and New Mexico: Bureau of Economic Geology, Report of Investigations No. 235, 86 p.

King, P. B., 1942, Permian of west Texas and southeast New Mexico: AAPG Bulletin, v. 26, p. 535–763.

King, P. B., 1948, Geology of the southern Guadalupe Mountains, Texas: U.S. Geological Survey Professional Paper P-215, 183 p.

Mutti, E., and W. R. Normark, 1987, Comparing examples of modern and ancient turbidite systems: problems and concepts, *in* J. K. Leggett and G. G. Zuffa, eds., Marine clastic sedimentology: concepts and case studies: London, Graham & Trotman, p. 1–38.

Oriel, S. S., D. A. Myers, and E. J. Crosby, 1967, West Texas Permian Basin region, *in* Paleotectonic investigations of the Permian System in the United States: U.S. Geological Survey Professional Paper P-515C, p. C17–C60.

Piper, D. J. W., 1970, A Silurian deep sea fan deposit in western Ireland and its bearing on the nature of turbidity currents: Journal of Geology, v. 78, p. 509–522.

Pray, L. C., 1988, Trail Guide: day one, Bone and Shumard Canyon area, *in* S. T. Reid et al., eds., Guadalupe Mountains revisited, West Texas Geological Society Publication, 88-84, p. 41–60.

Ross, C., 1986, Paleozoic evolution of southern margin of Permian Basin: Geological Society of America Bulletin, v. 97, p. 536-554.

Rossen, C., and J. F. Sarg, 1988, Sedimentology and regional correlation of a basinally restricted deep-water siliciclastic wedge, Brushy Canyon Formation, Cherry Canyon Tongue (Lower Guadalupian), *in* S. T. Reid et al., eds., Guadalupe Mountains revisited: West Texas Geological Society Publication, 88-84, p. 127–132.

Sageman, B. B., M. H. Gardner, J. M. Armentrout, and A. E. Murphy, 1998, Stratigraphic hierarchy of organic-rich siltstones in deepwater facies, Brushy Canyon Formation (Guadalupian), Delaware Basin, west Texas: Geology, v. 26, p. 451–454.

Vail, P. R., R. M. Mitchum, Jr., and S. Thompson III, 1977, Seismic stratigraphy and global changes in sea level: relative changes of sea level from coastal onlap, *in* C. E. Payton, ed., Seismic stratigraphy—Applications to hydrocarbon exploration: AAPG Memoir 26, p. 83–97.

Yang, K. M., and S. L. Dorobek, 1995, The Permian Basin of west Texas and New Mexico: flexural modeling and evidence for lithospheric heterogeneity across the Marathon Foreland, *in* S. Dorobek and J. Ross, eds., Stratigraphic evolution of foreland basins: SEPM Special Publication 52, p. 37–53.

Walker, R. G., 1975, Nested submarine-fan channels in the Capistrano Formation, San Clemente, California: Geological Society of America Bulletin, v. 86, 915–924.

Zelt, F. B., and C. Rossen, 1995, Geometry and continuity of deep-water sandstones and siltstones, Brushy Canyon Formation (Permian) Delaware Mountains, Texas, *in* K. T. Pickering, R. N. Hiscott, N. H. Kenyon, F. Ricci Luchi, and R. D. A. Smith, eds., Atlas of deep water environments—architectural style in turbidite systems: London, Chapman & Hill, p. 167–183.

Carr, M., and M. H. Gardner, 2000, Portrait of a basin-floor fan for sandy deepwater systems, Permian lower Brushy Canyon formation, west Texas, *in* A. H. Bouma and C. G. Stone, eds., Fine-grained turbidite systems, AAPG Memoir 72/SEPM Special Publication No. 68, p. 215–232.

Chapter 20

Portrait of a Basin-Floor Fan for Sandy Deepwater Systems, Permian Lower Brushy Canyon Formation, West Texas

Mary Carr
Michael H. Gardner
Department of Geology and Geological Engineering, Colorado School of Mines, Golden, Colorado, U.S.A.

ABSTRACT

The architecture of basin-floor deposits from unconfined flows, showing bed compensation and nonunique bed patterns, is stratigraphic in origin. Stratigraphic forcing is demonstrated in an upward coarsening succession of layered sheet sandstones that encase isolated distributary channel complexes deposited 30 km from a physiographic margin. This 70-m-thick basin-floor succession contains three fourth-order cycles that show organized changes in sediment body type, bed thickness, facies proportion, and percentage of sand. Siltstone-dominated Fan 1 (8% sandstone) drapes but incompletely fills underlying paleotopography. Layered sandstone sheets (~75%, 0.59-m-thick bed sets) dominate Fan 2 (93% sandstone) and encase the only channel-flanking wedge (~7.4%) and channelform (~7.35%) elements. Amalgamated sandstone sheets (~99%, 0.80-m-thick sandstone bed sets) dominate Fan 3 (87% sandstone) with minor channelform elements (0.3%).

INTRODUCTION

The stratigraphic organization of basin-floor deposits from unconfined flows has been a hotly contested and highly debated issue among deepwater workers (Ghibaudo, 1981; Hiscott, 1981; Anderton, 1995; Graham et al., 1996; Larue and Jones, 1997). The central issue regards compensational bedding (Mutti and Sonnino, 1981), whereby bed geometries of preceding flows control the position and geometry of subsequent flows to produce nonunique vertical bed patterns. Hence, upward sandstone bed thickening and thinning patterns, commonly interpreted to record cyclicity, are considered unreliable indicators of stratigraphic forcing. However, the question remains whether there are other sedimentologic parameters that reflect stratigraphic forcing and that can be used to make predictions of stratal architecture.

In the lower Brushy Canyon, siltstone intervals form the most laterally continuous deposits and define a hierarchy of three fourth-order stratigraphic cycles that consist of smaller-scale fifth-order cycles (Table 1). Although siltstone intervals of both fourth and fifth-order cycles can be correlated across the study area, only siltstone intervals of fourth-order cycles extend across the Delaware Mountains outcrop belt. This regional stratigraphic framework is discussed in Gardner and Borer (this volume). This stratigraphic framework is used to compare changes in bed and bed set thickness, facies diversity, frequency and magnitude of erosional surfaces, channelform to lobeform ratio, and lithology proportions. By examining systematic changes in these attributes, we can address the following attributes of basin-floor deposits: (1) degree of stratigraphic organization, (2) at what scale compensational bedding

Table 1. Summary of sedimentologic data within the stratigraphic framework for the lower Brushy Canyon at Colleen Creek.

Lithostratigraphy	4th-order cycles	5th-order cycles	Dominant Architectural Element	Thickness	Percent Sandstone	Stacking Pattern
Lower Brushy Canyon Fm	Fan 3	3	Tabular Sandstone	30 m	80%	landward
	Fan 2	2	Channelform	40 m	92%	basinward
Pipeline Siltstone Member	Fan 1	2?	Tabular Siltstone	30 m	8%	basinward

controls stratal architecture, and (3) examine the relationship between architecture and lithology of unconfined flow deposits.

GENERAL SETTING

Within Colleen Canyon, the Permian Cutoff Formation and Brushy Canyon Formation, including the basal Pipeline Shale Member (King, 1965), are exposed over an approximately 4 km^2 area (Figure 1). The north and south canyon walls expose normal faults trending N50°E that bound a central graben with approximately 60 m of displacement (Figure 1). This structural setting provides three-dimensional exposures where strike and dip changes are documented. The dominant paleoflow direction within the deposits of Fans 2 and 3 is to the ESE at 100° (Table 2).

METHODS AND DATA

Seventeen measured sections, showing decimeter resolution, are distributed at 200 to 800 m spacing around the canyon (Figure 1, Table 2). Scintillometer profiles were collected from 15 sections, and 10 sections were core sampled at least every meter for petrophysical analysis (Figure 1). These data are tied to computer-generated photomosaic panels of each canyon wall (Figure 1). Digital photomosaic panels, tied to measured section data, are used to map siltstone intervals bounding stratigraphic cycles from which facies architecture and architectural element distributions can be quantified. Architectural element percentages are calculated using the two-dimensional area of strike-oriented projections of the photomosaic panels (Figure 2).

STRATIGRAPHIC SETTING

Cutoff Formation

The Cutoff Formation was not studied in detail but several sections began in the upper portions of the unit. Although mapped as the Bone Spring Limestone, this unit contains fauna of the Cutoff Formation exposed along the Western Escarpment of the Guadalupe Mountains (Lance Lambert, personal communication, 1999). The Cutoff Formation in this area consists of two bench-forming units. The lower bench is composed of dilute lime mudstone turbidites. The upper bench consists of rotated submarine slide blocks that represent a mass-transport complex that can be traced northward across the Delaware Mountains outcrop belt. Overlying this mass transport complex are deepwater lime mudstone turbidites that filled some of the irregular topography produced by the mass transport complex deposit. The upper beds commonly contain large concretions that may be rotated perpendicular to bedding. The upper contact of the cutoff shows up to 28 m of relief within Colleen Canyon (Figure 2), and is commonly littered with ammonoids.

Pipeline Shale Member of the Brushy Canyon Formation

The Pipeline Shale (Fan 1) consists of interbedded siltstone, organic-rich siltstone, and minor thin sandstone beds, arranged in at least two fifth-order cycles (Table 1). The Pipeline Shale varies from 20.6 to 27 m in thickness and has an undulatory upper surface with up to 19 m of relief (Figures 2, 3). The upper contact is sharp to gradational, but there is no evidence of significant erosion by overlying Brushy Canyon sandstones. The relief at its upper surface indicates that Pipeline Shale deposits did not completely fill the topography from the Cutoff mass-transport complex.

The basal 4 m of the Pipeline Shale are the most organic-rich, show the greatest concentration and size of calcite and phosphatic concretions, have the greatest abundance of volcanic ash beds, and give the highest gamma-ray counts. This organic-rich interval is gradationally overlain by interbedded planar-laminated siltstone, organic-rich siltstone, volcanic-ash beds, and thin parallel-laminated very fine-grained sandstones. The upper 2–3 meters may contain parallel-laminated very fine-sandstones.

Lower Brushy Canyon Formation

The Brushy Canyon Formation in the Delaware Mountains has been informally divided into lower, middle, and upper members. The lower Brushy Canyon Formation, above the Pipeline Shale in Colleen Canyon, consists of two fourth-order cycles, each one containing three fifth-order cycles. These sand-rich deposits form a conspicuous physiographic mesa that can be traced across the Delaware Mountains outcrop belt. Fan 2 varies in thickness from 25 to 48 m and consists of three fifth-order cycles that show a trend from small single-story channelforms in Cycle 1 (F2C1), to multistory and multilateral channel complexes in Cycle 2 (F2C2), to siltstone in Cycle 3 (F2C3). Fan 2 Cycle 1 (F2C1) varies significantly in thickness (6–20m) because its deposits fill underlying topography on the Pipeline Shale and are eroded by the overlying channel complexes in F2C2 (Figures 3, 4). F2C1 has an average

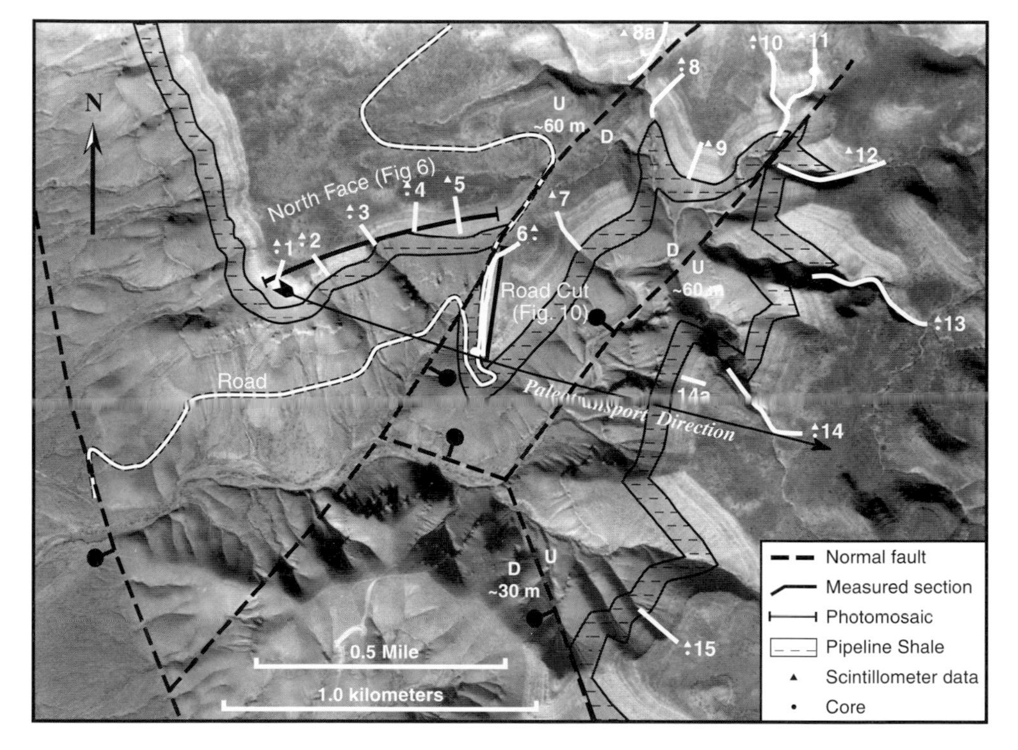

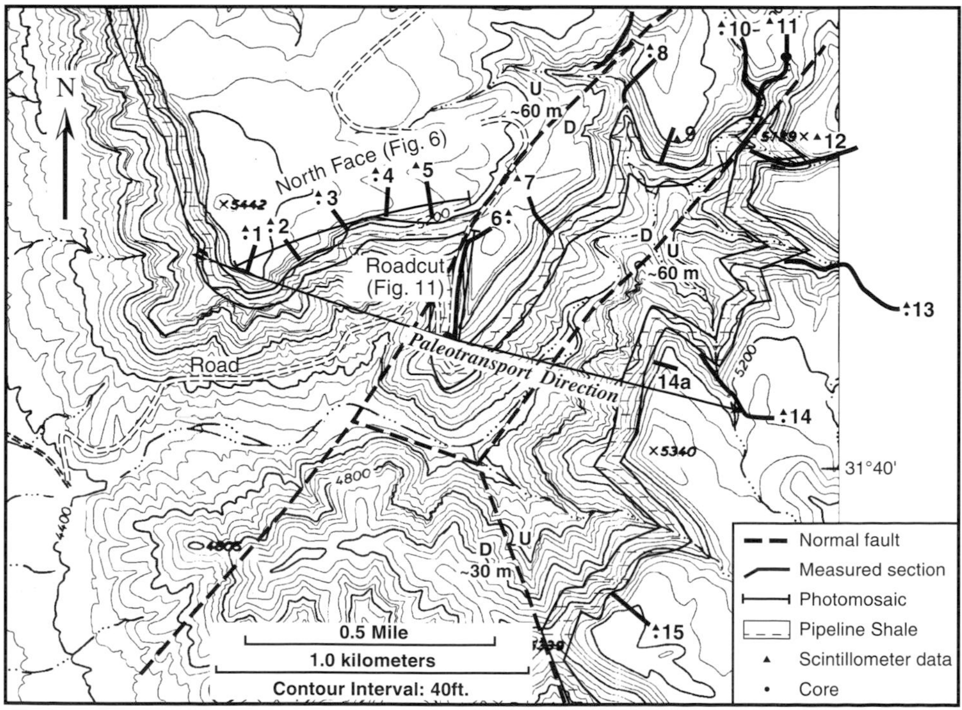

Figure 1—Airphoto and topographic base maps of the Colleen Canyon study area, showing locations of measured sections, photomosaics, scintillometer surveys, and core data. Large arrow indicates general trend and paleocurrent of the Fan 2 Cycle 2 channel complexes.

Table 2. Paleoflow data and sandbody architecture from the lower Brushy Canyon at Colleen Canyon.

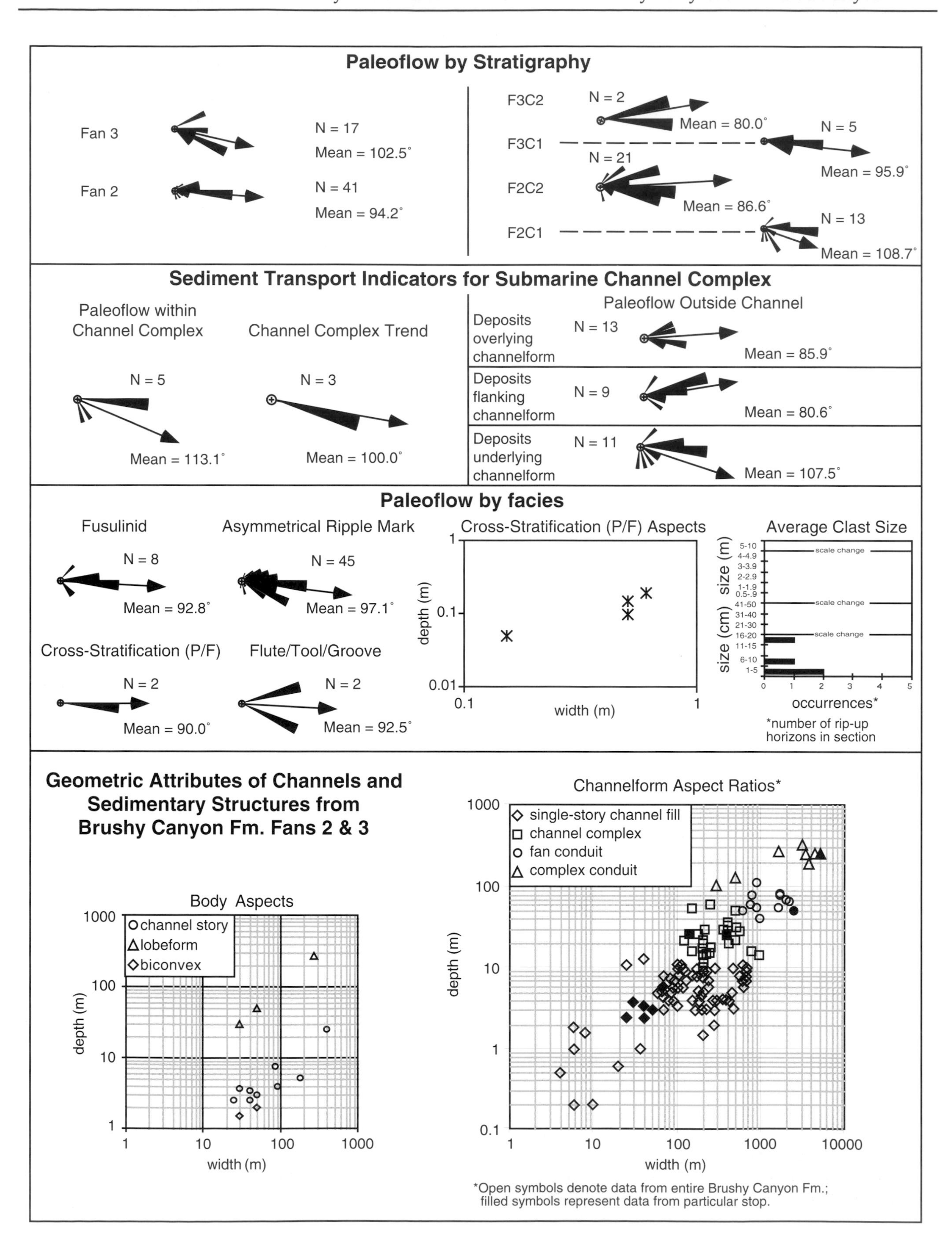

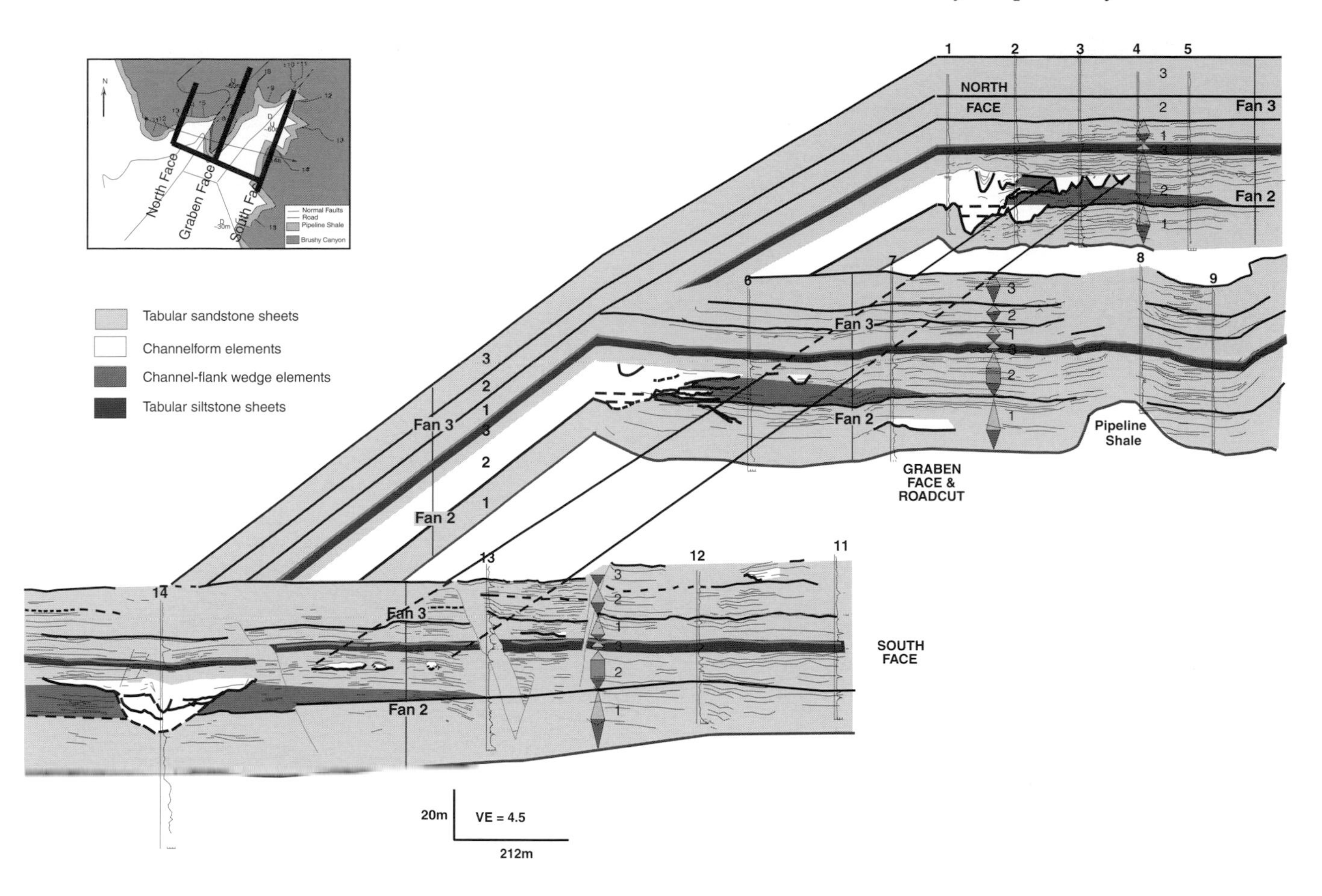

Figure 2—Strike-oriented fence diagram of Colleen Canyon outcrop faces, showing downprofile changes in architectural element distribution and bedding geometry. Gamma-ray profiles are overlain with line tracings from photomosaic panels. Two channel complexes can be seen in Fan 2 Cycle 2. The older complex is 25 m thick by 250 m wide. A dark pair of lines connects the younger complex (9 × 400 m) across the three faces. Note that the younger complex is not resolved on the distal panel. This channel complex may record the initiation of Fan 2 abandonment.

bedset thickness of 0.58 m and is dominated structureless and soft-sediment deformed to cross-stratified sandstones with a sandstone content (sandstone thickness/total cycle thickness) of 92% (Figure 5). Although they comprise only a very small percentage of the overall rock volume, numerous thin (2–5 cm) organic-rich siltstone interbeds vertically partition sandstone bed sets of F2C1 to produce layered or nonamalgamated bedding. Layered or nonamalgamated bedding is described as thin to medium sandstone intervals vertically partitioned by thin siltstone beds (Mahaffie, 1994). Isolated single story channels (~200m × 4m) represent approximately 5% of the total rock volume in F2C1. Sediment transport within these channel deposits was to the ESE. At the base of F2C1 immediately below the channel complex in F2C2 is a fusulinid-rich conglomerate (30–40% fusulinids). It is not possible to determine the geometry and lateral extent of this unit because of the nature of the outcrop, but the facies are consistent with other channels within F2C1.

Fan 2 Cycle 2 (F2C2) is 94% sandstone (Figure 5) and varies in thickness from 16 to 25 m (Figures 3,d 4). The interval is dominated by nonamalgamated structureless and stratified sandstone organized in layered sandstone sheets that encase two multistory to multilateral channel complexes with flanking overbank wedges (Figure 6). Average bedset thickness is 0.60 m, whereas within channel deposits it is 0.84 m and 0.54 m for deposits outside of the channel. Because the base of the channel complex downcuts into F2C1, F2C2 is thickest within and near this multistory channel complexes. Channel-flanking deposits show an overall symmetrical pattern of bed thinning, thickening, and thinning.

The percentage of sandstone for Fan 2 Cycle 3 (F2C3) is 15% and thickness varies from 2.2 to 5 m. Siltstone, organic-rich siltstone, and thin ripple-laminated

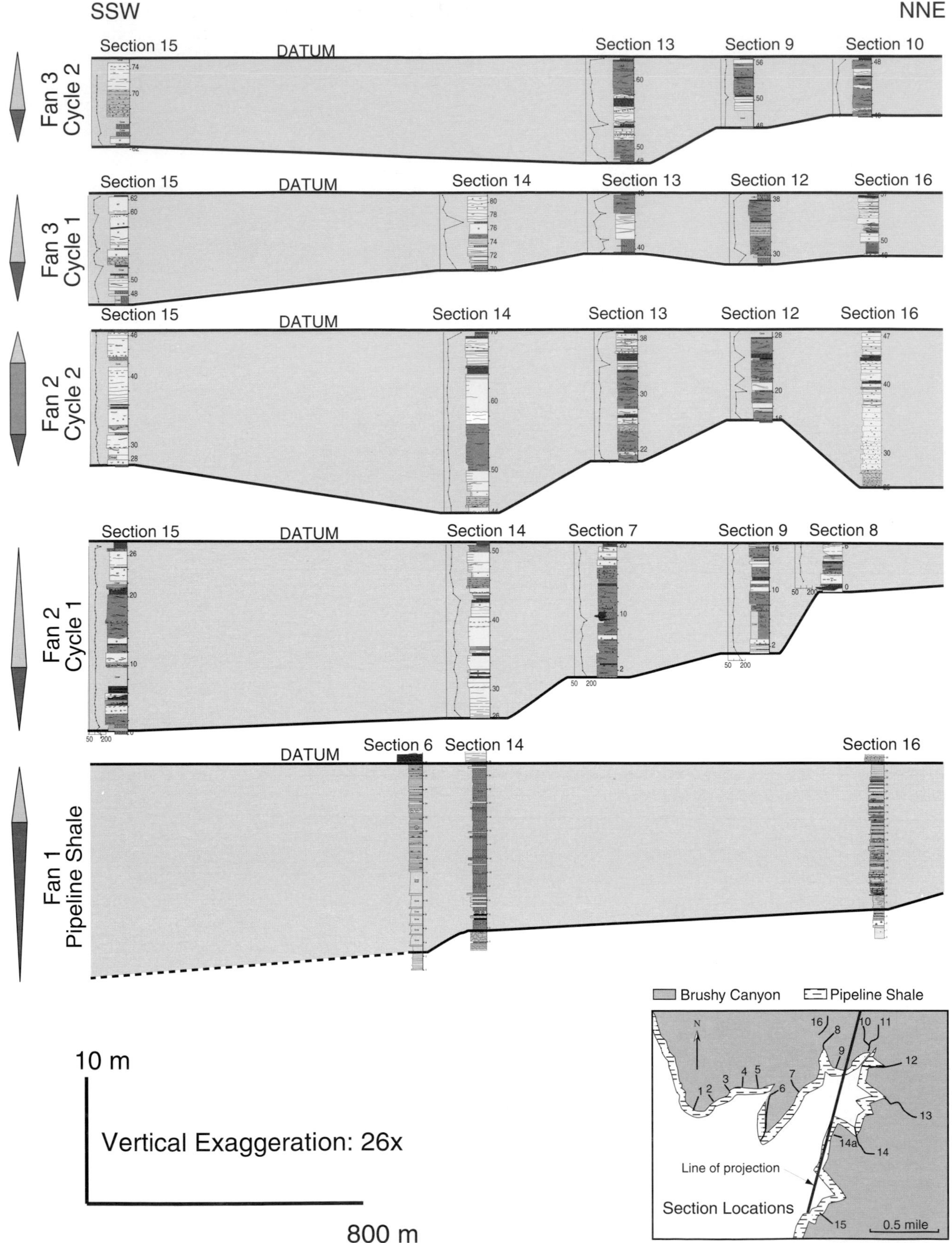

Figure 3—Multiple-data strike-oriented cross sections of Fans 1–3. Data for each fifth-order stratigraphic cycle is an upper organic-rich siltstone. This cross-section construction illustrates the effect of basal topography on depositional patterns in overlying deposits. The channel complex at the base of section 14 in Fan 2 Cycle 2 follows the low in Fan 1 and Fan 2 Cycle 1.

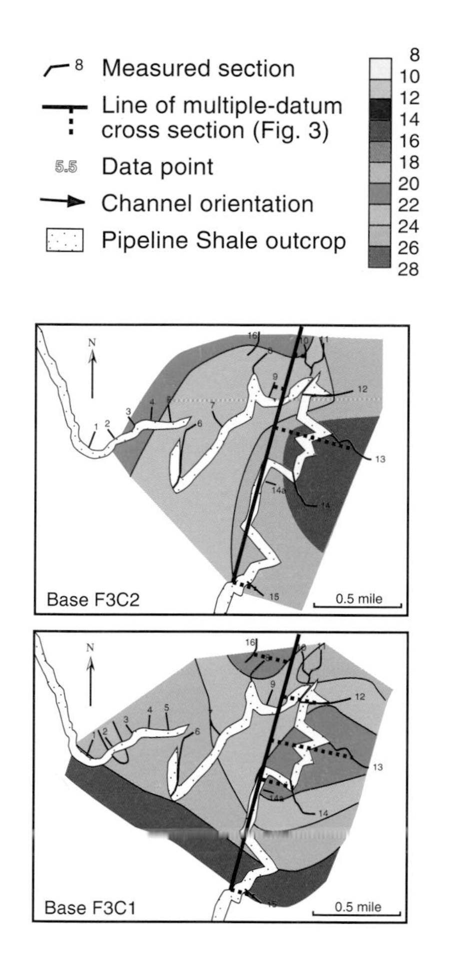

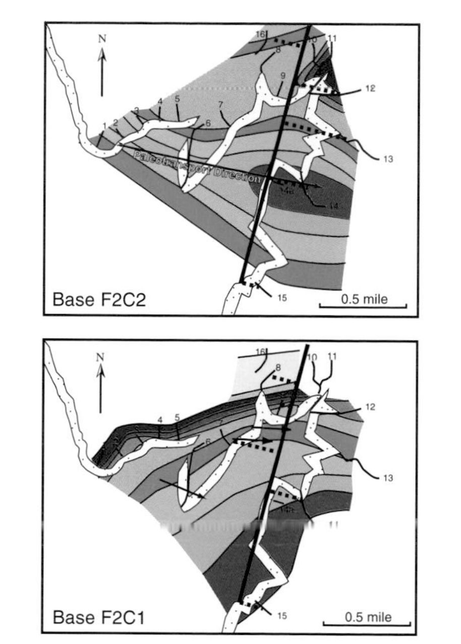

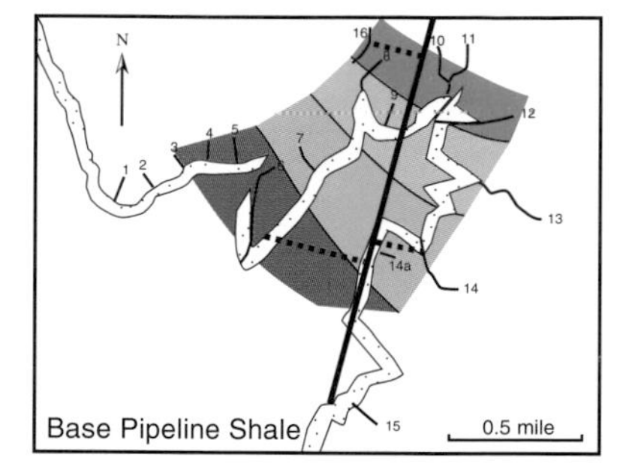

Figure 4—Isopach maps of each 5th-order stratigraphic cycle illustrating the initial topography affecting deposition in each cycle shown in Figure 3. Fan 2 Cycle 3 varies in thickness from 2 to 4 m and is not shown.

sandstone and thin structureless sandstones characterize this interval. The low percentage of sandstone records a significant reduction in sediment supply related to abandonment of Fan 2.

Fan 3 shows a uniform thickness of 30 m and contains three fifth-order cycles. These cycles show a trend of increasing average bedset thickness from 0.71 m in F3C1 to 0.84 m in F3C3. Fan 3 is dominated by amalgamated sandstone sheet elements and shows a significant increase in siltstone percentage and a corresponding drop in sandstone content (86% sandstone) (Table 1, Figure 5). Fan 3 Cycle 1 (F3C1) contains thick-bedded (avg. thickness 0.71 m) tabular sandstone bed sets. Although bed sets are thicker, the percentage of sandstone (86%) is lower relative to underlying Fan 2 cycles reflecting increased sandstone bed amalgamation (Figure 5). Fan 3 Cycle 1 is uniform in thickness, averaging 12 m, and shows an asymmetrical upward-bed–thickening pattern (Figures 3, 4). This fan is characterized by heavily cemented, nonamalgamated to amalgamated, structureless to cross-stratified sandstone. A silicified structureless sandstone with up to 10% fusulinid and crinoid debris caps F3C1. No channels were observed in this interval.

Fan 3 Cycles 2 and 3 (F3C2 and F3C3) generally are poorly exposed and form a recessive slope in Colleen Canyon, but clean gully exposures show thick-bedded, non-amalgamated to amalgamated tabular bed sets of dominantly structureless to cross-stratified sandstone that increases from an average thickness of 0.82 m in F3C2 to 0.84 m in F3C3. The average sandstone percentage for cycles 2 and 3 is 88% (Figure 5). Fan 3 Cycle 3 contains rare single-story channels; the channel margins are much lower relief than those in Fan 2, with about 1 m of erosional relief.

ARCHITECTURAL ELEMENTS

Four sediment body types, or architectural elements, were mapped within the high-resolution stratigraphic framework for the Lower Brushy Canyon Formation in Colleen Canyon (Figure 7). Architectural element percentages were calculated from three strike-oriented cross-sections that represent a two-dimensional estimate of architectural element abundance as opposed to the facies proportions calculated from one-dimensional sections (Figures 2, 5). Brushy Canyon architectural elements are listed in decreasing abundance:

1. Sandstone sheet element
2. Tabular siltstone sheet element
3. Channelform element
4. Channel-flanking wedge element

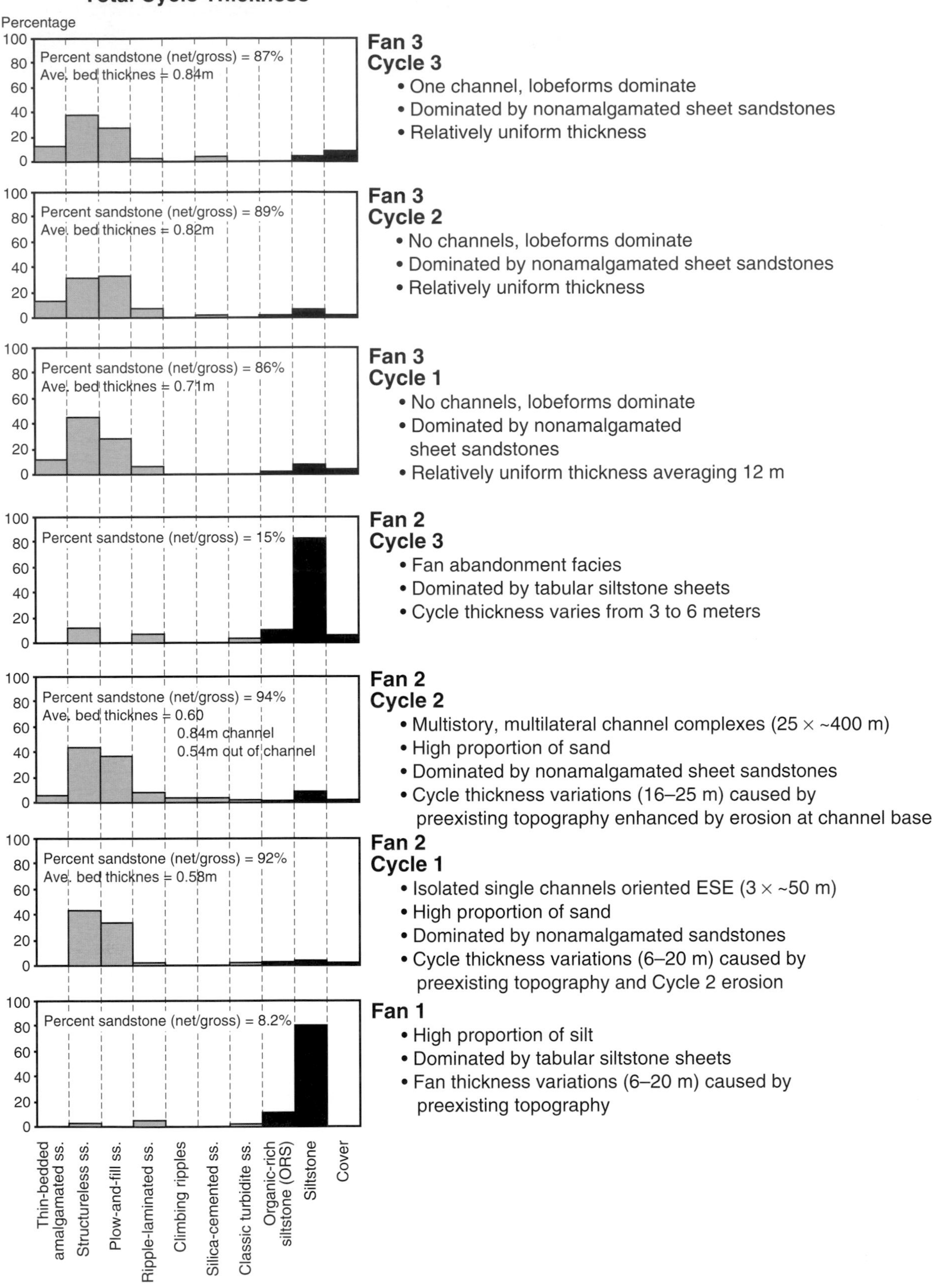

Figure 5—Facies type, distribution, and percentage of sand in fifth-order stratigraphic cycles. Data derived from measured sections.

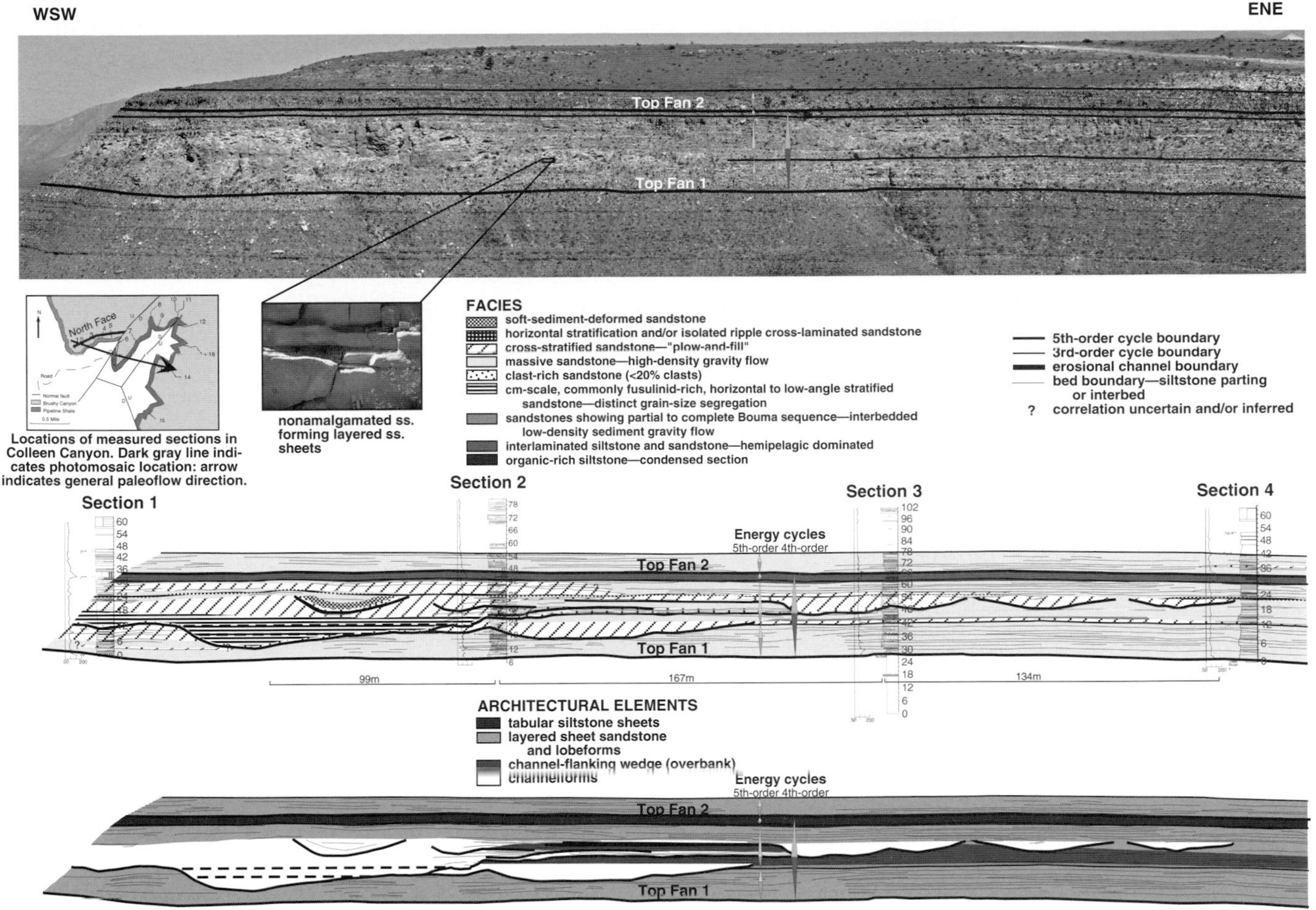

Figure 6—Photomosaic and line tracings of North Face of Colleen Canyon, illustrating bedding geometry, facies, and architectural elements. Outcrop trends N70°E. Locations of measured sections are indicated on the inset map.

Sandstone Sheet Element

Thin to thick sandstone bed sets vertically separated by very thin to thin siltstone interbeds and forming kilometer-scale layered sandstone sheets are by far the most abundant architectural elements in the lower Brushy Canyon Formation (85% of Fans 2 and 3) (Figures 7, 8). The sandstone sheets are composed of constructional, sandstone-rich lobeforms flanked by depressions that may contain more heterolithic deposits of structureless to dewatered and soft-sediment-deformed sandstone (Figure 8). These flanking lows are referred to as interlobe sediment bodies. Sandstones are deposited from unconfined high-density sediment-gravity flows that show compensation at the scale of lobe-interlobe sediment bodies.

Distinguishing individual beds as opposed to bed sets within deepwater deposits can be very difficult because of bed amalgamation patterns (Figure 8). Bed amalgamation can occur as a result of erosion, soft-sediment deformation, or the result of internal shear within high-density sediment gravity flows (i.e., vertical stacking of ungraded Bouma T_2 cycles) (Thorpe, 1969). An excellent example of bed amalgamation can be seen in one of the roadcuts at Colleen Canyon, where four distinct beds can be traced laterally along strike into one bed which is therefore a bed set (Figure 8).

Tabular Siltstone Sheet Element

Tabular siltstone sheets compose only 6% of the exposed outcrop in Colleen Canyon but form the most laterally continuous intervals in the study area (Figures 7, 9). This element is characterized gray to black siltstone and occasional thin turbidite sandstones. Claystone interbeds are common and are interpreted to represent altered volcanic ash beds (0.5–2 cm thick). This facies succession is interpreted as hemipelagic and low-density gravity flows that record reduced rates of sandstone deposition associated with periods of fan abandonment. The thickest and most continuous siltstone sheets (averaging 2–3 m thick) form recessive units that mark fourth-order fan boundaries. Thinner tabular siltstone sheets, forming fifth-order cycle boundaries, can be traced over several kilometers except where removed by erosion.

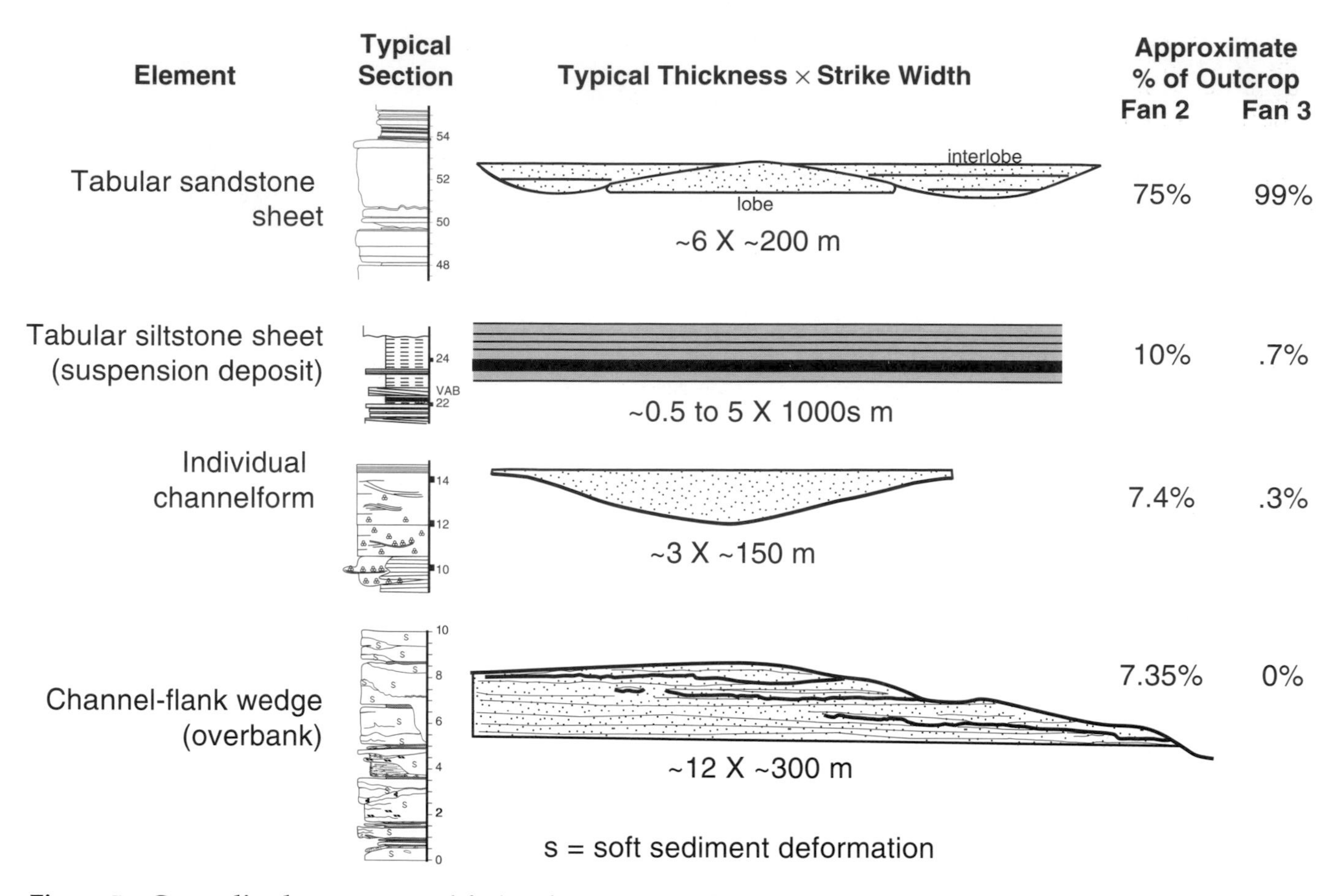

Figure 7—Generalized geometry and facies characteristics of architectural elements in Fans 2 and 3 of the Brushy Canyon Formation at Colleen Canyon.

Channelform Elements

Channelform sand bodies occur in a hierarchy of single-story channel sand bodies that stack to form larger multistory and multilateral channel complexes. This body type represents approximately 4.5% of the exposed outcrop. Individual channels have an average erosional relief of 3–4 m and are commonly filled with 10-cm-thick, spaced-stratified sandstones, or traction carpet deposits, overlain by cross-stratified to structureless and dewatered sandstone (Figures 7, 10). The lowest channel fills within a channel complex show the highest concentration of basal traction carpets and siltstone ripup clasts (2–10 cm). Individual channels vary from approximately 30 to 150 m wide.

Characteristics and Development of Channel Complexes

Two closely spaced multistory and multilateral channel complexes in Fan 2 consist of two to 10 erosive-based channels that stack to produce a serrated complex margin (Figures 2, 6). The lower complex is approximately 250 m wide by 25m thick at its maximum (Figures 6, 11). The basal channel story (5–7 m thick) erodes into F2C1 strata. The basal fill is thin-bedded, spaced-stratified sandstones interpreted as aggradational traction carpet deposits (Sohn, 1997). Overlying channel stories are offset laterally and interfinger with flanking overbank deposits. Within these multilateral channel stories (avg. 4 m thick) the fill changes upward from stratified sandstone to structureless sandstone and commonly show a horizontally laminated cap. The final phase of fill is dominated by soft-sediment-deformed sandstone.

This complex is offset 250 m to the ENE by a younger multilateral channel complex (400 × 9 m) composed of four broad and shallow single-story to multistory channel fills. The basal beds of the two deepest channel cuts are characterized by fusulinid and ripup clast conglomerate overlain by cross-stratified to soft-sediment-deformed sandstone. The multilateral nature of this channel complex is more distributary in character as opposed to the underlying dominantly multistory channel complex. Both channel complexes are overlain by approximately 6 m of structureless to stratified sandstone forming layered sandstone sheets that blanket the entire study area.

Channel-Flank Wedge Element

Channel-flank wedge elements consist of overbank deposits that flank, interfinger, and thin away from channel margins (~4.5% of the outcrop) (Figures 7, 12).

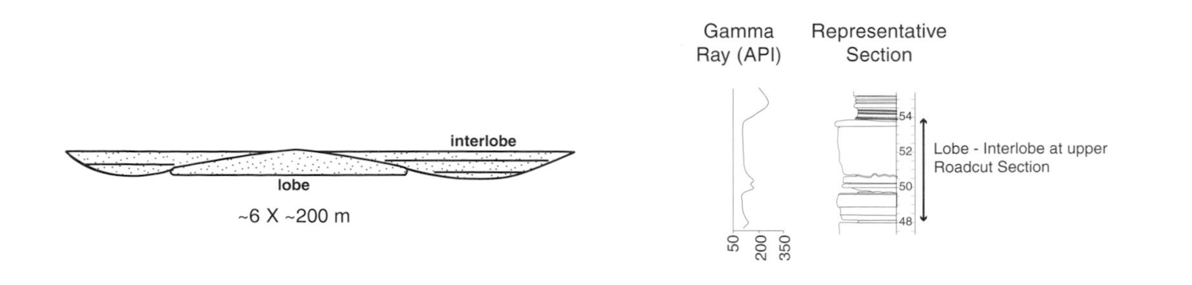

Figure 8—Sandstone sheets consist of lobeform and interlobe bodies. Representative section is from Fan 2 Cycle 2 in the upper roadcut measured section. Photo is from the eastern portion of the North Face not seen in Figure 6. Changes in bed amalgamation patterns can be seen as four beds on the left-hand side of the photo amalgamate to form one bedset toward the right-hand side of the photo.

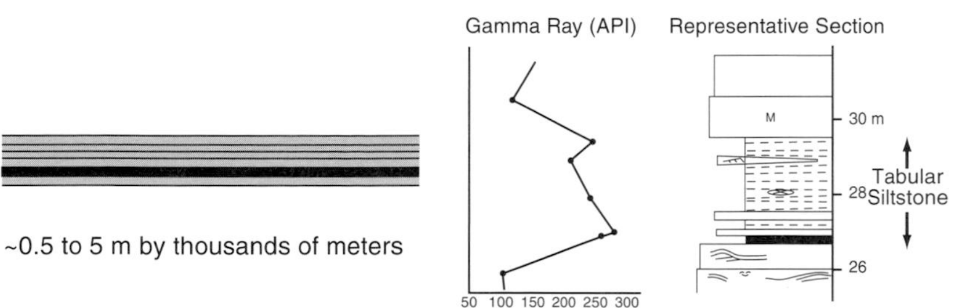

Figure 9—Architecture of tabular siltstone sheet element. Representative section is from Fan 2 Cycle 2 in measured section 10. Photo is from same interval in section 11. Jacob staff (1.8 m) for scale.

The overbank deposits show a 20° divergence in paleoflow away from the channel complex (Table 2). This element contains the greatest diversity of sandstone facies: structureless, cross-stratified, planar-laminated, and ripple-laminated sandstone. Sandstones are interbedded with thin laminated siltstone. The most distinguishing facies are abundant climbing ripple-laminated to horizontally laminated sandstone beds that thin away from channel margins. Stacked channel-flank wedges exposed in the Colleen Canyon roadcut (12 m thick) (Figure 11) also contain smaller channelform bodies interpreted to represent crevasse channels that interdigitated with channel-flanking deposits. A common vertical succession contains thicker inclined beds at the base that thin upward and show more siltstone interbeds and rippled sandstones reflecting an upward decrease in sandstone transport energy. These successions are commonly erosionally overlain by channel fills.

ANTECEDENT CUTOFF PALEOTOPOGRAPHY AND EFFECTS ON BRUSHY CANYON SEDIMENTATION

Paleotopography at the top of the mass transport complex in the upper Cutoff Formation is one of the most significant controls on lower Brushy Canyon (Fans 1 and 2) depositional patterns. It takes about 90 m of deposition, up to the base of Fan 3, to eliminate the effect of this topography on lower Brushy Canyon thickness and facies distributions. Additionally, the Fan 2 channel complexes and fusulinid conglomerates are restricted to paleotopographic lows (Figure 4). This paleotopographic control on lower Brushy Canyon depositional patterns and is relevant to reservoir producibility in the nearby Delaware Basin (Broadhead and Luo, 1996;

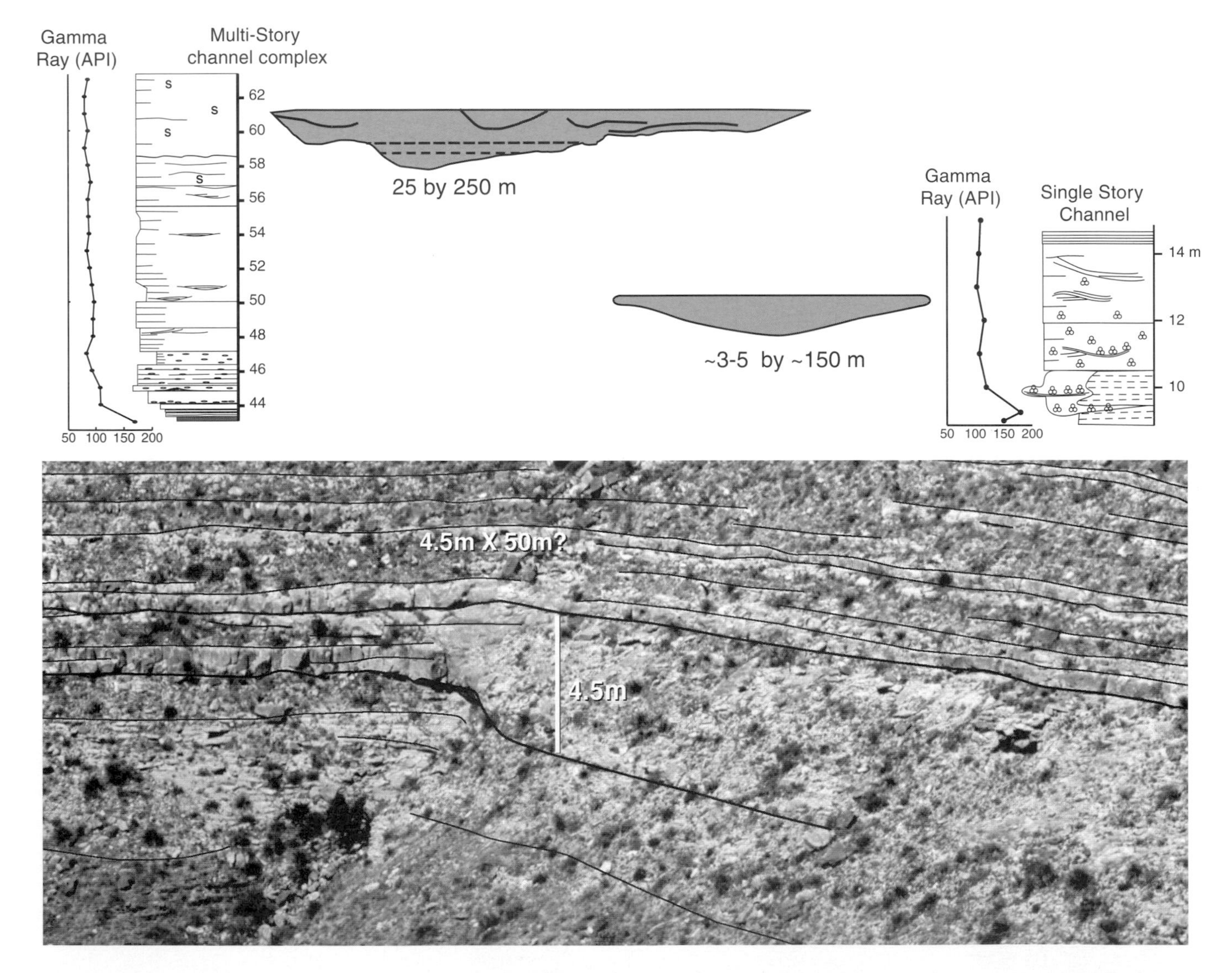

Figure 10—Architecture of single-story channel fill forming a channelform element. Representative section is from Fan 2 Cycle 1 in measured section 7. Photo is from same interval in section 7.

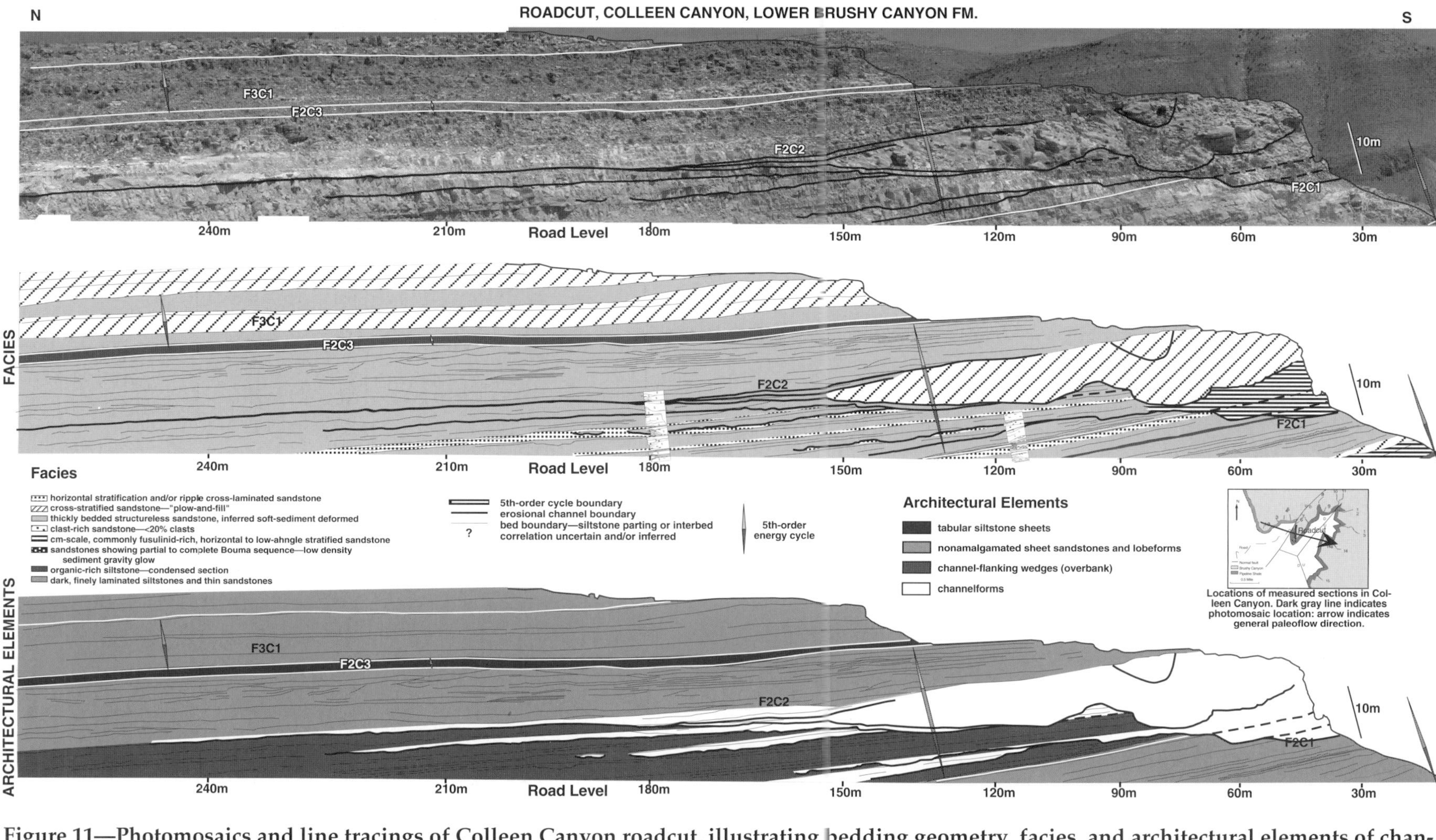

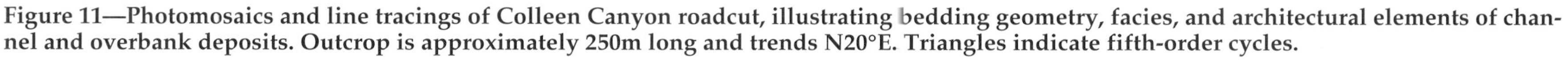

Figure 11—Photomosaics and line tracings of Colleen Canyon roadcut, illustrating bedding geometry, facies, and architectural elements of channel and overbank deposits. Outcrop is approximately 250m long and trends N20°E. Triangles indicate fifth-order cycles.

Thomerson and Catalano, 1996; Johnson, 1998). For example, at Cabin Lake Field the most productive wells in the lower Brushy Canyon are where sandstone percentage increases in lows in the underlying carbonate related to sediment ponding within these depressions.

Isopach maps of fifth-order cycles best illustrate the controls preceding topography had lower Brushy Canyon sedimentation patterns (Figures 3, 4). The variable but non-erosive Pipeline Shale (Fan 1) thickness trends suggest that low-energy siltstones did not completely infill topography developed at the top of the Cutoff Limestone mass-transport complex. Architectural element and facies distribution can be related to preexisting topography. For example, isopach maps show the Fan 2 channel complexes and their paleoflows lie within and follow the preexisting low in the Pipeline Shale (Figures 3, 4). By contrast, the paleotopographic maps of Fan 3 fifth-order cycles (F3C1 and F3C2) show broad uniform thickness patterns suggesting that paleotopography had been completely infilled by deposits of the upper cutoff and Fans 1 and 2.

STRATIGRAPHIC CHANGES IN FACIES ARCHITECTURE

Within the high-resolution stratigraphic framework outlined above, sedimentologic parameters such as frequency distribution and magnitude of erosional surfaces, bed and bedset thickness patterns, facies and architectural element proportions, and lithology ratios (net-to-gross sandstone) are compared to track systematic changes in architecture that may be related to stratigraphic position within the succession. Fan 1 has the highest siltstone proportion and most organic-rich deposits, whereas Fan 2 has the greatest frequency and magnitude of erosional surfaces, the highest proportion of channel and overbank deposits, highest sand content, and greatest facies and element diversity. These patterns suggest that Fan 2 records increased sediment bypass. Conversely, at the transition from Fan 2 to Fan 3 there is an increase in the siltstone proportion and a change from layered to amalgamated sandstone sheets. Although Fan 3 sandstone percentage decreases relative to Fan 2, the sandstone bedset thickness increases and the bounding siltstones increase in thickness. The increased bedset thickness in Fan 3 is primarily a function of bed amalgamation resulting from rapid deposition and soft-sediment deformation followed by long intervals of nondeposition that increased the siltstone percentage. These depositional patterns suggest that Fan 1 and 2 deposits step basinward and record increasing sediment supply, whereas the locus of deposition for Fan 3 either shifted landward or laterally (Figure 13). However, only an 8° shift in sediment transport is recorded between Fan 2 to 3 deposits, which does not support a lateral shift in deposition that could be related to fan scale compensation.

Unconfined gravity flow deposits in the lower Brushy Canyon Formation show organized depositional patterns that reflect depocenter migration along a slope-to-basin profile. Although compensational bedding patterns are recognized, they are best expressed at the scale of architectural elements that record the lateral offset of lobe-interlobe bodies that stack to form layered sandstone sheets (Figure 8). This trend decreases with increasing scale of cyclicity, such that fifth-order cycles show more sediment transport variation than deposits of fourth-order cycles (Table 2). In other words, the smaller-scale architectural elements are much more sensitive to compensational processes than the larger-scale stratigraphic cycles. Compensational patterns are not recognized in isopach maps of either fourth- or fifth-order cycles. Bed thickness patterns normally associated with progradation (i.e., upward bed thickening) are unreliable indicators of stratigraphic cyclicity because of nonunique vertical bed patterns. However, sandstone bed patterns of amalgamation and nonamalgamation do represent reliable indicators of changes in the frequency and magnitude of gravity flow events. Through lower Brushy Canyon deposition there are systematic changes in bedset thickness, lithology ratios, facies, and sediment body type that reflect the stratigraphic modulation of the frequency and magnitude of gravity flow events. Although compensation patterns larger than the 4 km^2 area cannot be ruled out, the stratigraphic framework captures important vertical and lateral variations in architectural patterns produced by deposits of unconfined gravity flows. These changes are systematic and unique and therefore must be incorporated into the characterization of basin-floor deposits. Hence, the presence of compensational bedding and nonunique bed thickness patterns does not preclude stratigraphic organization of unconfined gravity flow deposits.

SUMMARY

The basin-floor fan deposits at Colleen Canyon are dominated by kilometer-wide by decimeter-thick sandstone bed sets that encase multistory to multilateral channel complexes. Laterally extensive organic-rich siltstone intervals define a hierarchy of fourth- and fifth-order stratigraphic cycles that document temporal and spatial variations in facies architecture. Four basic architectural elements characterize these changes in the lower Brushy Canyon. Fan 1 is composed of tabular siltstone sheets (8% sandstone) that mark the initial progradation of clastic sediments into the basin, which partially infill antecedent topography of underlying carbonates. Fan 2 shows a major increase in sandstone deposition (93% sandstone) producing layered sandstone sheets (~75%), the highest proportion of channel-flanking wedge (~7.4%), and channelform (~7.35%) elements. Fan 2 channels change upward from scat-

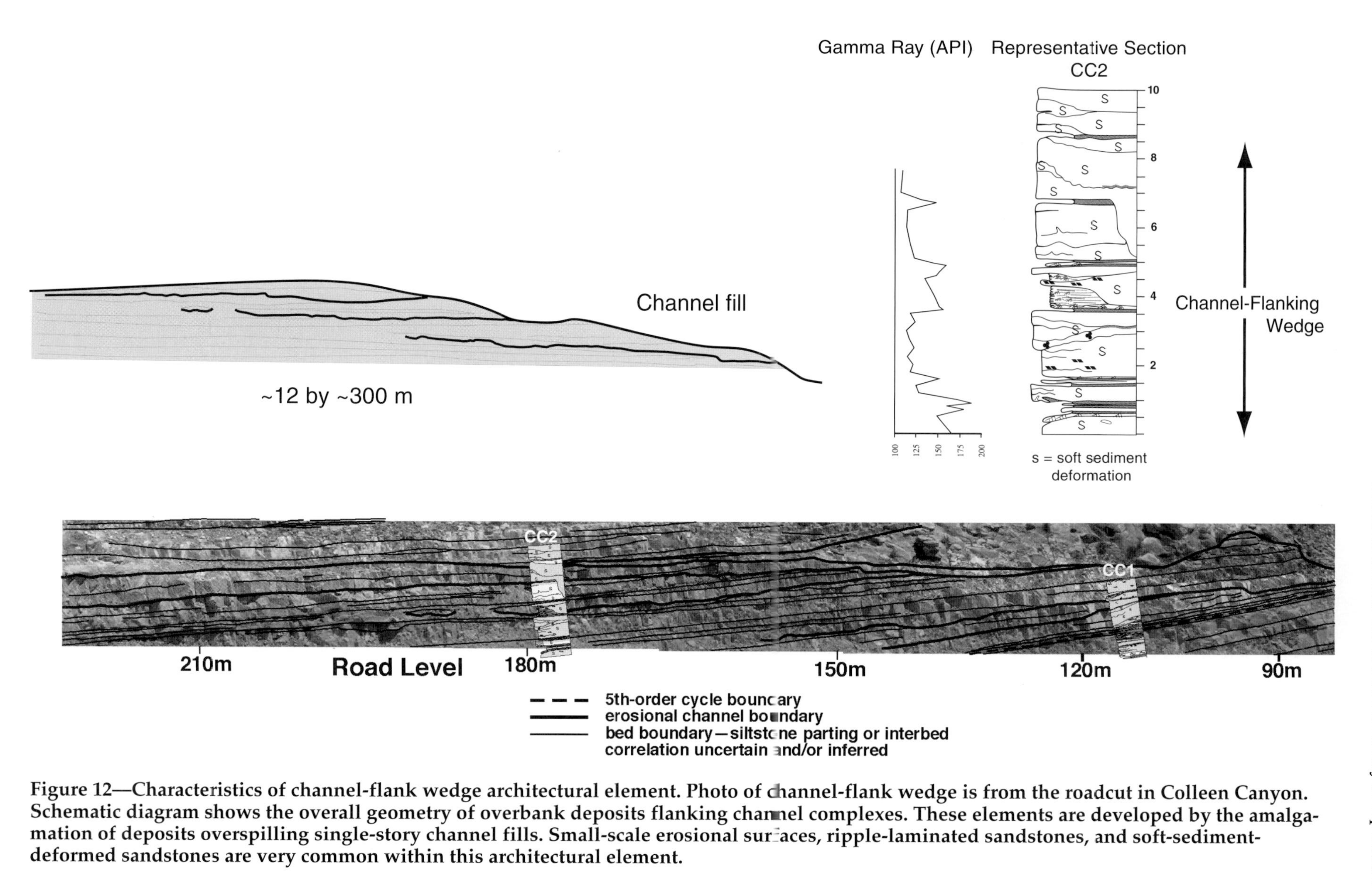

Figure 12—Characteristics of channel-flank wedge architectural element. Photo of channel-flank wedge is from the roadcut in Colleen Canyon. Schematic diagram shows the overall geometry of overbank deposits flanking channel complexes. These elements are developed by the amalgamation of deposits overspilling single-story channel fills. Small-scale erosional surfaces, ripple-laminated sandstones, and soft-sediment-deformed sandstones are very common within this architectural element.

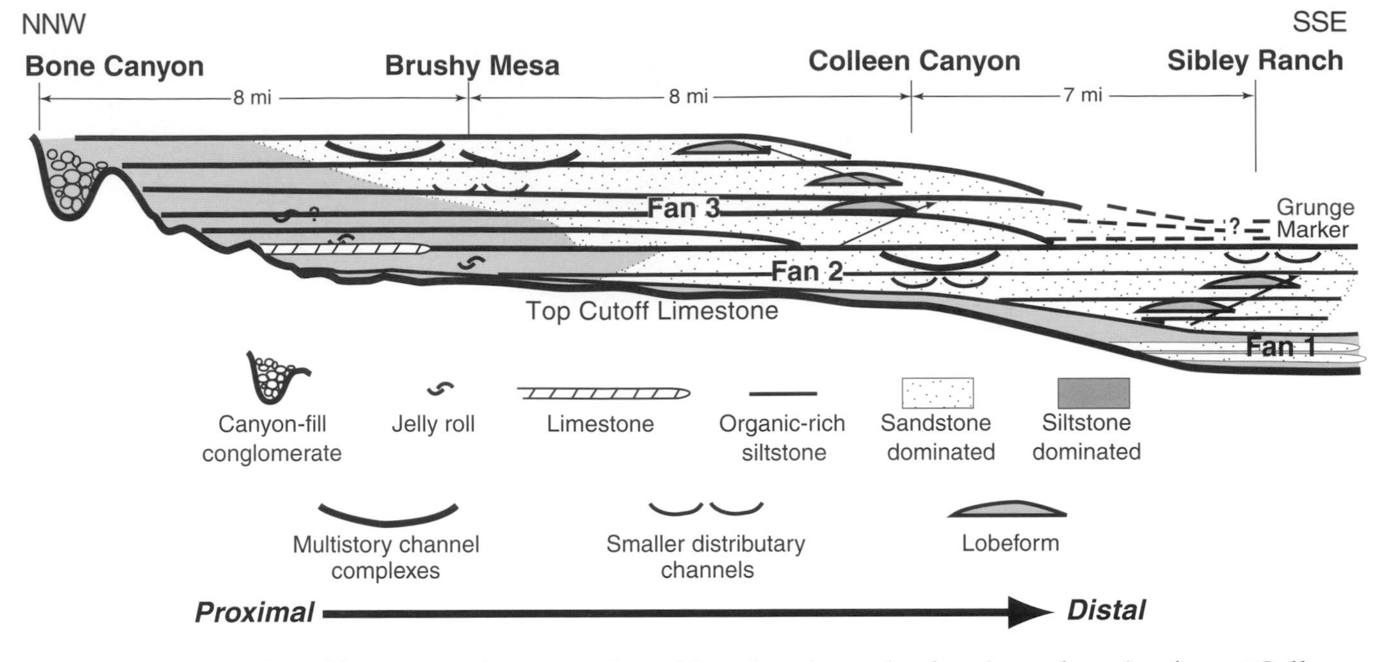

Figure 13—Proposed stacking pattern interpretation of fourth-order cycles forming submarine fans at Colleen Canyon. Cycles illustrate stratigraphic changes from sand-poor, silt-dominated Fan 1 deposits, to sand-rich Fan 2 channel deposits, to lobe-dominated Fan 3 deposits.

tered single-story channels (F2C1) to multistory and multilateral channel complexes (F2C2) that show a distributary pattern. Fan 3 deposits (87% sandstone) are dominated by amalgamated sandstone sheets (~99%), with only minor channelform elements (0.3%). The amalgamated sandstone sheets of Fan 3 record a change to more compensational bedding styles reflected in more divergent paleoflow indicators, thicker sandstone bedsets, and more siltstone.

Associated with these lithology and sediment body trends are significant lateral and vertical bedset thickness trends. Fan 2 has the highest sand percentage (93%) and an average bedset thickness is 0.59 m, whereas Fan 3 is 87% sandstone and has an average bedset thickness of 0.80 m. Hence, the greatest bedding and architectural complexity occurs within the most sandstone-rich interval within this basin-floor succession (Fan 2). Furthermore, the upward decreasing sand percentage corresponds to an increasing bedset thickness. These are significant architectural variations that can only be predicted from a robust framework based on a hierarchy of siltstone-bounded cycles that record the basin-floor evolution.

ACKNOWLEDGMENTS

Members of the Slope and Basin Consortium provided financial support. Members are listed in Gardner and Borer, this volume. Field assistance was provided by Mike Batzle, Jesse Melick, Conrad Woodland, Ted Playton, Jeremy Merrill, Marieke Dechesne, and Roger Wagerle. Charlie Rourke and Leann Wagerle provided invaluable office assistance. Linda Martin constructed all digital photomosaics. A special thanks goes to Mike and Ann Capron and Tony Kunitz for their unwavering support of geologists on the Six Bar Ranch. We would also like to thank Donna Anderson, Arnold Bouma, and Charles Stone for thoughtful reviews of the manuscript.

REFERENCES CITED

Anderton, R., 1995, Sequences, cycles and other nonsense: are submarine fan models any use in reservoir geology? *in* Hartley, A. J. and D. J. Prosser, eds., Characterization of deep marine clastic systems, Geological Society Special Publication 94, p. 5–11.

Broadhead, R. F., and F. Luo, 1996, Oil and gas resources in the Delaware Mountain Group at the WIPP site, Eddy County, New Mexico, *in* W. D. DeMis and A. G. Cole eds., The Brushy Canyon play in outcrop and subsurface: concepts and examples, guidebook: Permian Basin section, SEPM, No. 96–38, p. 119–130.

Ghibaudo, G., 1981, Deep-sea fan deposits in the Macigno Formation (Middle–Upper Oligocene) of the Godana Valley, Northern Apennines, Italy—discussion: Journal of Sedimentary Petrology, v. 51, no. 3, p. 1021–1026.

Graham, S. A., et al., 1996, Statistical analysis of bed-thickness patterns in a turbidite section from the Great Valley sequence, Cache Creek, northern California: Journal of Sedimentary Petrology, v. 66, no. 5, p. 900–908.

Hiscott, R. N., 1981, Deep-sea fan deposits in the Macigno Formation (Middle-Upper Oligocene) of the Godana Valley, Northern Apennines, Italy—discussion: Journal of Sedimentary Petrology, v. 51, no. 3, p. 1015–1020.

Johnson, K. R., 1998, Outcrop and reservoir characterization of the Brushy Mountain Formation, Guadalupe Mountain National Park, west Texas, and Cabin Lake field, Eddy County, New Mexico: Unpublished M.S. thesis, Colorado School of Mines, 256 p.

King, P. B., 1965, Geology of the Sierra Diablo region, Texas: U.S. Geological Survey Professional Paper P-480, 185 p.

Larue, D. K., and T. A. Jones, 1997, Object-based modeling of deep-water depositional system reservoirs. AAPG, Abstracts with Programs, v. 6, p. A67.

Mahaffie, J. J., 1994, Reservoir classification for turbidite intervals at the Mars discovery, Mississippi Canyon 807, Gulf of Mexico, *in* P. Weimer, A. H. Bouma, and B. F. Perkins, eds., Gulf Coast Section SEPM Foundation 15th Annual Research Conference, Submarine Fans and Turbidite Systems, p. 233–244.

Mutti, E., and M. Sonnino, 1981, Compensation cycles—A diagnostic feature of turbidite sandstone lobes, in International Association of Sedimentologists Second European Regional Meeting, Bologna, Italy: Abstracts Volume, p. 120–123.

Sohn, Y. K., 1997, On traction-carpet sedimentation: Journal of Sedimentary Research, Section A, v. 67, p. 502–509.

Thomerson, M. D., and L.E. Catalano, 1996, Depositional regimes and reservoir characteristics of the Brushy Canyon Sandstone, East Livingston Ridge Delaware field, Lea County, New Mexico, *in* W. D. DeMis and A. G. Cole, eds., The Brushy Canyon play in outcrop and subsurface: concepts and example: Permian Basin Section of SEPM, No. 96–38, p. 103–111.

Thorpe, S. A., 1969, Experiments on the instability of stratified shear flows: immiscible fluids: Journal of Fluid Mechanics, v. 39, p. 25–48.

Kirschner, R. H., and A. H. Bouma, 2000, Characteristics of a distributary channel-levee-overbank system, Tanqua Karoo, *in* A. H. Bouma and C. G. Stone, eds., Fine-grained turbidite systems, AAPG Memoir 72/SEPM Special Publication No. 68, p. 233–244.

Chapter 21

Characteristics of a Distributary Channel-Levee-Overbank System, Tanqua Karoo

Roland H. Kirschner[1]
Arnold H. Bouma
Department of Geology and Geophysics, Louisiana State University
Baton Rouge, Louisiana, U.S.A.

ABSTRACT

A complex of distributary channel fills (termed here "distributary channel system"), exposed on the Bloukop farm, shows a complex internal architecture, comprised of seven individual channel fills, levee, and crevasse splay deposits. This distributary channel system is embedded in overbank deposits that reflect times when distributary channels were active elsewhere in the basin. Two types of channel fill illustrate the downdip morphologic changes of the individual feeder channels within a distributary channel system. The vertical stacking of each channel type within the outcrops reflects both the updip/downdip shift of the center of sheet sand deposition and its lateral movement within a sheet sand lobe.

INTRODUCTION

A general introduction to the geology of the Tanqua Karoo subbasin and the location of the study area is illustrated in Wickens and Bouma (this volume). A distributary channel fill within Fan 5, the youngest fan of the Tanqua Karoo basin floor fan complex, is exposed in an approximately 1.5-km^2-large area on the Bloukop farm (Figure 1).

The internal architecture of such distributary channel fills has only been studied in a general sense. The model for fine-grained submarine fan systems, presented by Bouma (this volume), depicts this part of the fan as a series of channels that develop beyond the point of midfan channel bifurcation. Only one distributary channel is active at any given time, funneling sediment to the sheet sand lobes. As positive relief is created within a sheet sand lobe, as well as around its associated distributary channel, a new distributary channel is created that transports sediment to a different part of the basin. This study reveals the internal complexity of a single distributary channel, which is in fact composed of four different architectural elements: two types of channel fills, levee deposits, and crevasse splay deposits (Kirschner, 1999). Their distribution within the outcrops is presented in Figure 2. The association of these four architectural elements is termed here "distributary channel system."

This paper reviews the individual architectural elements of a distributary channel system and suggests an interpretation for the overall depositional patterns in the lower midfan to lower fan environment of fine-grained submarine fans. Furthermore, connectivity and fluid flow characteristics of sandstone bodies within distributary channel systems

[1]Present address: Phillips Petroleum Company, Bellaire, Texas, U.S.A.

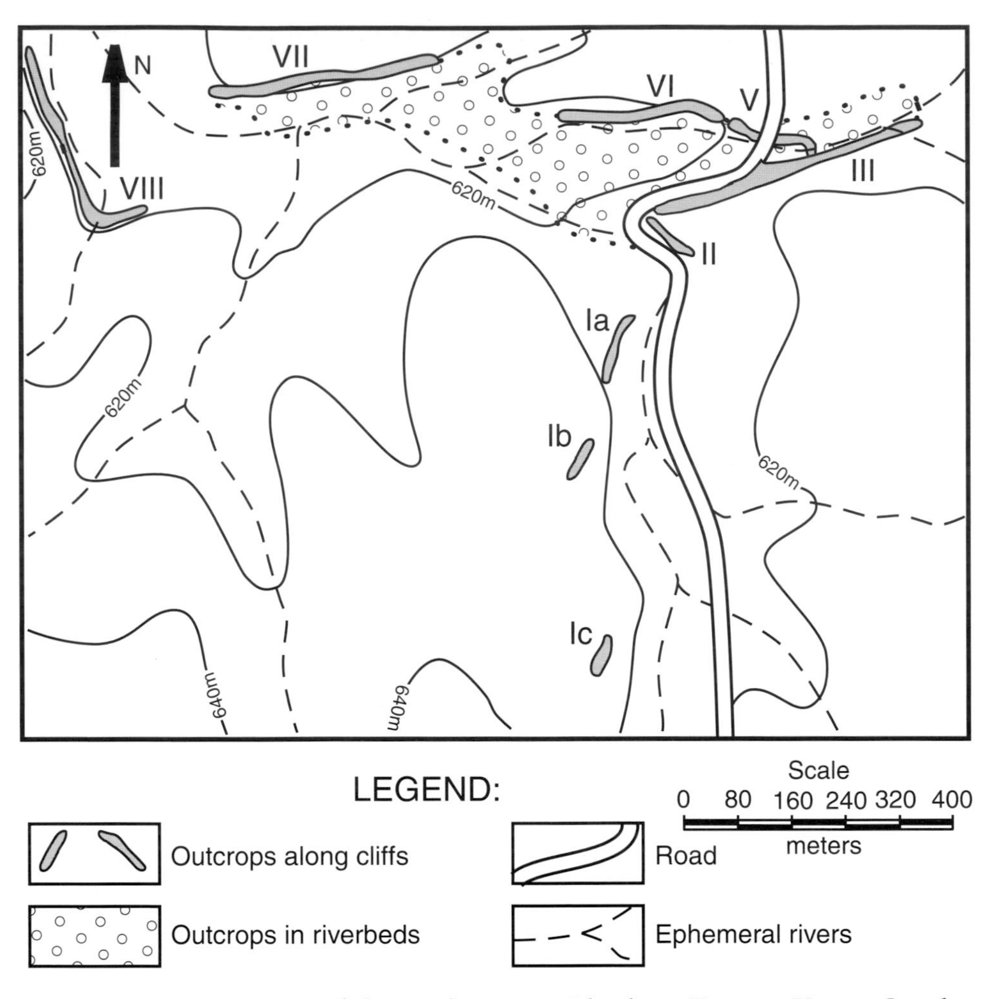

Figure 1—Outcrop map of the study area at Bloukop, Tanqua Karoo, South Africa. Two types of outcrops are present: cliffs and riverbeds.

will be discussed. The lithofacies types used in this study are based on the scheme presented by Wickens (1994). Facies I delineates fine-grained, massive sandstones; facies II describes very fine-grained climbing ripple cross-laminated sandstones; and facies III depicts laminated medium to fine-grained siltstones.

INTERCHANNEL DEPOSITS

Levee and Overbank Deposits

Levee and overbank deposits are laid down when turbidity currents confined within a submarine channel overtop the levees and deposit sediment beyond the channel form in a process termed overbank deposition by Mutti and Normark (1987). At Bloukop, the resulting deposits are packages of very thin-bedded facies III siltstones, interbedded with thin-bedded facies II climbing ripple cross-laminated sandstones. According to Walker (1985), these climbing ripples are indicative of interchannel deposits.

Figure 3 shows a cross section of levee deposits, LV1, in outcrops VII and VIII (Figures 1, 2). The term "levee" is used here for the part of the overbank area that borders the channel proper, even if the positive relief of the levee crest cannot be observed. The thickest sequence of levee deposits was recorded at measured section VII-1. They represent the paleolevee crest, where sedimentation rates were the highest during overbank flows. Westward from measured section VII-1, the levee deposits undergo several changes. In particular, a decrease in the facies II sandstone bed thickness, coupled with a decrease in their grain size, was observed. Pinching out of some thinner facies II sandstone beds in the lower third of the outcrop causes a decrease in the net:gross sand ratio from 8.1 at measured section VII-4 to 3.8 at measured section VIII-1. Furthermore, the percentage of amalgamation decreases from 35% at measured section VII-4 to 0% at measured section VII-2. Overall, these lateral changes indicate decreasing erosive power and sediment supply rates of the overbank flows in a westward direction.

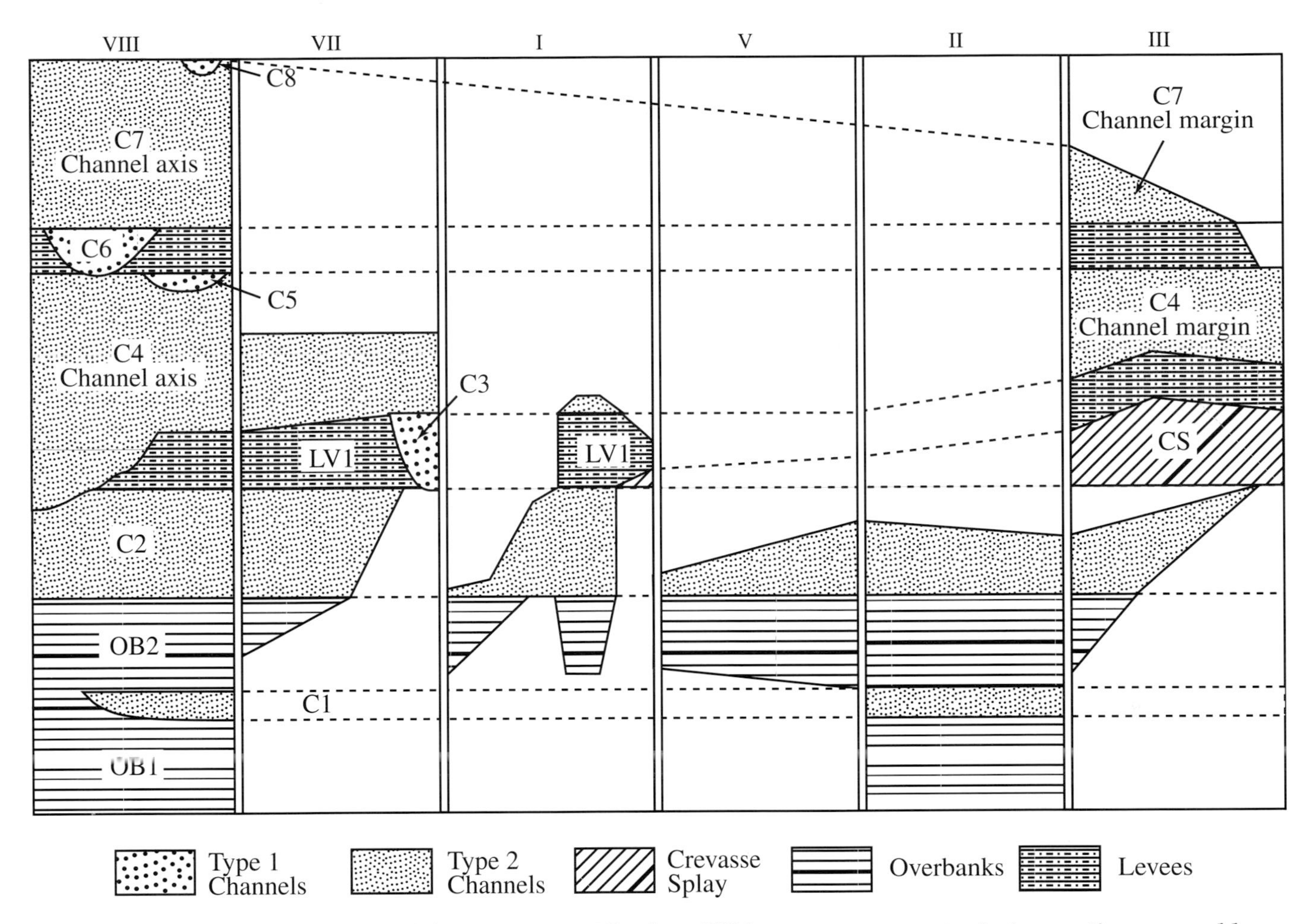

Figure 2—General stratigraphy of the outcrops at Bloukop. White areas represent missing section, caused by scree cover, erosion, or nonexposure. C = channel deposits, LV = levee deposits, OB = overbank deposits, CS = crevasse splay deposits. Different deposits of the same group are numbered, according to their stratigraphic position in the outcrops. Refer to Figure 1 for location of outcrops.

The overbank deposits at Bloukop (OB1 and OB2, Figure 2) are lithologically similar to the levee deposits, but their overall net:gross sand ratio is lower (0.67 to 1.2). The average facies II sandstone bed thickness (14 cm) is approximately half of that in the levee deposits (32 cm). Therefore, overbank deposits are interpreted as the lateral continuation of levee deposits.

Crevasse Splay Deposits

In the interchannel area to the east of channel C3, in outcrop area III (Figure 1), the lower beds of levee deposits LV1 are absent (Figure 2, CD-ROM Figures 1 and 2). In their place, a sequence of thick-bedded facies I and II sandstones interbedded with thin-bedded facies III siltstones was observed. A strong bilateral thinning trend was recorded both to the southwest and to the east (CD-ROM Figure 1). This trend is incongruent with an interpretation as levee or overbank deposits. Ball-and-pillow structures, indicative of high sedimentation rates at the time of deposition, were observed in the facies I sandstones. Overall, the facies association, bilateral thinning, and high sedimentation rates suggest an interpretation of this sequence as crevasse splay deposits.

CHANNEL DEPOSITS

Based on their aspect ratios (width/thickness ratio of channel fills), basal downcutting pattern, and internal architecture, the channel fills at Bloukop can be assigned to the two groups described below.

Type I Channels

This type of channel is characterized by a narrow channel form (aspect ratio <50) that is smoothly cut into the underlying and neighboring deposits. Its channel fill is entirely composed of amalgamated, massive facies I sandstones. Figure 4A and B shows a photomosaic and a line drawing that depict channel

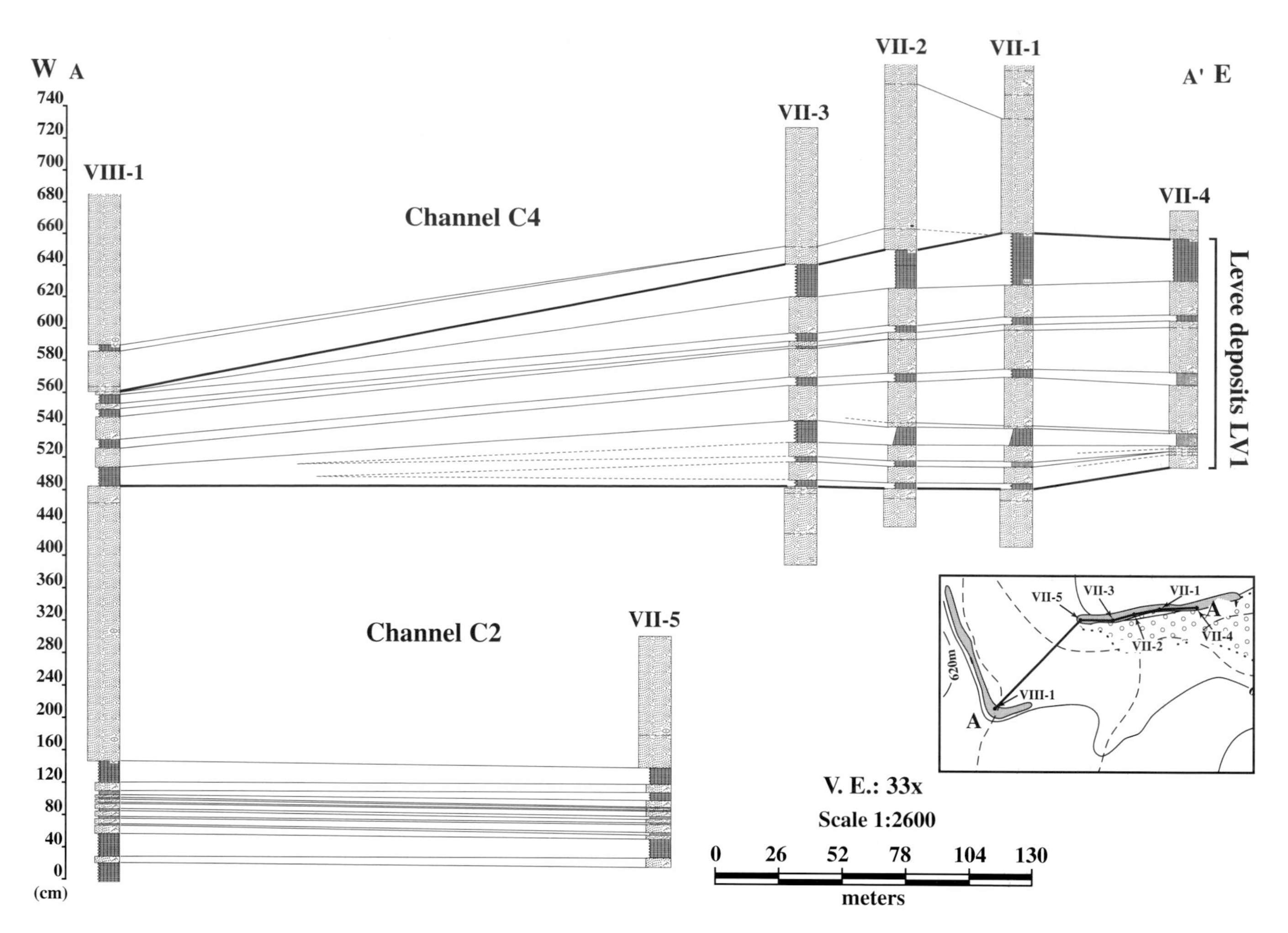

Figure 3—Correlation of measured sections in outcrops VII and VIII at the level of levee deposits LV1, outlined by the heavy black lines. Note the overall westward thinning trend of the levee deposits and the pinching out of the lowermost two sandstone beds in the levee deposits.

C3 in outcrop VII (Figures 1, 2). Levee deposits LV1 discussed above are exposed to the west of the channel fill. Two beds within the lower third of the levee deposits can be traced into the channel fill, where they gradually change from facies II and III, respectively, into facies I. In addition to the westward decrease in the strength of overbank flows, recorded in levee deposits LV1, this is evidence for the genetic relationship between the channel fill C3 and the adjacent levee deposits LV1. Four separate erosive contacts (e_1–e_4 in Figure 4) were observed at the margin of channel C3 that indicate a multiple-stage cut and fill history for type I channels.

Type 1 channel fills are the result of two kinds of gravity flows: rare, erosive gravity flows that cut out a narrow channel form into the underlying and neighboring deposits. The only remnants of these flows are silt layers at the channel margin (Figure 5, steps 1, 3, 5, 7). The created channel forms were later infilled by turbidity currents that deposited the amalgamated sheets of massive facies I sandstone beds. The progressive shallowing of the channel form led to spilling of sediment over the levee crests, which resulted in the formation of extensive levee deposits (Figure 5, steps 2, 4, 6, 8).

Type 2 Channels

Type 2 channels differ from type 1 channels due to their higher aspect ratio (minimum 150–200) and the difference in basal downcutting pattern. Compared with the smooth channel forms of type 1, type 2 channels expose a stepwise downcutting pattern as illustrated in Figure 6 (see CD-ROM Figure 3 for complete outcrop view). This suggests less erosive power of the heads of the turbidity currents. Two depositional subenvironments can be distinguished within type 2 channels: the channel axis deposits and the channel margin deposits (Mutti, 1977; Mutti and Normark, 1987).

Channel axis deposits for all three type 2 channels within the distributary channel system are exposed in

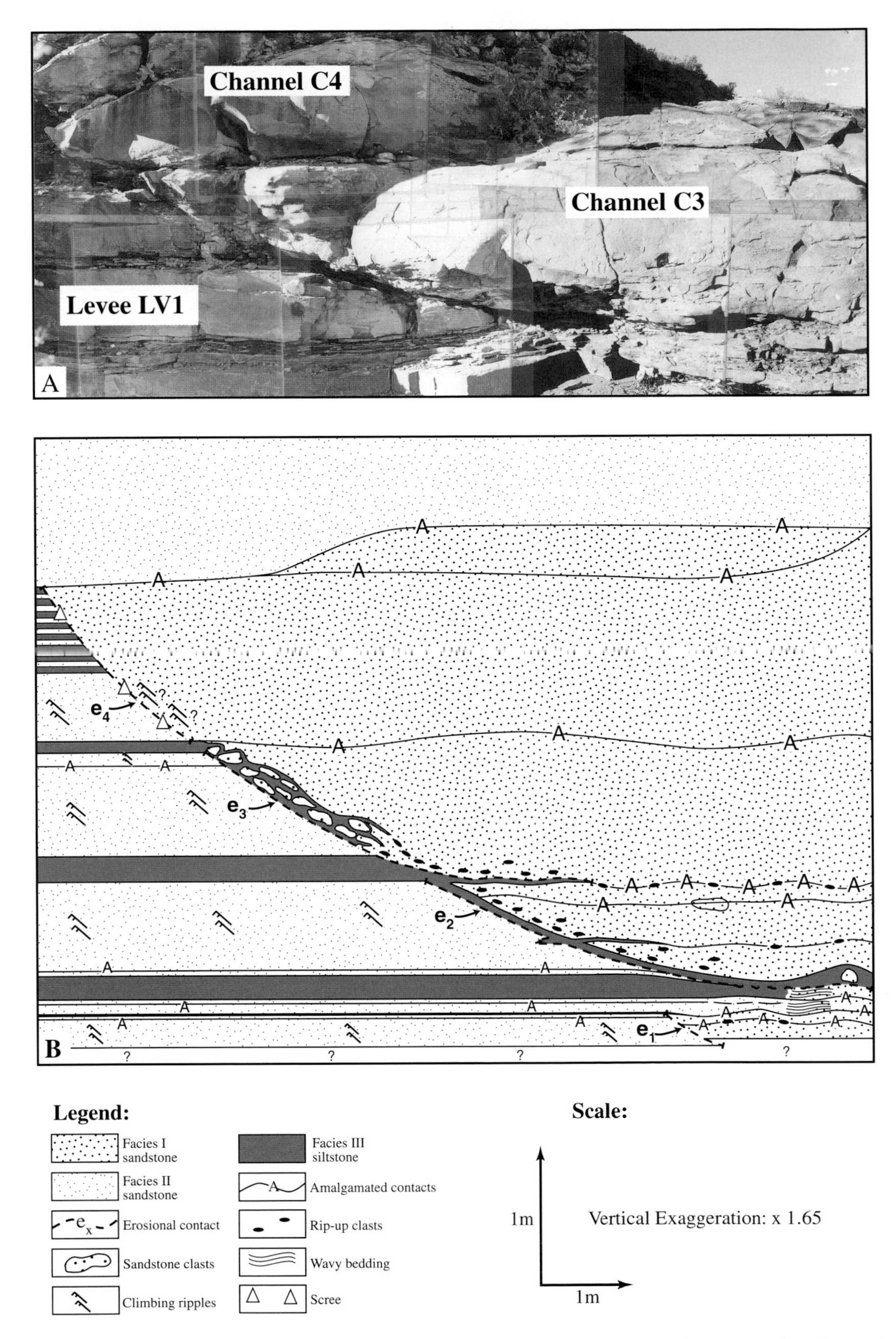

Figure 4—(A) Photomosaic of channel C3 in outcrop VII. Note the differences between the channel fill to the right, composed of massive facies I sandstones and the levee deposits to the left that consist of interbedded sheets of facies II and III. (B) Detailed sketch of the relationships between channel C3 and the adjacent levee deposits LV1. Note the two continuous beds in the lower third of the outcrop that cross the channel margin into the channel fill, and the four separate erosive contacts

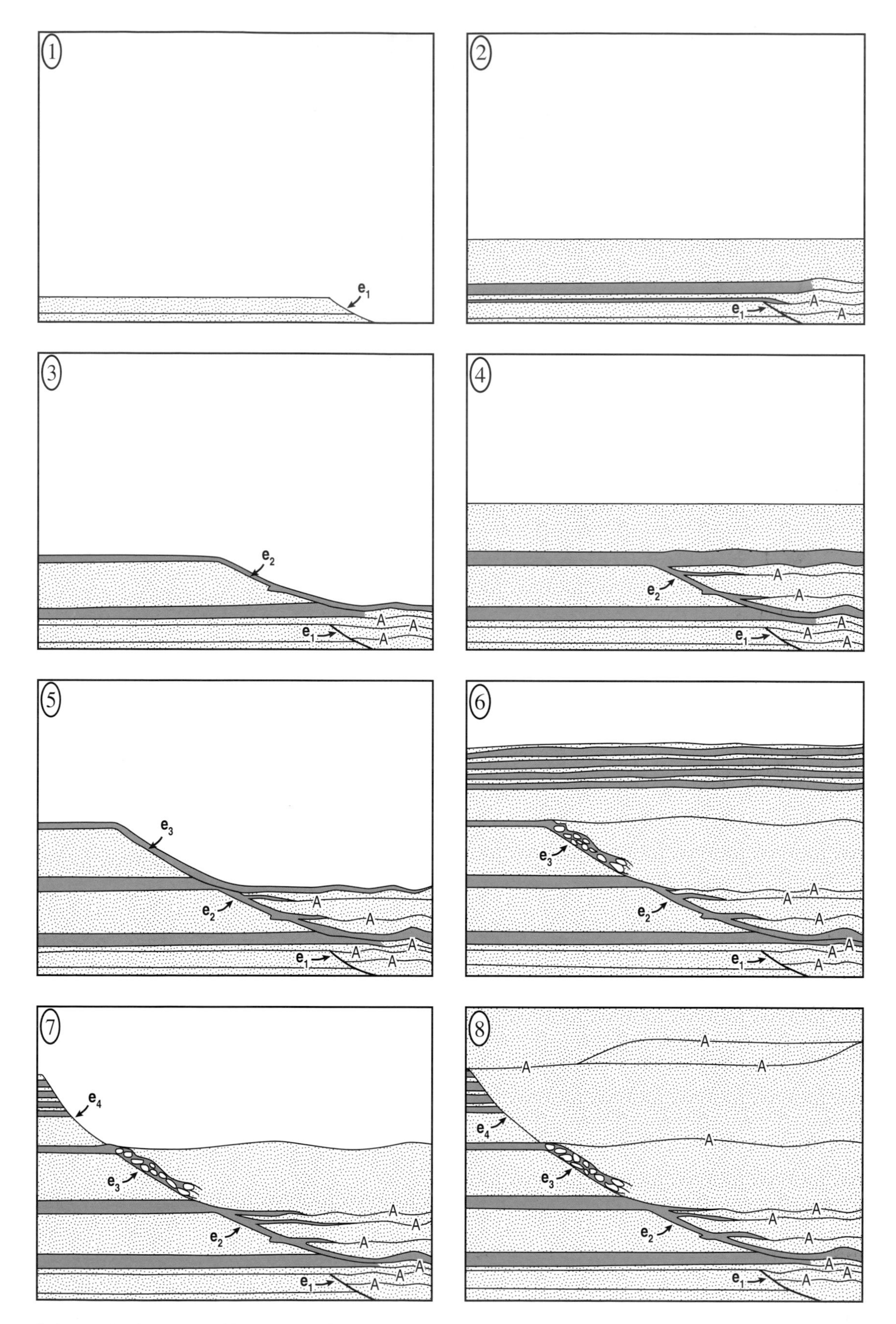

Figure 5—Schematic diagram, illustrating the multiple stages of evolution of the channel-levee complex shown in Figure 4. The set of sketches shows the complexity of alternating erosional and depositional phases. See text for detailed description.

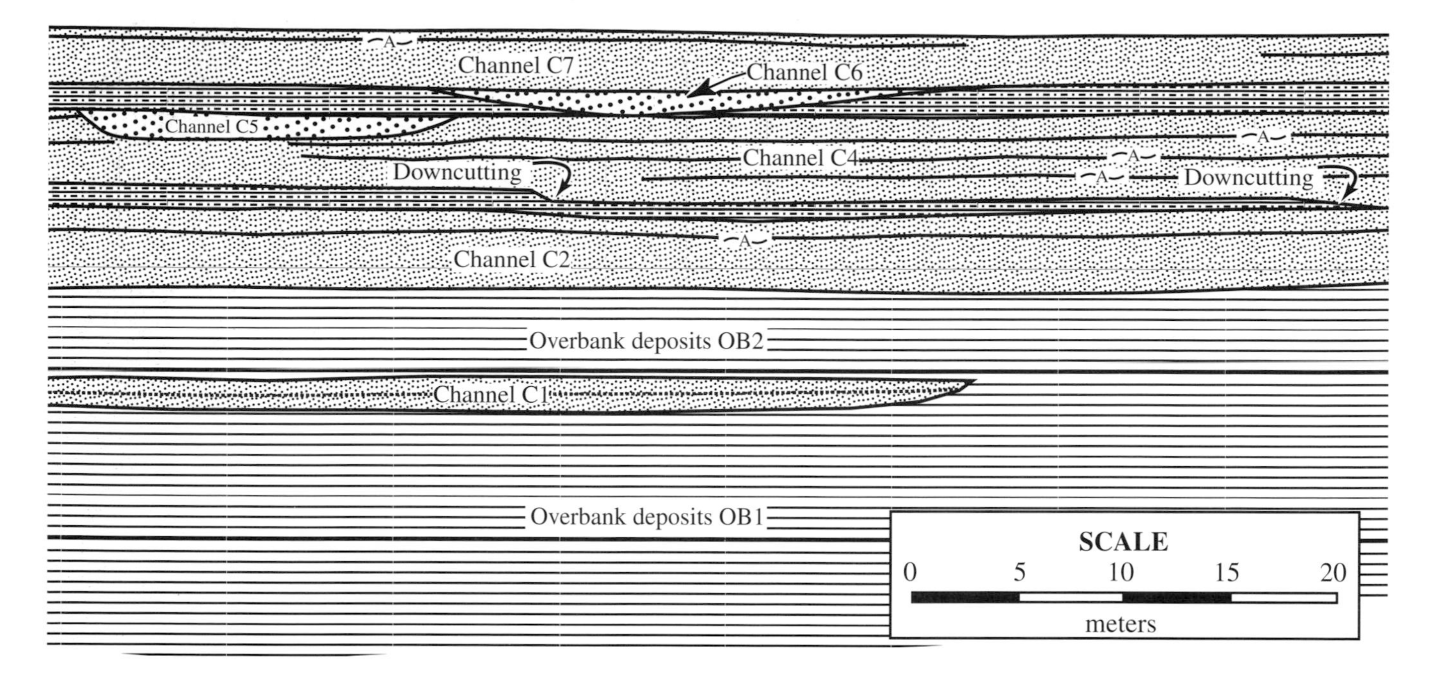

Figure 6—Line drawing of the north–south-trending outcrop VIII (Figures 1, 2, CD-ROM Figure 2). A thick sequence of overbank deposits at the base is overlain by the series of channel and levee deposits that construct the distributary channel system. In this outcrop, a series of type 1 channel fills and the channel axis deposits of three type 2 channels are exposed. These typically consist of sheets of amalgamated facies I sandstones. Also note the stepwise downcutting pattern at the base of channel C4.

outcrop VIII (Figures 1, 2, 5). They consist of amalgamated sheets of facies I sandstones, comparable with the channel fill of type I channels. Similar to the levee deposits, the channel margin deposits are composed of interbedded facies II and facies III beds. The primary difference between the levee and the channel margin deposits is the higher sedimentation rate recorded in the facies II beds of the channel margin deposits. This is evidenced by the thickness of the facies II beds (on average 108 cm, compared to 34 cm in the levee deposits) and the steeper angle of climb of the climbing ripples (largely >10°, compared to <10° in the levee deposits). Therefore, an environment of deposition within the channel rather than beyond the levee crest is suggested.

DISCUSSION

At the base of the outcrops, a 12-m-thick sequence of overbank deposits OB1 and OB2 dominates the exposures (Figure 6). It has been shown that overbank deposits, the lateral continuation of levees, may extend for several kilometers away from the channel-levee complex (Mutti, 1977; Kenyon et al., 1995). Therefore, this thick sequence is interpreted to reflect a time when distributary channels and their associated depositional lobes were located in different areas of the basin. Channel C1, which is sandwiched between the overbank deposits, reflects a short period of channel deposition at Bloukop. It is interpreted to be the most distal and short-lived of the type 2 channels, due to its shallowness and the absence of levee and channel margin deposits. The base of channel C2 (Figures 2, 6) marks an abrupt change in the depositional environment, when channel deposition started in the area. This change in depositional style most likely reflects the lateral switching of depositional lobes and their associated system distributary channel, as suggested by Bouma and Wickens (1991).

The channel association exposed above the base of channel C2 demonstrates the internal architecture of this "distributary channel system." It is apparent that a "distributary channel system" does not consist of a single, backfilled feeder channel, as depicted in the model shown in Bouma (this volume). It can be described more accurately as a channel sequence that comprises of seven individual channel fills at Bloukop (channels C2–C8). As shown earlier, this channel association is composed of two types of channel fills (types 1 and 2) and interchannel deposits related to type 1 channels.

A 3-D reconstruction of channels C2–C4, based on photomosaics and measured sections, revealed that all of the depicted channels have slightly different paleoflow directions. Each channel fill is thought to represent the feeder channel for a stack of sheet sand deposits. As positive relief is created, in response to the accumulation of sheet sands and channel fills, a lateral shift of the depocenter in the sheet sand lobe will occur. This shift is associated with the formation of a new feeder channel (Figure 7). The slight variance in the paleoflow directions of the channels is thought to reflect the lateral shift of the depositional center in the sheet sands.

Type 1 and 2 channels are interpreted to reflect the downdip morphologic changes within a feeder

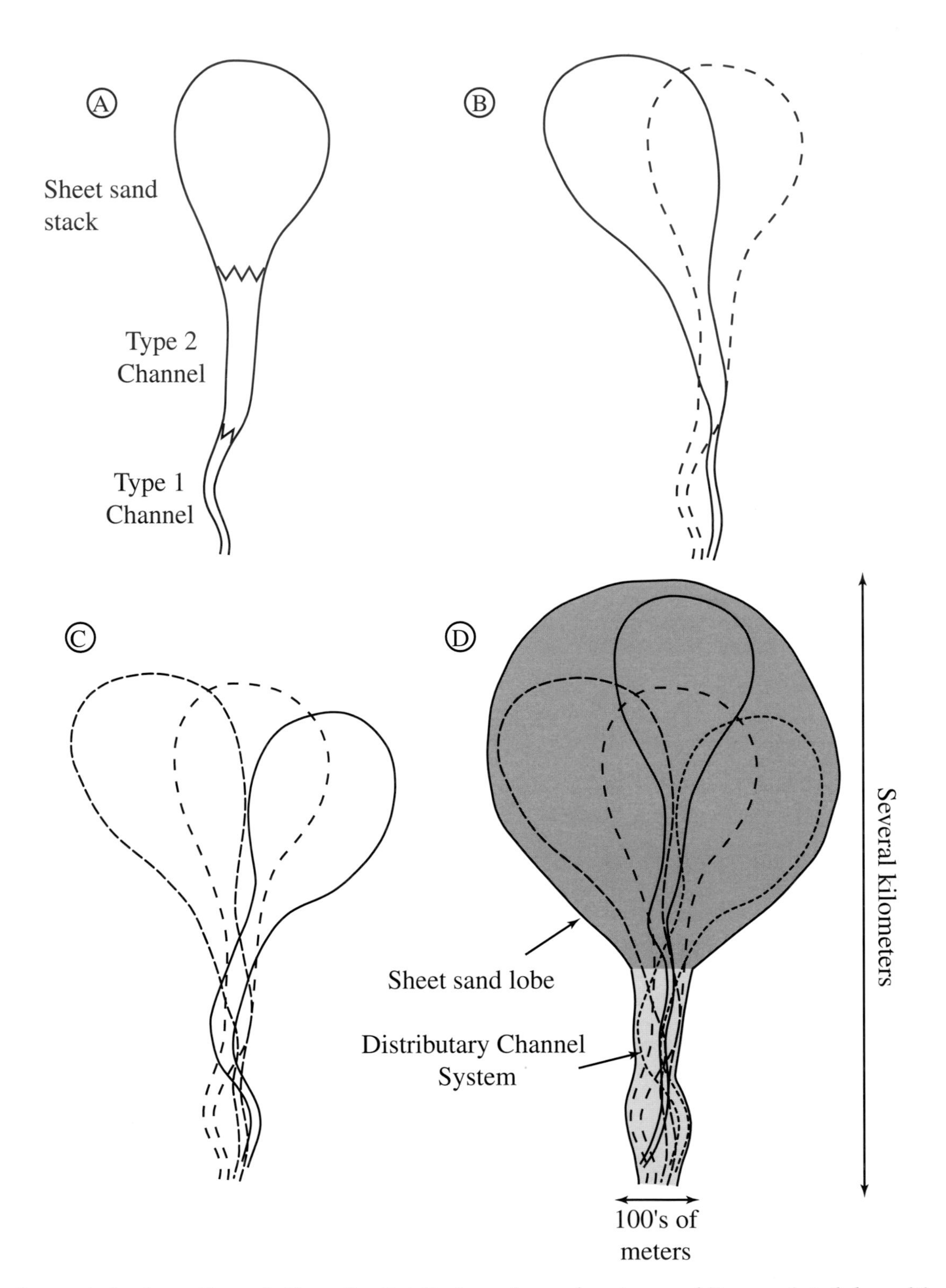

Figure 7—Suggested schematic evolution of a distributary channel system and its associated depositional lobe. (A) Distinction of the type 1 and type 2 morphologies of the feeder channel and its associated stack of sheet sands. (B) and (C) display the lateral switching of the center of sheet sand deposition and the associated development of new feeder channels over time, reflected in the outcrops by varying paleocurrent directions of the individual channel fills. (D) Downdip shift of the center of sheet-sand deposition, reflected in the outcrops by the vertical stacking of type 1 on top of type 2 channel fills.

channel. As turbidity currents start to spread out toward the lower fan, the type 1 channel morphology will gradually be replaced by the type 2 channel morphology. Therefore, the close vertical association of type 1 and 2 channels reflects the updip/downdip shift of the depocenter in the sheet sand lobes. The updip/downdip shift, is likely to be controlled by slight changes in the composition of the turbidity currents. Turbidity currents that carry a high load of fines remain in suspension for a longer period of time and, thus, can carry sediment further into the basin (Figure 7).

CONNECTIVITY AND FLUID FLOW

Connectivity is an interwell reservoir parameter that describes the potential for fluid flow between different reservoir sandstones. Two types of potential reservoir rocks were identified in the distributary channel system at Bloukop: the very fine-grained climbing ripple cross-laminated facies II sandstones in the levee deposits and the massive facies I and parallel-laminated facies II sandstones in the channel fills. In the deepwater Gulf of Mexico, levee-overbank reservoirs (e.g., Ram/Powell Field, Clemenceau, 1995; Tahoe Field, Shew et al., 1994), as well as reservoirs in channel fills, such as the N1 and O sands in the Auger Field (McGee et al., 1994) and the Green Canyon Block 205 (Rafalowski et al., 1994), have been drilled. The thin-bedded levee-overbank reservoirs were described as discrete sandstone beds with little or no vertical connectivity among each other (Shew et al., 1994). Within the channel fills, sandstone beds have a good connectivity, due to the high degree of amalgamation, especially in the channel axis deposits.

The difference in degree of connectivity among the different channel fills and the thin-bedded levee-overbank deposits at Bloukop is illustrated in Figure 8. Two factors contribute to the good vertical connectivity among the channel fills: first, type 1 channels often erode into the top of type 2 channels, creating an amalgamation surface between the two sandstone bodies. These sand-on-sand contacts facilitate unconstrained fluid flow between the two channel fills (Figure 8; Cook et al., 1994). Secondly, type 2 channels that cut into type 1 or 2 channels create connectivity for the same reason. Thus, channels C2-C8 (Figure 6) of the distributary channel system act as a single flow unit (Figure 8).

The connectivity between channels and the associated levee-overbank sandstones is usually hindered by a continuous silt drape that separates the channel fill from the levee-overbank deposits (Figure 5). This silt layer may be interpreted as a flow barrier; however, drill-stem tests performed in a similar setting in the N1 sand in Green Canyon Block 205 showed that flow restrictions between the channel fill and the levee deposits were only minimal (Rafalowski et al., 1994). Therefore, the silt layers are thought of merely as an obstruction to fluid flow between the channel and the levee deposits, not as a flow barrier.

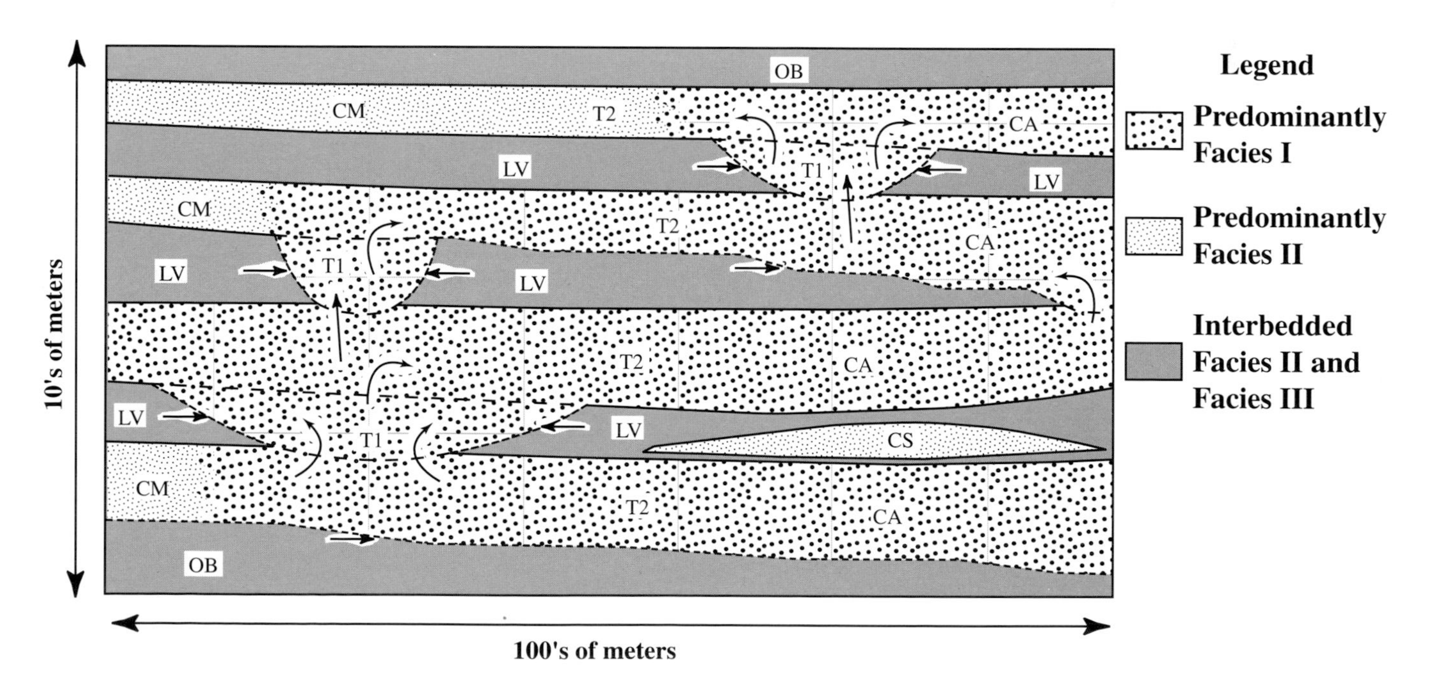

Figure 8—Proposed model of reservoir characteristics of distributary channel systems. Arrows indicate possible pathways for fluid flow between the different reservoir types. Length of arrows represents the relative strength of the possible fluid flow. T1 = type 1 channel fill, T2 = type 2 channel fill, CS = crevasse splay deposits, CA = channel axis deposits, CM = channel margin deposits, LV = levee deposits, OB = overbank deposits.

CONCLUSIONS

The present study focused on the examination of the lower midfan to upper lower fan environment, especially with respect to the architecture of a distributary channel system as exposed in Fan 5. It yielded the following main conclusions:

1. Two types of channels were recognized: type 1 is characterized by a relatively narrow, erosive channel form (aspect ratio <50) and develops as a channel-levee complex in a series of erosional and depositional events. Type 2 channels are broader (minimum aspect ratio ~150–200) and their channel form is cut into the underlying deposits in a stepwise fashion. Internally, they develop a channel axis facies that grades laterally into a channel margin facies. Overall, type 2 channels are less erosive in nature than type 1 channels. They are, therefore, interpreted as the downdip equivalent of type 1 channels.
2. Levee deposits stem from spilling of sediment from the body and wake of a turbidity current into the interchannel area as they pass through the channel forms. Levee deposits consist of a laterally continuous sequence of interbedded sandstones and siltstones. Away from the channel, a decrease in the sandstone bed thickness, percentage of amalgamation, number of sandstone beds, angle of climbing ripple contacts, grain size of the sandstones, and net:gross sand ratio was observed, which was interpreted to reflect the decrease in the strength of overbank flows.
3. Overbank deposits are the lateral continuation of levee deposits. They form a series of thinly interbedded fine-grained silt (facies III) with coarse-grained silt (facies II), which exhibits extreme lateral continuity over hundreds of meters.
4. The overall succession of the deposits at Bloukop suggests switching of channel location within the same fan system, which is indicative of the distributary channel environment. This avulsion occurs in response to the creation of positive relief during the deposition of a depositional lobe.
5. The exposed sequence of type 1 and 2 channels suggests that a depositional lobe is fed over time by several, spatially closely related feeder channels, termed here "distributary channel system." The feeder channel fills within a single distributary channel system evolve in close relationship with the lateral and updip/downdip switching of the center of sheet sand deposition.
6. Type 1 and type 2 channels reflect the downdip changes in the morphology of the feeder channels within a distributary channel system.
7. The degree of connectivity among the channel deposits is high, although it is moderate between the channel and levee-overbank deposits. All of these deposits are determined to represent one flow unit. Lack of exposure prevents connectivity analysis between the channel and crevasse splay deposits.
8. The inferred complexity of the two channel types, levee, overbank, and crevasse splay deposits clearly support the idea that successive density flows are of varying strength, composition, density, and volume. As a result, the transitions from type 1 to type 2 channels undergo an updip and downdip shift in location, within a distributary channel system, reflected here by the vertical stacking of type 1 and 2 channels.

ACKNOWLEDGMENTS

We wish to express our gratitude to the Tanqua Karoo Consortium of oil companies that sponsored this research project. We would also like to thank Khristina Kirschner for her critical comments that led to vast improvements of the original manuscript.

REFERENCES CITED

Bouma, A. H., and H. D. Wickens, 1991, Permian passive margin fan complex, Karoo Basin, South Africa: possible model to Gulf of Mexico: Gulf Coast Association of Geological Societies Transactions, v. 41, p. 30–42.

Clemenceau, G. R., 1995, Ram/Powell Field: Vioska Knoll Block 912, deepwater Gulf of Mexico: turbidites and associated deep-water facies, *in* R. D. Winn and J. M. Armentrout, eds., Turbidites and associated deep-water facies: SEPM Core Workshop No. 20, Tulsa, p. 95–129.

Cook, T. W., A. H. Bouma, M. A. Chapin, and H. Zhu, 1994, Facies architecture and reservoir characterization of a submarine channel complex, Jackfork Formation, Arkansas *in* P. Weimer, A. H. Bouma, and B. F. Perkins, eds., Submarine fans and turbidite system: sequence stratigraphy, reservoir architecture, and production characteristics: SEPM 15th Annunal Research Conference Proceedings, Houston, p. 69–81.

Kenyon, N. H., A. Amir, and A. Cramp, 1995, Geometry of younger sediment bodies of the Indus Fan, *in* K. T. Pickering, R. N. Hiscott, N. H. Kenyon, F. Ricci Lucchi, and R. D. A. Smith, eds., Atlas of deepwater environments, London, Chapman & Hall, p. 89–93.

Kirschner, R. H., 1999, Characteristics of a distributary channel-levee-overbank system, Fan 5, Tanqua Karoo, South Africa: unpublished Master's thesis, Louisiana State University, 200 p.

McGee, D. T., P. W. Bilinsky, P. S. Gary, D. S. Pfeiffer, and J. L. Sheiman, 1994, Geologic models and reservoir geometries of Auger Field, deepwater Gulf of Mexico, *in* P. Weimer, A. H. Bouma, and B. F. Perkins, eds., Submarine fans and turbidite systems: sequence stratigraphy, reservoir architecture, and production characteristics: SEPM 15th Annual Research Conference Proceedings, Houston, p. 245–256.

Mutti, E., 1977, Distinctive thin-bedded turbidite facies and related depositional environments in the Eocene Hecho Group (south-central Pyrenees, Spain): Sedimentology, v. 24, p. 107–131.

Mutti, E., and W. R. Normark, 1987, Comparing examples of modern and ancient turbidite systems: problems and concepts, *in* J. K. Leggett and G. G. Zuffa, eds., Marine clastic sedimentology, London, Graham & Trotman, p. 1–38.

Rafalowski, J. W., B. W. Regel, D. L. Jordan, and D. O. Lucidi, 1994, Green Canyon Block 205 lithofacies, seismic facies, and reservoir architecture, *in* P. Weimer, A. H. Bouma, and B. F. Perkins, eds., Submarine fans and turbidite systems: sequence stratigraphy, reservoir architecture, and production characteristics: SEPM 15th Annual Research Conference Proceedings, Houston, p. 293–306.

Shew, R. D., D. R. Rollins, G. M. Tiller, C. J. Hackbarth, and C. D. White, 1994, Characterization and modeling of thin-bedded turbidite deposits from the Gulf of Mexico using detailed subsurface and analog data, *in* P. Weimer, A. H. Bouma, and B. F. Perkins, eds., Submarine fans and turbidite systems: sequence stratigraphy, reservoir architecture, and production characteristics: SEPM 15th Annual Research Conference Proceedings, Houston, p. 327–334.

Walker, R. G., 1985, Mudstones and thin-bedded turbidites associated with the upper Cretaceous Wheeler Gorge conglomerates, California: a possible channel-levee comples: Journal of Sediment Petrology, v. 55, p. 279–290.

Wickens, H. D., 1994, Basin floor fan building turbidites of the Southwestern Karoo Basin, Permian Ecca Group, South Africa: unpublished Ph.D. thesis, University of Port Elizabeth (South Africa), 233 p.

Al-Siyabi, H. A., 2000, Anatomy of a type II turbidite depositional system: Upper Jackfork Group, Degray Lake area, Arkansas, *in* A. H. Bouma and C. G. Stone, eds., Fine-grained turbidite systems, AAPG Memoir 72/SEPM Special Publication No. 68, p. 245–262.

Chapter 22

Anatomy of a Type II Turbidite Depositional System: Upper Jackfork Group, Degray Lake Area, Arkansas

Hisham A. Al-Siyabi
Department of Geology and Geological Engineering, Colorado School of Mines Golden, Colorado, U.S.A.[1]

ABSTRACT

The upper Jackfork Group in the DeGray Lake area, Arkansas, represents a Type II turbidite depositional system with proximal lobes, distal lobes, and channel fills. Diagnostic facies assemblages and sand/shale ratios characterize each of these deposits. In addition, vertical facies arrangements, upward bed thickness trends, and downcurrent bed continuity vary among these deposits.

The depositional setting is divided into three areas: upper fan, transition zone, and middle to lower fan. In these settings, channel fills are envisioned to occur in the upper-fan region. While proximal lobes dominate the transition zone, distal lobes populate the middle- to lower-fan region.

INTRODUCTION

The Jackfork Group of the Ouachita Mountains of Arkansas and Oklahoma is part of a very thick 9120–13,680 m (30,000–45,000 ft) sequence of Carboniferous strata that is mostly of deepwater origin (Coleman et al., 1994). The Jackfork Group of south-central Arkansas and eastern Oklahoma has been examined by many geoscientists, and opinions about the depositional history and environments vary considerably.

Previous investigations have focused on the lithologic, sedimentologic, and facies variation in single vertical rock successions to arrive at three-dimensional depositional interpretations. In the DeGray Lake Spillway and Intake area, superbly exposed upper Jackfork rock sections can be integrated with descriptions of shallow cores to provide a detailed geologic framework of this interval. In this paper, a depositional model for the upper Jackfork is presented that takes into account not only the vertical variations in rock successions but also their variation in a downcurrent direction.

Six stratigraphic sections have been measured in the 1.6-km-long area between DeGray Lake Spillway and Intake (Figures 1, 2). The beds strike almost east–west and dip 55° toward the south. The 309 m (1017 ft) thick spillway section has been described by Morris (1977), Chester (1994), and Shanmugam and Moiola (1995). The intake section has been described by Jordan et al. (1993). All of these sections have been

[1] Present address: Exploration Directorate, Petroleum Development Oman LLC., Muscat, Sultanate of Oman.

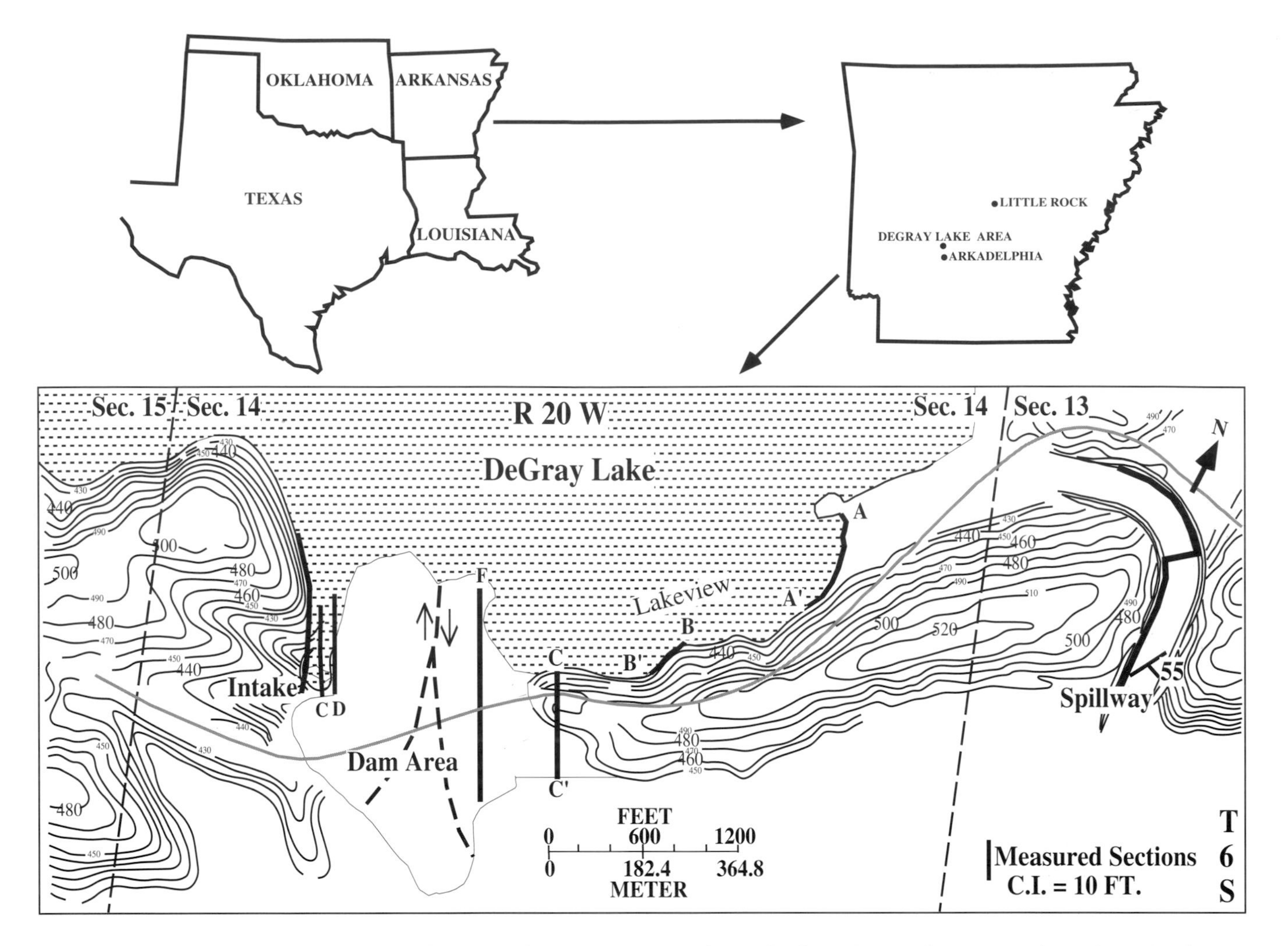

Figure 1—Approximate location of the study area. Depicted are the locations of measured sections in the DeGray Lake Spillway and Intake area (topographic map modified from U.S. Corps of Engineers, 1971). Outcrop exposures are: Spillway, Lakeview (divided into segments: A–A′, B–B′, and C–C′), and the Intake (Figure 2). Sections F, D, and C are constructed from borehole descriptions from the Dam area.

re-examined in detail, and an additional stratigraphic section termed "Lakeview" was measured along the shore of the DeGray Lake (Al-Siyabi, 1998). This section has been subdivided into three parts (A–A′, B–B′, and C–C′), which basically tie the spillway to the DeGray Dam along a continuous stratigraphic section. In addition, borehole descriptions from the U.S. Army Corps of Engineers (1971) are shown in Figure 2 as C, D, and F. Paleocurrent measurements indicate a general westward flow direction for these strata (Morris, 1974a; Al-Siyabi, 1998).

UPPER JACKFORK FACIES

Fifteen facies have been recognized in the DeGray Lake Spillway and Intake area (Table 1; Al-Siyabi, 1998). The classification follows that of Mutti (1992), even though not all of his facies are recognized in Jackfork strata.

A rather important and common feature in Jackfork strata is large- to medium-scale erosional scour-and-fill (Morris, 1974a). On the bases of sandstone beds, these features can resemble large load structures. However, when observed on the top surfaces of bedding, such as on an excellent exposure at the Murfreesboro quarry (west of the Spillway-Intake area), these features have a teardrop-shaped appearance typical of large flutes. These features are interpreted as being diagnostic of erosion/deposition in a region of change in slope gradient, such as the channel-lobe transition zone of Mutti and Normark (1987). Pebbly sandstone-filled scours are the product of high-energy, channelized flows, even though those filled with sandstones and mudstones are related to relatively lower-energy, nonchannelized flows (Al-Siyabi, 1998). The presence or absence of these scours is also interpreted as indicative of relatively proximal or distal settings, respectively.

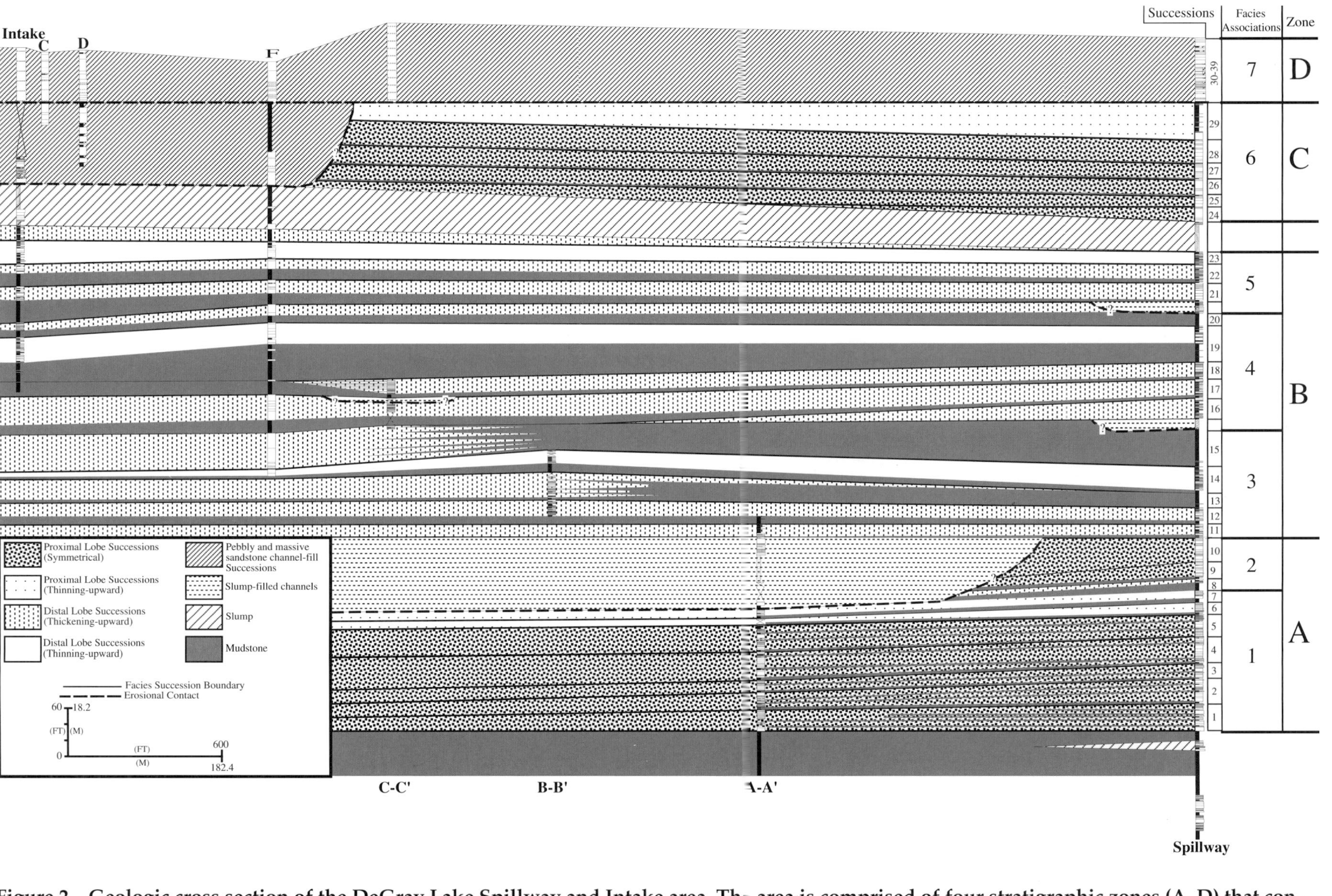

Figure 2—Geologic cross section of the DeGray Lake Spillway and Intake area. The area is comprised of four stratigraphic zones (A–D) that consist of seven facies associations and 39 facies successions.

Table 1. Various facies that comprise upper Jackfork strata in the study area (facies designations are modified after Mutti, 1992).

Facies	Facies Designation	Occurrence (%)*	Sediment Support Mechanism/ Depositional Mechanism	Interpretation
Mudstone with incorporated sandstone clasts	F1	1.1	Cohesive matrix strength/ cohesive freezing	Debris Flow Deposits
Clast-supported mudstones	F3a	0.2	Matrix buoyant lift and clast-to-clast contact/cohesive freezing	
Massive sandstones with incorporated mudstone clasts	F3b	4	Matrix strength and hindered settling(?)/cohesive freezing(?)	
Pebbly supported sandstones	F4a	1	Dispersive pressure and hindered settling/frictional freezing	Pebbly High-Density Sandstone Deposits
Massive sandstones with floating pebbles	F4b	0.4	Hindered settling and turbulence/ instantaneous collapse and freeze(?)	
Inversely graded pebbly sandstones	F4c	0.1	Dispersive pressure/instantaneous collapse and freeze	
Normally graded pebbly sandstones	F5	3	Turbulence and hindered settling/ deposition from suspension from a highly concentrated flow	
Sandstones with fluid escape structures	F8a	3.4	Upward movement of intergranular fluid/deposition from a liquified flow	Liquefied Flow Deposits
Sandstones with floating mudstone clasts	F8b	2.4	Turbulence and hindered settling/ deposition from a dense flow with clasts depositing during the final stage	Sandy High-Density Sandstone Deposits
Massive sandstones	F8c	57	Turbulence and hindered settling/ deposition from suspension from a highly concentrated flow	
Parallel-laminated sandstones	F9b	7.2	Waning turbulent flow/interaction of waning flow with bedload	Sandy Low-Density Sandstone Deposits
Rippled sandstones	F9c	20	Part of waning turbulent flow/ deposition from suspension from a low-density flow	
Convoluted sandstone beds	F9cc	0.2	Waning turbulent flow/deposition from suspension from a low-density flow followed by soft sediment deformation	
Laminated siltstones and mudstones	D	N/A	Waning turbulent flow/deposition from suspension clouds of a low-density flow	Suspension Deposits
Laminated mudstones	E	N/A	Suspension/deposition from water column	

* Percentages of sandstones calculated from Spillway, Lakeview, and Intake sections.

UPPER JACKFORK FACIES TRACT AND SUCCESSIONS

A facies tract is a continuum of genetically related facies (Mutti, 1992; Mutti et al., 1994). Following the method of Mutti, the previously described facies have been linked according to their presumed depositional flow path or facies tract (Figure 3). Along this path, a downdip change in the facies type can be observed from debris flows to pebbly high-density sandstones; to liquefied sandstones; to sandy, high-density sandstones; and then to sandy, low-density sandstones. In a few instances, such transitions can be seen over a relatively short

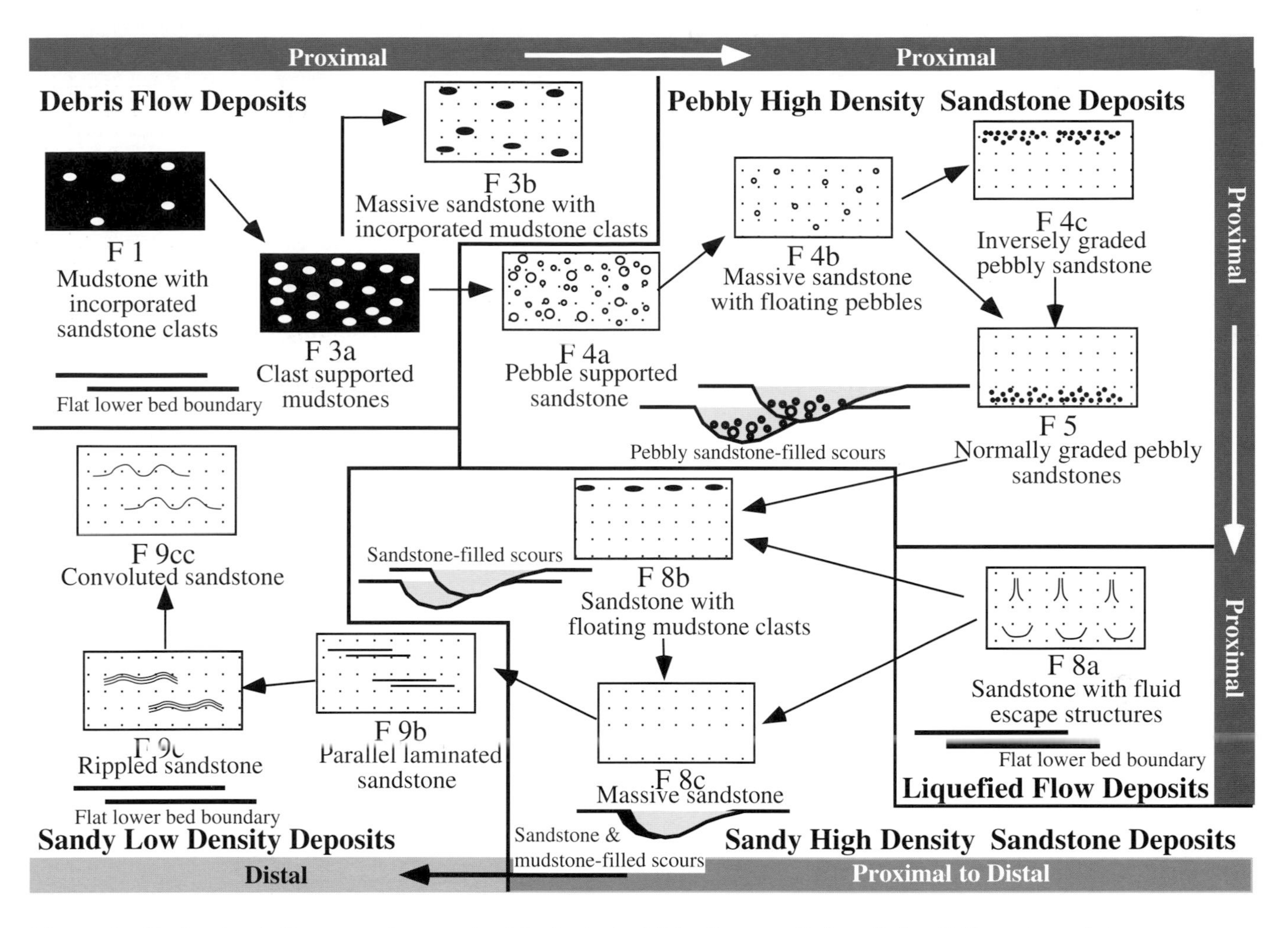

Figure 3—Facies tract diagram of upper Jackfork strata. This diagram links all the different sandstone facies in an evolutionary continuum. Bed boundaries that characterize each facies category are also depicted (facies designations are modified after Mutti, 1992).

distance (Al-Siyabi, 1998). Other such facies continua have been published by Middleton and Hampton (1973), Walker (1978), and Lowe (1982).

The flow path depicted in Figure 3 is a complete turbidite continuum. However, in most outcrops, only parts of this continuum will be recognized laterally or vertically. The delineation of facies tracts is critical in deciphering vertical facies arrangements into facies successions. A facies succession is defined as the progressive upward change of one or all the following parameters: sandstone/mudstone ratio, grain size, and sedimentary structures (Walker, 1992).

Mutti et al. (1994) regarded facies successions, termed elementary facies tracts (EFTs), as the building block for turbidite stacking pattern analysis. The EFT concept is a useful tool for detailed analysis of turbidite deposits. Aside from bed scale correlations, the EFT presents another scale, by which high-resolution correlations can be achieved within turbidite strata. In this paper, the concept of EFT is substituted by the facies succession. This term is less rigid and can be applied to other depositional settings, making it more comprehensible. The second scale of observation used is the facies association, which consists of multiple facies successions.

The combination of vertical facies successions, downcurrent facies relationships, and depositional geometries are critical in the interpretation of depositional environments. Based on depositional environments, the upper Jackfork in the DeGray Lake Spillway and Intake area consists of four stratigraphic zones within which a total of 39 facies successions that comprise seven facies associations have been identified (Figure 2). Three types of facies associations are recognized: proximal lobes, distal lobes, and channel fills.

PROXIMAL LOBE FACIES ASSOCIATIONS

Proximal lobe facies associations, represented by zones A and C, are comprised of successions that are dominated by facies F3b, F8a, F8b, and F8c (Figure 4; Table 2). Two types of facies successions are recognized: symmetrical and thinning-upward. In a symmetrical succession, bed thickness is recognized to

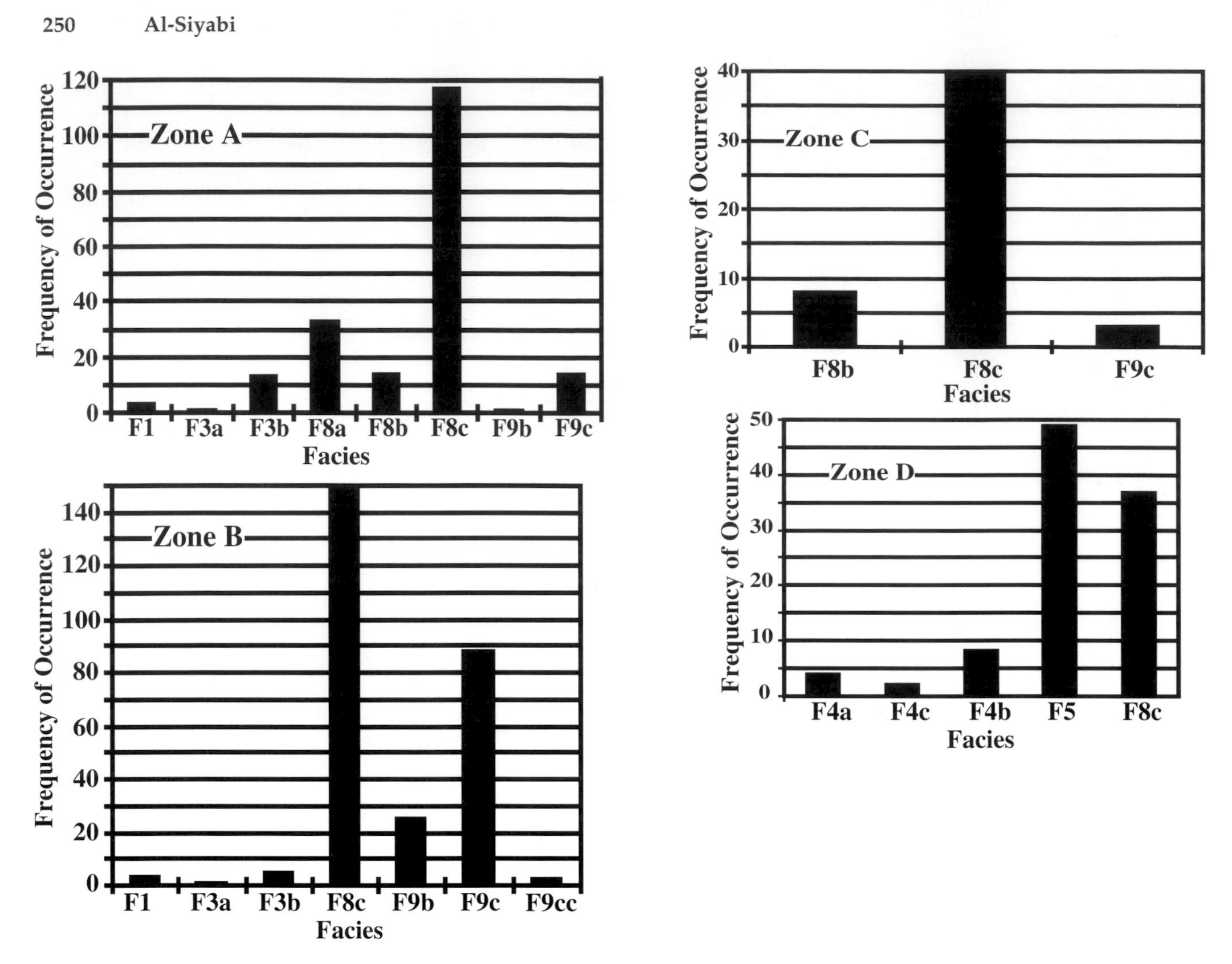

Figure 4—Proportions of the different facies that comprise zones A, B, C, and D.

Table 2. Depositional interpretations, net to gross (N/G), percentage of sand-on-sand contacts (Sd-Sd), and continuity of beds for the four stratigraphic zones as calculated from various outcrop exposures and borehole descriptions.

Section	Zone	Interpretation	Total Thickness	SSThickness	N/G	$\overline{N/G}$	Sd-Sd (%)	Continuity
Spillway	A	Proximal lobes	245.5	207.25	0.84	0.84	72	Good
	B	Distal lobes	363.5	153.5	0.42	0.45	53	Moderate
	C	Proximal lobes	114.5	111	0.97	0.92	88	Good
	D	Channel	81.5	78	0.96	0.99	90	Poor
Lakeview	A	Proximal lobes	N/A	N/A	N/A			
	B	Distal lobes	195.5	69.5	0.36			
	C	Covered	Covered	Covered	N/A			
	D	Channel	100	100	1.00			
F–F'	B	Distal lobes	212	108	0.51			
	C	Channel	50	50	1.00			
	D	Channel	60	60	1.00			
Intake	B	Distal lobes	147	72	0.49			
	C	Channel	31.5	25	0.79			
	D	Channel	69	69	1.00			

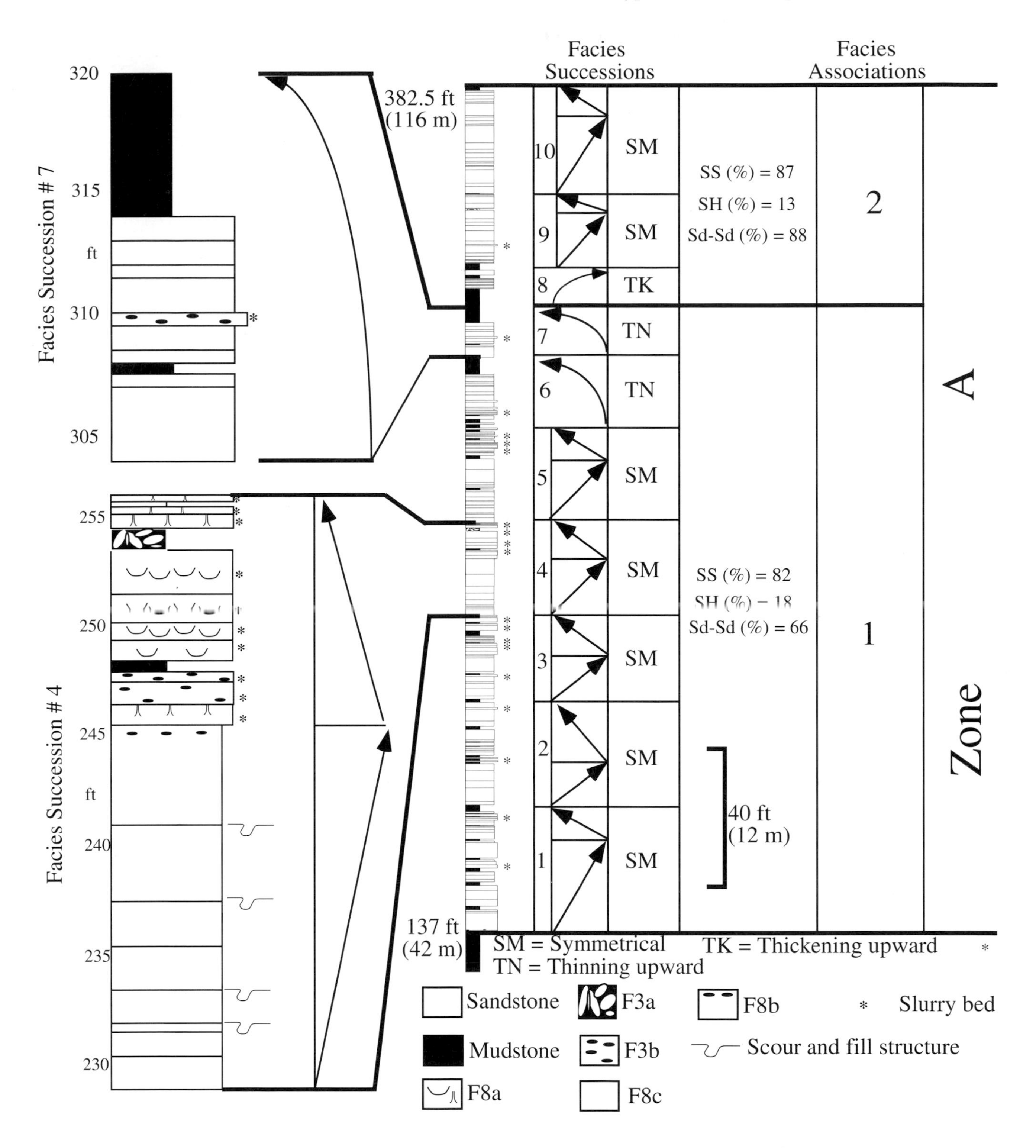

Figure 5—Examples from zone A of a symmetrical (#4) and a thinning-upward (#7) facies successions. Zone A consists of two proximal lobe facies associations #1 and 2. Facies association #1 is composed of symmetrical facies successions that change upward to thinning-upward successions. Association #2 is composed dominantly of symmetrical facies successions.

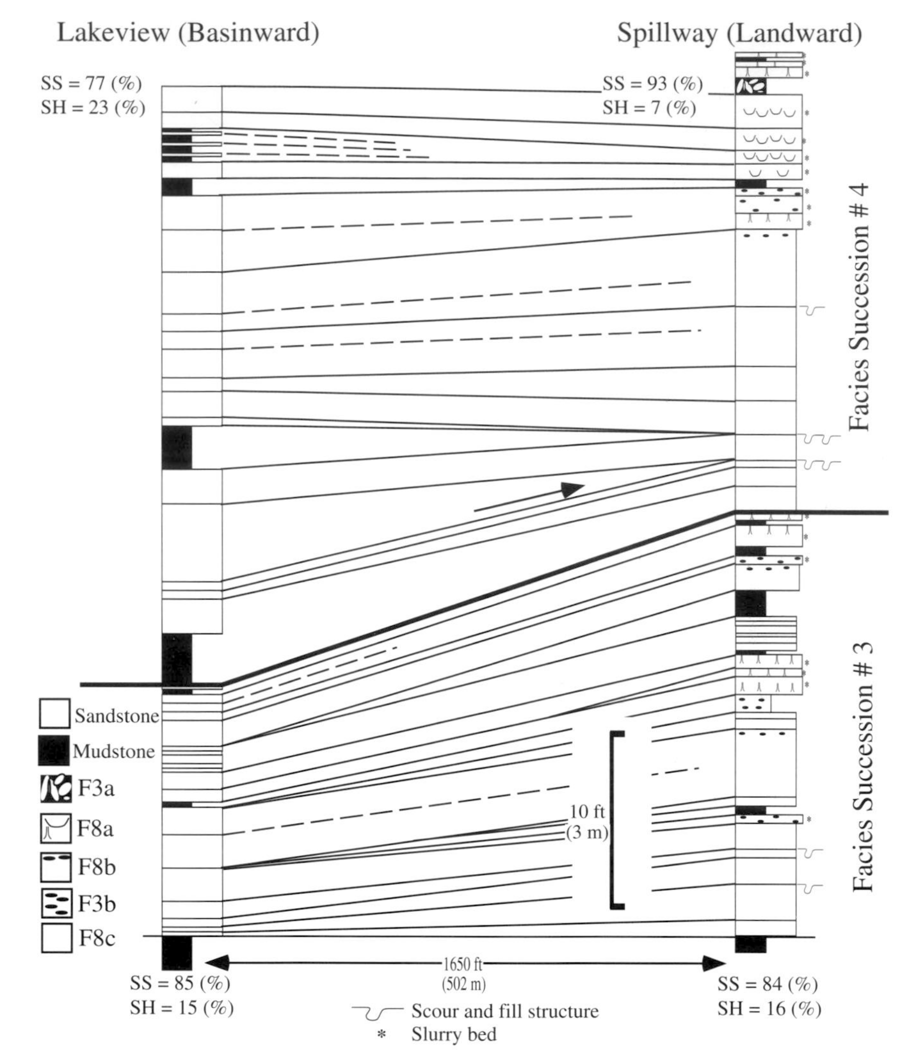

Figure 6—Correlations of symmetrical facies successions #3 and #4. In facies succession #3, 39% of the beds can be correlated between the Spillway and Lakeview sections. For succession #4: 44%. Note also the basinward change in facies type from proximal to distal. In addition, the amount of mudstone contained in succession #4 increases in a basinward direction.

generally increase and then decrease upward (Figure 5). A typical succession starts with amalgamated massive sandstone beds (F8c), which are then overlain by beds of facies F8a and/or facies F3b. This type of succession is sometimes capped by a relatively thin mudstone (facies D). Sandstone-filled scour-and-fill features are prevalent in these successions.

Thinning-upward successions may record proximal to distal facies transitions or a thinning-upward bed trend with no vertical facies change. An example of a facies succession that thins upward with no vertical facies change is presented in Figure 5. Both types are capped by relatively thick mudstone intervals. Sandstone-filled scour-and-fill features are conspicuously absent from these successions, but mudstone-filled scours are present.

Based on vertical stacking of facies successions, zone A can be divided into two facies associations (Figures 2, 5). Facies association #1 is composed of symmetrical facies successions (#1–#5 in Figure 5), which are capped by two thinning-upward facies successions (#6, # 7). Facies association #2 consists of thickening-upward and symmetrical facies successions (#8–#10). Analysis of mudstone proportions between the two associations reveals that #1 is slightly muddier than #2 (Figure 5). Sandstone- and mudstone-filled scour-and-fill features are common.

Correlations of symmetrical and thinning-upward successions reveal the following attributes. Bed scale correlations are possible, as illustrated in Figures 6 and 7. Symmetrical successions (Figure 6) exhibit a basinward proximal to distal facies change, with liquefied flow deposits (F3b) and massive sandstone beds with floating mudstone clasts (F8b) grading to massive sandstone beds (F8c). In addition, scour-and-fill structures are not recognized basinward. Mudstone proportions increase in a basinward direction (Figures 6, 7). Proximal lobe deposits are incised by large, deep channels. Examples are the upper part of zone A (facies association #2) and all of zone C (facies association #6 in Figure 2). Table 3 summarizes lateral changes within the seven facies successions in zone A over a distance of approximately 502 m. Total thickness and percentage sandstone of five of the seven facies successions decrease in the downcurrent direction. Moreover, facies succession scale correlations reveal a general unidirectional thinning in the downcurrent direction.

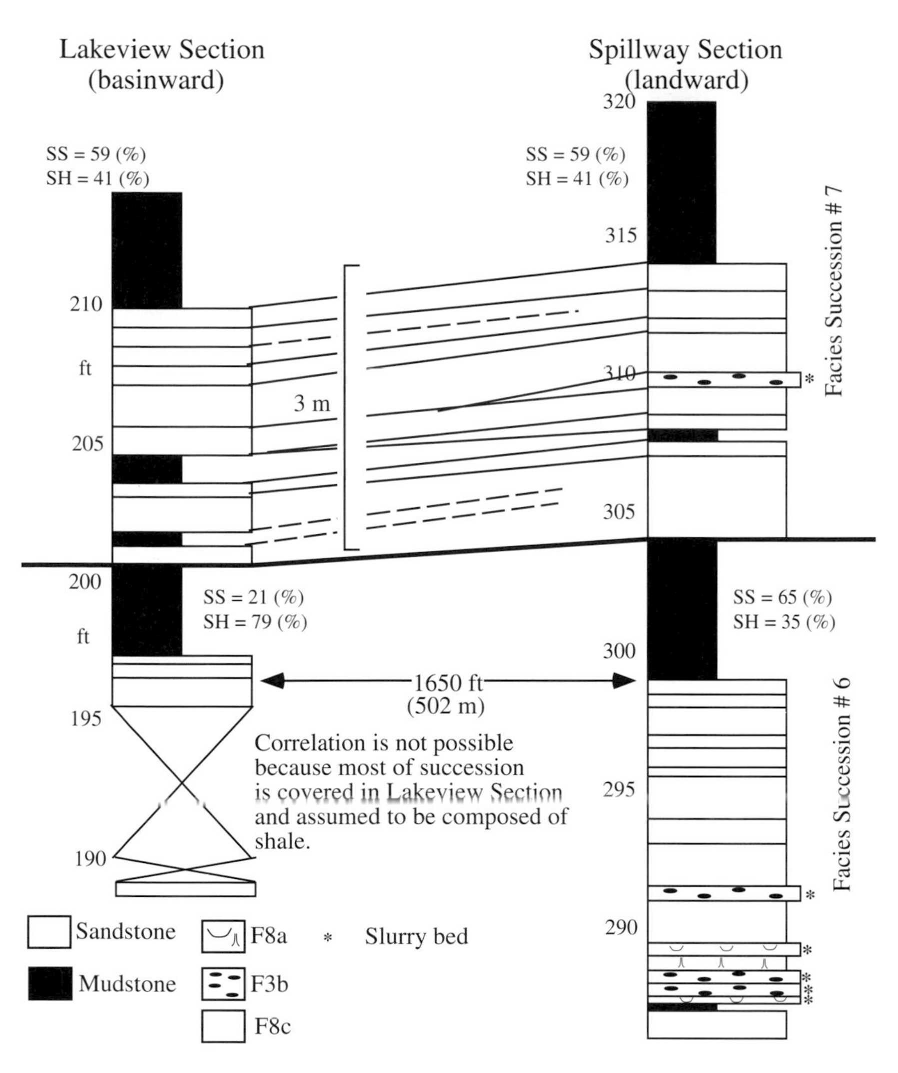

Figure 7—Correlations of thinning-upward facies successions #6 and #7. Bed scale correlations are possible, especially for succession #7. Succession #6, in the Spillway section, exhibits an upward facies change from proximal to distal; succession #7 shows a decrease in bed thickness in the same direction.

Proximal lobe facies association #6, represented by zone C, is dominantly comprised of stacked symmetrical facies successions with high sandstone-to-shale ratio (successions #24–#28 in Figure 8). The last facies succession in this association is thinning-upward and is capped by a thick mudstone unit (succession #29). In addition, all the scour-and-fill features are sandstone-filled.

DISTAL LOBE FACIES ASSOCIATIONS

Distal lobe facies associations, represented by zone B, are dominated by both high- and low-density turbidity flow deposits (Figure 4). The high-density facies occur as massive sandstones (F8c), whereas the low-density facies are in the form of parallel-laminated (F9b), ripple-laminated (F9c), and convoluted beds (F9cc). Sandstone percentages and other data are presented in Table 2. Two types of facies successions are recognized: thickening- and thinning-upward.

A typical thickening-upward facies succession starts with either parallel (F9b) or ripple-laminated beds (F9c) and progresses upward to a relatively thick, massive sandstone (F8c) (Figure 9). These thickening-upward successions are comprised of higher frequency thickening-upward cycles. A typical thinning-upward facies succession starts with massive sandstones (F8c) that become progressively overlain by parallel (F9b) and/or rippled laminated (F9c) beds (Figure 9). These thinning-upward successions, which are commonly capped by relatively thick mudstones, are comprised of higher frequency thinning-upward cycles.

Based on vertical stacking of facies successions, zone B can be divided into three facies associations: #3, #4, and #5 (Figures 2, 9). Facies associations #3 and #4 start with thickening-upward successions, which are then capped by thinning-upward units (Figure 9). Facies association #5, on the other hand, starts with a symmetrical facies succession that is overlain by thickening- and thinning-upward units (successions #21–#23 in Figure 9). A general upward increase in the sandstone content and the degree of bed amalgamation is also recognized among the three associations, with #5 being the most sandy and highly amalgamated. In addition, associations #4 and #5 contain slumped-filled channels.

Table 3. Correlations, at the facies succession scale, between two stratigraphic sections of proximal lobe deposits (zone A). Most of the successions exhibit unidirectional thinning to the west. The direction of thinning is inferred from the amount of sandstone present in a particular succession.

Direction of Thinning	Sections: Basinward Lakeview	Sections: Landward Spillway	Successions
No Change	13.5 SS(59), SH(41)	16 SS(59), SH(41)	7
←	12 SS(21), SH(79)	20.5 SS(65), SH(35)	6
←	32.5 SS(65), SH(35)	27.5 SS(89), SH(11)	5
←	35 SS(77), SH(23)	27 SS(93), SH(7)	4
→	16.5 SS(85), SH(15)	25.5 SS(84), SH(16)	3
←	22 SS(75), SH(25)	32 SS(86), SH(14)	2
←	28 SS(73), SH(27)	34.5 SS(88), SH(12)	1

Top number = thickness of successions (in feet); SS = sandstone percentage; SH = shale percentage.

Table 4 summarizes downcurrent changes of the 13 facies successions comprising the distal lobes of zone B. Facies successions generally thin in opposite (east–west and west–east) directions in a compensatory fashion over a 1.6 km area of correlation. In general, the percentage of sandstone decreases in the direction of thinning of individual facies successions. Individual beds within facies successions #16–#18 also thin in alternating directions, in a compensatory fashion (Slatt et al., 1997). Details of facies successions #19, #21, and #22 are shown in Figures 10 and 11. Bed scale correlations of thickening- and thinning-upward successions are difficult to achieve.

CHANNEL-FILL FACIES ASSOCIATIONS

Two types of channel fills are recognized. The first type, represented by zone D (facies association #7 in Figure 2), is dominated by normally graded pebbly sandstones (F5) and massive sandstones (F8c) (Figure 4). Other pebbly sandstone facies comprise the rest of this facies association. A typical succession begins with normally graded pebbly sandstones, which are overlain by massive sandstones and/or mudstones in an overall fining- and thinning-upward pattern (Figure 12). Pebble-filled scour-and-fill features are prevalent in these successions (Table 2). Figure 13 illustrates details of downcurrent correlations of this channel-fill association. Owing to the lenticularity of individual beds, it is not possible to correlate them from section to section; however, the percentage of sandstone is uniformly high (>95%) across the interval correlated.

The second type of fill is composed of amalgamated, slumped, sandy debris flow deposits interlayered with cohesive debris flows (F1) (Figure 14). This type of channel fill has no systematic vertical trend. In the study area, these types of channels occur only in association with zones A and B (Figure 2).

DEPOSITIONAL INTERPRETATION

The studied part of the Jackfork turbidite depositional system is interpreted as a Type II turbidite system of Mutti (1985, 1992) for the following reasons: abundance of sandstone-filled scours, high sandstone content, moderate to poor bed scale correlations exhibited by distal lobes, tectonic setting, and type of turbidite deposits (Table 5).

According to Mutti and Normark (1987, 1991), physiographic changes characterizing the channel-lobe transition area induce turbulent flows to expand and, in turn, lead to an increase in their erosive

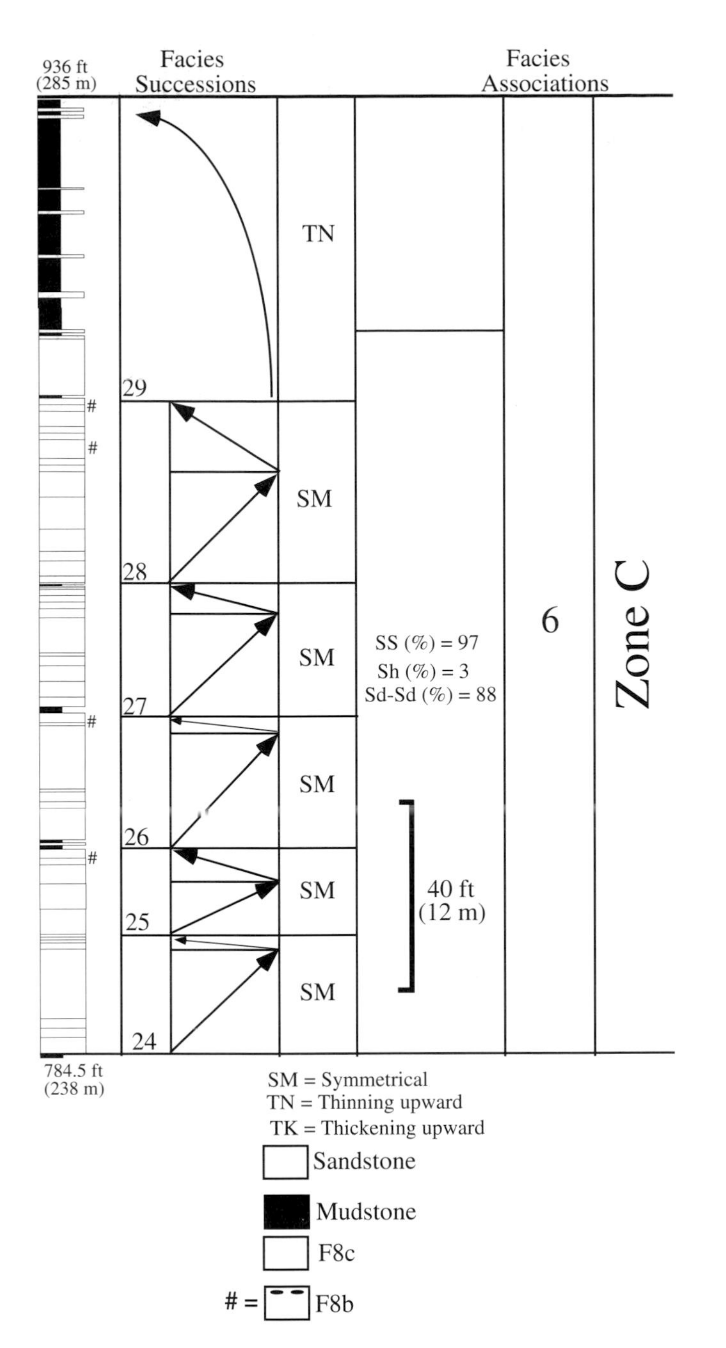

Figure 8—Facies association #6, represented by zone C, is dominated by stacked symmetrical facies successions that are capped by a thinning-upward unit with an associated thick mudstone.

capacity. They attributed the change in flow behavior to the existence of a hydraulic jump. According to the same authors, flows that are dominated by sandstones and coarser sediments undergoing a hydraulic jump will experience increased turbulence that will lead to rapid dilution and an immediate deposition in the transition area. These sandstone-rich flows are termed "poorly efficient" by Mutti and Normark and are characteristic of a Type II turbidite system.

The dominance of sandstone-filled scours in thick sandstone packages that represent proximal lobe facies associations, in upper Jackfork strata, indicate that most of the sandstone deposition occurred in the channel-lobe transition area. Moreover, the scarcity of mudstone-filled scours and the absence of cross-stratified and lenticular sandstone units substantiate the Type II turbidite system interpretation. Furthermore, Type II systems are characterized by high proportions of sandstone and coarser grained sediments (Mutti and Normark, 1987). The total sandstone content of the upper Jackfork is calculated at 71% on average, which qualifies it as a sandstone-rich system in the classification of Reading and Richards (1994).

Unlike Type I turbidite systems, lobes in a Type II system are less developed (Mutti and Normark, 1987). The quantification of lobe developments is based on their aerial extent. Distal lobes of the upper Jackfork, which represent the most basinward sandy deposits in this system, exhibit moderate to poor correlations at the bed scale (Figures 10, 11). This implies a limited aerial extent and, in turn, poor development. Additional supporting evidence is the occurrence of associated shallow channels with these lobes (Figure 2). Type I lobes, in contrast, are not channelized (Mutti, 1985, 1992).

An additional supporting evidence is related to the tectonic setting during the deposition of the Jackfork. These sediments were deposited during a contractional phase that involved the subduction of the southern margin of the North American craton beneath an island arc or microcontinent (Morris, 1974b; Blythe et al., 1988; Viele and Thomas, 1989). According to Mutti and Normark (1987), Type II depositional systems occur in basins associated with active margin settings of all forms that include extensional (rifting), strike-slip (transform), and contractional (collisional) regimes.

From Mutti's (1985, 1992) division of turbidite systems, it can be inferred that the dominance of a particular turbidite deposit(s) (i.e., lobes, channels, etc.) is an indication of the type of turbidite system involved. It was illustrated in the previous sections that upper Jackfork strata are comprised of the following turbidite facies associations: proximal lobes, distal lobes, and channel fills (Figures 2, 15). Consequently, these deposits are the architectural elements of a Type II system.

Coleman et al. (1994) envisioned the Jackfork fan complex to consist of the following depositional areas: upper fan, transition zone, middle fan, lower fan, and abyssal plain. According to this model, turbidites of the DeGray Dam area were deposited in the transition zone. The author concurs with that interpretation because of the presence of various types of scour-and-fill features in upper Jackfork strata. The author interprets sandstone- and slump-filled channels to have been deposited in the upper fan region (Figure 15). The deposition of proximal lobes occurred in the transition zone, whereas that of distal lobes occurred beyond the transition zone or the middle- to lower-fan region (Figure 15).

Figure 9—Examples, from zone B, of thickening-upward (#18) and thinning-upward (#14) facies successions. Zone B consists of three distal lobe facies associations: #3, #4, and #5. Facies associations #3 and #4 start with thickening-upward successions that are then capped by thinning-upward units. Association #5 starts with a symmetrical succession that is capped by thickening- and thinning-upward units.

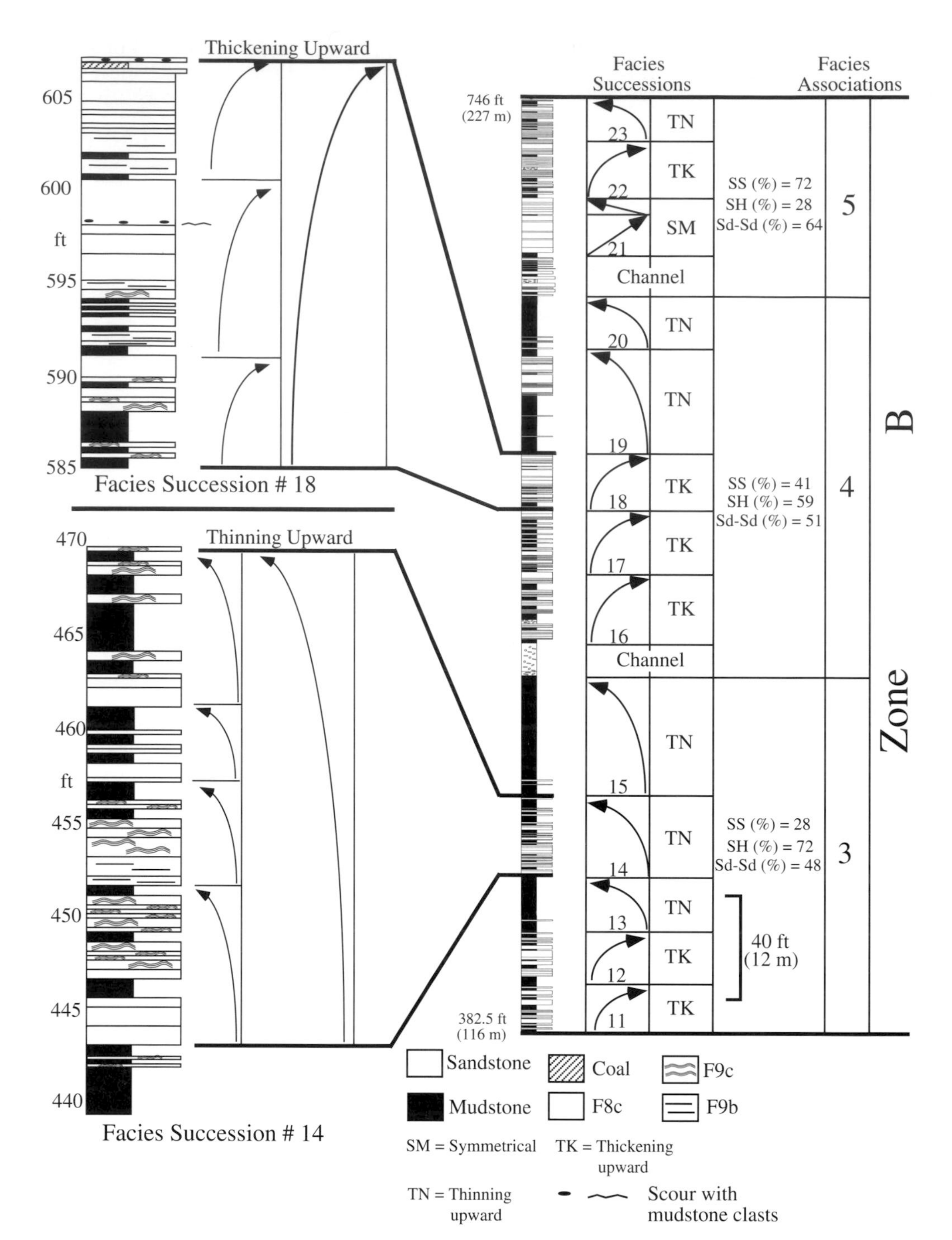

CONCLUSIONS

The upper Jackfork Formation in the DeGray Lake area is representative of a Type II turbidite depositional system. In this system, the following turbidites deposits are recognized: proximal lobes, distal lobes, and channel fills.

Proximal lobe deposits are dominated by facies F3b, F8a, F8b, and F8c. The percentage of sandstone ranges from 84 to 97%. Sand-on-sand contacts range from 72 to 88%. Proximal lobe facies associations are comprised of symmetrical and thinning-upward successions that exhibit unidirectional thinning. Sandstone-filled scours are prevalent in these deposits. Good bed scale correlations are attained among these lobes.

Distal lobes consist of facies F8c, F9b, F9c and F9cc. The percentage of sandstone in these lobes ranges from 36 to 51%. The percentage of sand-on-sand contacts is 53%. Distal lobe facies associations consist of thickening- and thinning-upward successions that display complex thickness variations induced by compensation style deposition. Scour-and-fill features are rare in these units. Moderate to poor bed scale correlations are attained among distal lobes.

Depending on their fill, channels occur in two types. Pebbly and massive sandstone-filled channels

Table 4. Correlations, at the facies succession scale, of distal lobe deposits of zone B over the DeGray Lake Spillway and Intake area. Alternating thinning directions are exhibited by successions #11–#20. In contrast, successions #21–#23 show unidirectional thinning to the west. The direction of thinning is inferred from the amount of sandstone present in a particular succession.

Direction of Thinning	(Basinward)		Sections			(Landward)	Succession
	Intake	F	C–C'	B–B'	A–A'	Spillway	
←	32.5 SS(55), SH(45)	22.5 SS(64), SH(36)	N/A	N/A	N/A	15.75 SS(62), SH(38)	23
←	27.5 SS(36), SH(64)	27 SS(49), SH(51)	N/A	N/A	N/A	23.25 SS(70), SH(30)	22
←	45.5 SS(40), SH(60)	31 SS(52), SH(48)	N/A	N/A	N/A	23.5 SS(89), SH(11)	21
←	19 SS(50), SH(50)	23.75 SS(63), SH(37)	N/A	N/A	N/A	25.5* SS(7), SH(93)	20
↔	56.5 SS(36), SH(64)	72.5 SS(40), SH(60)	N/A	N/A	N/A	37 SS(33), SH(67)	19
←	13 SS(4), SH(96)	18.75 SS(7), SH(93)	22.5 SS(78), SH(22)	N/A	N/A	22.5 SS(78), SH(22)	18
↔	N/A	50 SS(67), SH(33)	34.25* SS(28), SH(72)	N/A	N/A	24 SS(53), SH(47)	17
←	0	0	0	0	0	27.5 SS(55), SH(45)	16
↔	N/A	46.25 SS(97), SH(3)	4.75 SS(74), SH(26)	26 SH(100)	N/A	47.5 SS(2), SH(98)	15
←	N/A	8.75 SS(100)	N/A	28.5 SS(23), SH(77)	N/A	33.5 SS(60), SH(40)	14
→	N/A	N/A	N/A	35.5 SS(51), SH(49)	N/A	19 SS(3), SH(97)	13
→	N/A	N/A	N/A	30.5 SS(43), SH(57)	N/A	21 SS(55), SH(45)	12
←	N/A	N/A	N/A	N/A	17 SS(30), SH(70)	17.5 SS(53), SH(47)	11

Top number = thickness of successions (in feet); SS = sandstone percentage; SH = shale percentage.

are dominated by facies F5 and F8c. The average percentage of sandstone comprising this type of fill is 99%. Beds are highly amalgamated with a calculated sand-on-sand contact of 90%. This type of channel-fill facies association is exclusively composed of thinning-upward successions that display abrupt thickness variations. Pebbly sandstone-filled scours are prevalent in these deposits. Poor bed scale correlations are attained among these channels. The second type is composed of slump-fills consisting of amalgamated sandy and cohesive debris flow deposits.

The depositional setting for this turbidite system is divided into three areas: upper fan, transition zone, and middle to lower fan. Upper-fan environments are populated by large sandstone- and slump-filled channels. Slumped horizons can occur in the upper-fan region and may extend to/or beyond the transition zone area. Proximal lobes occur in the

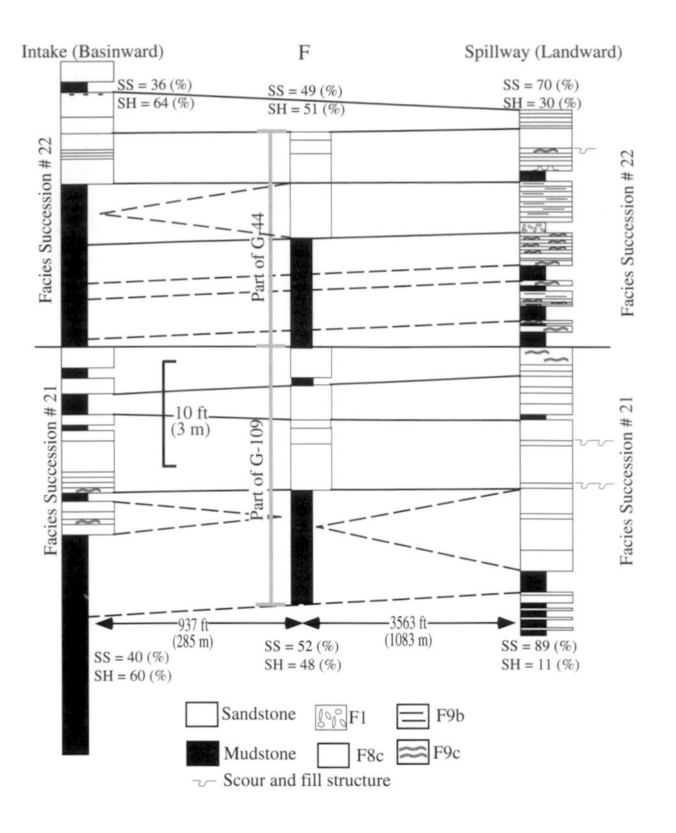

Figure 10—Correlations of symmetrical (#21) and thickening-upward (#22) facies successions, from zone B, over the DeGray Lake Spillway and Intake area. Bed scale correlations are moderate to poor.

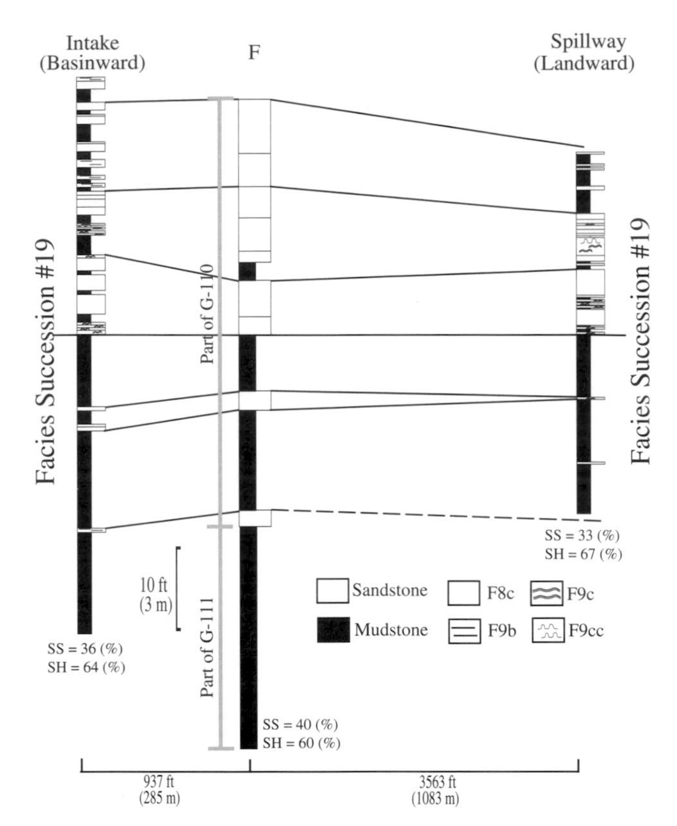

Figure 11—Correlations of thinning-upward facies succession #19 over the DeGray Lake Spillway and Intake area. Bed scale correlations are moderate to poor.

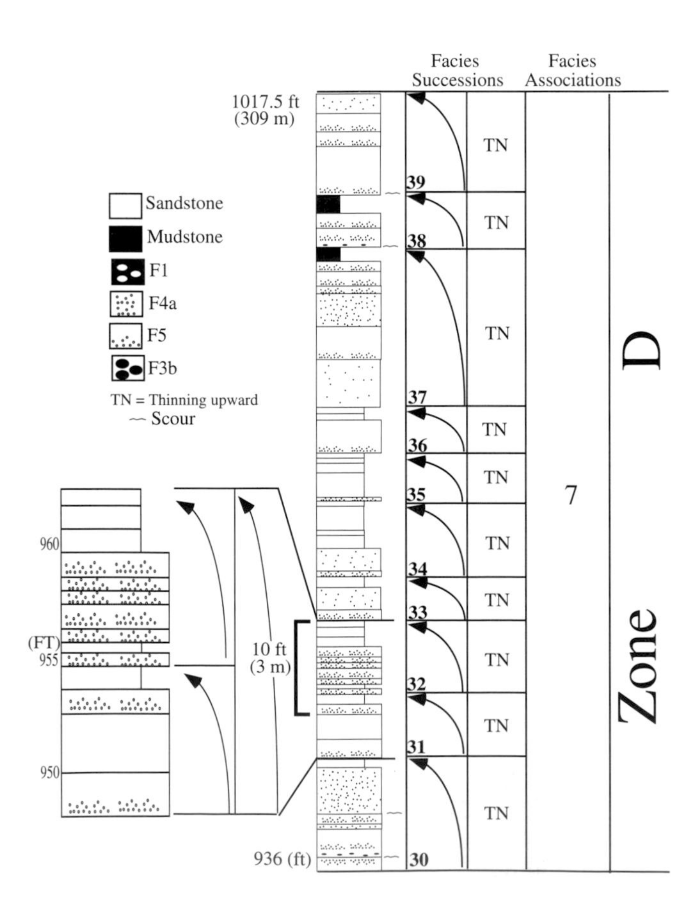

Figure 12—Zone D is composed of stacked thinning-upward facies successions that represent multiple filling events. Each succession records an upward transition from proximal to distal facies coupled with an upward decrease in bed thickness.

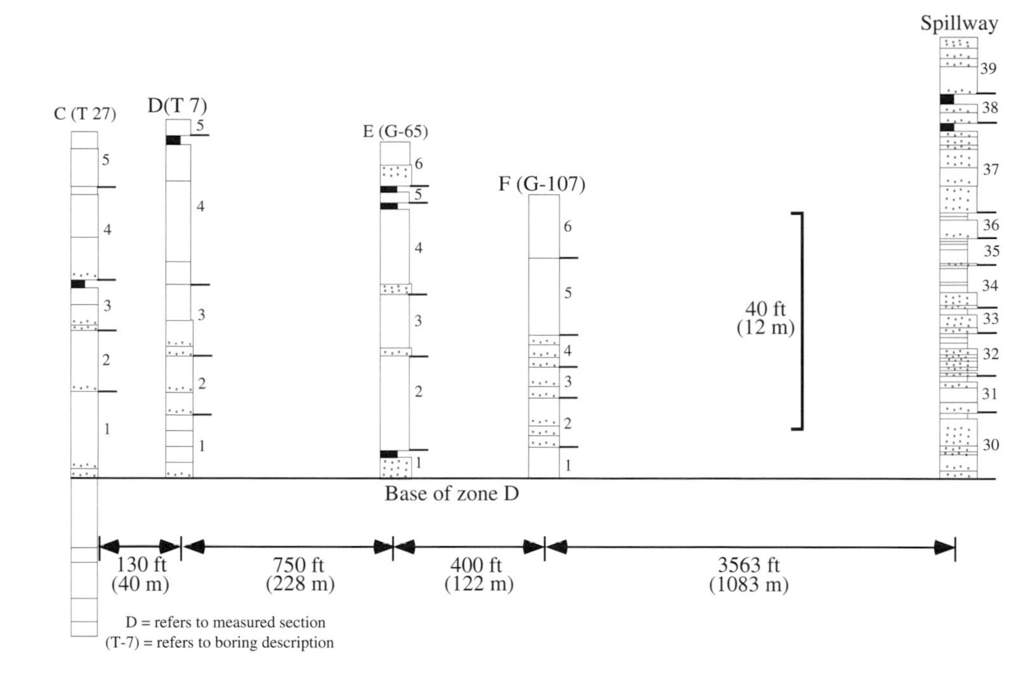

Figure 13—Correlations of facies successions comprising zone D, over the DeGray Lake Spillway and Intake area. Variations in thickness and sedimentological features make it very difficult to correlate at the bed scale.

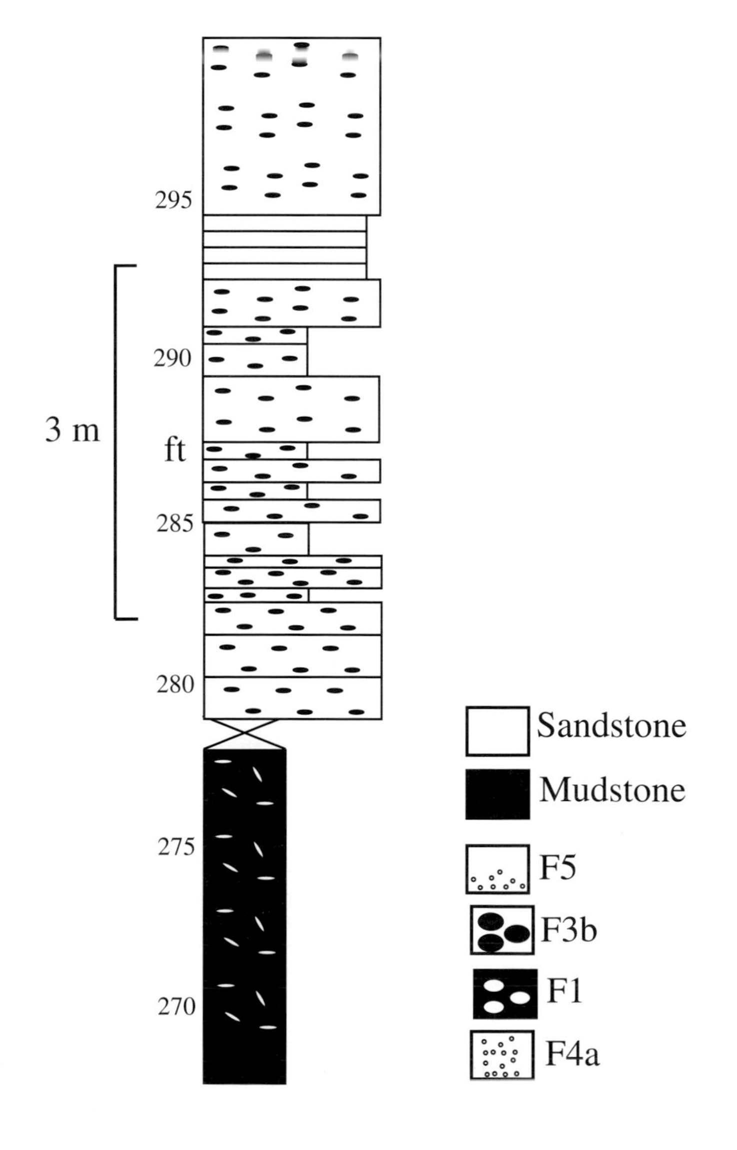

Figure 14—An example of a channel-fill that consists of sandy (F3b) and cohesive debris flow (F1) deposits with no discernible vertical facies trend. This example is part of the channel interval recognized in the Lakeview stratigraphic column that incises into facies association #2 in zone A (see Figure 2).

Table 5. Comparison of geological attributes of a Type II turbidite depositional system as documented by Mutti and Normark (1987, 1991) and the features recognized in upper Jackfork strata in the study area. Similarities between the two qualify the upper Jackfork as a Type II turbidite system.

Comparison	Type II Turbidite System	Upper Jackfork Strata
Turbidite deposits	Channels with depositional and mixed fills; lobes	Channels with sandstone- and slump-fills; proximal and distal lobes
Stratigraphy	Rapid downcurrent change from extensively scoured and thick-bedded sandstone facies into thinner-bedded sandstone facies.	Channel-fills that evolve to proximal lobes, which, in turn, evolve to distal lobes
Sandstone content	Poorly efficient = sandstone-rich	Total sandstone content is 71%* = sandstone-rich†
Sedimentological attributes	• Extensively scoured and amalgamated sandstones and pebbly sandstone facies • Out-sized ripup mudstone clasts • Mud-draped scours are not common • Lenticular and cross-stratified sandstone units are scarce • Associated lobes are less well developed	• Extensively scoured and amalgamated sandstones and pebbly sandstone facies • Out-sized ripup mudstone clasts • Sandstone- and pebbly sandstone-filled scours are common • Lenticular and cross-stratified sandstone units do not occure • Correlation of distal lobe deposits at the bed scale is moderate to poor
Tectonic setting	Associated with different forms of active margins: extensional, strike-slip, and contractional	Deposited during a contractional tectonic phase

*Sandstone content was calculated from the entire upper Jackfork interval.
†Classification is based on sandstone content according to Reading and Richards (1994).

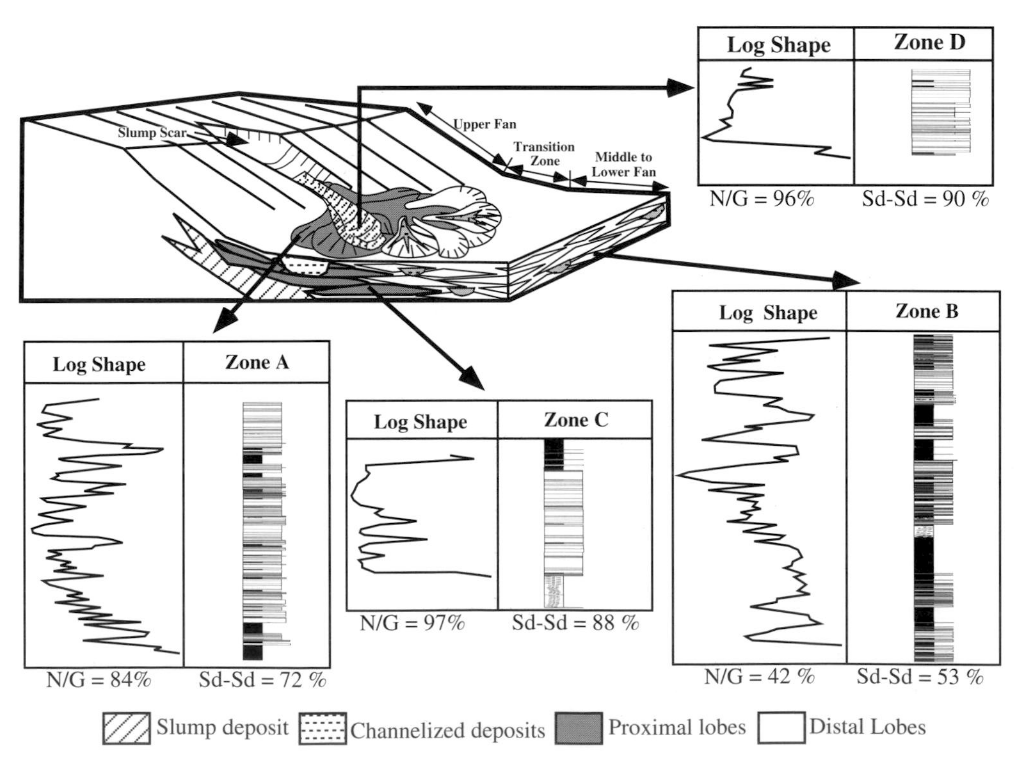

Figure 15—Depositional interpretations, well log signature (gamma-ray), net to gross, and percentage of sand-on-sand (Sd-Sd) contacts (percentages of sandstone and sand-on-sand contacts are from the Spillway section only) for the four zones recognized in the study area (gamma-ray profile modified from Jordan et al., 1993).

transition zone area. In contrast, distal lobes occur beyond the transition area or the middle- to lower-fan regions.

ACKNOWLEDGMENTS

I thank Roger M. Slatt, Arnold H. Bouma, Peter Osterloff, and two anonymous reviewers for critically reviewing this paper. Opinions expressed in this paper, however, are the sole responsibility of the author. I also thank Charles G. Stone for sharing his geologic knowledge of these rocks with me, and Brian C. Westfall who assisted in part of the fieldwork. Cindy Castle and Federico Cordovez assisted in the drafting. Funding was provided by Petroleum Development Oman LLC (PDO).

REFERENCES CITED

Al-Siyabi, H. A., 1998, Sedimentology and stratigraphy of the early Pennsylvanian upper Jackfork interval in the Caddo Valley Quadrangle, Clark and Hot Spring Counties, Arkansas: unpublished Ph.D. dissertation, Colorado School of Mines, Golden, Colorado, 272 p.

Blythe, A. E., A. Sugar, and S. P. Phipps, 1988, Structural profiles of the Ouachita Mountains, western Arkansas: AAPG Bulletin, v. 72, p. 810–819.

Chester, T. L., 1994, Analysis of vertical cyclicity patterns in two sediment gravity flow sequences: unpublished Master's thesis, Stanford University, 169 p.

Coleman, J. L., M. Van Swearingen, and C. E. Breckon, 1994, The Jackfork Formation of Arkansas: a test for the Walker–Mutti–Vail models for deep sea fan deposition: South-Central Section of the Geological Society of America Annual Meeting Field Trip Guidebook, 67 p.

Jordan, D. W., D. R. Lowe, R. M. Slatt, A. E. D'Agostino, M. H. Scheihing, R. H. Gillespie, and C. G. Stone, 1993, Scales of geological heterogeneity of Pennsylvanian Jackfork Group, Ouachita Mountains, Arkansas: application to field development and exploration for deepwater sandstones: Arkansas Geological Commission, Guidebook 93-1, 141 p.

Lowe, D. R., 1982, Sediment gravity flows, II: depositional models with special reference to the deposits of high density turbidity current: Journal of Sedimentary Petrology, v. 52, p. 279–297.

Middleton, G. V., and M. A. Hampton, 1973, Sediment gravity flows: mechanics of flow and deposition, *in* G. V. Middleton and A. H. Bouma, eds., Turbidites and deep-water sedimentation, SEPM Short Course 1, p. 1–38.

Morris, R. C., 1974a, Carboniferous rocks of the Ouachita Mountains, Arkansas: a study of facies patterns along the unstable slope and axis of a flysch trough, *in* G. Briggs, ed., Carboniferous of the southeastern United States: The Geological Society of America Special Paper 148, p. 241–279.

Morris, R. C., 1974b, Sedimentary and tectonic history of the Ouachita Mountains, *in* W. R. Dickinson, ed., Tectonics and sedimentation: SEPM Special Publication No. 22, p. 120–142.

Morris, R. C., 1977, Flysch facies of the Ouachita trough—with examples from the Spillway at DeGray Dam, Arkansas, *in* C. G. Stone, ed., Symposium on the geology of the Ouachita Mountains: Arkansas Geological Commission, v. 1, p. 158–169.

Mutti, E., 1985, Turbidite systems and their relations to depositional sequences, *in* G. G. Zuffa, ed., Provenance of arenites: NATO-ASI Series, Dordrecht, Reidel, p. 65–93.

Mutti, E., and W.R. Normark, 1987, Comparing examples of modern and ancient turbidite systems: Problems and concepts, *in* Leggett, J.K., and G.G. Zuffa, eds., Marine clastic sedimentology: Concepts and case studies, London, Graham & Trotman, p. 1–38.

Mutti, E., and W. R. Normark, 1991, An integrated approach to the study of turbidite systems: *in* P. Weimer, and M. H. Link, eds., Seismic facies and sedimentary processes of submarine fans and turbidite systems: New York, Springer–Verlag, p. 75–106.

Mutti, E., 1992, Turbidite sandstones, Agip S.p.A., Milan, 275 p.

Mutti, E., D. Giancarlo, M. Stefano, and P. Lorenzo, 1994, Internal stacking patterns of ancient turbidite systems from collisional basins: *in* P. A. Weimer, H. Bouma, and B. F. Perkins, eds., Submarine fans and turbidite systems: sequence stratigraphy, reservoir architecture, and production characteristics, Gulf of Mexico and international: Gulf Coast Section SEPM Foundation 15th Annual Research Conference, p. 257–268.

Reading, H. G., and M. Richards, 1994, Turbidite system in deep-water basin margins classified by grain size and feeder system: AAPG Bulletin, v. 78, p. 792–822.

Shanmugam, G., and R. G. Moiola, 1995, Reinterpretation of the depositional processes in the classic flysch sequence (Pennsylvanian Jackfork Group), Ouachita Mountains, Arkansas and Oklahoma: AAPG Bulletin, v. 79, p. 672–695.

Slatt, R. M., P. Weimer, and C. G. Stone, 1997, Reinterpretation of the depositional processes in the classic flysch sequence (Pennsylvanian Jackfork Group), Ouachita Mountains, Arkansas and Oklahoma: Discussion: AAPG Bulletin, v. 81, p. 449–459.

U.S. Army Corps of Engineers, 1971, DeGray Dam and Dike-Caddo River, Clark County, Arkansas, DeGray Dam Foundation Report, 42 p.

Viele, G. W., and W. A. Thomas, 1989, Tectonic synthesis of the Ouachita orogenic belt, *in* R. D. Hatcher, W. A. Thomas, and G. W. Viele, eds., The Appalachian-Ouachita orogen in the United States: Geological Society of America, The Geology of North America, v. F2, p. 695–728.

Walker, R. G., 1978, Deep-water sandstone facies and ancient submarine fans: models for exploration for stratigraphic traps: AAPG Bulletin, v.62, p.932–966.

Walker, R. G., 1992, Facies, facies models and modern stratigraphic concepts, *in* R. G. Walker and N. P. James, eds., Facies models, response to sea level change: Geological Association of Canada, p. 1–14.

Basu, D., and A. H. Bouma, 2000, Thin-bedded turbidites of the Tanqua Karoo: physical and depositional characteristics, *in* A. H. Bouma and C. G. Stone, eds., Fine-grained turbidite systems, AAPG Memoir 72/SEPM Special Publication No. 68, p. 263–278.

Chapter 23

Thin-Bedded Turbidites of the Tanqua Karoo: Physical and Depositional Characteristics

Debnath Basu
Schlumberger GeoQuest
Muscat, Sultanate of Oman

Arnold H. Bouma
Louisiana State University
Baton Rouge, Louisiana, U.S.A.

ABSTRACT

Thin-bedded turbidites (TBTs, 5–60 cm in bed thickness) constitute a common lithofacies in fine grained deepwater clastic environments. Field and laboratory data from the present study reveal systematic differences between the various deepwater depositional settings. Grain-size and bed-thickness distribution in the TBTs are the result of varying flow velocity and sediment load. The finest grain-size and thinnest beds are developed in the midfan levee-overbank and midfan passive channel-fill deposits. The coarsest grains and thickest beds are observed in the distal sheet sandstones. The data indicate bedform type and bedform stacking, in several scales, to vary with subfacies. Levee-overbank and passive channel-fill TBTs are dominated by base truncated Bouma Sequences (T_c, T_{cd}, T_{cde}), but channel-sheet transition and distal sheet deposits are characterized by top-truncated Bouma Sequences (T_a, T_{ab}, T_{abc}). The TBTs were deposited as a result of "bodyspill" from channelized turbidity currents or as deposits from the lagging tails of degenerating turbidity currents.

INTRODUCTION

This study is a detailed description and documentation of thin-bedded deposits from outcrops with the purpose of differentiating thin-bedded turbidites (TBTs) developed in various subenvironments. The deposits comprise the upper midfan channel levee-overbank, midfan passive channel fill, lower midfan channel-sheet transition, and lower fan distal sheet environments of the fine-grained submarine fans in the Tanqua Karoo. Thin-bedded turbidites have been neglected in terms of detailed scientific analyses primarily due to their subtle physical characteristics and partly due to the general belief about their lacking economic potential. Conventional petrophysical and geophysical data cannot resolve their internal architecture, and extensive outcrops for analog studies are rare. Recently with the advent of high-resolution electrical logging techniques (borehole images and high-resolution log data), TBTs have been demonstrated to be economic in a variety of deepwater basins around the world. Proper characterization and delineation of these deposits require high-resolution log and seismic interpretations as well as detailed core analyses and analog studies. The Tanqua Karoo deposits (Figure 1) were selected because they are tectonically undeformed and traceable for multikilometers. This subbasin has well-developed TBTs from

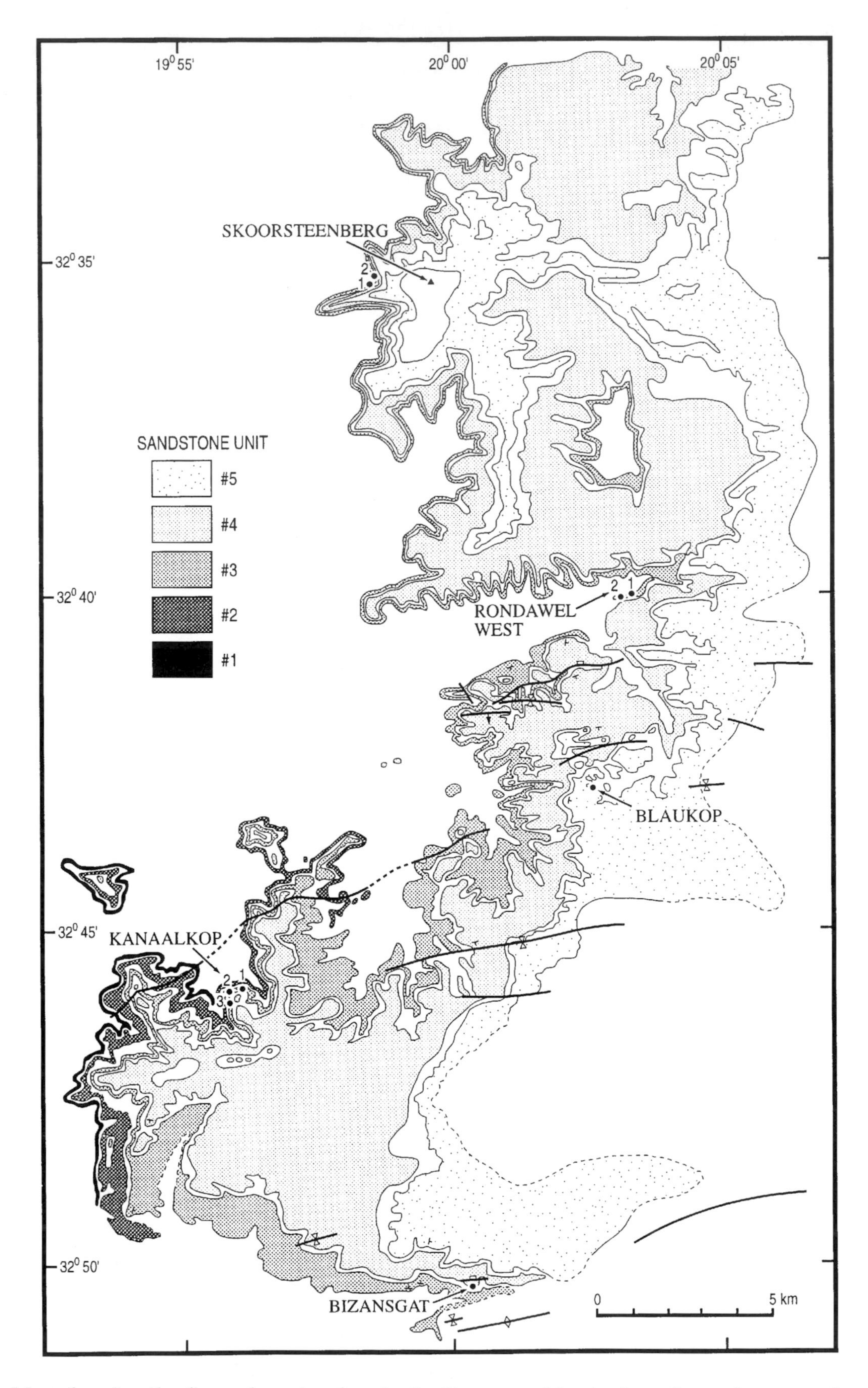

Figure 1—Map showing the five submarine fans in the Tanqua subbasin. Locations of measured profiles are indicated, Kanaalkop (Fan 3 midfan), Skoorsteenberg (Fan 3 lower fan), Rondawel West (Fan 3 lower midfan), Blaukop (Fan 5 midfan), and Bizansgat (Fan 4 midfan).

a suite of depositional elements allowing detailed facies analysis (Wickens and Bouma, this volume; refer to this paper for general information about the Tanqua Karoo turbidite system).

A multipronged approach involved fieldwork documenting variations and relationships between different depositional facies, detailed measurement of sedimentary profiles, paleocurrent analysis, measurement of gamma-ray profiles, and collection of samples for laboratory analyses. Laboratory analyses entailed grain-size and mineralogical studies, grain-orientation analysis using anisotropy of isothermal remanent magnetization (IRMA), and X-ray diffraction analyses on extracted clay fractions. Attributes that pertain to the sedimentary characteristics of the TBTs, within the context of the genetic and temporal evolution of a submarine fan, are presented in this paper. The TBTs from the various settings are characterized by diverse physical attributes (grain size, grain orientation, sedimentary structures, bed-thickness distributions, and relationship with associated facies), which can be related to their respective depositional styles.

The Tanqua submarine fan complex is composed of five submarine fan units. The outcrop locations on which this study is based are indicated in Figure 1. The fans are composed of siltstone and fine sandstone, and are separated stratigraphically by basinal shales and silty shales (Bouma and Wickens, 1994). The five fans reflect characteristics of middle and lower fan settings (see Bouma, this volume).

GENERAL UNDERSTANDING OF THIN-BEDDED TURBIDITES

Thin-bedded turbidites comprise a common facies in submarine fan deposits in the Recent and ancient geological record in both active and passive margins, but are more extensively developed in the latter (Bouma et al., 1985). Outcrop studies on TBTs include the Ordovician Cloridorm and Tourelle formations (Quebec; Hiscott, 1980), Cambrian St. Roch Formation, Gaspe Peninsula (Quebec; Strong and Walker, 1981), Precambrian Kongsfjord Formation, (Norway; Pickering, 1982), and Pennsylvanian Jackfork Formation, Arkansas (Jordan et al., 1991). Similar thin-bedded analogs from the Recent and Tertiary submarine fans are commonly encountered from deepwater reservoirs in the Gulf of Mexico, North Sea, Indonesia, Alaska, and West Africa. From the subsurface these TBT deposits are commonly referred to as low-resistivity, low-contrast (LRLC) thin-bedded sands in the Gulf of Mexico (Darling and Sneider, 1992). Examples of well-documented TBTs from the Gulf of Mexico include the Ram Powell, Spirit, and Tahoe prospects (Viosca Knoll; Shew et al., 1994) Einstein prospect (DeSoto Canyon; Hackbarth and Shew, 1994), and Auger prospect (Garden Banks; McGee et al., 1994). Well-developed TBTs from the flank of fan lobes and overbank complexes from the Mississippi Fan are reported from the Leg 96 DSDP efforts (Bouma et al., 1985). Thin-bedded turbidites are extensively developed in the outer Bengal Fan as reported from ODP Leg 116 (Stow et al., 1989). These studies from the Recent and the ancient record addressed general issues such as the occurrence of TBTs as part of a regional study, as well as applied studies on lateral continuity and reservoir characterization. This paper highlights the characteristics of TBTs from the various depositional subenvironments displaying these deposits and documents properties useful for differentiating between these subfacies.

We follow the McKee and Weir (1953) bed-thickness classification scheme, in which 1- to 5-cm-thick strata are classified as very thin bedded, 5- to -60-cm-thick strata are thin bedded, and 60- to 120-cm-thick strata are thick bedded. In the Tanqua Karoo system, the very thin beds (1–5 cm) and thin beds (5–60 cm) are commonly associated with each other. It is to be noted that the very thin beds and thin beds less than 30 cm are below the resolution of standard logging tools.

Bouma et al. (1995) introduced a submarine fan depositional model that is applicable to systems with a low sand/shale ratio (Bouma, this volume; (Figure 1) (CD Figure 1) like the Tanqua Karoo. Depositional settings that develop under relative tectonic quiescence representing or mimicking a passive margin are better described using this nomenclature.

DESCRIPTION AND INTERPRETATION OF MEASURED SECTIONS

Specific localities in the Tanqua Karoo deposits were selected where thin-bedded intervals are well developed and exhibit clear relationships with associated facies. Five sites were selected (Figure 1), from which nine sections presented in this paper comprise the following subenvironments: (1) levee-overbank in an upper midfan setting (Fan 3, Kanaalkop); (2) distal sheet sandstone (depositional lobes) in a lower-fan setting (Fan 3, Skoorsteenberg); (3) channel-sheet transition deposits in a mid- to lower-fan setting (Fan 3, Rondawel West); (4) levee-overbank deposits in a lower midfan setting (Fan 5, Blaukop); and (5) passive channel fill in a midfan setting (Fan 4, Bizansgat).

Interpretations are based on observations made on grain-size variations, sedimentary structures, paleocurrent patterns, bed geometry, gamma-ray profiles, and relationship with associated deposits. These observations enabled division of the sections into genetic intervals.

Kanaalkop (Fan 3)

Profiles KP 1, KP 2, and KP 3 are measured sections of Fan 3 at Kanaalkop (Figure 1), which are 48, 37 and 39 m thick, respectively. A partial view of the outcrop is presented in Figure 2 (CD Figure 2). These measured

Figure 2—An overview of the central part of the north face of the Kanaalkop (KP) outcrop with the locations of measured sections KP 1 (left, east) and KP 2 (right, west). The "Kanaalkop Channel" is exposed at the top with a conspicuous erosional base. The channel has a maximum thickness of 20 m, as indicated by the scale bar.

profiles exhibit a wide array of bed thickness and sedimentary structures but are fairly restricted in grain-size variations (Figure 3). The salient physical and depositional characters of these profiles are described as follows.

Base to Level A

The siltstones and silty shales are characterized by very thin and thin beds (~3–8 cm) showing parallel and ripple cross-laminations (Figure 3). The ripples indicate an easterly paleocurrrent direction. They are interpreted to represent thin, base-truncated Bouma Sequences (T_{cd}) having been deposited in an overbank setting as spillovers from minor flows from a channel. This channel is inferred to have existed to the west of Kanaalkop and is not present in outcrop ("West Channel," Figure 4). The easterly directed paleocurrent flow in this interval, considered within the regionally established N–NE directed downfan direction, supports this interpretation. This is considered to be part of a levee-overbank complex referred in its entirety as levee-overbank complex (LOC 1) in Figures 3 and 5.

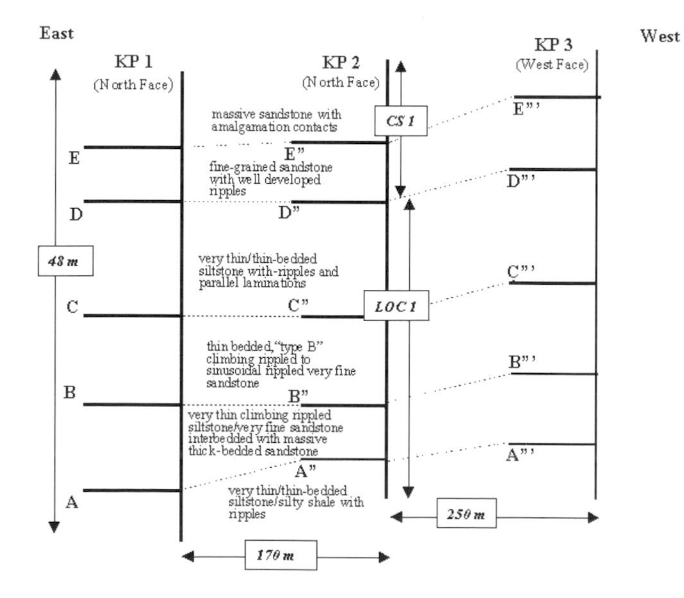

Figure 3—Summary diagram demonstrating the depositional characteristics of Fan 3 at Kanaalkop (KP). The vertical profiles are approximately 35–50 m thick. See Figure 2 and text for details.

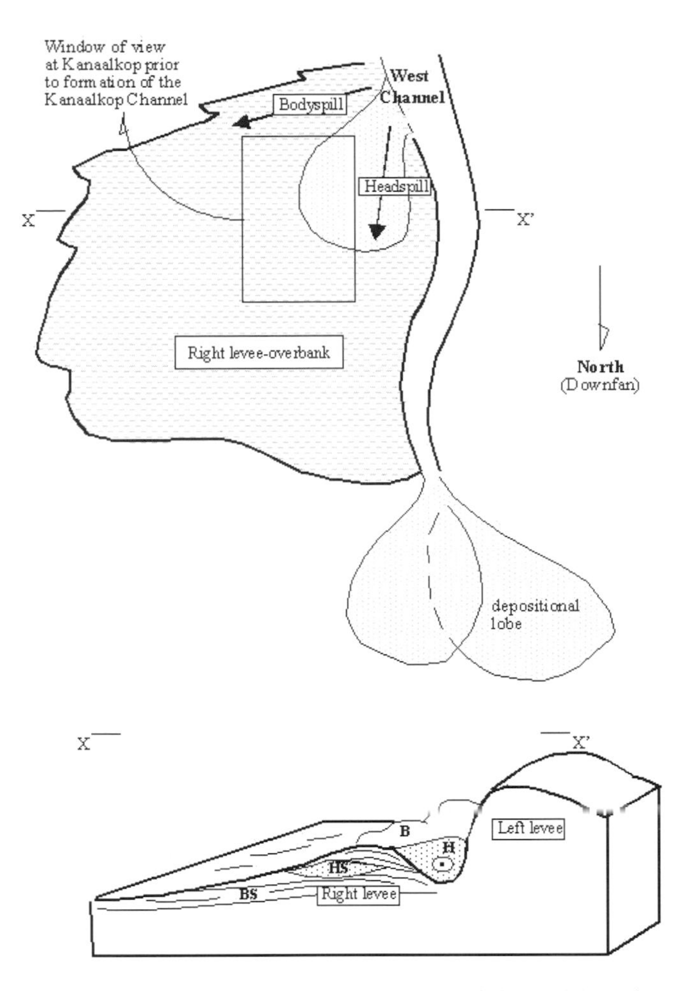

Figure 4—Schematic representation of depositional regime in the Kanaalkop area showing its position on the right, eastern levee-overbank complex (LOC 1, in text) with respect to the West Channel. Map and cross-sectional view demonstrates the style of deposition of the headspill and bodyspill deposits (H = head, B = body, HS = headspill, and BS = bodyspill from turbidity currents). The paleocurrent direction in the cross-sectional representation of the channel is toward the viewer.

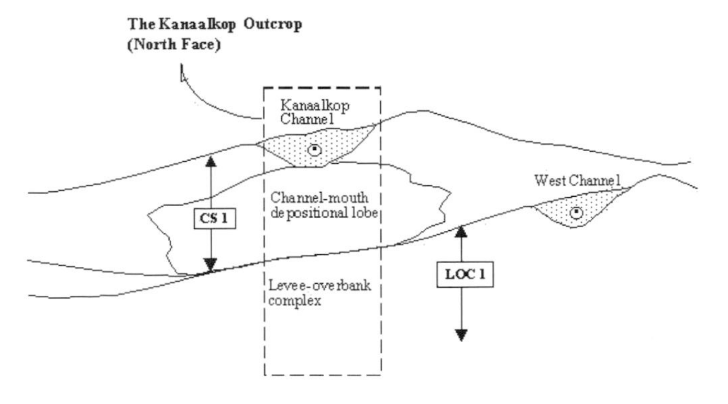

Figure 5—The inferred spatial relationship between the older levee-overbank complex (LOC 1) and the younger channel system (CS 1). The window of view of the Kanaalkop outcrop is represented. Paleoflow in the channels is toward the viewer.

Interval A to B

This interval is composed of 10–40 cm thin beds interbedded with 50–130 cm thick beds (Figure 3), often with very thin beds (15 cm) in between. The 10–40 cm thin beds are characterized by very fine-grained sandstone fining upward to siltstone. They have Type A and B climbing ripple laminations (Type B dominant), gradational to sinusoidal ripple laminations as defined by Jopling and Walker (1968), representing T_c beds. The paleocurrent direction from the ripples is E–ESE. The less numerous, 50–130 cm thick beds, are very fine-grained sandstones that represent T_{ac} beds (thickness of $T_a >> T_c$). Groove marks indicate a NE–SW flow azimuth. These thicker beds (50–130 cm) do not show any evidence of erosion at their base and are conformable with the interbedded thinner (10-40 cm) units. The thick (50–130 cm), predominantly massive beds, often pinch out laterally within the extents of the outcrop (~1 km). The variations in sedimentary structures, paleocurrent pattern, and thickness indicate that the two groups, with 10–40 cm and 50–130 cm beds, had considerably different depositional styles. A "headspill" vs. a "bodyspill" mechanism from channelized turbidity currents is invoked to explain the fewer massive, thicker beds (50–130 cm) interbedded with the more numerous, climbing ripple laminated thinner (10–40 cm) beds (Figures 3, 4). The appearance of headspill and bodyspill deposits is shown in (CD Figure 3). The 10–40 cm beds, with Type A and B ripple-drift cross-laminations and sinusoidal aggradational ripple laminations (CD Figure 4), were deposited from spillover flows with very high suspended-load to bed-load ratio. It is suggested that these 10–40 cm beds are lateral spillovers from the turbulent suspended sediment cloud that rides as part of the body, above and behind the coarser-grained head of a turbidity current flowing downfan along a channel (West Channel; Figures 4, 5). Deposition sets in because of flow expansion once the body spills across the levee-crest. In contrast, the 50- to 130-cm T_{a-c} beds (T_a thickness >> T_c; T_b is lacking or not observable) are possibly spillovers from larger flows when the confines of the channel fail to restrict the upper reaches of the head of turbidity currents at bank-full stages (Figure 4). Once across the levee crest, deposition occurs from these headspill flows by a flow-freezing mechanism (Lowe, 1982) resulting in the T_a division. The thin T_c divisions at the top are the result of deposition from the lagging tail of the same spillover current, as well as possible reworking of the newly deposited bed by current drifts. This interpretation is supported by the angular divergence in the direction of paleoflow, eastward from the thinner (10–40 cm) beds with climbing ripples and northeastward from the thicker (50–130 cm) massive beds with sole marks (Figure 4). The headspills, due to their higher state of inertia, emanate into the overbank area at a shallower angle (30°–45°, headed NE) with respect to the sourcing channel (headed N–NE) than the body-spill deposits (80°–90°, headed E; Figure 4). This interval is part of the same levee-overbank complex (LOC 1, Figures 3, 5) as the first interval.

Interval B to C

The western section KP 3 (Figure 1) is characterized by a thinning upward trend (20–40 cm grading to 10–25 cm) and the correlatable eastern sections KP 1 and KP 2 display a thickening upward trend (10–20 cm to 30–45 cm). The reversal of thickness patterns in the correlated intervals from west to east is possibly due to depositional topography leading to compensation in a levee-overbank setting. This interval (B to C) is also part of the same levee-overbank complex (LOC 1). Presence of Type B ripple drift cross-laminations and sinusoidal ripples indicate an easterly paleocurrent. This interval also suggests a bodyspill mode of deposition from the suspended sediment cloud in channelized flows from the inferred channel located to the west (West Channel, Figure 5). The high suspended-load to bed-load ratio of the density flows form the Type B climbing ripples grading toward sinusoidal (aggradational) climbing ripple laminations (see CD-ROM Figure 4).

Interval C to D

This interval is composed of very thin (1–5 cm) siltstone beds characterized by ripples and occasional parallel lamination (Figure 3). The ripples indicate an eastward paleocurrent. This interval is interpreted to represent the abandonment of the levee-overbank complex LOC 1 (Figures 3, 5), which also comprises the intervals previously described. The transition from active levee-overbank to a passive overbank fill probably records an upfan channel avulsion in the sourcing West Channel, rendering a shift in the local depocenter (Figure 5).

Interval D-E

Datum E corresponds to the erosional base of the overlying "Kanaalkop Channel" (Figure 2, CD-ROM Figure 2). The Kanaalkop Channel is thicker in KP 1 than in KP 2 and KP 3 because of erosional relief at its base. This interval is characterized by poorly developed bedding and is composed of very-fine to fine-grained sandstone that have well-developed ripples, with rare climbing ripples (Type A, B, and sinusoidal ripples). Paleocurrent directions are to the NE (N30°–45°E) in KP 1 and NW (N25°–65°W) in KP 2 and KP 3, suggesting an approximately symmetrical and radially diverging flow with respect to the overlying Kanaalkop channel axis (N–NE). The presence of current ripples (suggesting a lower suspended-load to bed-load ratio, contrasting with stratigraphically underlying deposits) and poorly developed bedding indicate that these beds formed in a more active depositional setting than the underlying units. The Kanaalkop Channel is stratigraphically younger and located farther east of the inferred West Channel (Figure 5). The latter was the westerly source of the levee-overbank complex sediments LOC 1, up to level D (Figures 3, 5). The location of the overlying Kanaalkop Channel was most likely prompted by an avulsion process in the inferred lower channel (the West Channel), farther upfan than the Kanaalkop location. The avulsion process initially led to the deposition of the channel terminus depositional-lobe (interval DE, lower part of CS 1 in Figures 3, 5) to the east of the preexisting West Channel. The younger Kanaalkop Channel subsequently entrenched itself into its own depositional lobe and later filled up the erosional cut with a compound fill (Figures 2, 5). The spatial offset of the Kanaalkop Channel course with respect to the West Channel is most likely guided by postdepositional differential compaction in the older channel/levee-overbank complex (LOC 1). The preexisting course of the West Channel effectively became a sandy submarine ridge due to differential compaction, and the younger Kanaalkop Channel levee-overbank lens onlapped onto the flanks of the older levee-overbank wedge (LOC 1; Figure 5). The characteristics of the TBTs in this locality are summarized in Table 1.

The Paleoposition of South Africa and the Kanaalkop Deposits

The common occurrence of the 50- to 130-cm-thick beds in a levee-overbank setting requires some justification because it is an exception to a general observation from well-documented modern fans.

The paleoposition of South Africa in the Permian was in high latitudes, ~65°S (Smith et al., 1973). It has been observed that late Tertiary to Recent submarine fan channels in high latitudes (>45°N or S) have a profound difference in levee heights at any transverse channel profile (Hesse and Rakofsky, 1992; Carter and Carter, 1988). The levee asymmetry is attributed to the Coriolis Effect. Right-hand levees are higher in the Northern Hemisphere due to a pull in sediment gravity flows to the right. In the Northern Hemisphere this results in preferential aggradation of right levees with fine-grained sediments from the body of channelized flows. The North Atlantic Mid-Ocean Channel (NAMOC), in the Laurentian Fan, has levee asymmetries up to 90 m (Hesse and Rakofsky, 1992). The converse is true in the Southern Hemisphere: the right-hand levees are lower. In the channels in the Otago Fan Complex, offshore New Zealand, the left levees are higher than their right-hand counterparts (Carter and Carter, 1988).

Coriolis Force = $2\Omega\mu \sin\Phi$, where Ω is the angular velocity of the earth's rotation, μ the velocity of flow, and Φ is latitude. Here Ω is a constant and μ has a

Table 1. Physical and depositional characteristics of thin-bedded turbidites (TBT) from different depositional subenvironments of the Tanqua Karoo turbidite system.

Depositional Environments	Bed Thickness	Grain Size	Sedimentary Structures	Depositional Mechanism
Kanaalkop (upper/midfan levee-overbank)	Bimodal ~<40 cm ~50–130 cm	Coarse siltstone ~43 μm mean	~<40 cm: ripple-drift cross-laminations (Type B), T_c beds; ~50–130 cm: massive (T_a), very thin T_c tops	~<40 cm: bodyspill from channelized turbidity currents; ~50–130 cm: headspill from channelized turbidity currents
Bizansgat (midfan passive channel fill)	1–12 cm	Medium siltstone ~31.5 μm mean	Lower plane bed parallel and rare current ripple laminations with nonlaminated silty shale tops	Base truncated Bouma Sequence ~weak turbidity currents
Blaukop (midfan levee-overbank)	Trimodal ~<9 cm 10–15 cm 35–45 cm	Medium siltstone ~40 μm mean	~<9 cm & 10–15 cm: ripple cross-laminations; ~10–45 cm: climbing ripple laminations (Type C)	Suspended sediment spillover from channelized flows
Rondawel West (lower midfan channel-sheet transition)	Trimodal ~<15 cm 2–25 cm 30–40 cm	Coarse siltstone vfL* sandstone ~56 μm mean	~<25 cm: T_{ab}, T_{abc} beds; ~30–40 cm: T_a, T_{ab} beds; large flutes, ripup clasts, large scours	Deposition at the transition of confined and unconfined flows
Skoorsteenberg (lower fan distal sheet sandstone)	Trimodal ~<9 cm, 10–15 cm 35–45 cm	Coarse siltstone, vfL* sandstone ~70 μm mean	~<9 cm: T_{cd}; ~10–45 cm: T_{ab}, T_{abc} beds	Deposition from fully unconfined flows

*vfL = very fine L.

restricted range of values in submarine channels; hence, the high paleolatitudinal position (~65°S) of the submarine fan complex of the Tanqua Karoo is likely to have a profound impact on channelized flows due to Coriolis Effect. The position of the Kanaalkop outcrop is to the right-hand (lower levee) side with respect to the north directed sourcing channel (West Channel, Figures 4, 5). This generalized set of information supports the previous discussion regarding the deposition in the Kanaalkop area (up to level D in Figure 3) as being characterized by frequent bodyspill and occasional headspills across the right-hand lower levee (CD Figure 3). This happened when the body and head of flows exceeded the levee relief at bankfull stages. The higher left-handed levee of the West Channel would have acted as a barrier to spillover, especially from the heads of turbidity currents. The lower right-handed levee (LOC 1 in the eastern levee, Figures 4, 5), on the contrary, is the reason not only for the relative easy spillover of sediment clouds from the body of the flows but also the occasional headspills. Sand beds of similar thickness (100+ cm) have been reported from overbank sites across the lower left levee in the NAMOC setting (being in the Northern Hemisphere), but similar sand bodies have not been found in the right overbank across the higher levee (Hesse et al., 1996).

A detailed analysis of medium- to thick-bedded sandstones in a levee-overbank setting is necessary to evaluate their depositional affinities. Conventional interpretation of these massive sandstones is channeling in some form, but the full spectrum of variability in processes revealed through the study of modern fans (Bouma et al., 1985; Piper and Stow et al., 1991; Hesse et al., 1996) warrant more than such casual interpretation from outcrops.

Skoorsteenberg (Fan 3)

Two sections were measured from Fan 3 at Skoorsteenberg (SK 1 and SK 2), separated by 200 m. Profile SK 1 is located on the depositional strike section and SK 2 on the depositional dip section of Fan 3 (Figure 1). A summary diagram with tie-lines delineating genetic intervals is shown in Figure 6 and described as follows (CD Figure 5).

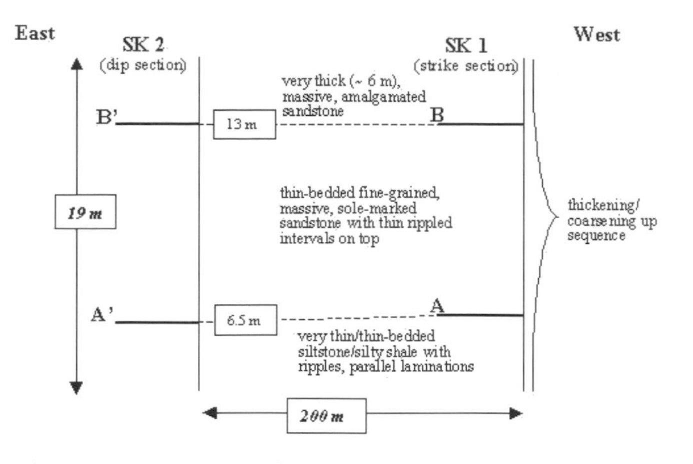

Figure 6—Summary diagram demonstrating the depositional characteristics of Fan 3 at Skoorsteenberg (SK). The measured profiles SK 1 and SK 2 are approximately 19 m thick.

Interval Base to A

This interval is composed of very thin siltstone and silty-shale beds that are 0.5–5 cm thick (Figure 6) characterized by parallel lamination and ripple cross-lamination (Bouma T_{cde} and T_{cd} divisions). These very thin beds were deposited by minor, dilute turbidity currents at the distal end of the fan system. These dilute turbidity flows were the local precursors of major ensuing flows.

Interval A to B

This interval is characterized by thin-bedded very fine-grained sandstone (8–30 cm, avg. 20 cm) that are flat-based and tabular in geometry (Figure 6) (CD Figure 6) representing T_{ab} or T_{abc} divisions with dominant T_a units. The bases of these beds commonly exhibit groove marks with an azimuth of NNE–SSW. Interbedded with these beds are rare thin-bedded (4–10 cm) siltstone units characterized by small ripup clasts, fine degraded plant debris, and convolute laminations. These represent debris-flow deposits or "slurry beds" (Morris, 1973). They could have originated upfan or even farther updip indicated by the common presence of plant matter. Alternatively, the debris flows could have a more immediate source, for example, resedimentation of fine-grained material derived from levee instabilities in the midfan. The observed features pointing to the turbidite origin for fine-grained sandstones and debris-flow origin for the interbedded convoluted siltstones are similar to observed facies in depositional lobes of the outer Mississippi Fan (Schwab et al., 1996).

The sudden increase in bed thickness and grain size from datum A to datum B is most likely in response to a variety of possible forcing mechanisms. A progradational shift in deposition farther to this downfan depocenter is likely. This possibly resulted from major sediment influx related to renewed slope instabilities or a further fall in relative sea level. Upfan channel avulsion, favoring the onset of active deposition in this locality, is also a possibility. That this thin-bedded interval may be the marginal or lateral equivalent of thicker beds developed at the axis of depositional lobes represents another possible scenario. The thickening upward could then be due to depositional topography resulting in compensation style lateral switching of lobes (Mutti and Sonnino, 1981; Bouma et al., 1995).

Interval B to the Top

The uppermost interval is characterized by a single, very thick, amalgamated, fine-grained sandstone bed (Figure 6). This bed is 6 m thick and is devoid of any observable sedimentary structures. The base of this bed is sharp and slightly undulatory but does not show any evidence of erosion. This sudden increase in bed thickness is interpreted to be in response to continued downfan progradation or sudden lateral shift of Fan 3. This bed possibly represents a sandy debris flow. For the summarized characteristics of the TBTs in this locality, see Table 1.

The Skoorsteenberg Outcrop in a Regional Perspective

The characteristics of the uppermost part of Fan 3 at Kanaalkop (upper midfan, Figures 2, 5) and at Skoorsteenberg (lower fan, Figure 6), (CD Figure 5) points to an interesting correlation. The top of Fan 3 at Kanaalkop is marked by the erosional Kanaalkop Channel and its compound fill. At Skoorsteenberg the top of Fan 3 is characterized by a 6-m-thick amalgamated sandstone sheet. It appears that the flows that caused the scoured base of the Kanaalkop Channel could have transported sediments that built the uppermost 6-m-thick massive bed at Skoorsteenberg. The midfan erosional surface at Kanaalkop could then be the time-correlative channel-cut that effectively remained as a sediment bypass horizon in an upfan location for an extended period of time while the 6-m-thick amalgamated bed was deposited at Skoorsteenberg in a distal setting. This temporal and spatial relationship is depicted in Figure 7. The bulk of the sediments passing down the Kanaalkop Channel conduit (*erosional vacuity*) is interpreted here to be deposited in the upper reaches of Fan 3 in the Skoorsteenberg area seen as the 6-m thick amalgamated bed (*correlative deposit*). Most of the well-documented modern fans represent this style of deposition, sequestering coarser sediments in the distal depositional lobes during the initial progradational phase. The sediments get progressively finer in an upfan direction with consequent backstepping of the system. Presence of 35–50% sand in the lower fan depositional lobes of the Mississippi Fan as compared to 4–12% in the middle and lower fan channel and channel-levees (Stow et al., 1985) is a typical example.

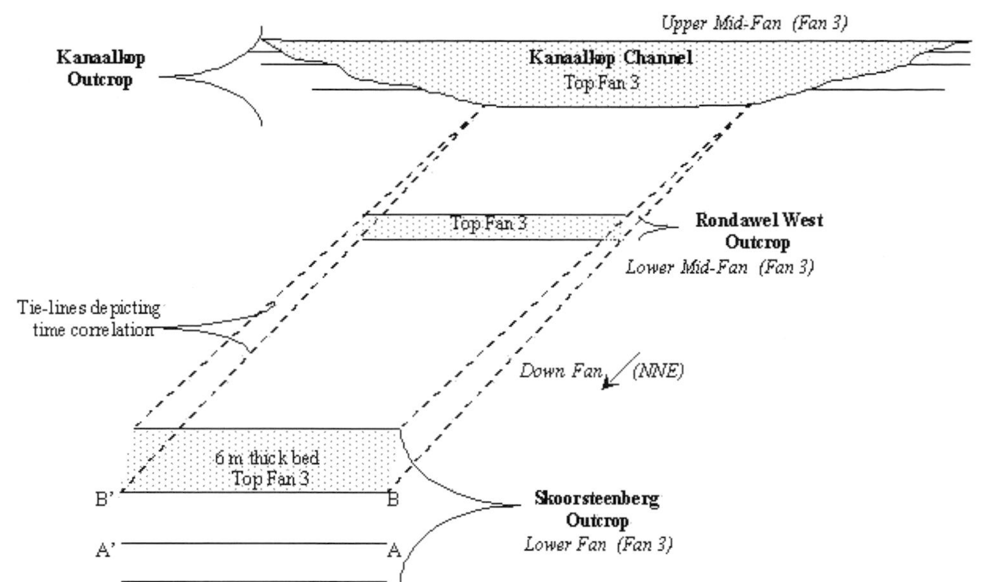

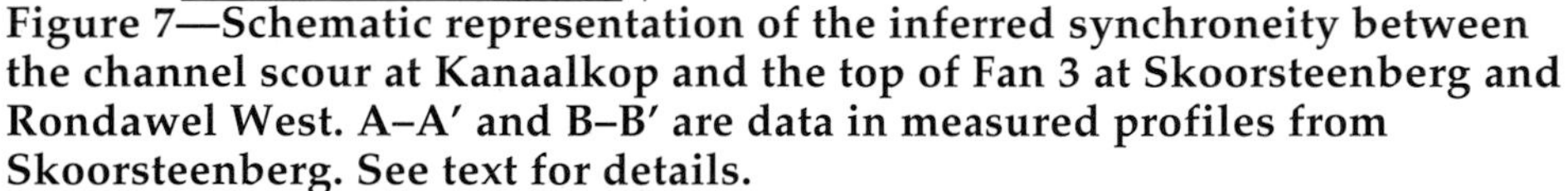

Figure 7—Schematic representation of the inferred synchroneity between the channel scour at Kanaalkop and the top of Fan 3 at Skoorsteenberg and Rondawel West. A–A′ and B–B′ are data in measured profiles from Skoorsteenberg. See text for details.

Subsequent to deposition of the 6-m-thick bed (see CD-ROM Figure 5) at Skoorsteenberg, Fan 3 is interpreted to represent a backstepping mode of deposition most likely in response to a rise in relative sea level or upfan channel avulsion (Figure 7). The channel avulsion would have resulted in consequent decrease in the volume of density currents reaching this location and the sediments got directed elsewhere in the lower fan. Gradual rise in relative sea level shifted the main depocenter farther upfan and filled the channelized conduit in the updip location at Kanaalkop. The latter was already partially filled by incremental deposition from turbidity currents that traveled past this site during active deposition farther downfan. The extrapolated trend of the Kanaalkop Channel axis northward and the location of both the intervals (at Kanaalkop and Skoorsteenberg, Figure 1) at the very top of Fan 3 supports the interpretation that the Kanaalkop Channel acted as the sediment dispersal fairway for depositing the very thick interval at Skoorsteenberg.

Rondawel West (Fan 3)

Two profiles were measured in Fan 3 at Rondawel West, separated by 50 m (Figure 1). The base of Fan 3 is not exposed in this locality. Figure 8 represents part of the outcrop, and the summary diagram is shown in Figure 9.

Base to A

This interval is characterized by thin siltstone/silty-shale beds that are interbedded with very fine-grained thin-bedded sandstones. The siltstones (2–5 cm thick) are commonly ripple laminated on a very fine scale with occasional parallel laminations. The very fine sandstones are 5–25 cm thick and are mostly massive T_a beds with or without T_b and T_c intervals. These thicker beds are predominantly tabular, but have considerable variation in thickness within the length of the outcrop (~120 m). As a result they appear lens shaped. Amalgamation surfaces commonly divide beds internally into subunits. The subunit thickness and lengths vary due to the wavy and/or discontinuous nature of the amalgamation contacts. Ripup clasts are commonly associated with the upper surfaces of the beds and are distributed on certain bedding planes. The meter-plus interval just below level A is characterized by thin-bedded siltstones. A minor proportion of siltstone beds have disrupted/convolute bedding and outsized clay-clasts.

This interval is interpreted to have been under the effect of active deposition, resulting in beds with minor but observable variations in thickness, scours in the form of wavy bases, amalgamated contacts, and residual lags as ripup clasts. These features are most likely due to increased flow turbulence commonly encountered in channel-sheet transition areas (Mutti and Normark, 1991; Chapin et al., 1994). The meter-plus thick siltstone bed just below datum A probably represents the waning activity of the feature that deposited most of this interval (a sourcing channel farther updip). The siltstone beds with disrupted/convolute bedding are interpreted to have been deposited by minor debris flows.

Interval A to B

This is characterized by thick beds with common thin interbeds (Figure 9). The beds are 30–300 cm

Figure 8—Outcrop photograph of the Rondawel West section. The scale bar to the left is 1.5 m high.

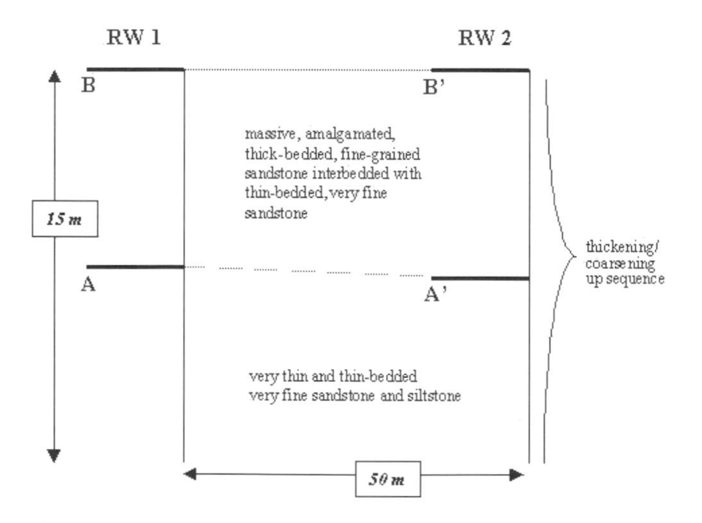

Figure 9—Summary diagram depicting the general depositional characteristics of Fan 3 at Rondawel West (RW).

thick. Bed thickness is commonly seen to change drastically within the length of the outcrop (~120 m), implying lobate bed geometries. They are primarily massive T_a beds with common amalgamation surfaces, separating single beds into two or more subunits traceable laterally along the outcrop. Large flutes are commonly present. The presence of large scours (megaflutes, 1.5–2.0 m wide) acting as short-term sediment bypass features, common ripup clasts occurring as residual lag deposits, lobate geometry of the medium to thick beds, and location between known channel-fill and distal sheet deposits support the interpretation that Rondawel West represents channel-sheet transition facies. The characteristics of the TBTs in this locality are summarized in Table 1.

The Rondawel West Outcrop in a Regional Perspective

The thicker beds at the top of the Rondawel West outcrop (interval A–B, in Figures 8, 9) are inferred to be related to the upfan and downfan expressions of the upper reaches of Fan 3 studied in the Kanaalkop (midfan channel-levee overbank) and Skoorsteenberg (lower fan distal sheet) areas, respectively. The thicker beds at the top of Fan 3 at Skoorsteenberg and at Rondawel West can be considered as time equivalents of the interval through which the Kanaalkop Channel site acted as a sediment bypass feature (Figure 7).

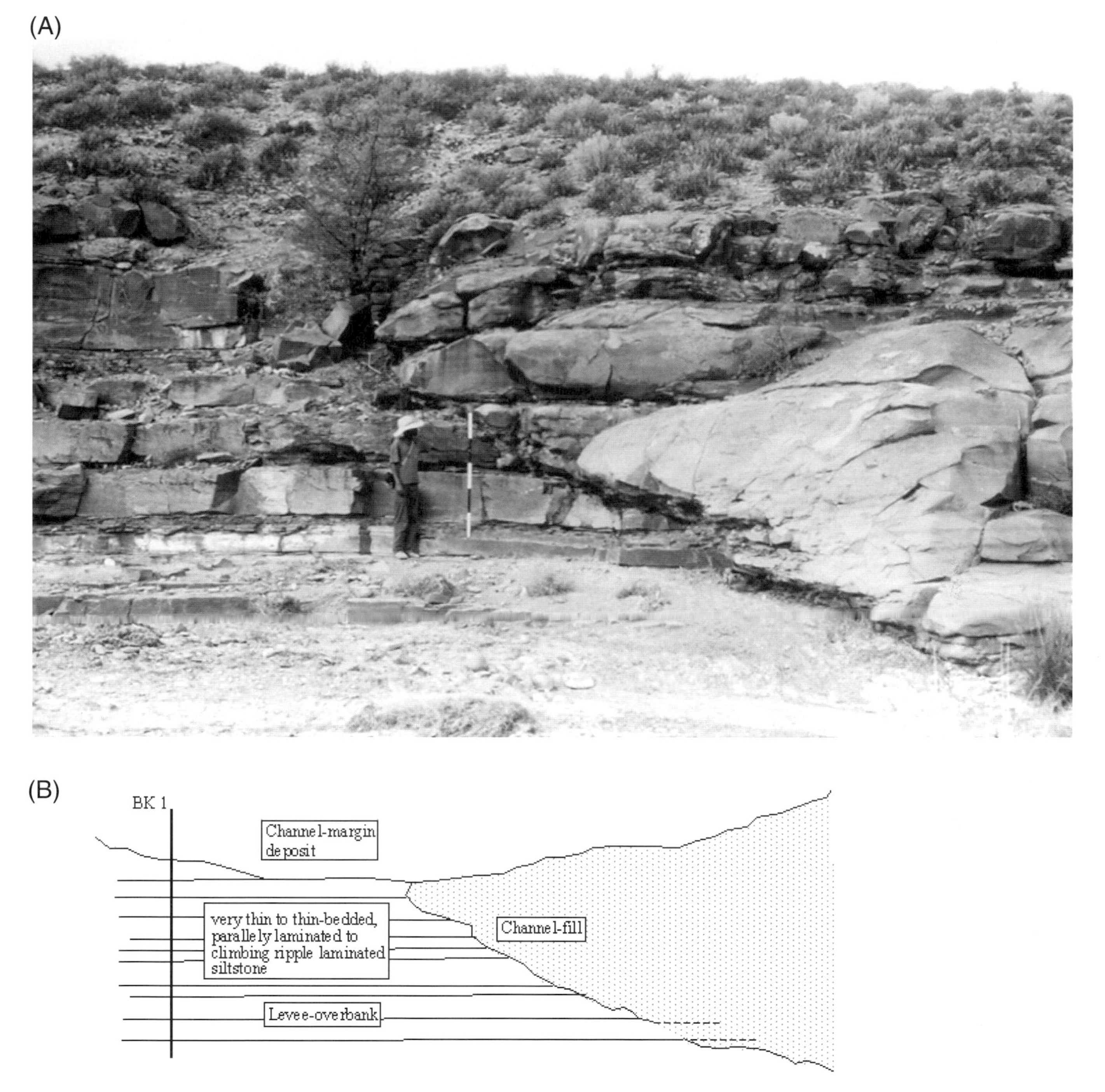

Figure 10—The spatial relationship between the levee-overbank and channel-fill deposits of Fan 5 at Blaukop (A) and the interpreted line drawing (B). The location of profile BK 1 and its depositional elements are depicted in the line drawing. The scale bar is 1.5 m high.

The underlying deposits of Fan 3 at Rondawel West and Skoorsteenberg can be related to the inferred older West Channel in the Kanaalkop area (Figure 5). The lower part of Fan 3 at these two downfan sites are possible time equivalents of the levee-overbank wedge exposed in the lower part of Fan 3 at Kanaalkop (LOC 1 in Figures 3, 5), which was sourced by the inferred West Channel. The depositional break seen at Kanaalkop (thin-bedded interval from datum C to D, top of LOC 1 in Figure 3), representing the waning activity of the older West Channel, is interpreted to be represented at Rondawel West by a 1-m-thick, thin-bedded, sediment starved interval just below datum A (Figure 9).

Blaukop (Fan 5)

The relationship between thin-bedded siltstones and a massive channel fill is well exposed in Fan 5 at

Figure 11—The Bizansgat outcrop showing well-developed thin-bedded turbidites in Fan 4. The scale is 1.5 m high.

Blaukop (Basu, 1997; Kirschner and Bouma, this volume). The genetic link between a channel fill and its laterally equivalent thin-bedded levee-overbank deposits is observed from the relationship exposed at this location. A few thin beds are seen as continuations from within the confines of the channel going into the levee-overbank area (Figure 10, B). A similar relationship between channel and levee-overbank deposits is exposed in the Permian Brushy Canyon Formation, West Texas (Basu, 1995). One representative profile from this site, BK 1 (Figure 1), is discussed in detail.

Base to ~2.75 m

The base of Fan 5 is not exposed at this locality. The very thin- to thin-bedded interval in profile BK 1 (Figure 10) comprises medium to coarse siltstones. The bed thickness distribution has three modes, less than 9 cm, 10–15 cm, and 20–45 cm. The two thinner-bedded populations are characterized by parallel laminations and ripple cross-laminations, the latter being abundant in the 10–15 cm group. The 35- to 45-cm-thick group is characteristically climbing ripple laminated (Type A climbing ripples of Jopling and Walker, 1968). These three bed-thickness groups are interbedded with each other. The less than 9 cm and 10- to 15-cm bed-thickness populations often have wavy lower and upper bedding contacts. The ripple cross-laminations in the thin-bedded intervals (less than 9 cm and 10–15 cm groups) are directed N–NE, whereas the climbing ripples in the medium bedded units (35–45 cm) are directed NE. The associated channel fill has its margin exposed, and the paleocurrent direction observed from flute and prod marks is E–NE.

The two thinner-bedded populations (<9 cm and 10–15 cm) of TBTs exhibit parallel and ripple cross-laminations but are devoid of any climbing ripples. This suggests deposition from minor spillover flows in a levee-overbank setting. The greater deviation (~60°) between the channel trend and the paleoflow direction from these thin beds is also suggestive of spillover of suspended sediment clouds from relatively smaller flows with a lower inertial state passing down the channel. The wavy bedding contacts in these thin beds are attributed to deposition over a ripple field in a levee-overbank setting—a local and subtle paleotopographic influence. The bedding contacts mimic the depositional topography of the ripple field. The presence of climbing ripples (Type A, Jopling and Walker, 1968) in the 20- to 45-cm TBT group suggests that these beds were deposited from larger flows charged with a higher suspended sediment load. The lower angular deviation (~40°) in respective paleocurrent directions between the channel and these thin-bedded (20–45 cm) siltstones is attributed to greater inertia of the spillover plumes having originated from larger and faster flows passing down the channel.

2.75 m to the Top

The layers are thin-bedded very-fine sandstones. These are massive beds with few parallel laminations seen toward the tops. Very thin interbeds are rare in this interval. The sandstone beds are commonly amalgamated and have occasional ripup clasts. The bases are mostly flat to slightly undulatory with the exception of a 1.5-m-thick bed with erosional base. This interval is inferred to be channel margin deposits related to renewed flow along the older channel (Figure 10B). For the summarized characteristics of the TBTs in this locality, see Table 1.

Bizansgat (Fan 4)

Profile BZ 1 at Bizansgat is measured from a short but well-exposed section of Fan 4 (Figure 1). This section represents a passive fill of the upper part of a channel in Fan 4 (Bouma et al., 1995; Figure 11; see also CD-ROM Figure 7).

The profile is 3.5 m thick and is characterized by thin-bedded siltstone and silty shale units with base-truncated Bouma sequences (T_{cde} and T_{de}) as shown in CD-ROM Figure 11. Beds are on an average 3 cm thick (range ~0.5–10 cm), but beds having up to 1 cm thickness are common (Figure 11, CD-ROM Figure 7). The most commonly occurring sedimentary structure is parallel lamination, which is often difficult to observe due to lack of variation in grain size. Structureless siltstones and silty shales are also common. Some incipiently developed ripple cross-laminations are often observed. Each of these beds is fining upward. Siltstones are capped by organic-rich silty shale and shale partings. Degraded plant fragments, very common in the millimeter-scale partings, are thought to represent deposition of the very fine-grain-size fraction from suspension. These thin beds may be deposited from the tail of turbidity currents that have passed through a particular location or result from small turbidity currents that represent the waning phases of lowstand deposition. Organics are fairly well dispersed in the siltstone beds but with a lower density of occurrence than the top of each bed. A few thin to medium bedded units, composed of very fine-grained sandstones, occur toward the top of the section. They are mostly massive with or without convolute bedding and ripup clasts. They are interpreted to represent renewed activity in the channel that ensued after a period of relative dormancy. The characteristics of the TBTs in this locality are summarized in Table 1.

SUMMARY

Thin-bedded turbidites studied from various depositional settings in the Tanqua Karoo submarine fans reveal sufficient differences to enable distinction between them. The thin-bedded deposits are from the following depositional subenvironments: upper midfan levee-overbank (Kanaalkop), midfan passive channel fill (Bizansgat), midfan levee-overbank (Blaukop), channel-sheet transition deposits (Rondawel West), and distal sheet sandstones (Skoorsteenberg). Table 1 highlights the differences between the thin-bedded turbidite subenvironments. The key findings are:

- The thinnest beds are developed in the midfan levee-overbank and midfan passive channel-fill deposits. Lower fan sheet sandstone and upper midfan levee-overbanks also have well-developed thin beds.
- Grain size is finest (medium silt) in the midfan levee-overbank and passive channel-fill deposits. The coarsest grain-size population is observed from the distal sheet sandstones (coarse siltstone to very fine sandstone).
- Sedimentary structures range from stacked ripple to climbing ripple laminated T_c beds from levee-overbank environments to top-truncated Bouma sequences (T_a or T_{ab} beds) in the distal sheet sandstones and "headspill" deposits in overbank settings. Bottom-truncated Bouma Sequences (T_{cd} or T_{cde} beds) are common in the passive channel fills.
- The thin beds were deposited as a result of "bodyspill" from channelized turbidity currents or as deposits from the lagging tails of degenerating turbidity currents.

Associated with the study of TBTs two important conceptual insights were gained:

- The Tanqua fans sequestered coarser sediments in the distal depositional lobes during the initial progradational phase and the sediments got progressively finer in an upfan direction with consequent back-stepping of the system.
- The Tanqua subbasin by virtue of its high-latitude paleoposition (~65°S) in the Permian represents an ancient analog of TBT deposits controlled by asymmetric levees resulting from Coriolis forcing. The Coriolis Effect is invoked to explain the common occurrence of 50- to 130-cm-thick, laterally discontinuous, non-erosional massive sandstone beds in a levee-overbank setting. These beds are emplaced by easy spillover from heads of channelized turbidity currents across the right-handed lower levee into overbank areas.

REFERENCES CITED

Basu, D., 1995, Rock magnetic determination of angular deviation in grain alignment between channel and levee sandstones within submarine fan channel-levee complex, Brushy Canyon Formation, west Texas: Gulf Coast Association Geological Societies Transactions, v. 45, p. 47–51.

Basu, D., 1997, Characterization of thin-bedded turbidites from the Permian Tanqua Karoo submarine fan deposits, South Africa: unpublished Ph.D. dissertation, Louisiana State University, 159 p.

Bouma, A. H., C. E. Stelting, and J. M. Coleman, 1985, Mississippi Fan Gulf of Mexico, *in* A. H. Bouma, W. R. Normark, and N. E. Barnes, eds., Submarine fans and related turbidite systems: New York, Springer–Verlag, p. 143–150.

Bouma, A. H., and H. deV. Wickens, 1994, Tanqua Karoo ancient analog for fine-grained submarine fans, *in* P. Weimer, A. H. Bouma, and B. F. Perkins, eds., Submarine fans and turbidite systems: sequence stratigraphy, reservoir architecture and production characteristics, Gulf of Mexico and international: Gulf Coast Section of SEPM Foundation 15th Annual Research Conference Proceedings, p. 23–34.

Bouma, A. H., H. deV. Wickens, and J. M. Coleman, 1995, Architectural characteristics of fine-grained submarine fans: a model applicable to the Gulf of Mexico: Gulf Coast Association of Geological Societies Transactions, v. 45, p. 71–76.

Carter, L., and R. M. Carter, 1988, Late Quaternary development of left-bank–dominant levees in the Bounty Trough, New Zealand: Marine Geology, v. 78, p. 185–197.

Chapin, M. A., P. Davies, J. L. Gibson, and H. S. Pettingill, 1994, Reservoir architecture of turbidite sheet sandstones in laterally extensive outcrops, Ross Formation, western Ireland, *in* P. Weimer, A. H. Bouma, and B. F. Perkins, eds., Submarine fans and turbidite systems: sequence stratigraphy, reservoir architecture and production characteristics, Gulf of Mexico and international: Gulf Coast Section of SEPM Foundation 15th Annual Research Conference Proceedings, p. 53–68.

Darling, H. L., and R. M. Sneider, 1992, Production of low resistivity, low contrast reservoirs offshore Gulf of Mexico Basin: Gulf Coast Association of Geological Societies Transactions, v. 42, p. 73–88.

Hackbarth, C. J., and R. D. Shew, 1994, Morphology and stratigraphy of a mid-Pleistocene turbidite leveed channel from seismic, core and log data, northeastern Gulf of Mexico, *in* P. Weimer, A. H. Bouma, and B. F. Perkins, eds., Submarine fans and turbidite systems: sequence stratigraphy, reservoir architecture and production characteristics, Gulf of Mexico and international: Gulf Coast Section of SEPM Foundation 15th Annual Research Conference Proceedings, p. 127–133.

Hesse, R., and A. Rakofsky, 1992, Deep-sea channel/submarine-Yazoo system of the Labrador Sea: a new deep-water facies model: AAPG Bulletin, v. 76, p. 680–707.

Hesse, R., et al., 1996, Imaging Laurentide Ice Sheet drainage into the deep sea: impact on sediments and bottom water: GSA Today, September, p. 3–9.

Hiscott, R. N., 1980, Depositional framework of sandy midfan complexes of Tourelle Formation, Ordovician, Quebec: AAPG Bulletin, v. 64, p. 1052–1077.

Jopling, A. V., and R. G. Walker, 1968, Morphology and origin of ripple-drift cross-lamination, with examples from the Pleistocene of Massachusetts: Journal of Sedimentary Petrology, v. 38, p. 971–984.

Jordan, D. W., D. R. Lowe, R. M. Slatt, C. G. Stone, A. D'Agostino, M. H. Sheihing, and R. H. Gillespie, 1991, Scales of geological heterogeneity of Pennsylvanian Jackfork Group Ouachita Mountains, Arkansas: applications to field development and exploration for deepwater sandstones: Guidebook Dallas Geological Society Field Trip #3 (AAPG Annual Convention), 142 p.

Lowe, D. R., 1982, Sediment gravity flows: II. Depositional models with specific reference to the deposits of high-density turbidity currents: Journal of Sedimentary Petrology, v. 52, p. 279–297.

McGee, D. T., P. W. Bilinski, P. S. Gary, D. S. Pfeiffer, and J. L. Sheiman, 1994, Geologic models and reservoir geometries of Auger Field, deepwater Gulf of Mexico, *in* P. Weimer, A. H. Bouma, and B. F. Perkins, eds., Submarine fans and turbidite systems: sequence stratigraphy, reservoir architecture and production characteristics, Gulf of Mexico and international: Gulf Coast Section of SEPM Foundation 15th Annual Research Conference Proceedings, p. 245–256.

McKee, E. D., and G. W. Weir, 1953, Terminology for stratification and cross-stratification in sedimentary rocks: AAPG Bulletin, v. 64, p. 381–390.

Morris, R. C., 1973, Flysch facies of the Ouachita trough—with examples from the Spillway of DeGray Dam, Arkansas, *in* C. G. Stone, B. R. Haley, and G. W. Viele, eds., A guidebook to the geology of the Ouachita Mountains, Arkansas: Arkansas Geological Commission Guidebook, p. 158–168.

Mutti, E., and M. Sonnino, 1981, Compensation cycles: a diagnostic feature of turbidite sandstone lobes: Abstracts Volume, 2d European Regional Meeting of International Association of Sedimentologists, Bologna, Italy, p. 120–123.

Mutti, E., and W. R. Normark, 1991, An integrated approach to the study of turbidite systems, *in* P. Weimer and M. H. Link, eds., Seismic facies and sedimentary processes of submarine fans and turbidite systems: New York, Springer-Verlag, p. 75–106.

Pickering, K. T., 1982, Middle-fan deposits from the Late Precambrian Kongsfjord Formation Submarine fan in north-east Finnmark, northern Norway: Sedimentary Geology, v. 33, p. 79–110.

Piper, D. J. W., and D. A. V. Stow, 1991, Fine grained turbidites, *in* G. Einsele, W. Ricken, and A. Seilacher, eds., Cycles and events in stratigraphy: New York, Springer–Verlag, p. 360–376.

Schwab, W. C., H. J. Lee, D. C. Twichell, J. Locat, C. H. Nelson, W. G. McArthur, and N. H. Kenyon, 1996, Sediment mass-flow processes on a depositional lobe, outer Mississippi Fan: Journal of Sedimentary Research, v. 66, p. 916–927.

Shew, R. D., D. R. Rollins, G. M. Tiller, C. J. Hackbarth, and C. D. White, 1994, Characterization and modeling of thin-bedded turbidite deposits from the Gulf of Mexico using detailed subsurface and analog data, *in* P. Weimer, A. H. Bouma, and B. F.

Perkins, eds., Submarine fans and turbidite systems: sequence stratigraphy, reservoir architecture and production characteristics, Gulf of Mexico and international: Gulf Coast Section of SEPM Foundation, p. 327–334.

Smith, A. G., J. C. Briden, and G. E. Drewry, 1973, Phanerozoic world maps: Special Papers in Palaeontology, v. 12, p. 1–42.

Stow, D. A. V., et al., 1989, The Bengal Fan: some preliminary results from ODP Drilling: Geo-Marine Letters, v. 9, p. 1–10.

Stow, D. A. V., et al., 1985, Mississippi Fan sedimentary facies, composition, and texture, *in* A. H. Bouma, W. R. Normark, and N. E. Barnes, eds., Submarine fans and related turbidite systems: New York, Springer–Verlag, p. 259–266.

Strong, P. G., and R. G. Walker, 1981, Deposition of the Cambrian continental rise: the St. Roch Formation near St. Jean–Port-Joli, Quebec: Canadian Journal of Earth Sciences, v. 18, p. 1320–1325.

Rozman, D. J., 2000, Characterization of a fine-grained outer submarine fan deposit, Tanqua-Karoo Basin, South Africa, *in* A. H. Bouma and C. G. Stone, eds., Fine-grained turbidite systems, AAPG Memoir 72 / SEPM Special Publication No. 68, p. 279–290.

Chapter 24

Characterization of a Fine-Grained Outer Submarine Fan Deposit, Tanqua-Karoo Basin, South Africa

Daniel J. Rozman
BP Amoco
Houston, Texas, U.S.A.

ABSTRACT

Fan 2 deposits of the late Permian Tanqua-Karoo subbasin in South Africa are sheet sandstones typical of an outer submarine fan. The fan consists of three sand-rich units, separated by several meters of shale, that thin to the north and northeast into the Tanqua subbasin. The stacking pattern is progradational, with lateral offset to the east, suggesting that the lower units influenced the depositional location of those above.

Broadly lens-shaped beds, random vertical thickness trends, and variable paleocurrent directions suggest that sedimentation occurred through relatively broad, shallow pathways whose position changed frequently during deposition of the outer fan. The resulting sheet sand deposits have estimated dimensions of 3–10 km in a dip direction and 100–2000 m in a strike direction. These individual sheet sands, which laterally have relatively uniform sedimentary characteristics, stack to form depositional lobes.

Detailed outcrop-scale observations and measurements reveal that amalgamation and scouring are more prevalent than is initially apparent, enhancing sandstone connectivity both vertically and laterally. Shale clasts, bed-scale scours, and the three sand-rich units of Fan 2 represent three scales of geologic heterogeneity that would affect fluid flow in this type of reservoir.

INTRODUCTION

At the seismic or wellbore scale, characteristics of lower or outer fine-grained submarine fan sheet sand deposits are readily obtained; however, much less information is available regarding the depositional characteristics of these deposits at the subseismic or interwell scale. Such information is needed to make realistic reservoir models that will optimize the recovery of hydrocarbons and improve the understanding of transport and depositional processes.

The Fan 2 submarine fan outcrops in the Tanqua-Karoo basin were studied because of the high quality of exposures over relatively long distances (Wickens and Bouma, this volume). The dimensions, bedding character, and lithofacies distribution could be studied in detail at various locations in the area. It is believed that this deposit is an appropriate analog for some turbidite reservoirs in the Gulf of Mexico (Wickens and Bouma, 1995). The objectives of this study were to (1) characterize the large-scale (external) geometry of the outer submarine fan, (2) characterize the outcrop-scale (internal) bedding style, (3) describe changes in the sedimentary character of the fan in both a lateral and vertical sense, and (4) describe depositional features of this deposit that are significant for its performance as a hydrocarbon reservoir.

GEOLOGIC BACKGROUND, METHODS, AND LITHOFACIES DESCRIPTION

Much of the geologic background for Fan 2, which is situated stratigraphically between Fans 1 and 3, is given by Wickens and Bouma (this volume). The Fan 2 outcrops are located in the southern portion of the Tanqua subbasin and are exposed for approximately 9 km, oblique to the general depositional dip direction (Figure 1). Fan 2 is separated from the Fan 1 deposits by about 20 m of silty shale, and from Fan 3 by about 40 m of silty shale (Figure 2).

Field methods included the measurement of vertical sections and detailed paleocurrent measurements from groove casts. This information was used to make cross sections, isopach maps, and paleoflow diagrams. A more detailed characterization of Fan 2 was based on photomosaics, outcrop sketches, numbering and physical tracing of beds, and sampling of lithofacies. Along some outcrops vertical sections were measured every 5 m to calibrate photomosaics of the outcrops. The resulting profiles were used to determine lithofacies proportions and bed lengths in a manner similar to that described by Arnot et al. (1997).

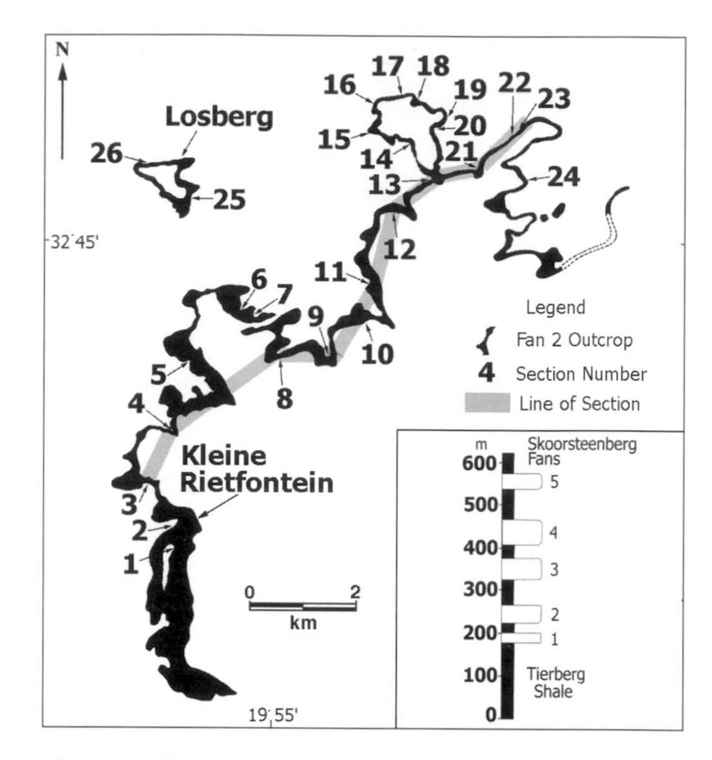

Figure 2. Outcrop map of Fan 2 outcrop showing locations of measured sections. See Figure 1 for location of Fan 2 outcrops. Stick log shows relative thickness and position of Fan 2 in the Skoorsteenberg Formation.

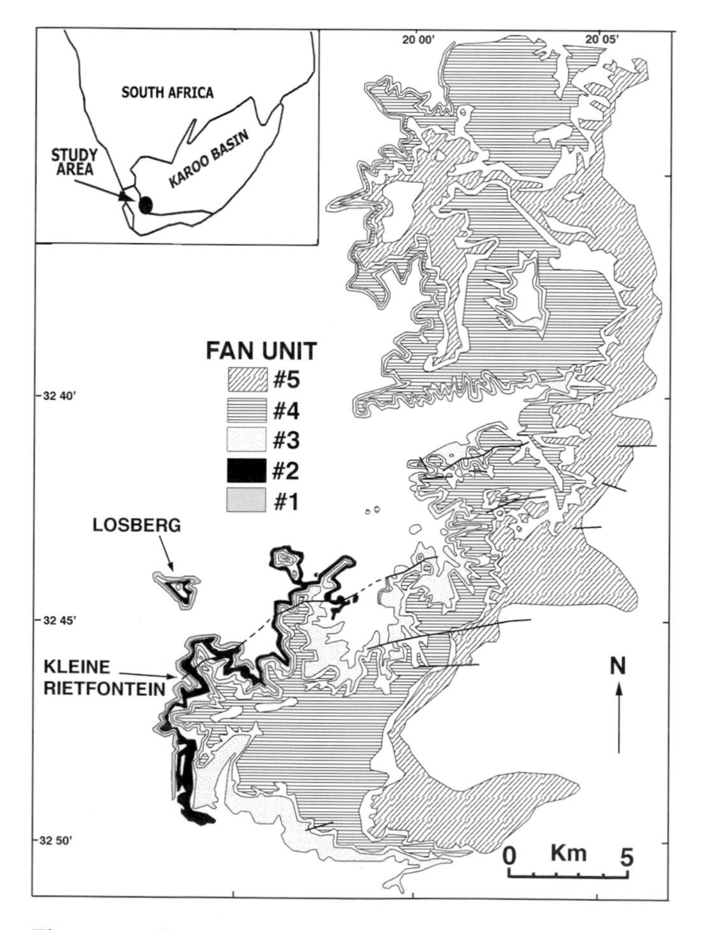

Figure 1—Outcrop map of Tanqua–Karoo fan deposits showing location of Fan 2 outcrop. Location of study area is indicated by the inset map.

Wickens (1994) categorized five lithofacies for the entire Tanqua-Karoo submarine fan complex, including massive (structureless) sandstone, parallel and/or ripple cross-laminated sandstone, siltstone, shale, and carbonaceous/micaceous siltstone. All of these facies were identified in the Fan 2 deposits; however, the massive sandstone and siltstone facies comprise more than 95% of the deposit. A typical vertical section through Fan 2 consists of a basal thickening and coarsening upward section of siltstone. This is overlain by a section that consists of doublets of massively bedded, 10- to 100-cm-thick, very fine-grained sandstone (T_a), and 2- to 20-cm parallel-laminated siltstone. Two silty shale beds approximately 3 m thick separate the sand-rich sections into three parts.

ARCHITECTURE AND GEOMETRY

At the southern end of the outcrop area the Fan 2 deposit has three units 5- to 15-m-thick that consist of alternating medium-bedded sandstones and thinly bedded siltstones. Each sand-rich unit is separated from the other by about 3 m of shale. This division can be seen across the exposure area; however, the middle and upper units extend much farther into the basin than the lower unit, which pinches out halfway across the area (Figure 3). These sandstone units thin fairly constantly across the area, but pinch out dramatically at the end.

Isopach maps and paleocurrent trends for the sandstone units show that each unit thins generally to the north and northeast (Figure 4). Additionally, the source of the clastic input appears to have shifted from the west to the south during the time Fan 2 was deposited. In addition to the progradational stacking pattern illustrated by Figure 3, the three sandstone units have a laterally offset pattern. Figure 5 shows the

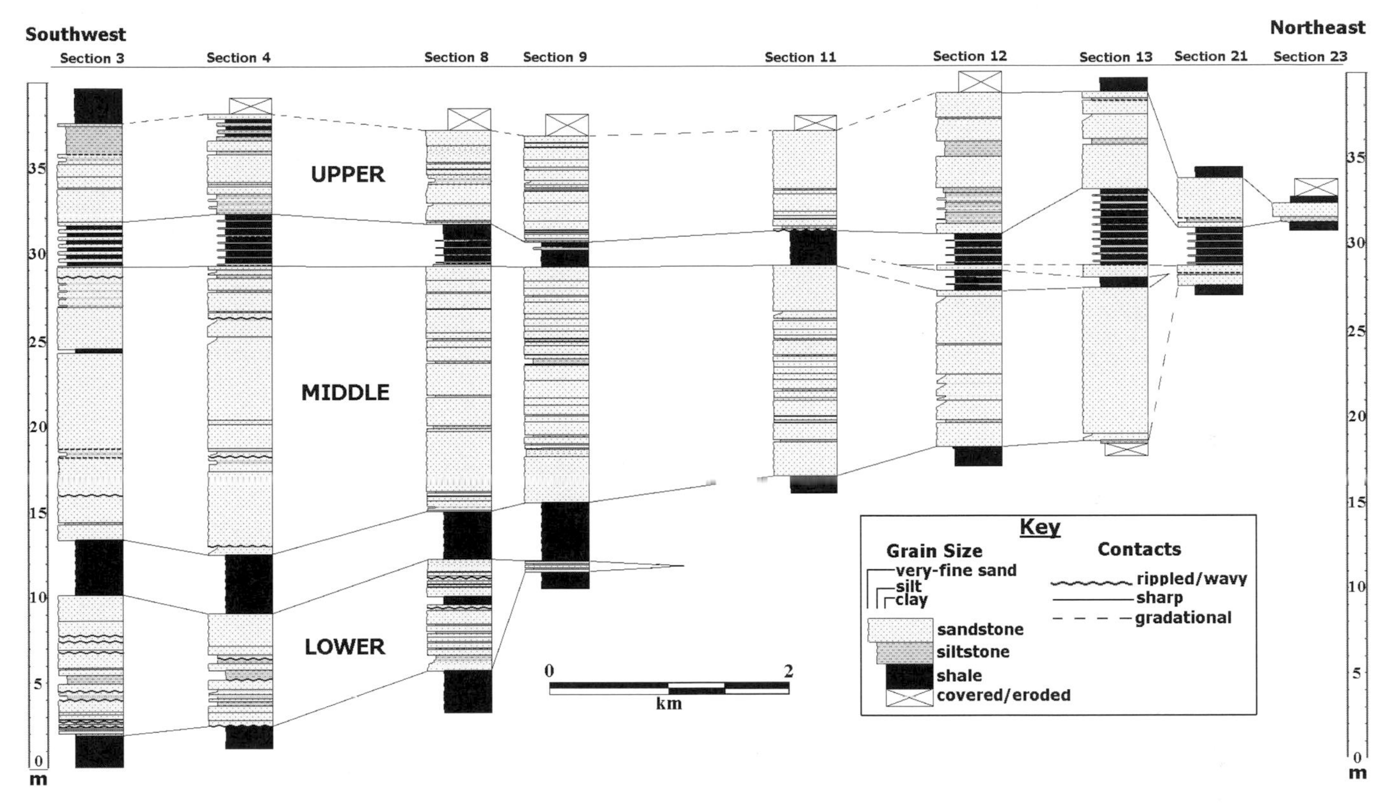

Figure 3—Cross section through Fan 2 in general dip direction showing division of deposit into three sand-rich units that thin to the northeast. Line of section is indicated on Figure 2.

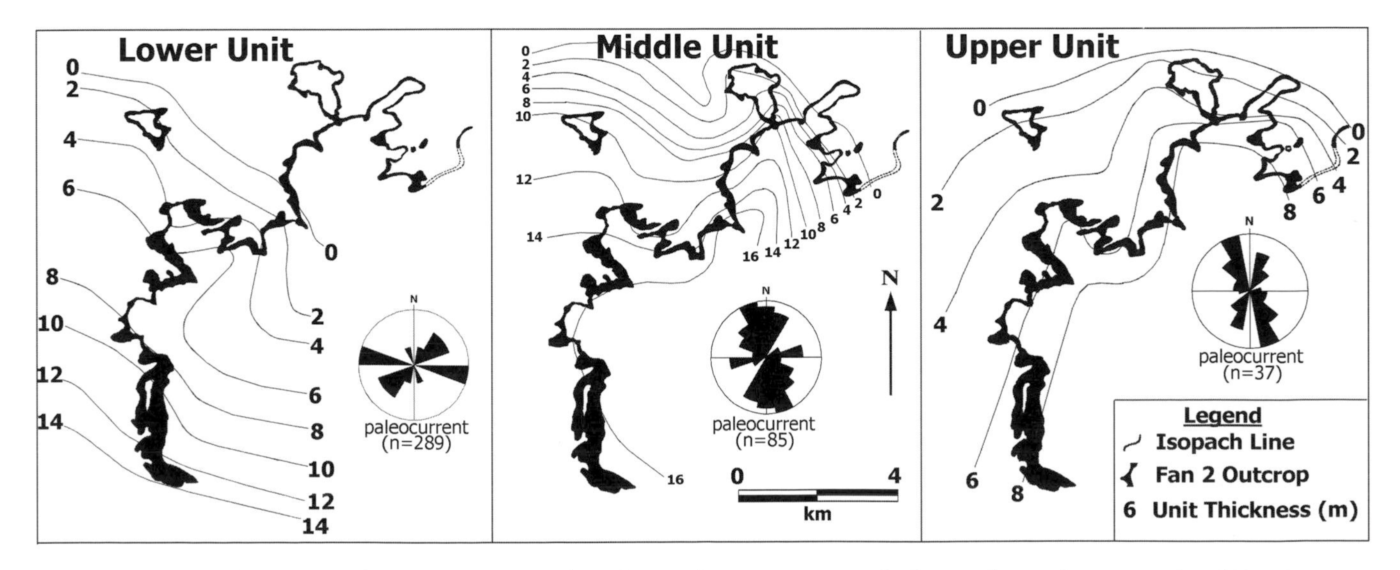

Figure 4—Isopach maps and rose diagrams showing paleocurrent trends for each sandstone unit of Fan 2.

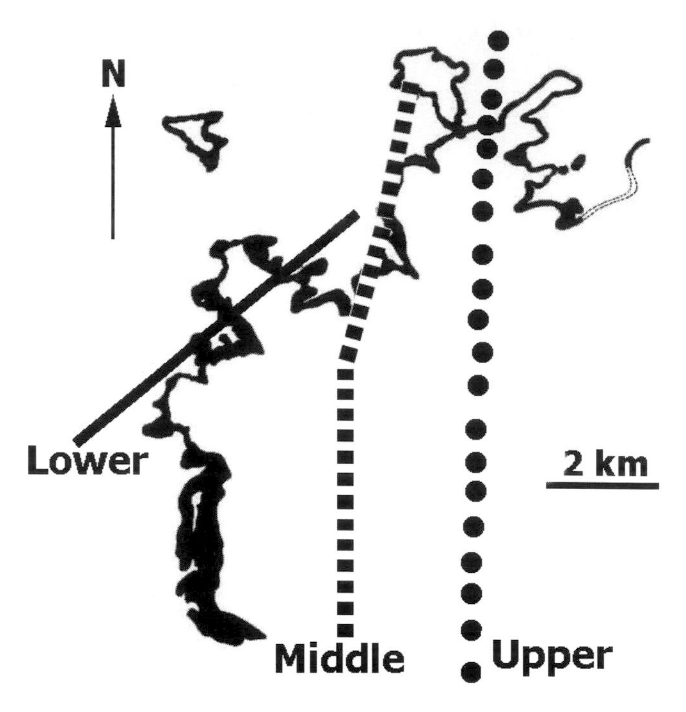

Figure 5—Estimated depositional axes of the three sand-rich units in Fan 2 superimposed over the outcrop map. Note the progressive movement of the depositional axes to the east, most likely due to the topographic expression of the underlying unit.

estimated depositional axis of each unit. The location of the middle and upper axes are shifted slightly east of the lower and upper axes, respectively.

In a vertical sense, the differences between the three sandstone units are relatively minor (Table 1). The lower unit has local scours and wedge-shaped beds, some of which onlap the basal shale. Consequently, the average bed length in this unit is lower than that of the thick, amalgamated beds of the middle unit (73 cm average) and those of the regularly bedded upper unit. The upper unit consists of some thick sandstone (58 cm average) and thick sections of thinly interbedded sandstone, silt, and shale. The upper unit is the only one that has a significant amount of ripple cross-lamination. The 78% net-to-gross of sand is the lowest of the three units.

Thickening- or thinning-upward cycles in fan deposits have often been cited as evidence of fan progradation (Mutti et al., 1978). Bed thickness analysis of Fan 2 reveals that there is no consistent thinning or thickening trend. Both types of cyclicity can be seen at different places in the outcrop area. Most commonly, a thinning trend in one bed coincides with thickening of the unit above.

To examine depositional changes in a spatial sense, several values were plotted against the distance across the outcrop from southwest to northeast in the general flow direction. Figure 6 shows graphs for sandstone percentage, or net-to-gross, average sandstone bed thickness, and percentage of amalgamation. Percentage of amalgamation is defined as the percentage of sand-on-sand contacts in a vertical section that were identifiable (Chapin et al., 1994) and can be an indication of the vertical connectivity between sheet sands. In general, the sandstone percentage and average sandstone bed thickness do not show a consistent downdip trend. However, percentage of amalgamation does seem to decrease slightly, suggesting that the erosive nature of the sediment-gravity flows decreased somewhat toward the end of the fan. The high degree of variability in the plots is due to differences in the weathered character of various outcrops. Also, because the fan consists of only a few beds at the end of the outcrop, the presence of one amalgamated contact or one thick sand greatly skews the graph. Despite this variability, the plots suggest that most of

Table 1. Depositional characteristics of the lower, middle, and upper sand-rich units of Fan 2.

Unit	Facies	Average Net-to-Gross Ratio (%)	Average Sandstone Thickness (cm)	Bedding Styles	Scours and Amalgamation
Upper	Alternating massive sandstones, bedded sandstones, and thin silts and shales; some traction strucures	78	58	More interbedded sequences relative to lower units	Few
Middle	Alternating massive sandstones and silty shales	85	73	Thick, amalgamated units of high lateral extent	Long amalgamated contacts
Lower	Alternating massive sandstones and silty shales; sole markings and load casts	80	39	Medium beds of limited lateral extent due to scours and pinch-outs	Locally common

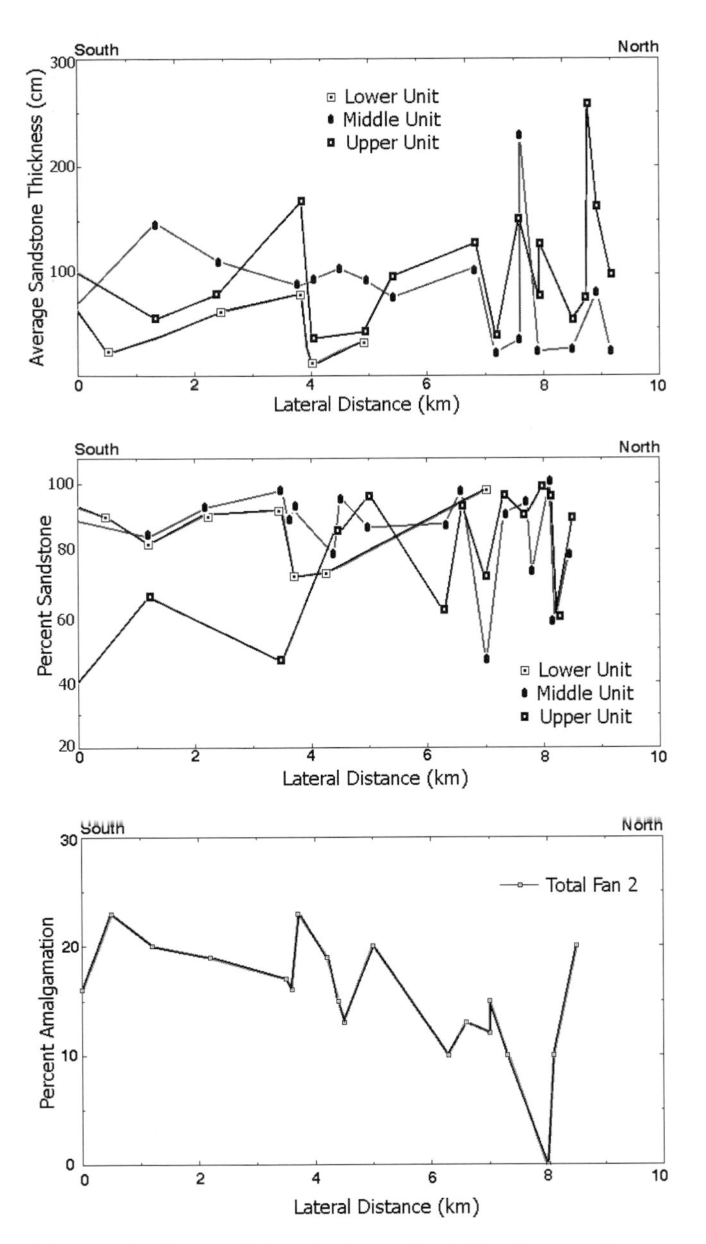

Figure 6—Depositional characteristics for each sand-rich unit of Fan 2 as a function of distance. There is no consistent downdip change in average sandstone bed thickness or in the percentage of sandstone (net-to-gross); however, the amount of amalgamation decreases slightly toward the end of the deposit.

the depositional characteristics of Fan 2 are laterally uniform to the very outer portion of the deposit.

Lithologically the sandstone beds themselves do not change significantly downdip; rather, they remain structureless and very fine grained to the outer 100 m. For example, at one location in the northeast of the Fan 2 outcrop area an individual sandstone bed thins over a distance of about 100 m, grading from a structureless 50-cm bed into a 5-cm siltstone containing clay clasts and plant fragments (Figure 7).

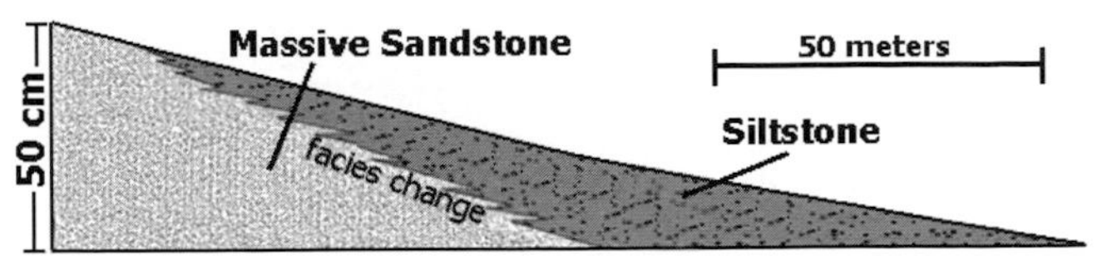

Figure 7—Lithologic change from a 50-cm structureless sandstone to thin siltstone over a distance of approximately 100 m at the end of Fan 2, upper unit.

Interpretation of Architecture and Geometry

Relatively slow, muddy sedimentation into the Tanqua subbasin was interrupted several times during the Late Permian by periods of sediment-gravity flow deposition. The flow character changed very little over this portion of the fan, resulting in beds that have uniform sedimentary characteristics to the end of the deposit. Slight basin-floor topography confined the flows and affected the bedding style of the lower sandstone unit, producing more pinch-outs and local scouring. The volume and energy of the sediment-gravity flows that deposited the middle unit were much higher; therefore, the beds are frequently amalgamated, forming thick, laterally extensive beds. The interbedded nature of the upper unit, its lower net-to-gross ratio, and the presence of lower flow regime traction structures suggests that it represents a gradual decline in the amount of sand-size sediment that was delivered to the Tanqua subbasin at this time.

The absence of age control for the intervening shales of Fan 2 makes it difficult to determine the time span of deposition. Because the three sandstone units are separated stratigraphically by relatively thin shales they have previously been considered to be parts of one outer-fan deposit (Wickens, 1994). However, the difference in paleocurrent directions between each unit suggests that they were separated by enough time for the source area to switch, or for an avulsion event to occur in the middle or upper fan area. Work by Scott et al. (this volume) suggests that sedimentation into the southwestern part of the Karoo basin actively shifted around the basin margin.

The stacking pattern and geometry of Fan 2 appears to have been controlled by both basin-floor topography and the character of each sediment-gravity flow. Topography on the basin-floor confined some of the early sedimentation events; additionally, the topographic expression of the lower sandstone units forced subsequent flows farther to the east. Relatively random thickening trends that are seen in Fan 2 suggest that the depositional style was aggradational rather than progradational at the bed scale, and produced compensation cycles as described by Mutti and Sonnino (1981).

BEDDING-SCALE DESCRIPTION: INTERNAL GEOMETRY

At the scale of the outcrop many of the sandstone beds appear to be very parallel and laterally continuous. However, detailed lateral measurements reveal that many beds change thickness over hundreds of meters, and some pinch out altogether. Although lithologically the sandstone beds are virtually identical, paleocurrent trends vary by as much as 90 degrees from bed to bed in a vertical sequence.

Profiles were made of two particularly high-quality outcrops to show bedding-scale characteristics. The first profile was made at Kleine Rietfontein (2 on Figure 2). Here a 220-m-wide oblique dip exposure of the lower sandstone unit is accessible. Twenty-two sandstone beds, illustrated in Figure 8, are present in the outcrop, two of which can be seen to pinch-out into the shale below. Less than a 1% change in total thickness was measured across the exposure. The second profile was made from an outcrop at Losberg (25, Figure 2), and is a 160-m-wide strike exposure of the middle sandstone unit. No change in thickness exists across this exposure.

One of the most notable features that these profiles illustrate is the abundance of amalgamation. Very few intervening siltstone beds are continuous across the entire exposure. Also, these profiles illustrate the presence of pinch-outs, truncations, and local scours that are often unrecognized on initial inspection. Analysis of the profile images showed that there is approximately 86% sandstone exposed at Kleine Rietfontein and 90% sandstone at Losberg. These net-to-gross values are typical for the entire Fan 2 outcrop.

Interpretation of Bedding-Scale Observations

The two outcrop profiles, combined with observations from other locations across the Fan 2 outcrop area, suggest a particular depositional style. The broadly lenticular sandstone beds, highly variable paleocurrent trends, and absence of typical overbank deposits suggest that deposition occurred via broad sediment "pathways." The term "channel" is not used because it implies a relatively long-term conduit for multiple sediment-gravity flows (Mutti and Normark, 1991). Fan 2 sheet sands are envisioned to have been deposited in highly transitory, shallow, broadly unconfined pathways that were active for a relatively short period of time. The highly variable paleocurrent trends suggest that these pathways crisscrossed the outer fan, and were probably controlled by switching of midfan channels (Pickering, 1981). These sheet sands stack primarily in a compensational to aggradational pattern to form individual depositional lobes.

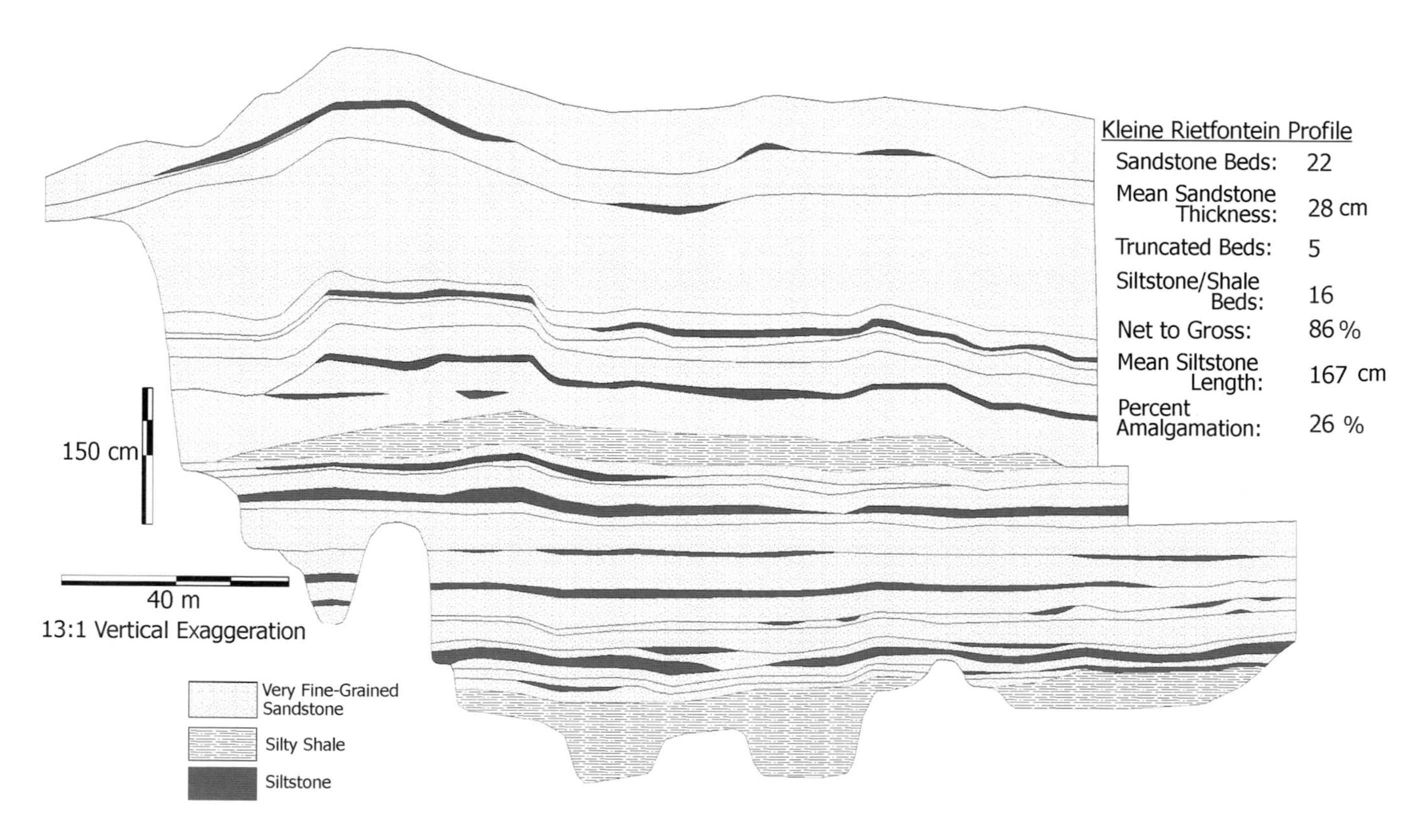

Figure 8—Outcrop profile at Kleine Rietfontein (see Figure 2 for location), lower sand-rich unit. The outcrop faces north-northeast. Note that there is only one silty-shale bed that is continuous across the exposure.

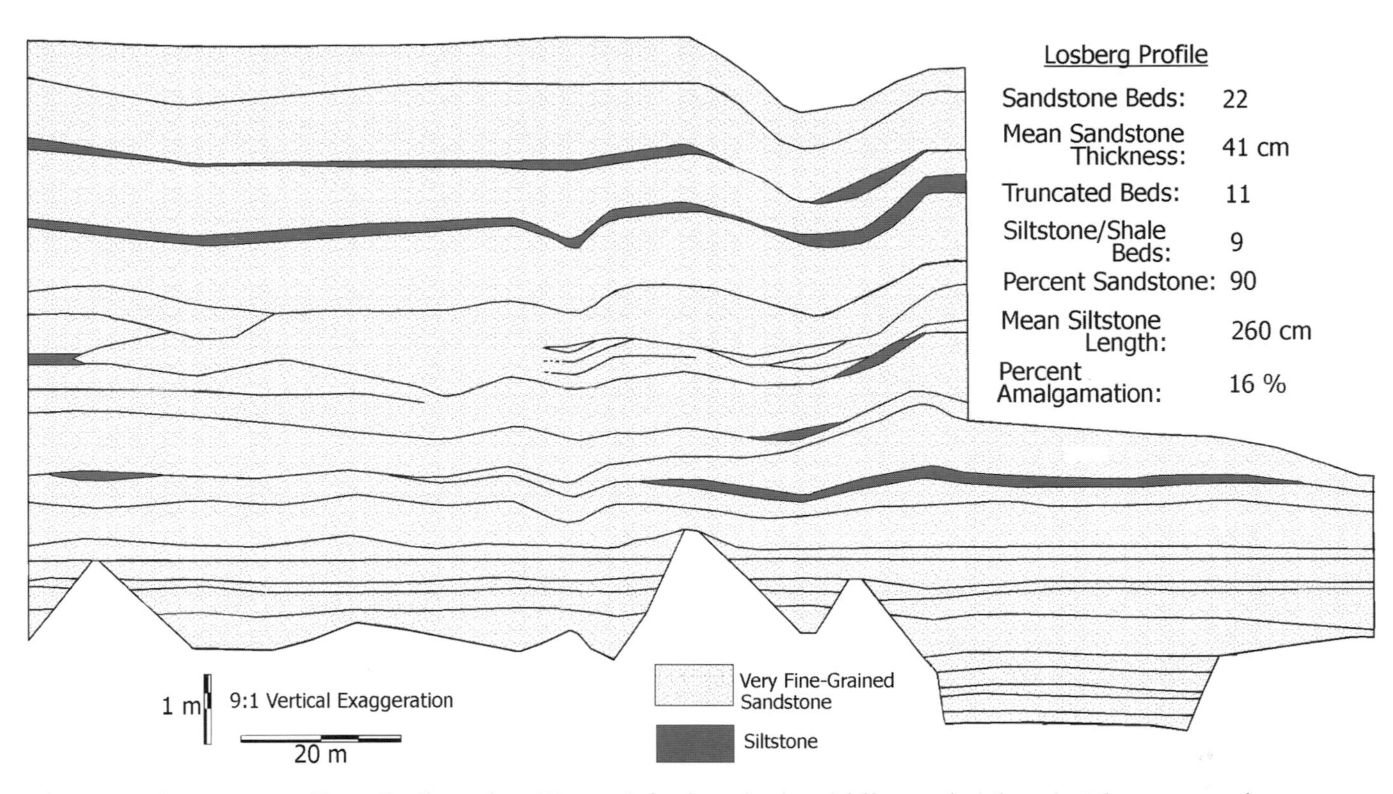

Figure 9—Outcrop profile at Losberg (see Figure 2 for location), middle sand-rich unit. The outcrop faces north. There are two fairly continuous silty shales in this exposure.

RESERVOIR IMPLICATIONS

Slatt and Galloway (1992) described three scales of reservoir heterogeneity, or geologic complexity, that affect fluid flow: (1) wellbore, (2) interwell, and (3) fieldwide. The Fan 2 deposit has depositional features that fit into each of these three categories.

Wellbore Scale

Parameters such as primary porosity and permeability, grain size, type of sedimentary structures, and other sedimentary characteristics control fluid flow at this scale. Generally, these features can be seen in cores. The lack of variation in grain size, sorting, and absence of prominent sedimentary structures in the Fan 2 deposit is notably consistent. For this reason relatively little time is devoted to describing grain-size variations and other features at this scale. Nevertheless, the occurrence of shale clasts throughout Fan 2 was noted to have significant potential as a barrier or baffle to fluid flow (Weber, 1982). Several different types of shale clasts were identified. The most common were flat, subround, 1–10 cm clasts that occurred either "floating" in isolation within a structureless sandstone bed or clustered together along a specific layer within or on top of the sandstone bed. Figure 10 shows a sketch of an interval that corresponds with the "brecciated shale clasts" of Johansson and Stow (1995), interpreted to be the result of disruption of a silty bed in the channelized area updip.

Interwell Scale

Heterogeneity on the scale of tens of meters thick by hundreds of meters wide is most difficult to assess in the subsurface because it must be extrapolated to the interwell region from the wellbore; therefore, outcrop analogs serve as a useful guide to interpretation (Slatt and Galloway, 1992).

In Fan 2, the features that have the most profound influence on bed to bed connectivity and potential fluid flow are scours. Note that, as used here, a scour is broadly defined and includes the amalgamation of two sandstones. Based on observations made at Kleine Rietfontein, Losberg, and other outcrops, four types were classified (Figure 11).

Type I: This type occurs when a depositional unit cuts through a lower unit at a high angle but then flattens out to become basically parallel to the unit below. Both lateral and vertical connectivity are enhanced.

Type II: This is the most common type of scour in the Fan 2 outcrops. The upper unit removes the intervening fine sediment over large areas, leaving thin lenses of relatively

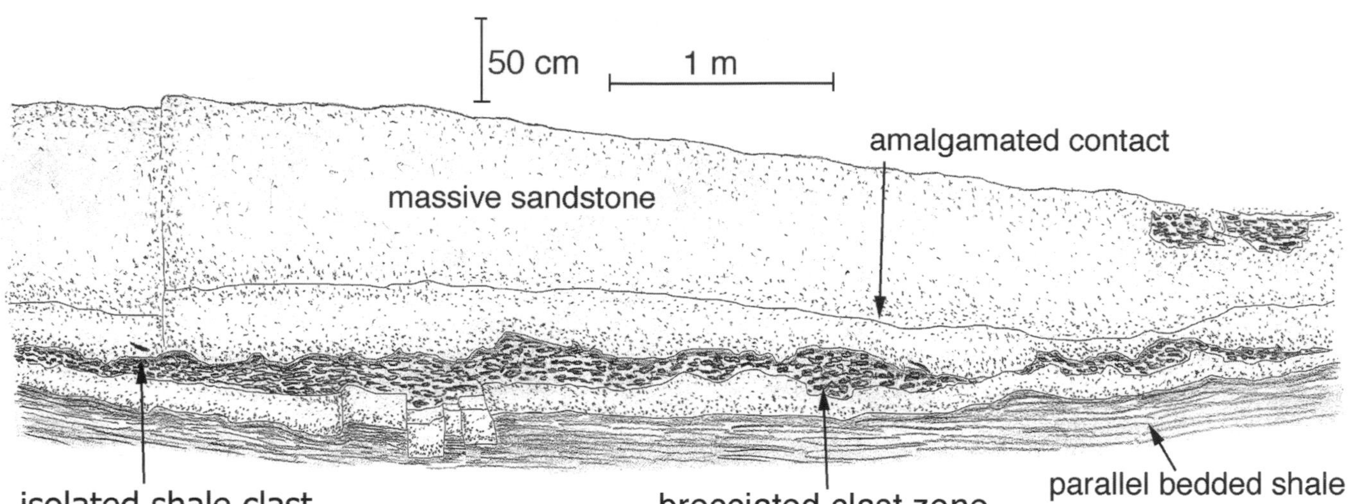

Figure 10—Sketch of a portion of the lower sandstone unit near Kleine Rietfontein (Figure 2). The brecciated clast zone between massive sandstones is interpreted as the result of disruption of a muddy deposit in a more proximal location on the fan by a coarser-grained sediment-gravity flow.

impermeable sediment between amalgamated contacts. Vertical communication between sandstone beds is enhanced.

Type III: Termed "compensational scour" because one unit thins and the one above thickens, this type produces greater lateral continuity within the reservoir. The sandstones at the margins of each unit are amalgamated, creating a sheet sand much longer than an individual one.

Type IV: This type is relatively uncommon in Fan 2. Scouring by the upper unit is fairly sharp, cutting into two or more units that would otherwise have been unconnected. Connectivity is enhanced in all directions.

An example of the Type II scour was seen in an outcrop at the location of Section 8 (Figure 2). Here, thick massive sandstones are inconsistently amalgamated along a particular contact such that discontinuous lenses of shale and silt occur with "pillars" of sandstone (1–4 on the diagram) in between (Figure 12). Such sand on sand contacts have obvious value as potential fluid flow paths. Their origin is interpreted to be due to local removal of the fines by a sediment-gravity flow. However, it is also possible that these features represent preexisting topography.

Interwell Scale Shales

As potential barriers to flow, shale beds need to be adequately accounted for in a reservoir. The reservoir engineer can create stochastic models based on parameters such as the minimum, maximum, and modal shale width and thickness (Haldorsen and Chang, 1986).

Qualitatively, it appears that shaly units thinner than 10 cm in Fan 2 are very likely to be scoured away at some point over their three-dimensional area, with the likelihood that the interval is continuously increasing proportionally to their thickness. Quantitative thickness data were obtained from vertical measured sections. Figure 13 shows the thickness distribution of siltstone and shale beds in Fan 2, as well as some statistical parameters. The average thickness is 11 cm, a value that is positively skewed by uncommonly thick shales such as that seen in the middle of middle unit at Kleine Rietfontein (Figure 8).

Quantitative measurements of siltstone and shale bed lengths are considerably more difficult to obtain. The majority of these units cannot be traced for their full extent because of erosion or cover. An estimate can be made from the two detailed profiles previously described. The mean siltstone lengths are 260 cm at Losberg and 167 cm at Kleine Rietfontein, suggesting the majority of Fan 2 siltstones are laterally continuous for less than several tens of meters. Because the scours have only been observed in two dimensions, it is difficult to estimate what the three-dimensional extent of the shales might be.

Field Scale

Depositional bodies that are 1- to 10-km-wide and hundreds of meters thick represent fieldwide features, and determine the volume of in-place hydrocarbons and their spatial distribution (Slatt and Galloway, 1992). In Fan 2 the three sandstone units can together or separately be considered small fieldwide features. A feature of this scale might also be considered to be one flow unit, that is, a sedimentary

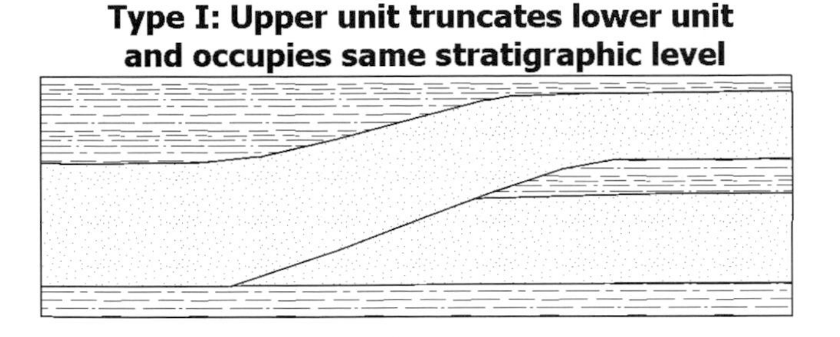

Type II: Upper unit scoured through intervening silt or shale but remains parallel to lower unit

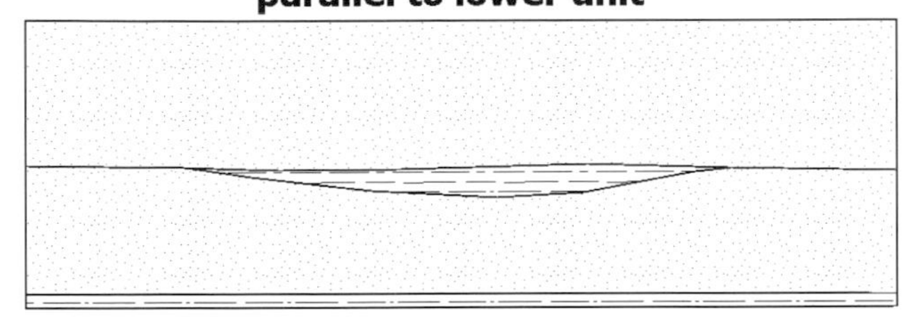

Type III: Compensational scour at unit margin increases lateral connectivity

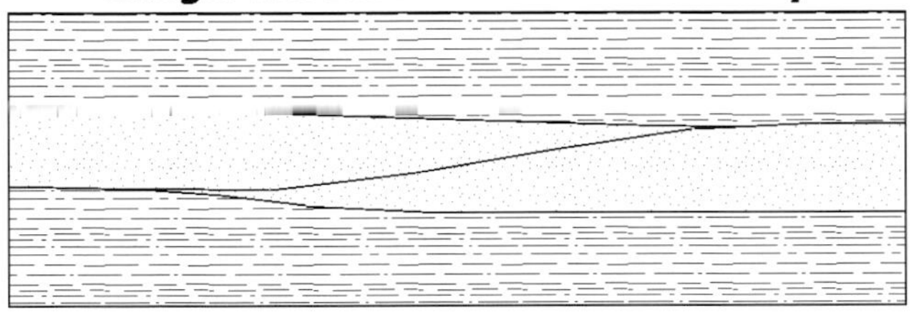

Type IV: Vertical scouring connecting multiple units

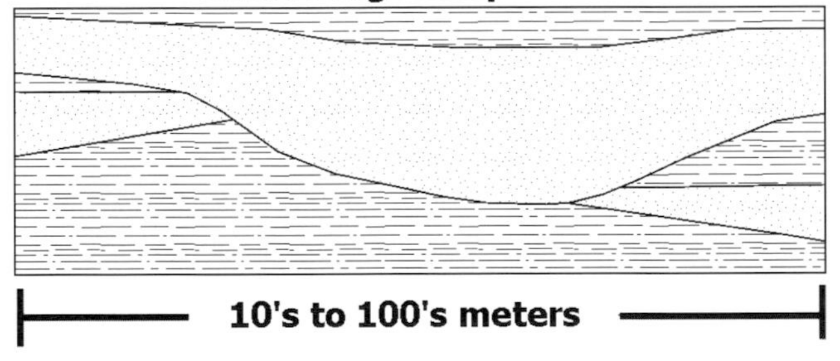

Figure 11—Generalized classification of four scour types observed in the Fan 2 deposit. The vertical scale is that of an individual sandstone bed.

body that is consistent with respect to properties that affect fluid flow and is mappable at the interwell scale (Ebanks et al., 1992). Dividing a reservoir into flow units can be a useful way to model and predict its performance. Fluid communication between flow units similar to the three sandstone units of Fan 2 would be expected only if the reservoirs were faulted and/or fractured.

Several scales of heterogeneity have been demonstrated in the Fan 2 deposits, all of which would influence its performance as a hydrocarbon reservoir. Figure 14 illustrates the different scales identified in this study.

CONCLUSIONS

Fan 2 is an outer submarine fan sheet sand deposit consisting of three sand-rich units, separated by silty shales, that thin to the north and northeast. The overall stacking pattern is progradational and offset to the east, suggesting that each sandstone unit influenced the location of the one above. Paleocurrent indicators suggest that the sediment source shifted from the west to the south during periods of reduced sedimentation, which were probably related to stillstands in relative sea-level fall or avulsion events. Variable paleocurrent directions and the bedding style suggest that sedimentation occurred through broad, shallow, highly transitory transport paths resulting in coalesced sheet sands. The three sheet sand units can be considered to be individual depositional lobes.

Few changes in lithology or depositional style were identified in either a lateral or vertical sense. Differences between the three sandstone units of Fan 2 are believed to be due to differences in basin-floor topography and sediment-gravity flow volume. Random bed thickness trends also indicate highly variable sediment pathways.

Two outcrop-scale profiles show that bedding is not as even and continuous as it initially appears, but that pinch-outs and truncations are relatively common. At a scale of hundreds of meters, most thin silt or shale beds observed in these profiles are scoured by overlying sandstones at some point within the two dimensional exposure; therefore, considerable connectivity in three dimensions is expected. Characterization of bedding in these profiles supports the idea of deposition through shallow, broadly unconfined channels of short lifetime. Individual sheet sands have general dimensions of 3–10 km in a dip direction and 100–2000 m in a strike direction.

Three scales of reservoir heterogeneity are identified in Fan 2 deposits. Wellbore-scale shale clasts are common and would act as barriers or baffles to fluid flow. Four types of interwell-scale amalgamation and scours are identified that would enhance connectivity in at least one dimension. Also, the average siltstone and shale beds within the sand-rich units have a thickness of 11 cm and an estimated lateral extent of 1–10 m, implying that they would not, in general, act as interwell scale barriers to flow. Fieldwide features such as Fan 2, or its three constituent parts, may be treated as individual flow units with relatively homogenous fluid flow character.

ACKNOWLEDGMENTS

I wish to thank Arnold Bouma for his guidance and support throughout this study. The work also benefited from the comments of various Louisiana State University graduate students and the AAPG editors.

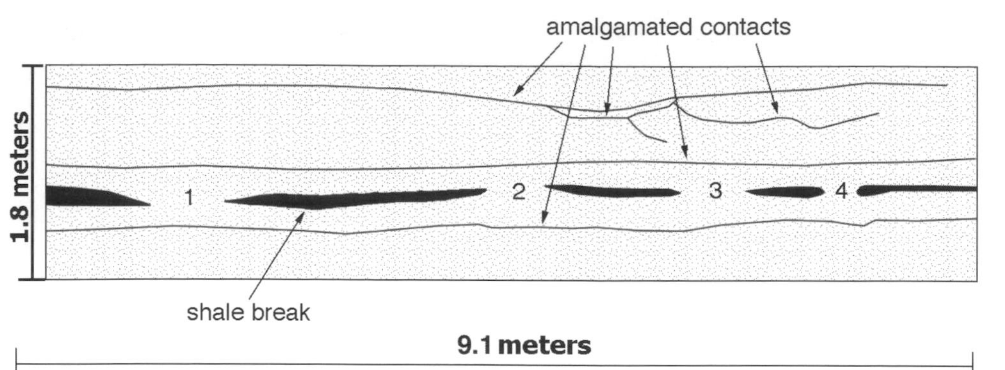

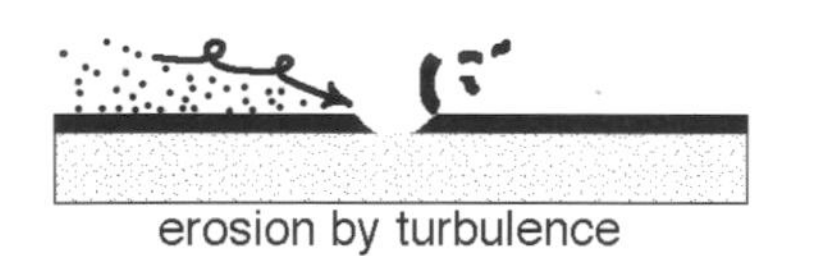

OR

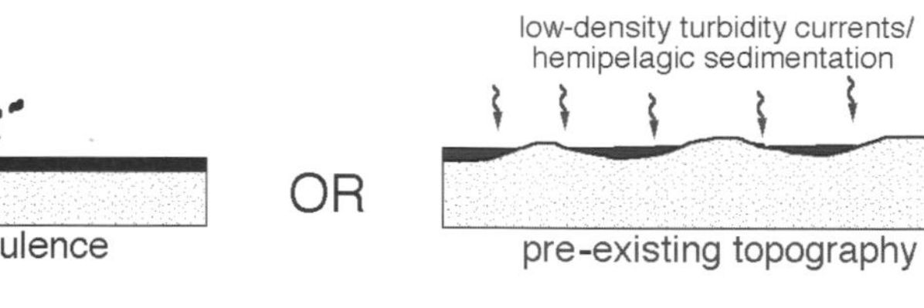

Figure 12—Example of Type II scour at the location of Section 8 (Figure 2). Massive sandstone is inconsistently amalgamated, leaving "pillars" of sandstone (1–4) separated by lenses of siltstone. The origin of these features could be erosional or depositional.

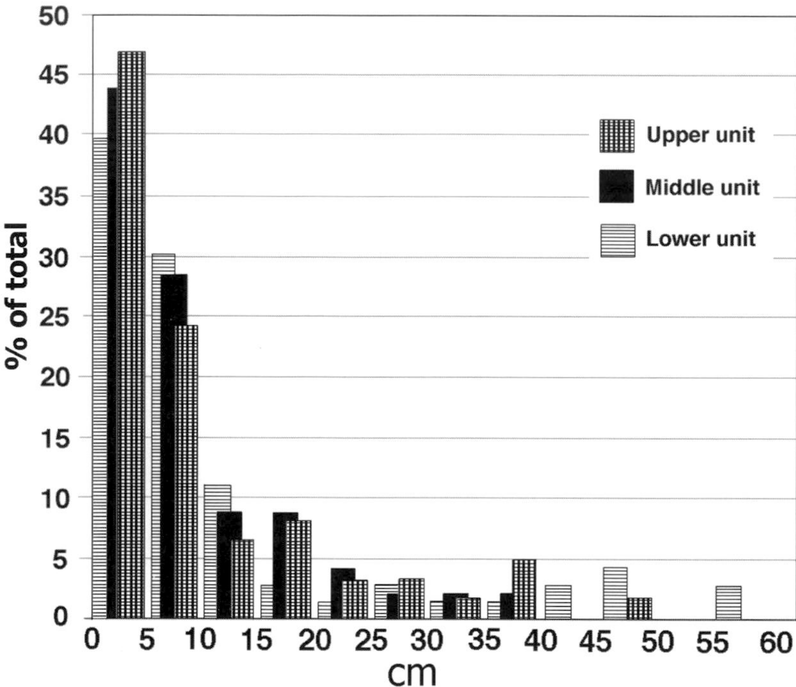

Value (in cm)	Lower Unit	Middle Unit	Upper Unit	Fan 2 Total
Mean	0.5	8	29	11
Median	2	8	25	7
Mode	2	10	20	10
Standard Deviation	7	3.2	11.4	11.4
Maximum (excluding 3 m shales)	60	38	48	60

Figure 13—Graph of thickness distribution of siltstone and shale beds in Fan 2. The table indicates size statistics for the siltstone and shale beds within each sandstone unit, and for the total Fan 2 deposit.

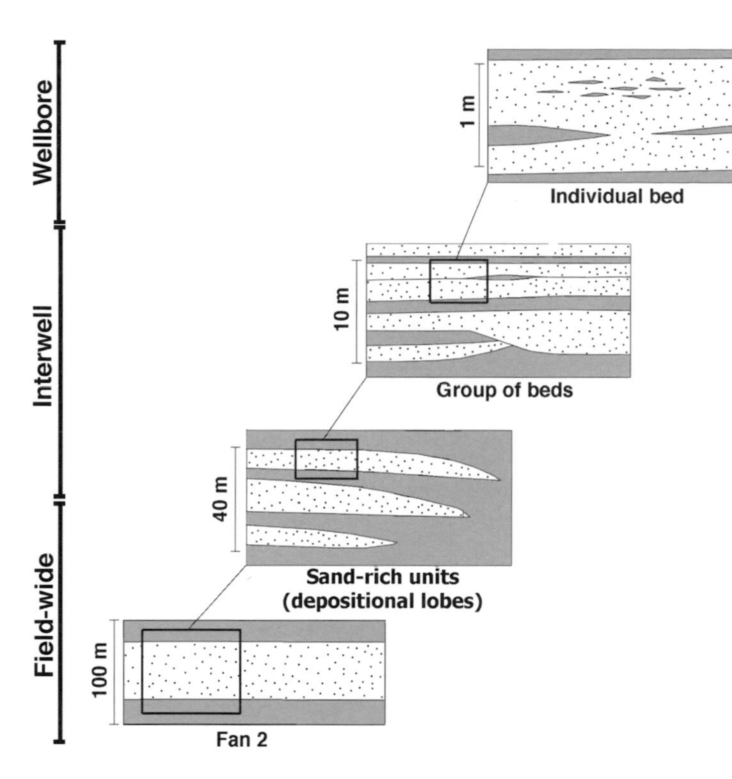

Figure 14—Various scales of geologic heterogeneity in Fan 2 that would affect fluid flow. The dark pattern represents shale.

REFERENCES CITED

Arnot, M. J., T. R. Good, and J. M. Lewis, 1997, Photogeological and image-analysis techniques for collection of large-scale outcrop data: Journal of Sedimentary Research, v. 67, p. 984–987.

Chapin, M., P. Davies, J. L. Gibson, and H. S. Pettingill, 1994, Reservoir architecture of turbidite sheet sandstones in laterally extensive outcrops, Ross Formation, western Ireland, *in* P. Weimer, A. H. Bouma, and B. F. Perkins, eds., Submarine fans and turbidite systems: sequence stratigraphy, reservoir architecture, and production characteristics, Gulf of Mexico and international: Gulf Coast Section SEPM Foundation 15th Research Conference, p. 53–68.

Ebanks, W. J., M. H. Scheihing, and C. D. Atkinson, 1992, Flow units for reservoir characterization, *in* D. Morton–Thompson and A. M. Woods, eds., Development geology reference manual: AAPG Methods in Exploration Series 10, p. 282–285.

Haldorsen, H. H., and D. M. Chang, 1986, Notes on stochastic shales: from outcrop to reservoir simulation model, *in* L. W. Lake and H. B. Carroll Jr., eds., Reservoir characterization: San Diego, Academic Press, p. 445–485.

Johansson, M., and D. A. V. Stow, 1995, A classification scheme for shale clasts in deepwater sandstones, *in* A. J. Hartley and D. J. Prosser, eds., Characterization of deep marine clastic systems: Geological Society Special Publication 94, p. 221–241.

Mutti, E., T. H. Nilsen, and F. Ricci Lucchi, 1978, Outer fan depositional lobes of the Laga Formation (upper Miocene and lower Pliocene), east-central Italy, *in* D. J. Stanley and G. Kelling, eds., Sedimentation in submarine canyons, fans, and trenches: Stroudsbourg, PA, Dowden, Hutchinson, and Ross, p. 210–225.

Mutti, E., and M. Sonnino, 1981, Compensation cycles: a diagnostic feature of turbidite sandstone lobes: International Association of Sedimentologists, 2d European Regional Meeting, Bologna, 1981, Abstracts, p. 120–123.

Mutti, E., and W. Normark, 1991, An integrated approach to the study of turbidite systems, *in* P. Weimer and M. L. Link, eds., Seismic facies and sedimentary processes of submarine fans and turbidite systems: New York, Springer–Verlag, p. 75–106.

Pickering, K. T., 1981, Two types of outer fan lobe sequence from the late Precambrian Kongsfjord Formation submarine fan, Finnmark, North Norway: Journal of Sedimentary Petrology, v. 51, p. 1277–1286.

Slatt, R. M., and W. E. Galloway, 1992, Geological heterogeneities, *in* D. Morton-Thompson and A. M. Woods, eds., Development geology reference manual: AAPG Methods in Exploration Series 10, v. 10, p. 278–281.

Weber, K. J., 1982, Influence of common sedimentary structures on fluid flow in reservoir models: Journal of Petroleum Technology, v. 34, p. 665–672.

Wickens, H. deV., 1994, Submarine fans of the Ecca Formation: unpublished Ph.D. dissertation, University of Port Elizabeth, South Africa, 350 p.

Wickens, H. deV. and A. H. Bouma, 1995, The Tanqua basinfloor fans, Permian Ecca Group, western Karoo Basin, South Africa, *in* K. T. Pickering, R. N. Hiscott, N. H. Kenyon, F. Ricci Lucchi, and R. D. A. Smith, eds., Atlas of deepwater environments, architectural style in turbidite systems: London, Chapman & Hall, p. 317–322.

Bouma, A. H., and D. J. Rozman, 2000, Characteristics of fine-grained outer fan fringe turbidite systems, *in* A. H. Bouma and C. G. Stone, eds., Fine-grained turbidite systems, AAPG Memoir 72/SEPM Special Publication No. 68, p. 291–298.

Chapter 25

Characteristics of Fine-Grained Outer Fan Fringe Turbidite Systems

Arnold H. Bouma
Daniel J. Rozman[1]
Department of Geology and Geophysics, Louisiana State University
Baton Rouge, Louisiana, U.S.A.

ABSTRACT

Comparing the fringes of modern and ancient fans seemingly results in different interpretations, probably caused by incompatible datasets. Studies on the Mississippi Fan suggest that the entire fan is controlled by a special form of branching of channelized flow. The fan-fringe sand bodies are finger shaped and considered to be overbank deposits. Observations from the Tanqua Karoo outcrops suggest that sand transport starts either as narrow fingers at the end of many distributaries or as one "wide" sheet sand that breaks up into fingers. The two different interpretations, derived from modern and ancient deposits, have significant implications on transport processes of sand and on sand-body geometries, volumes, and connectivities.

INTRODUCTION

The purpose of this paper is an attempt to compare modern and ancient submarine fan fringes and to highlight that both are studied in a seemingly incomparable 2-D mode. Although there may be legitimate comparison of the two sets of data, the interpretations could also have been influenced negatively by different scales or lithologies.

Two examples are discussed: (1) the surface depositional patterns of the Mississippi Fan and (2) vertical sections through the Permian Tanqua Karoo in South Africa. Modern fan fringes can best be observed with high-resolution equipment, such as side-scan sonar and high-resolution, shallow-penetration seismic reflection instruments, such as 3.5 kHz and Minisparker. Additional coring provides ground truth, but the observations are basically restricted to a surface picture of the waning phase of a submarine fan. When studying outcrops, the obtained information reflects a vertical 2-D section that represents the terminal area of the fan. It is not known how much is eroded away or whether the present outcrops represent a valid picture.

Comparing both the modern and the ancient examples, using different data sources with different resolutions, and not knowing whether both areas differ somewhat in depositional location, may already cause dissimilarities. However, this type of comparison study will help to better understand the processes and the deposits one may encounter in the subsurface or any outcrop.

A complex of submarine fans/turbidite systems, consisting of two or more stratigraphically succeeding

[1]Present address: BP Amoco Exploration, Houston, Texas, U.S.A.

fans, is called mud-rich when the individual fans are separated by shales/mudstones of equal or greater thickness than most of the fans (Bouma, 1997, this volume). The opposite is called sand-rich (Reading and Richards, 1994). Confusion in this terminology may exist because people tend to consider a fan by itself and exclude the intervening shales. Typically, most of these shales are thinly layered and are composed of very thin to thin beds of coarse silts or very fine sands, separated by laminae that basically consist of clay minerals and organic matter.

When dealing with fine-grained deposits, other confusions may arise in the terminology of lithologies. Sand/sandstone are terms applied to average grain sizes coarser than 4ϕ (62.5 μ). If the bulk of the sediment is finer than 4ϕ, the term mud/mudstone is used with an additional word to emphasize the major component, such as silty mud or clayey mud. However, in most cases the term shale is used rather than mudstone when the estimated API value is greater than 70 counts, even when the sediment does not show fissility.

Fine-grained submarine fan/turbidite systems are often bypass systems, meaning that on the middle fan the deposition of the coarsest sediment is restricted to the channel itself, whereas it is finer on the levees (the part of the overbank area that borders the channel) and becomes even finer in the real overbank areas. Much of the sand is transported to the outer fan and deposited there as sheet sands/depositional lobes (Bouma, this volume). It must be kept in mind that the sand-rich or coarse-grained and the mud-rich or fine-grained turbidite systems are end members of a continuum and most submarine fans fall between these extremes. Nevertheless, it is important to keep the end members in mind when establishing where a certain field or outcrop should be placed because that indicates the mode of sand distribution.

FAN FRINGES

Very little study has been conducted on ancient fan fringes, the transition area between the depositional lobes and basin plain deposits (Mutti and Ricci Lucchi, 1972). This is a zone that becomes thinner in downdip direction with a decrease in the sand/shale ratio. When dealing with outcrops, these decreases in layer thickness and sand/shale ratio enhance the erosional breakdown of the rocks. For this reason one seldom sees the transition to the basin plain in its entirety but only some updip part of it.

Subsurface observations of fan fringes are very difficult because seismic resolution does not permit definition of the fan fringe in adequate detail. Therefore, trap sizes are restricted to seismically identified sandstones or the overall sand thickness trend, and wells are targeted within that polygon.

Because of the seemingly incompatible results obtained from modern fans and outcropping turbidite systems, no comparison is foolproof; hence, there is a need for descriptive kinematic models to predictively "fill in the gaps" (R.L. Phair, personal communication,1999). Information from modern fans basically describes the two-dimensional basin-floor surface that may include a thickness of about 2–10 m, depending on penetration. The outcrop provides a vertical 2-D picture with sometimes another 2-D panel nearby if outcrops in different locations are close together, or boreholes are available (Slatt, this volume). Another difficulty is determining the original location of the eroded-away outside of the fan fringe. It is known that thinning can be gradual over most of the fan but dramatic at the very end (Wickens, 1994; Bouma and Wickens, 1994; Rozman, 1998). In Fan 2 of the Tanqua Karoo one can observe within the outer 20 m of an individual bed a rapid thinning and a grain-size change from sandstone to siltstone to basin shale (Rozman, this volume).

MISSISSIPPI FAN

High-resolution, shallow-tow and deep-tow side-scan sonar, and subbottom surveys across a part of the outer/lower Mississippi Fan revealed an intriguing pattern (Twichell et al., 1991, 1992, 1995; Nelson et al., 1992). Using GLORIA long-range side-scan sonar images, that part of the fan was divided into eight elongate deposits or depositional sublobes. These sublobes can be as long as 200 km and less than 35 m thick. Small channel shapes and lineations on the surface of many of those depositional sublobes radiate from a single, larger main channel. The high-resolution seismic profiles show that adjacent depositional sublobes overlap one another (Twichell et al., 1991).

The depositional sublobes on the lower (outer) Mississippi Fan show a series of smaller linear features, 5–40 km in length and <5 m deep, interpreted as small channels (Twichell et al., 1991, 1992). About half of the deposits in the cores showed clasts supported by matrix. Those authors concluded that small channelized turbidity and debris flows were the dominant transport processes in constructing that depositional sublobe. An interpretation of a GLORIA image of one of the depositional sublobes that originates from a sharp bend in the Mississippi Channel is presented in Figure 1A. Figure 1B shows the interpretation of a medium-resolution SeaMARC IA image at the seaward end of a feeder channel, where small distal lobe channels are fringed by high backscatter areas. Twichell et al. (1992) suggest that sands and debris flow sediments move through those small channels as confined flows (Schwab et al., 1996). Their studies on all side-scan imagery led to the suggestion that these flows are confined to these small feeder channels for the first 100+ km. Although high backscatter deposits fringe the last 20–40 km of these small channels, the deposits extend laterally only a few hundreds of meters from

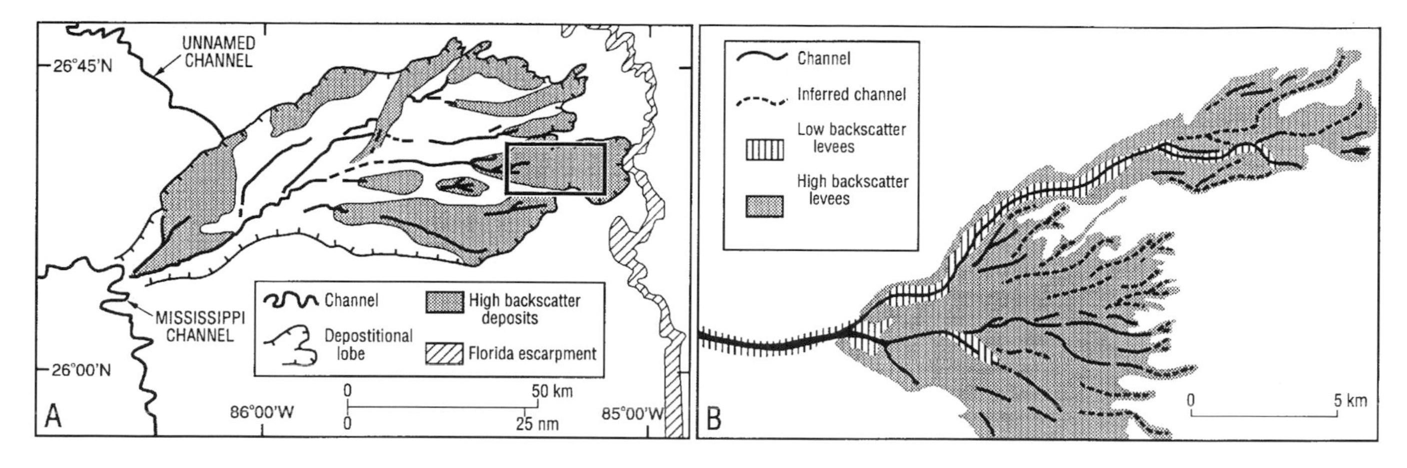

Figure 1—Interpretation of a depositional lobe (A) and detail of one of the sublobes (B) of the Mississippi Fan as seen in GLORIA (A) and SeaMARC (B) side-scan sonar images; for location, see rectangle. (A) Many distributary channels and high backscatter sandy deposits (redrawn after Twichell et al., 1991). (B) Interpretation of a higher-resolution side-scan image of (A), indicating a network of previously unrecognized channels and that the sand-rich area is found at the end of long feeder channel and that the sand deposits are closely associated with these small channels (redrawn after Twichell et al., 1992).

the channels. The images suggest that flows do not travel more than several hundred meters from the channel (D. C. Twichell, personal communication, 1999). In summary, overbanking sheet flows are not an important transport process. These very valuable images provide a good insight into the complexity of the modern fan surface. How representative this surface is for the entire fan and how representative it is of fans in general is unknown. We interpret it to represent the waning phase of active fan deposition (Bouma et al., 1985, 1986).

The channels in the distal part of the lobe are less than 100 m wide and have a relief <2 m. Lack of topographic relief in the area indicates that variations in backscatter strength reveal differences in lithologies (Twichell et al., 1992; Figure 1B). Cross-cutting of channels within a depositional sublobe suggests that only one channel was active at a time (Twichell et al., 1991) similar to the conclusions reached by Weimer (1989). Low to moderate backscatter is characteristic for the channels in those fingers, whereas the feeder channel fill shows high backscatter. It is not known whether these differences relate to sediment types or to lack of resolution of the narrow channels (Twichell et al., 1992). The high-backscatter areas shown in Figure 1B have abrupt, irregular, fingerlike outer boundaries. The 3.5-kHz profiles across the side-scan study area show no bathymetric relief that can be resolved with a surface-towed echo sounder. The side-scan imagery shows channels and stratigraphic relationships that are below the resolution of the high-resolution seismic profile techniques (D. C. Twichell, personal communication, 1999). The lateral edges of the coarsest grained sediments (high backscatter) are remarkably abrupt (Twichell et al., 1992). Those authors suggest that only confined flow is capable of transporting the sand-rich sediments and that as soon as the channel terminates, the sand is transported as an unconfined flow for only a few hundreds of meters (Twichell et al., 1992). The bathymetry is inadequately resolved to indicate whether there is a local sediment pile directly outside the channel mouth (D. C. Twichell, personal communication, 1999).

TANQUA KAROO

The Permian Tanqua Karoo subbasin fill provides an excellent opportunity to study fan fringe areas of fine-grained turbidite systems in outcrop (Wickens, 1994; Bouma and Wickens, 1994; Rozman, 1998; Wickens and Bouma, this volume). Initial, rather general studies on the distal parts of Fans 1 through 4 have been reported on thus far. At the end of Fan 2 lateral thickness changes are relatively dramatic in a strike orientation. Rozman (1998) observed that a 2-m-thick massive sandstone thins to less than 50 cm over a distance of approximately 50 m. Measured vertical sections reveal a pinch-and-swell cross-sectional shape of the upper surface of a major sandstone, observed only in the distal area. No scouring or erosion into the underlying shales could be observed. The sandstone layer appears to be one continuous sheet and no amalgamated contacts are visible, which may be the result of the well-sorted fine grain size. The interpretation represents "fingers" of sand that prograded, whereby the thick portions of the sheet sand comprise the elongate finger parts and the thin parts comprise the thinner sections of the areas in between the fingers (Rozman, 1998; Figure 2). This is remarkably similar to the

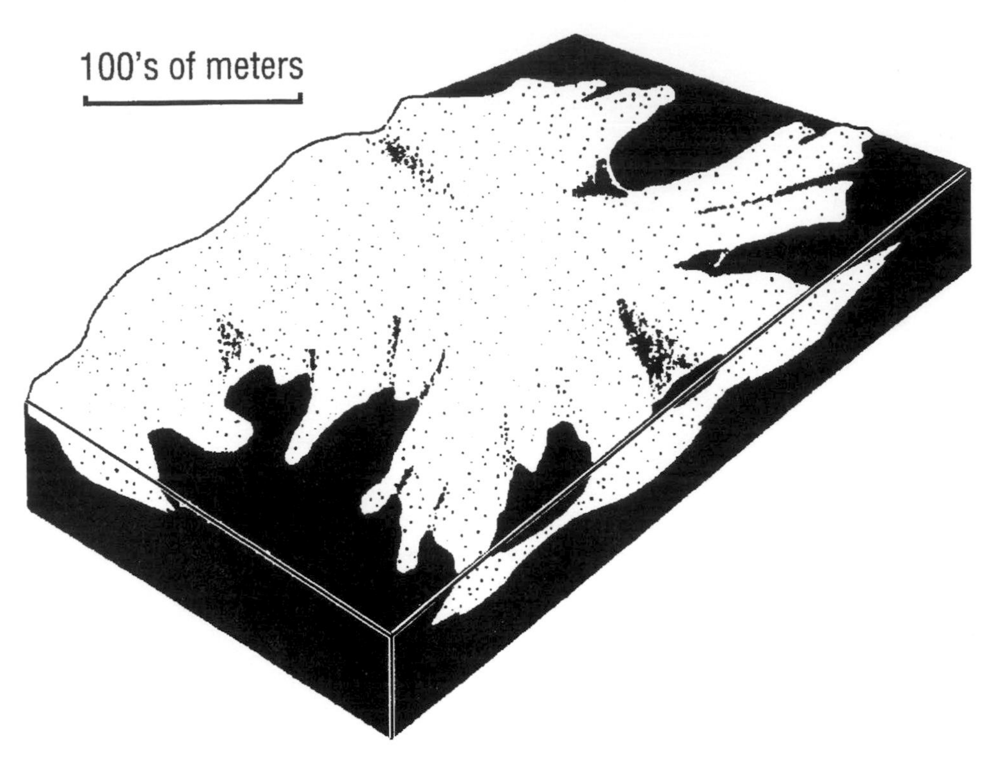

Figure 2—Conceptual diagram of prograding lobe fingers in the fan fringe area (after Rozman, 1998). The light stippled areas represent the elongated fingers, and the narrower densely stippled ones show the much thinner zones where fingers just onlap to one another.

observations on the Mississippi Fan, only such small vertical variations are below the resolution of the 3.5-kHz seismic system (D. C. Twichell, personal communication, 1999).

The distal part of Fan 3 is well exposed in the northern area (downdip part) of the Tanqua Karoo. West of Skoorsteenberg is an L-shaped outcrop, the sides of which are somewhat oblique to the general paleocurrent of that fan (Bouma and Wickens, 1994; Wickens, 1994; Basu, 1997; Kirkova, 1998; Wickens and Bouma, this volume). The short N–S arm is a "dip" line, about 960 m long, whereas the E–W strike line measures close to 1600 m. The thickness of the fan at this location is nearly 15 m. A vertical section through Fan 3 can be divided into four units: (1) lower 3.7 m comprised of thin-bedded (4–10 cm), very fine-grained, flat-based sandstones and siltstones, each overlain by <2-cm thick silty shales; (2) middle 6.2 m of thin- to medium-bedded (10–100 cm), flat-based, tabular sandstones (T_a, T_{ab}, $T_{a\text{-}c}$) with groove casts running SSW–NNE; (3) upper 4.7-m-thick massive sandstone; and (4) top meter of thin- to medium-bedded sandstones alternating with thin silty shales. Fan 3 also shows pinch and swell in the distal area of the 4.7-m-thick layer (J. Wyble and S. K. Huisman, personal communication, 1998). The top of this thick, flat-based sandstone is basically flat in both outcrops, except that the strike section shows at irregular distances (250–450 m), shallow (50–100 cm), narrow (30–50 m) depressions (see figures on CD-ROM). Aside from some possible amalgamation, no sedimentary structures could be observed that would reveal some type of erosion. Again, such small vertical variations could not be observed on the Mississippi Fan.

Fan 4 is the thickest of the five fans in the Tanqua Karoo (Wickens and Bouma, this volume). Its overall paleocurrent direction is from the west. In the Klipfontein area, just west of Skoorsteenberg, Fan 4 is nearly 60 m thick and maintains that thickness at least over a distance of 10 km in a northeastern direction. Continuing in that direction for about 4 km toward Katjiesfontein (Wickens and Bouma, this volume), the top of Fan 4 is eroded away for about 4 km and the next 5 km is removed by the Tanqua River. At Katjiesberg Fan 4 is complete again, but is only 14 m thick. Along a WNW–ESE section about 1200 m long, one can observe tremendous pitch and swell structures with thicknesses ranging from 42 cm to 7 m (Figure 3). The bottom of that sandstone is flat (D. E. Rehmer and P. R. C. Dudley, personal communication 1998).

The Brushy Canyon Formation in the Delaware Basin shows outer fan terminations that have a similar pinch and swell structure as reported here from the Tanqua Karoo (M.E. Gardner, personal communication, 1998).

CONCLUSIONS

The different interpretations of the distal terminations (fan fringe) of modern and ancient fine-grained submarine fans can indicate two different transport processes, different scales, different types of sediments, a misinterpretation, or a combination of those. The discussion on the "modern" Mississippi Fan is based on deep-tow, side-scan sonar and high-resolution reflection seismics, together with more than 30 piston and gravity cores. In contrast, the discussion on ancient fan fringe characteristics comes from outcrop observations.

Figure 3—Upper layer of Fan 4 at Katjiesberg. Fan 4 at this location has an average thickness of 14 m. The portion of this layer shown in the photograph varies in thickness from 90 cm (flag on SE side), to 7 m in the center, to 42 cm at the flag on the NW side. Distance between the flags is approximately 150 m.

The observations on the Mississippi Fan deal with the upper part of upper Pleistocene fan fringe deposits, accumulated during the waning stage of fan construction. Whether the characteristics are representative for the entire fan in both stratigraphic and areal direction cannot be answered from the present datasets. Lithologically, we notice a decrease in the volume of the sediment of a depositional sublobe and in the sand/shale ratio. It is unknown whether that is controlled by changes in transport and depositional processes or by a decrease in sediment volume.

The described outer fan deposits of the Tanqua Karoo represent systems with less discharge than the Mississippi Fan. It is likely that these may not be a major factor. It is important to realize that the resolution of the observation types differs, that is, the resolution of side-scan sonar and of high-frequency seismics versus detailed outcrop observations differs. Another unknown is whether we are dealing with the same parts of a fan fringe. The fringes reveal a total thinning in downdip direction, often a grain-size fining, and a decrease in layer thickness and in sand/shale ratio. Fracturing resulting from tectonics, as well as the easier weathering of shales compared with sandstones, commonly removes the final termination of the outcropping fan very easily. However, we know enough about these outcrops to define their general location within the overall setting of a fine-grained submarine fan.

The Mississippi Fan interpretations by Twichell et al. (1991, 1992) and Nelson et al. (1992) are completely acceptable, considering the data. The same can be said from the deductions made by the present authors, even if the vertical outcrop exposures are thicker than 3.5 kHz records can reveal.

At the present we can distinguish two types of interpretations: (1) the construction of fine-grained submarine fans/turbidite systems is completely governed by channel flow and the depositional sublobes/sheet sands are very limited in size, and (2) only the base-of-slope and middle fan are governed by leveed channels, whereas the outer fan deposition is controlled by the transition from confined to unconfined flows through distributaries, followed by sheet sand transport (Figure 4). The two interpretations may not be as different as they seem. However, both datasets lack sufficient information to fully answer that question. The answer can be very important with regard to flow processes on the outer fan, especially in the fan fringe area, and to the distribution and connectivity of the sands.

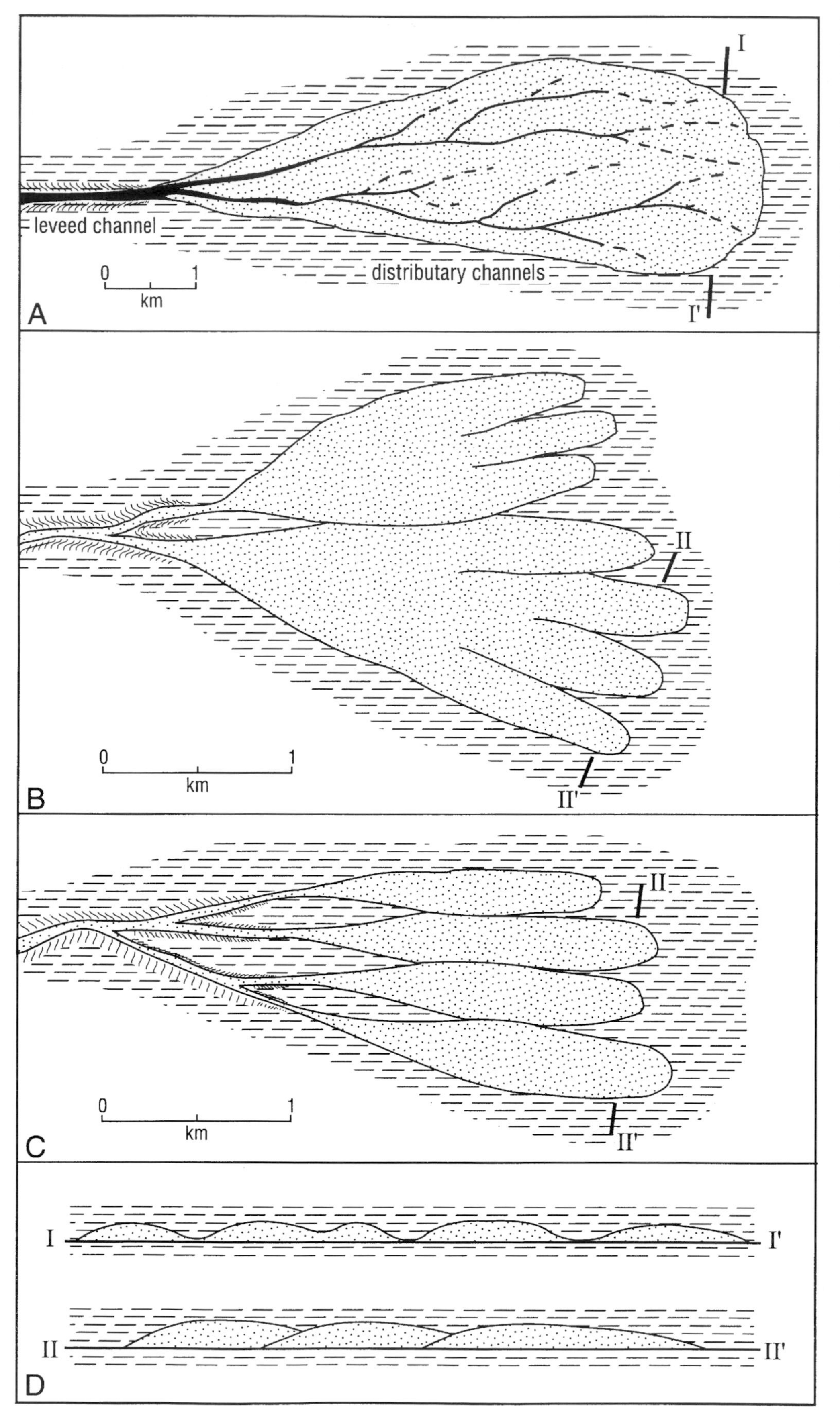

Figure 4—Schematic representations of fan fringe deposits as discussed in this paper: (A) represents Figure 1, showing the channel dominance; (B) shows sheet sand coverage, the sheet flows break up into fingers in the fan fringe area; (C) variety on type B in which no broad sheet sands exist but fingers are formed directly (more or less a combination of A and B); and (D) schematic cross sections: I–I′ through the channelized example showing that most of the fingers are not in communication with one another, II–II′ through sheet sand fingers showing onlap of fingers with excellent connectivity.

Important questions to be answered are the sizes of the observed features. Are they really different and is the difference in scale a critical issue? Comparison of both sets of observations is limited because both studies present a 2-D picture. The lithologies between both sites vary. Are such lithologies important? The large amounts of debris flow deposits, observed in cores from the near-surface deposits of one of the sublobes of the Mississippi Fan, suggest that highly efficient channelized channel flows transported large clasts. The general core description indicates a series of thin-bedded turbidites, debris-flow deposits, and hemipelagic clays (Schwab et al., 1996). The pixel size of the side-scan image for the Mississippi Fan sublobe is 12.5 m and the vertical resolution of the 3.5-kHz seismic instrument is about 2–3 m in a water depth of 3000 m. We do not observe all of those lithologies in the Tanqua Karoo.

The concepts in Figures 1 and 2 suggest channelized transport with overbanking, especially in the outer parts. The side-scan sonar backscatter images certainly support such an interpretation. Nelson et al. (1992) present a conceptional block diagram in which a major feeder channel starts at the base-of-slope and transports sediment to a point where the sublobe (depositional lobe) starts. Frequent bifurcations result in a large number of sublobe channels that are, in cross section, composed of a set of lenticular turbidite and debris flow deposits. No mention is made of the fan fringe.

Most of the Tanqua Karoo examples may well be located a little more inside the edge of the fan. However, the deposits do not look channel-fill shaped but sheet shaped with a thick layer near the top, which shows an irregular pinch-and-swell upper surface. Although no amalgamation contacts could be detected, which may be a result of the well-sorted fine-grain size, the deposits certainly would not be identified as channel fills.

At this time there are too many uncertainties to be able to draw hard conclusions and to suggest that the schematics in Figure 4 are completely acceptable. It is important, however, that more observations are being conducted in different areas because of the different types of datasets. The outcomes of these studies will help to better understand the involved transport and depositional processes, and the distribution and connectivity characteristics of the sands.

ACKNOWLEDGMENTS

The authors are thankful to Dave Twichell and to Jack Thomas for their in-depth reviews and suggested additions to make this contribution more valuable and readable. Ronald Phair suggested the addition of the application of diffusion limited aggregates to distal fan deposition, but we left it out because of length. Thanks to Lieneke Bouma for typing the various versions of the manuscript.

REFERENCES CITED

Basu, D., 1997, Characterization of thin-bedded turbidites from the Permian Tanqua Karoo submarine fan deposits, South Africa: Unpublished Ph.D. dissertation, Louisiana State University, Baton Rouge, 159 p.

Bouma, A. H., W. R. Normark, and N. E. Barnes, eds., 1985, Submarine fans and related turbidite systems: New York, Springer-Verlag, 351 p.

Bouma, A. H., et al., 1986, Initial reports of the Deep Sea Drilling Project Leg 96: Washington, D.C., U.S. Government Printing Office, 824 p.

Bouma, A. H., and H. DeV. Wickens, 1994, Tanqua Karoo ancient analog from fine-grained submarine fans, *in* P. Weimer, A. H. Bouma, and B. F. Perkins, eds., Submarine fans and turbidite systems: sequence stratigraphy, reservoir architecture and production characteristics, Gulf of Mexico and international: Gulf Coast Section SEPM Foundation 15th Annual Research Conference Proceedings, p. 23–34.

Bouma, A. H., 1997, Comparison of fine-grained, mud-rich and coarse-grained, sand-rich submarine fans for exploration-development purposes: Gulf Coast Association of Geological Societies Transactions, v. 47, p. 59–64.

Kirkova, J. T., 1998, Architectural changes within a fine-grained submarine fan, Tanqua Karoo, South Africa: Unpublished M.S. thesis, Baton Rouge, Louisiana State University, 130 p.

Mutti, E., and F. Ricci Lucchi, 1972, Le torbiditi dell'Appennino settentrionale: Introduzione all'analisi di facies: Memorie dell Societa Geologica Italiana, v. 11, p. 161–199.

Nelson, C. H., D. C. Twichell, W. C. Schwab, H. J. Lee, and N. H. Kenyon, 1992, Late Pleistocene turbidite sand beds and chaotic silt beds in the channelized distal outer fan lobes of Mississippi Fan: Geology, v. 20, p. 693–696.

Reading, H. G., and M. Richards, 1994, Turbidite systems in deep-water basin margins classified by grain size and feeder system: AAPG Bulletin, v. 78, p. 792–822.

Rozman, D. J., 1998, Characterization of a fine-grained outer submarine deposit, Tanqua-Karoo Basin, South Africa: Unpublished M.S. thesis, Louisiana State University, Baton Rouge, 147 p.

Schwab, W. C., H. J. Lee, D. C. Twichell, J. Locat, C. H. Nelson, W. C. McArthur, and N. H. Kenyon, 1996, Sediment mass flow processes on a depositional lobe, outer Mississippi Fan: Journal of Sedimentary Research, v. 66, p. 916–927.

Twichell, D. C., N. H. Kenyon, L. M. Parson, and B. A. McGregor, 1991, Depositional patterns of the Mississippi Fan surface: Evidence from GLORIA II and high resolution seismic profiles, *in* P. Weimer and M. H. Link, eds., Seismic facies and sedimentary processes of submarine fans and turbidite systems: New York, Springer-Verlag, p. 349–363.

Twichell, D. C., W. C. Schwab, C. H. Nelson, N. H. Kenyon, and H. J. Lee, 1992, Characteristics of a sandy depositional lobe on the outer Mississippi Fan from SeaMARC IA sidescan sonar images: Geology, v. 20, p. 689–692.

Twichell, D. C., W. C. Schwab, and N. H. Kenyon, 1995, Geometry of sandy deposits at the distal edge of the Mississippi Fan, Gulf of Mexico, *in* K. T. Pickering, R. N. Hiscott, N. H. Kenyon, F. Ricci Lucchi, and R. D. A. Smith, eds., Atlas of deep water environments: architectural style in turbidite systems: London, Chapman & Hall, p. 282–286.

Weimer, P., 1989, Sequence stratigraphy of the Mississippi Fan (Plio–Pleistocene), Gulf of Mexico: Geo-Marine Letters, v. 9, p. 185–272.

Wickens, H. DeV., 1994, Submarine fans of the Ecca Formation: Unpublished Ph.D. dissertation, University of Port Elizabeth, South Africa, 350 p.

Wynn, R. B., D. G. Masson, D. A. V. Stow, and P. P. E. Weaver, 2000, Turbidity current sediment waves in subsurface sequences, *in* A. H. Bouma and C. G. Stone, eds., Fine-grained turbidite systems, AAPG Memoir 72/SEPM Special Publication No. 68, p. 299–306.

Chapter 26

Turbidity Current Sediment Waves in Subsurface Sequences

Russell B. Wynn
Douglas G. Masson
Dorrik A. V. Stow
Phillip P. E. Weaver
Southampton Oceanography Centre, Southampton
Hampshire, United Kingdom

ABSTRACT

Two sediment wave fields on the submarine slopes of the Canary Islands display wave heights up to 70 m and wavelengths up to 2.4 km. Wave sediments consist of fine-grained turbidites and pelagic/hemipelagic sediments. The sediment waves are formed beneath unconfined turbidity currents, and are similar to sediment waves found on channel-levee backslopes.

Sediment wave morphology is resolvable on high-resolution seismic profiles. In areas lacking high-resolution seismic data, analysis of dipmeter readings may provide a useful tool for recognizing buried sequences of migrating waves. Thick sequences of sediment waves will impart a marked heterogeneity to a potential reservoir, leading to complications during reservoir production.

INTRODUCTION

Sediment waves formed beneath turbidity currents are large-scale depositional bed forms that occur in a variety of deepwater turbiditic environments. They can be broadly subdivided into two groups, on the basis of their grain size, environment, and type of depositional flow:

1. Gravel waves have been described from a number of modern fan valleys (Normark and Piper, 1991) and are interpreted as having been formed beneath channelized turbidity currents. They typically display wavelengths of less than 100 m and wave amplitudes of less than 10 m.
2. Fine-grained turbidity current waves occur on channel-levee backslopes and on the flanks of islands and seamounts (Normark et al., 1980; Carter et al., 1990; McCave and Carter, 1997; Nakajima et al., 1998). They typically display wavelengths of 1–6 km and wave heights of 5–70 m and are formed beneath unconfined turbidity currents. The wave crests are up to 60 km long and are aligned parallel to the slope. There is generally a regular downslope decrease in wave dimensions and the waves often display upslope migration. Wave sediments consist of thin-bedded sand/silt turbidites interbedded with pelagic/hemipelagic sediments. Individual

turbidites are thicker on the upslope face of the wave and are thinner, or eroded, on the downslope face.

Many previous studies of turbidity current–deposited sediment waves are restricted because of low-quality or sparse data. The present study uses a variety of high-quality data, including 3.5-kHz profiles, single-channel seismic profiles, GEOSEA 3-D imagery, and sediment cores. The principal objectives of this paper are to (1) summarize the data presented by Wynn et al., describing two sediment wave fields on the submarine slopes of the Canary Islands; (2) compare these data with other published data on fine-grained turbidity current waves; and (3) discuss how the results may be applied to hydrocarbon exploration, in particular the recognition and interpretation of sediment wave sequences in the subsurface.

STUDY AREA

The two sediment wave fields investigated are located on the submarine slopes of the Canary Islands (Figure 1). The La Palma wave field lies to the northwest of La Palma Island and is centered on 18°30′W, 29°30′N. The Selvage wave field is situated on the lower rise south of the Agadir Basin, at 15°45′W, 31°10′N. The continental slope in this region is 100–200 km wide and passes to the continental rise in water depths of 4000–4500 m. The rise is 400–600 km wide and displays gradients of about 1° on the lower slope/upper rise to 0.1° on the lower rise (Masson et al., 1992).

PREVIOUS WORK

The sediment waves on the submarine slopes of the northern Canary Islands were first described by Jacobi et al. (1975). They mapped a nearly continuous zone of sediment waves on the lower rise, lying parallel to the regional bathymetric trend. Those authors found that the present-day bottom current circulation in the area is too weak to have formed the waves, and suggested that relatively stronger bottom currents in the recent past may have been responsible for wave formation. However, they also discussed the possibility that turbidity currents may have influenced wave distribution. Masson et al. (1992) and Jacobi and Hayes (1992) mapped the La Palma wave field and concluded that bottom currents were probably responsible for wave formation. However, recent work by Wynn et al. (in press a, b), using the newly obtained dataset, suggest that the sediment waves in this area have been formed and maintained by unconfined turbidity currents.

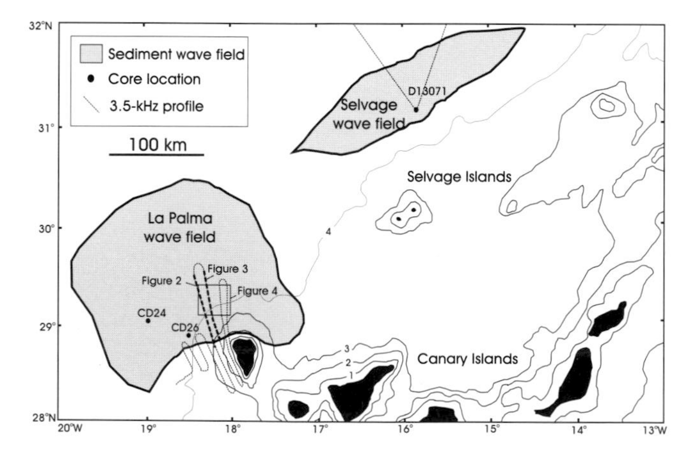

Figure 1—Location map of the study area, highlighting the position of the La Palma and Selvage wave fields, and relevant core positions. Locations of Figures 2–4 are indicated. Bathymetric contours shown in kilometers.

DATA COLLECTION

The dataset used was collected during RRS *Charles Darwin* cruises 56 and 108, and RRS *Discovery* cruise 225. A variety of data were collected to facilitate detailed surface and subsurface characterization of the waves.

3.5-kHz Profiles

We obtained 3.5-kHz profiles across both sediment wave fields (Figure 2). Analysis of the bathymetry, and previously obtained GLORIA sidescan sonar images, ensured that the profiles were taken along a line roughly perpendicular to the wave crests.

Seismic Profiles

A series of single-channel seismic profiles, using a single 300 in^3 airgun, were collected from the region of the La Palma wave field (Figure 3). The profiles used in this study were taken roughly perpendicular to the regional bathymetric gradient and the wave-crest lines.

Multibeam Bathymetry

The La Palma wave field was also imaged using the Simrad EM12 multibeam echo sounder. These data were then processed using GEOSEA modeling software, which produces a high-resolution 3-D image of the waves (Figure 4).

Sediment Cores

Cores were recovered from both wave fields. Three short (<2 m) Kasten cores were obtained from the southern margin of the La Palma wave field, and a 12 m piston core was obtained from the Selvage wave field (Figure 5). The core locations are shown in Figure 1.

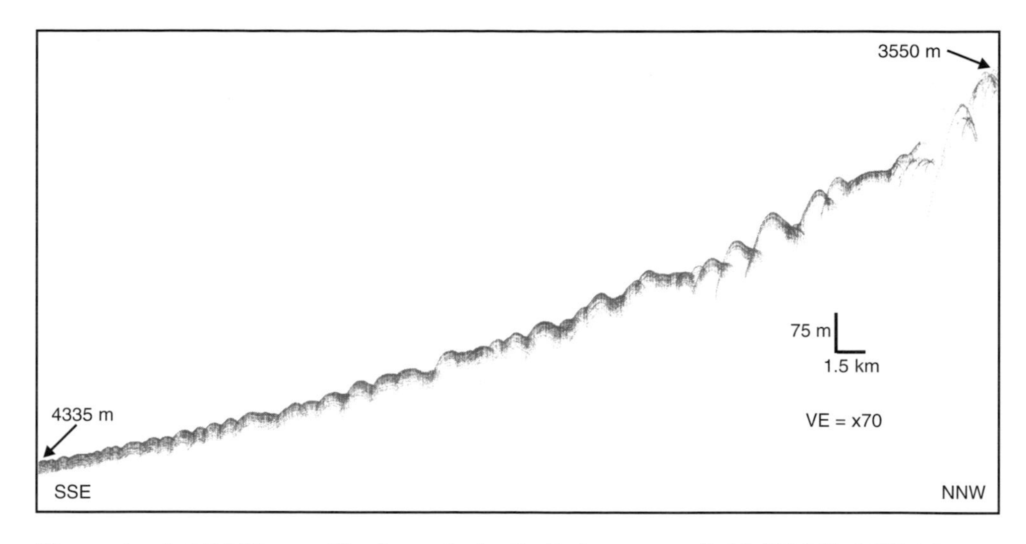

Figure 2—A 3.5-kHz profile through the La Palma wave field (NNW–SSE). Note the regular downslope decrease in wavelength and wave height, and the gradual increase in penetration depth. Waves show clear evidence of upslope migration and, in some cases, show evidence of increased sedimentation on the stoss face. Location is shown in Figure 1.

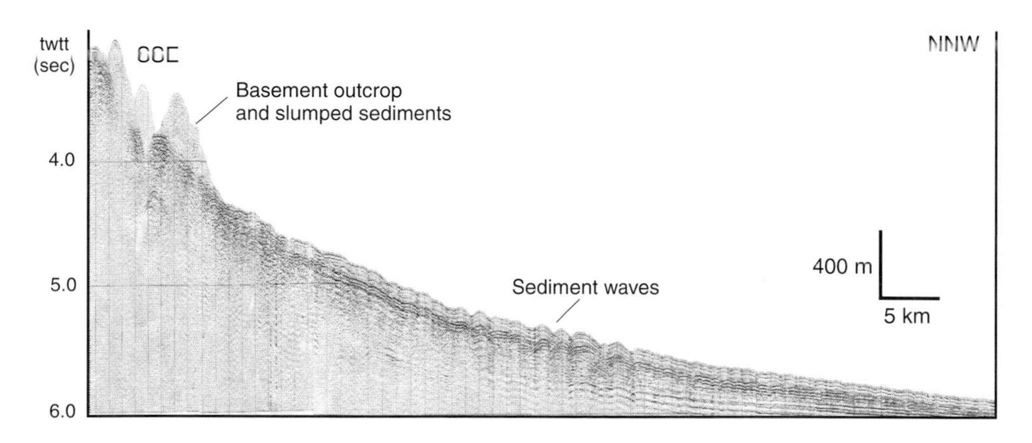

Figure 3—Single-channel seismic profile through the La Palma wave field. Location is shown in Figure 1.

RESULTS

The La Palma Wave Field

Surface Morphology

3.5-kHz and single-channel seismic profile data across the La Palma wave field reveal that the sediment waves at the surface display heights of less than 5–70 m and wavelengths of 0.4–2.4 km (Figures 2, 3). There is generally a slight asymmetry across the waves, with a steeper downslope face. There is also a regular downslope decrease in wave dimensions (Figure 2). GEOSEA images of the waves indicate that the wave crests are aligned roughly parallel to the regional gradient and are laterally continuous for up to 40 km (Figure 4). The crestline morphology is complex, with common bifurcation and well-developed sinuosity.

Subsurface Morphology

Single-channel seismic profiles through the upper wave field (Figure 3) reveal that the sediment waves are still present to a subsurface depth of at least 470 m (assuming a sediment velocity of 1600 m/s). Downslope, the sediment wave sequence gradually decreases to a thickness of 125 m. The waves appear to have been migrating upslope throughout deposition of the sequence. Sediment cores recovered from the upper 2 m of the wave field contain a number of thin volcaniclastic turbidites (≤20 cm thick) interbedded with bioturbated pelagic marls and oozes.

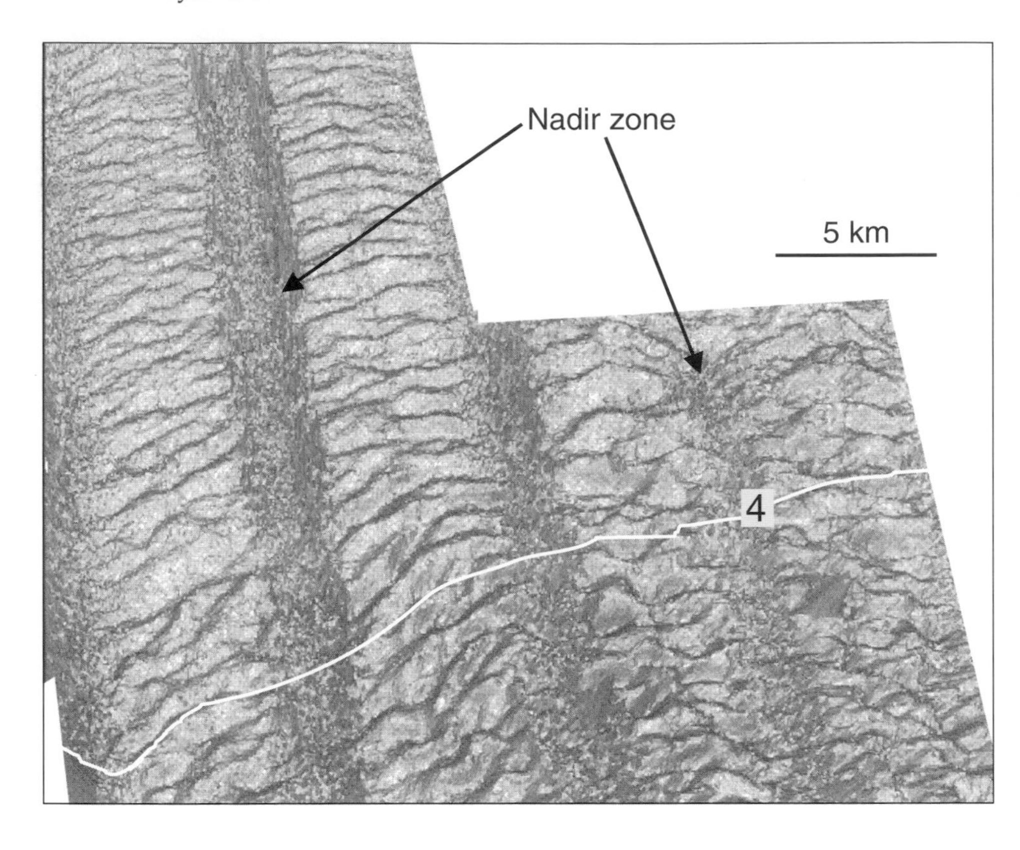

Figure 4—A GEOSEA 3-D image of the upper reaches of the La Palma wave field (oblique plan view). Note the common bifurcation and well-developed sinuosity of the wave crests. White dashed line with number is depth contour in kilometers. Vertical exaggeration (VE) = ×50. View is looking NNW toward the top of the page. Illumination direction is from the south. Nadir zone is a zone of "noise" beneath the ship. Location is shown in Figure 1.

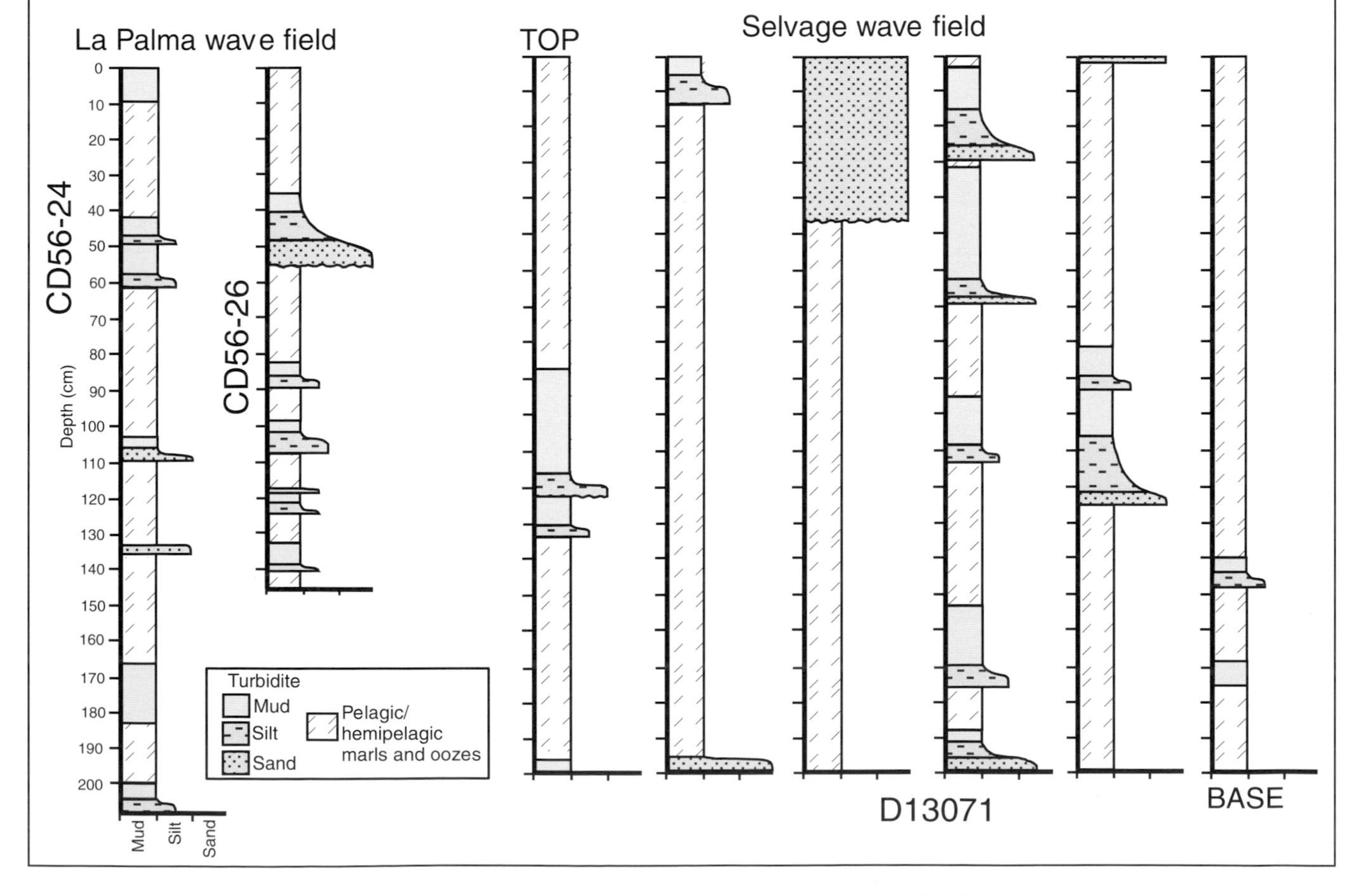

Figure 5—Sediment core descriptions from the La Palma and Selvage wave fields. Note the maximum sand-bed thickness of 50 cm. Core locations are shown in Figure 1.

The Selvage Wave Field

Surface morphology

A 3.5-kHz profile across the wave field reveals that the surface waves have a wavelength of 0.4–1.1 km (Figure 2). The wave height reaches a maximum of 7 m at 4300 m water depth, and gradually decreases as the waves die out downslope at 4370 m.

Subsurface morphology

The wave morphology is present to a depth of at least 30 m, and the waves have migrated upslope throughout deposition of the sequence (Figure 2). Core D13071 is 12 m long and was collected from the wave field at a water depth of 4311 m. It contains a series of 12 turbidites interbedded with bioturbated pelagic marls and oozes (Figure 5). Some of the turbidites have thin sandy bases, with sand-bed thicknesses reaching 50 cm. Overall, the core contains 5% sand.

DISCUSSION

Wave-Forming Processes

Detailed study of the La Palma and Selvage sediment wave fields (Wynn, Masson, et al., in press; Wynn, Weaver, et al., in press) revealed that the sediment waves have been formed and maintained by unconfined turbidity currents originating on the flanks of the Canary Islands. Evidence for this interpretation includes:

1. The large number of turbidites seen in cores recovered from the wave fields. In addition, the sediments show no evidence of bottom current sedimentation or reworking.
2. The regular downslope decrease in wave dimensions is typical of sediment waves formed by unconfined turbidity currents, e.g., Monterey Fan levee-backslope (Normark et al., 1980). Bottom current waves typically display more irregular dimensions.
3. The bottom current flow in this area is not strong enough at present to maintain the waves, although flow velocities may have been stronger in the past and/or subject to episodic fluctuations (Jacobi et al., 1975; Lonsdale, 1982; Jacobi and Hayes, 1992).
4. The crestline orientation of the waves is parallel to the bathymetric contours. This is compatible with a turbidity current origin. Recent models of mudwave dynamics (Blumsack and Weatherly, 1989) have revealed that sediment waves formed by bottom currents flowing alongslope will generally develop at an angle to the flow direction and the regional slope.

Most previous descriptions of fine-grained turbidity current waves are from channel-levee backslopes. The La Palma and Selvage wave fields differ because they occur on unchannelized open slopes on the continental slope and rise. However, the dimensions of the La Palma and Selvage sediment waves are very similar to those described from channel-levee backslopes (Table 1). In addition, the single-channel seismic section through the La Palma wave field (Figure 3) reveals that the overall wave package displays very similar characteristics to those taken through wave sequences on channel-levee backslopes (Carter et al., 1990; Piper and Savoye, 1993; Savoye et al., 1993; Nakajima et al., 1998).

These results therefore strongly suggest that the sediment waves on the flanks of the Canary Islands were formed by unconfined turbidity currents and have very similar characteristics to those described from channel-levee backslopes around the world.

Recognition of Sediment Waves in the Subsurface

Turbidity current sediment waves can be recognized in the subsurface at a variety of scales:

On 3.5-kHz profiles, surficial sediment waves can typically be resolved to subsurface depths of 20–60 m. Migration direction and angle are clearly visible, as are individual wave dimensions. The seismic character is typically one of low to moderate amplitude, more or less continuous, subparallel wavy reflectors. The higher amplitude reflectors most likely correlate with more sandy sections. They can be distinguished from compressional features such as creep folds (Hill et al., 1982; Mulder and Cochonat, 1996) by the fact that they migrate upslope and they show a regular downslope decrease in wave dimensions. Compressional features show no evidence of migration, and the varying dimensions are clearly not related to changes in slope angle.

On single-channel seismic profiles, relatively thick and laterally extensive successions of sediment waves can be resolved within the upper few hundred meters of section. The seismic character is one of regular wavy reflectors of moderate amplitude. Individual reflectors appear discontinuous, although the whole seismic facies is more continuous. Apparent migration direction can be resolved in some cases, but wave dimensions are not readily determined.

On lower-resolution seismic profiles it is generally not possible to recognize subsurface sediment waves on the basis of their morphology. It is therefore likely that sediment wave sequences in turbidite environments, such as occur on channel-levee backslopes and on unconfined slope aprons, are frequently overlooked during hydrocarbon exploration. This has important implications for the reservoir production of thin sands in interpreted levee sequences, because a stacked sequence of migrating waves on a channel levee is likely to impart a marked vertical and lateral heterogeneity to any potential reservoir.

Subsurface Applications

As this study has shown, the La Palma and Selvage wave fields have similar characteristics to wave fields described from channel-levee backslopes. Therefore, a

Table 1. Dimensions and key features of the La Palma and Selvage wave fields, compared with published data on wave fields in channel-levee environments.

Wave Field	Length (km)	Height (m)	Features	Sediments	Reference
La Palma	0.4–2.4	<5–70	Wave length + height decrease downslope. Waves migrate upslope.	Thin volcaniclastic turbidites interbedded with pelagics	Wynn, Masson, et al., in press
Selvage	0.4–1.1	0–7	Wave length + height decrease downslope. Waves migrate upslope.	Sand/silt turbidites interbedded with pelagic marls/oozes	Wynn, Weaver, et al., in press
Monterey Fan Levee	0.3–2.1	2–37	Wave length + height decrease downslope. Waves migrate upslope.		Normark et al., 1980
Bounty Channel Levee	0.6–6	2–17	Wave length + height decrease downslope. Waves migrate upslope.	Thin-bedded sand/silt turbidites and hemipelagites	Carter et al., 1990
Var Fan Levees	1–7	<50	Waves migrate upslope.		Savoye et al., 1993
Toyama Channel Levee	0.5–3	<70	Wave length + height decrease downslope. Waves migrate upslope.		Nakajima et al., 1998

series of measurements taken from these wave fields may help to increase our understanding of subsurface levee sequences where sediment waves are present. Nevertheless, how do we determine the presence or absence of sediment waves if they are not resolvable on standard seismics? One possible method involves the use of dipmeter readings. Theoretically, a dipmeter log taken through a wave sequence on a levee backslope should show fluctuating values, because a core taken through such a sequence will penetrate a series of wave crests, upslope faces, wave troughs, and downslope faces (Figure 6). Analyses of measured dip readings from the La Palma wave field, based on high-resolution seismic profiles, have shown that in the

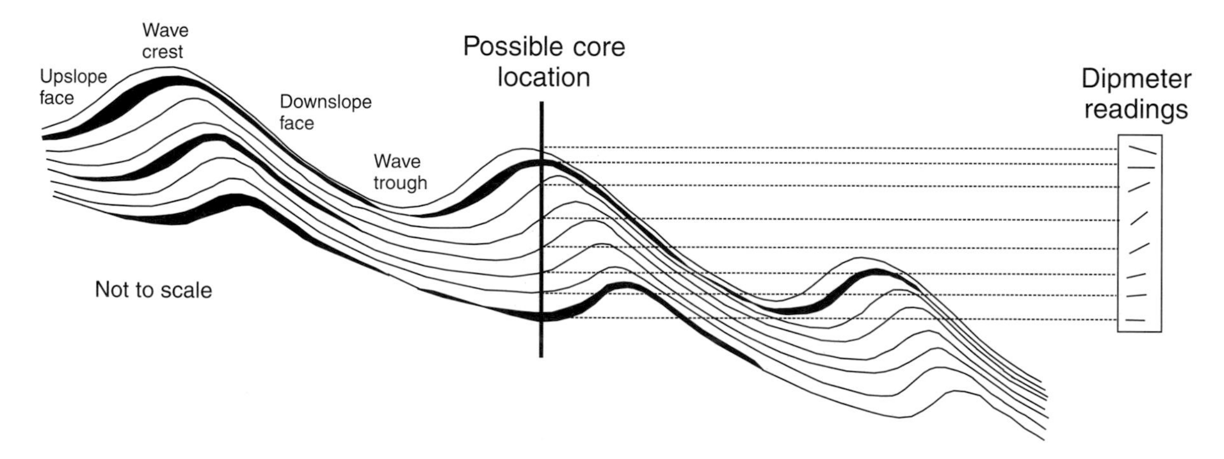

Figure 6—Schematic diagram illustrating a hypothetical sequence of migrating sediment waves in the subsurface. A potential series of dipmeter readings is given for a vertical core taken through the waves. Note the fluctuations in dip readings downcore, and the possible lateral discontinuity and variation of any sandy units within the sequence. Black areas represent sandy units (≤50 cm thick).

upper wave field the overall slope is 1.2°. The upslope faces of the waves dip upslope at an average of 1.5°, although the downslope faces dip downslope at an average of 3.2°. There is, therefore, an average difference of almost 5° in the angle between the upslope and downslope faces. In the lower wave field this value decreases to 1.6°. These results suggest that a dipmeter log taken through a series of stacked migrating waves will display dip readings that show regular fluctuations in the region of 1°–5°. In addition, subsurface sequences that preserve the original paleoslope (typically 1°–2° on channel levees) may show regular reversals in the dip azimuth. Unfortunately, existing cores from known sediment wave fields are not long enough to test this hypothesis.

There are, however, many examples from the deep subsurface in which regular fluctuations of dipmeter angle have been noted, but no generally accepted interpretation of these patterns exists. One striking example is from the Upper Jurassic fine-grained turbidite successions seen in several wells in the Brae Field area of the North Sea. According to Stow et al. (1982), these were deposited in an open slope setting distal to the gravels and sands of a coarse-grained slope apron system. They interpret the fluctuating dip motif in these wells in terms of prograding mud lobes, but we suggest here that an alternative model is one of a relatively thick sequence of turbidity current waves deposited by unconfined flows on an open slope.

If sediment waves are recognized within a subsurface sequence, it becomes necessary to understand their reservoir characterization properties. Although the sand/mud ratios in fine-grained sediment wave fields are generally low, the sands may have high poroperm values and be economically viable (A. H. Bouma, personal communication, 1998). In a stacked sequence of migrating sediment waves, these sandy units will be unevenly distributed. There will theoretically be a marked vertical and lateral heterogeneity in sand-bed thickness and grain size, which hinders fluid flow through the reservoir (Figure 6). Generally, the upslope faces of the waves comprise thicker beds than the downslope faces, and in some cases beds may actually pinch out on the downslope face. Therefore, the thin sandy units within a sediment wave sequence will vary in thickness laterally, and may pinch out downslope. In addition, the grain size may also vary laterally. An individual bed that is laterally continuous across a wave crest may consist of silt-size sediment on the upslope face, that grades into sand-size sediment on the downslope face. This heterogeneity will be most marked in the proximal areas of the wave field, where the largest variations in bed thickness and grain size occur.

CONCLUSIONS

This study describes two extensive sediment wave fields that formed beneath unconfined turbidity currents derived from the flanks of the volcanic Canary Islands. The morphology and distribution of the waves is similar to that described from a number of wave fields on channel-levee backslopes. This allows us to infer that the turbidity current processes acting on the continental slope and rise north of the Canary Islands are very similar to those occurring in channel-levee backslope environments.

In the subsurface, turbidity current sediment waves can be recognized at a variety of scales. On 3.5-kHz and single-channel seismic profiles, they can be recognized and distinguished from gravitational creep features on the basis of their morphology and upslope migration. On lower-resolution seismic profiles individual sediment waves are not generally resolvable in cored sequences; however, they may be detected using detailed dipmeter analysis. Where they do occur, subsurface sequences of migrating sediment waves are likely to impart a marked vertical and lateral heterogeneity to any potential hydrocarbon reservoir. This will lead to complexities at the reservoir production stage.

ACKNOWLEDGMENTS

The authors gratefully acknowledge the contribution of the masters and crew of the RRS *Charles Darwin* and RRS *Discovery*. Andy Pulham is thanked for helpful comments on various elements of this study. The manuscript benefited greatly from the reviews of Arnold Bouma, John Bratton, and John Southard. RBW acknowledges the Ph.D. research grant supplied jointly by the University of Southampton and the Southampton Oceanography Centre Challenger Division.

REFERENCES CITED

Blumsack, S. L., and G. L. Weatherley, 1989, Observations of the nearby flow and a model for the growth of mudwaves. Deep-Sea Research, v. 36, p. 1327-1339.

Carter, L., R. M. Carter, C. S. Nelson, C. S. Fulthorpe, and H. L. Neil, 1990, Evolution of Pliocene to Recent abyssal sediment waves on Bounty Channel levees, New Zealand: Marine Geology, v. 95, p. 97–109.

Hill, P. R., K. M. Moran, and S. M. Blasco, 1982, Creep deformation of slope sediments in the Canadian Beaufort Sea: Geo-Marine Letters, v. 2, p. 163–170.

Jacobi, R. D., P. D. Rabinowitz, and R. W. Embley, 1975, Sediment waves on the Moroccan continental rise: Marine Geology, v. 19, p. M61–M67.

Jacobi, R. D., and D. E. Hayes, 1992, Northwest African continental rise: effects of near-bottom processs inferred from high-resolution seismic data, *in* C. W. Poag and P. C. de Graciansky, eds., Geologic evolution of Atlantic continental rises: Berlin, Reinhold, p. 293-325.

Lonsdale, P., 1982, Sediment drifts of the northeast Atlantic and their relationship to the observed abyssal current: Bulletin Institute de Geologie Bassin d'Aquitaine, Bordeaux, v. 31, p. 141-150.

McCave, I. N., and L. Carter, 1997, Recent sedimentation beneath the Deep Western Boundary Current off northern New Zealand: Deep-Sea Research I, v. 44, p. 1203-1237.

Masson, D. G., R. B. Kidd, J. V. Gardner, Q. J. Huggett, and P. P. E. Weaver, 1992, Saharan continental rise: Facies distribution and sediment slides, *in* C. W. Poag and P. C. de Graciansky, eds., Geologic evolution of Atlantic continental rises: Berlin, Reinhold, p. 327-346.

Mulder, T., and P. Cochonat, 1996, Classification of offshore mass movements: Journal of Sedimentary Research, v. 66, p. 43–57.

Nakajima, T., M. Satoh, and Y. Okamura, 1998, Channel-levee complexes, terminal deep-sea fan and sediment wave fields associated with the Toyama Deep-Sea Channel system in the Japan Sea: Marine Geology, v. 147, p. 25-41.

Normark, W. R., G. R. Hess, D. A. V. Stow, and A. J. Bowen, 1980, Sediment waves on the Monterey Fan levee: A preliminary physical interpretation: Marine Geology, v. 37, p. 1-18.

Normark, W. R., and D. J. W. Piper, 1991, Initiation processes and flow evolution of turbidity currents: Implications for the depositional record: SEPM Special Publication 46, p. 207-230.

Piper, D. J. W., and B. Savoye, 1993, Processes of late Quaternary turbidity current flow and deposition on the Var deep-sea fan, north-west Mediterranean Sea: Sedimentology, v. 40, p. 557-582.

Savoye, B., D. J. W. Piper, and L. Droz, 1993, Plio-Pleistocene evolution of the Var deep-sea fan off the French Riviera: Marine and Petroleum Geology, v. 10, p. 550–571.

Stow, D. A. V., C. D. Bishop, and S. J. Mills, 1982, Sedimentology of the Brae oilfield, North Sea: fan models and controls: Journal of Petroleum Geology, v. 5, p.129–148.

Wynn, R. B., D. G. Masson, D. A. V. Stow, and P. P. E. Weaver, in press, Turbidity current sediment waves on the submarine slopes of the Canary Islands: Marine Geology.

Wynn, R. B., P. P. E. Weaver, G. Ercilla, D. A. V.Stow, and D. G. Masson, in press, Sedimentary processes in the Selvage wave field, NE Atlantic: a new model for the formation of turbidity current sediment waves: Sedimentary Geology.

Thomas, J. B., 2000, Lithology-driven rock calibration: "laboratory" modeling, *in* A. H. Bouma and C. G. Stone, eds., Fine-grained turbidite systems, AAPG Memoir 72/SEPM Special Publication No. 68, p. 307–316.

Chapter 27

Lithology-Driven Rock Calibration: "Laboratory" Modeling

J. B. Thomas
BP Amoco (retired)
Houston, Texas, U.S.A.

ABSTRACT

The critical step necessary to model seismic response for composite deepwater fan sequences is creation of "pseudowell" logs using high-quality outcrop data. The "laboratory" outcrop model can be built to generate "pseudologs" of gamma ray, resistivity, density, and velocity from outcrop descriptions. Synthetic acoustic impedance, reflectivity, and synthetic seismogram curves then are generated on the same scale as measured thicknesses. Extrapolation from outcrop to seismic data is completed by resorting to geophysical data from depositionally similar basins under study, such as the deepwater of the Gulf of Mexico. By varying fluid and rock properties iteratively, it is possible to "forward-model" hydrocarbon presence and its influence on seismic attributes (e.g., AVO) calibrated with rock and log data.

The Permian Skoorsteenberg Formation of the Tanqua Karoo Basin in South Africa is a natural "laboratory" in which to demonstrate how detailed outcrop data can be used to improve rock-calibrated seismic analysis of prospective deepwater fan exploration targets.

INTRODUCTION

The extensive fine-grained fan sequence exposed in the Tanqua Karoo subbasin, South Africa, is an ideal "natural laboratory" in which to demonstrate how lithology-based geophysical models on a seismic wavelet scale can be created. The outcrop-driven approach helps geoscientists prospecting for deepwater fan hydrocarbon traps if limited to a seismic dataset and insufficient local well control for rock property calibration. Apparent amplitude anomalies can then be tested for response due to contrasts in lithology, fluid properties, or their combination.

Lithologically based seismic and log response models can greatly improve the level of understanding of a prospect before drilling. A "normal" sequential work flow to identify a lead seismically is acquisition, processing, migration, and isochron mapping. Depth-converted maps are used to estimate the trap area, thickness, and relationship between hydrocarbon source and trap. Reservoir "quality" estimates such as porosity, permeability, and internal architecture are constrained using well control (e.g., logs, cuttings and/or core, flow tests). Finally, attribute analysis completes "definition" of the prospect. The accuracy of rock/fluid assessment ultimately impacts the exploration risk level assigned to the prospect. Oftentimes that assessment is the last and least analyzed information in prospect risk evaluation. It is this at this point when the rock-driven modeling can have its greatest value.

Basic Assumptions for Model Building

1. Outcrop documentation of the geometry, depositional setting, and internal architecture of reservoir and seal lithologies is accurate.
2. The total stratigraphic thickness of outcrops (e.g., 200 m) exceeds the calculated seismic tuning thickness.
3. "Pseudologs" can be generated based on the ideal log responses for minerals in the outcrops and thicknesses for each lithology; these data are available from logging service companies.
4. Interpretations of depositional conditions (i.e., grain size, basin type) must be similar for the model and the prospect areas.
5. Synthetic seismograms can be created on the same scale as the lithologic sections through acoustic impedance curves and assuming reasonable velocities for each lithology.
6. Fluid substitution data can be added to simulate oil-, gas-, or water-filled reservoir cases.
7. Iterations of rock/fluid combinations can be done on a workstation allowing direct integration of seismic, log, and outcrop models.

Based on observable lithologic associations, the assumptions outlined allow the explorationist and engineer to build seismic models for different types of reservoirs. More sophisticated forward and backward modeling of reservoir systems calibrated between seismic, log, and rock/fluid data are the result. The Permian Skoorsteenberg Formation in the Tanqua Karoo subbasin of South Africa (Figure 1) serves as a calibration model for deepwater fan reservoir sandstones (Wickens and Bouma, this volume). Similar procedures can be used to develop models for nearshore marine, estuarine, fluviodeltaic, and other nonmarine sandstone reservoirs and carbonates. For any outcrop to be useful in such modeling, the exposure thicker than the length of the seismic wavelet and outcrop gamma-ray measurements must be available for calibration purposes.

DISCUSSION OF DATA GATHERING

Outcrop studies led by Bouma and Wickens (1991, 1994) and graduate students from Louisiana State University and Stellenbosch University provide detailed documentation of the Permian Ecca Group, Skoorsteenberg Formation stratigraphic succession (Wickens and Bouma, this volume). They have divided the 1300-m-thick sequence into five mappable, fine-grained sand-rich deepwater fan sand bodies separated by interfan shales and siltstones. Bed thicknesses, sedimentary structures, and types of lithologic contacts help confirm the processes forming the depositional sequence; outcrop gamma-ray measurements taken at fixed vertical distances (Figure 2) provides the link to pseudolog creation. Sandstones in the five fans are the result of confined (in-channel) or unconfined (levee-overbank and sheet deposits) flow built out onto a stable basin slope and floor and derived from an igneous-metamorphic upland (Scott et al., this volume). At the Skoorsteenberg locality (Figures 1, 3), a continuous exposure of Fans 3, 4, and 5 are ideal for developing the rock-driven seismic model. Detailed discussion of each fan and the Skoorsteenberg section appear in Wickens (1994), Bouma and Wickens (1994), and Wickens and Bouma (this volume).

The fan sequences are interpreted by Wickens and Bouma (this volume) to have been deposited by processes similar to those forming the Tertiary fans in the deepwater Gulf of Mexico. Fans in the two basins contain sediments with similar grain-size variation, sedimentary structures, and depositional geometry. However, the Pangean source terrain for Skoorsteenberg sediments was probably syndepositionally deformed (Wickens, 1994; Scott, 1997; Scott et al., this volume) and did not involve salt tectonism in developing combination structural-stratigraphic hydrocarbon traps as in the Gulf of Mexico.

Petrography/Mineralogy of Fans

Mineral abundances are reported from thin section, X-ray diffraction, and FTIR (Fourier Transform Infrared Spectrometry) analyses. Fan sandstones of the Tanqua Karoo subbasin are enriched in quartz and albite feldspar compared with the shales. The shales are also feldspathic but relatively enriched in illite. Finely disseminated pyrite and organic material are concentrated in the shales, which are darker than associated sandstones and siltstones.

Table 1 reports mineralogy for shale and sandstone samples collected by Marot (1992), Rozman (1998), and Thomas (this paper) from Fans 1, 2, 3, and 4 of the Skoorsteenberg Formation. Soekor data (Marot, 1992) are normative thin section estimates. Using McBride's classification scheme (1963), all samples are feldspathic litharenites (10–50% feldspar, >25% lithic fragments, and generally less than 25% quartz). Rozman's (1998) data from Fan 2 at Losberg are visual estimates but from siltstone and very fine-grained samples, making identification difficult. Quantitative mineralogy determinations by the FTIR technique differ somewhat because polymineralic rock fragments (e.g., granites, phyllites) are reported by single mineral species. In all cases, the accessory minerals such as calcite and dolomite are interpreted as late-stage replacement or cements and therefore are not indicative of the sediment provenance. Original pore space has been reduced by burial compaction or formation of quartz and feldspar overgrowths during pressure solution or burial.

Mineralogy vs. Log Response

Assuming that shales are relatively higher in radiogenic illite, gamma-ray scintillation counts at the

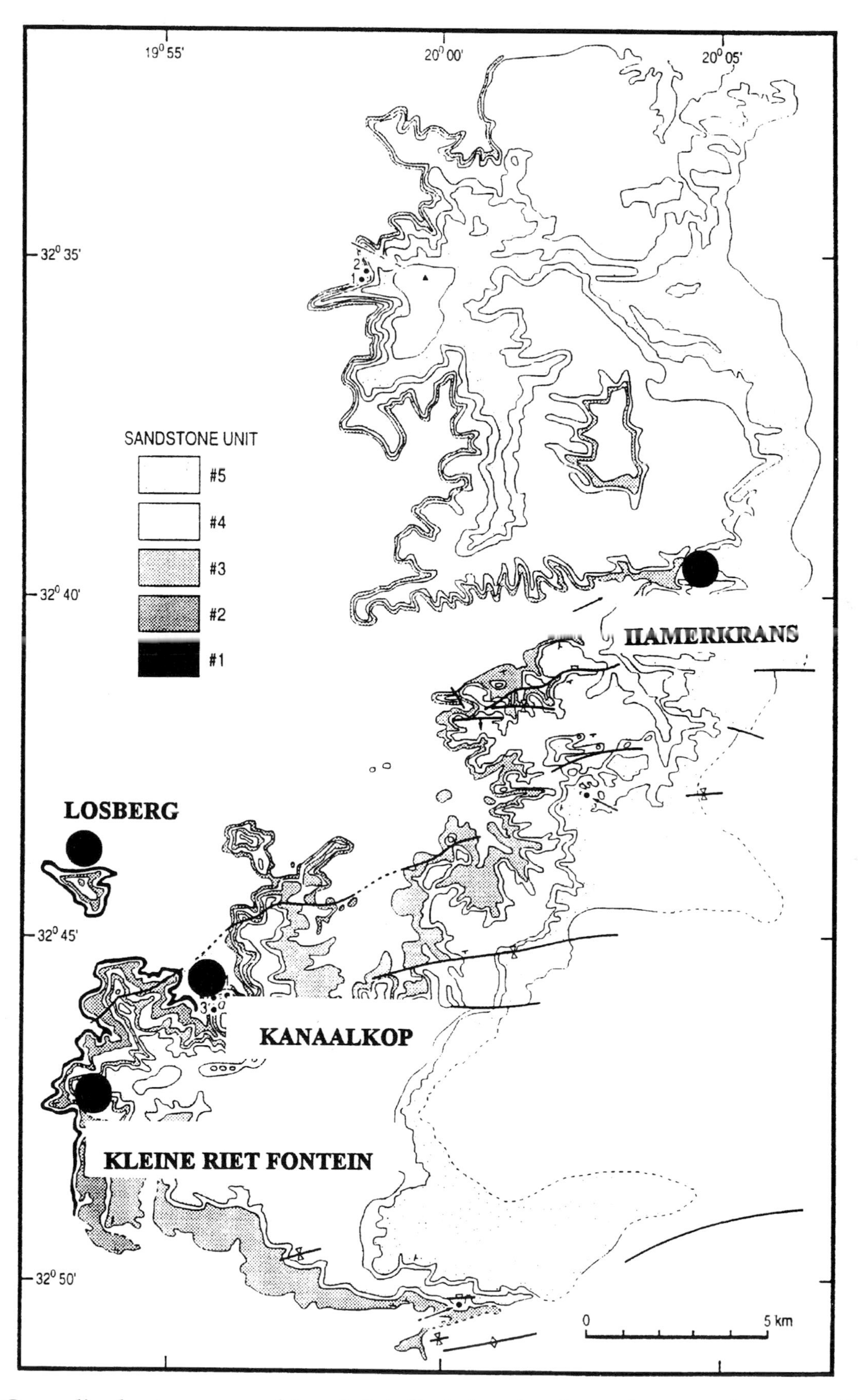

Figure 1—Generalized outcrop map of Fans 1–5. Indicated are locations where samples were collected for mineralogic analyses (modified from Wickens and Bouma, this volume).

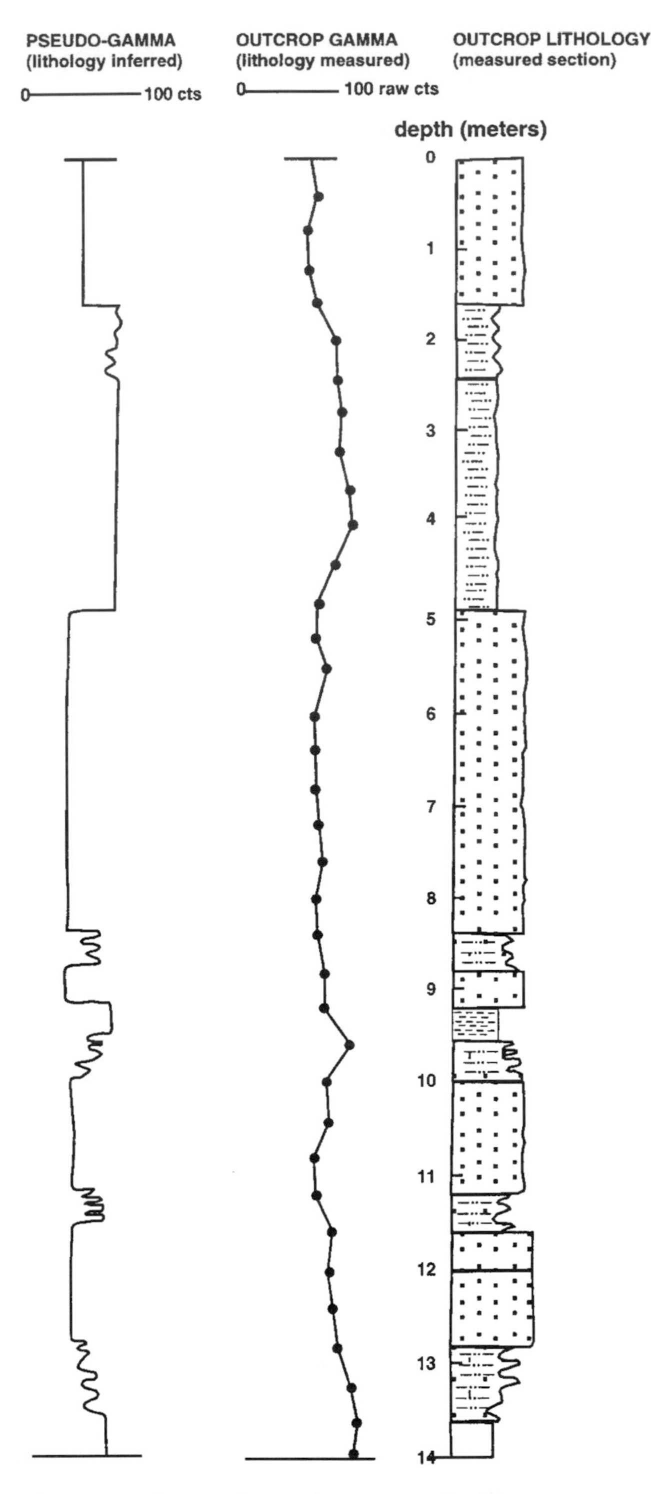

Figure 2—Comparison of outcrop scintillometer readings (raw counts) and pseudo gamma-ray curve hand-drafted to major lithologies; Kleine Rietfontein, Fan 2, lower, (modified from Rozman, 1998). Shale >70, siltstone >40–70, bedded sandstone >10–40, massive sandstone = 10 counts on gamma ray. Depth is reported in meters.

outcrop face should reflect changes in mineralogy and, therefore, lithology. It is also interpreted that intrastratal solution removal of chemically unstable minerals such as feldspars from the shales has occurred less completely than in the sandstones because of lower initial permeability in the shales. If the shales described in outcrop contain both illite and radiogenic feldspars, they will be assumed to respond as a shale on a gamma-ray curve.

Comparison work between Rozman (1998, this volume) and Thomas (this paper) shows that independently measured outcrop gamma count data compare very favorably with assumed responses (Figure 2). The outcrop gamma and "pseudo/gamma" logs for Fan 2 differ in character because the field data were measured at equal spacings up the outcrop and the other curve changes abruptly at the lithology boundaries as shown. Numerous other scintillometer surveys completed across other Tanqua Karoo outcrops (e.g., Rozman, 1998) show the same variations in natural gamma ray counts corresponding to lithology variations. The "pseudo-gamma" curves were constructed assuming the dominant mineral yielding radioactivity count was illite (contains radiogenic potassium isotope K^{40}). Therefore, shales are high in gamma-ray count, sandstones low, and siltstones intermediate reflecting differences in the illite abundance.

"Pseudodensity curves" were constructed assuming quartz and feldspars with grain densities of about 2.65 g/c^3 dominate the sandstones. Shale densities are greater than that because of the more dense clays (e.g., illite at 2.72 g/c^3) and pyrite. Siltstones have grain densities intermediate between the two (e.g., 2.68 g/c^3). More thickly bedded sandstones are assumed to be coarser and more quartz-rich, with a lower average grain density (2.65–2.64 g/c^3). The representative values were corroborated with laboratory measured grain densities (Table 2).

The calculated acoustic impedance curve was generated using the densities and matrix velocities assumed for each lithology. Expressed on the same vertical scale as the outcrop, the acoustic impedance curve can be broken down further into a "pseudo-reflectivity curve" for each of the lithologic interfaces.

Finally, resolution differences between outcrop and logging tools must be considered. Description of the outcrops has been much finer than typical logging tools. Bedding and lithology changes across vertical distances of 5 cm (2 in.) were recorded at the outcrop. Pseudologs assumed a vertical resolution of 46 cm (~18 in.) for the gamma ray, 100 cm (~39 in.) for the deep-resistivity tool, and about 61 cm (~24 in.) for the density and acoustic tools. The abrupt, blocky shape of the pseudolog curves (Figures 2, 4) decreased as the bed thicknesses increased. Thus, gradational contacts of interlaminated lithologies could be expressed with a gradational signature on deep resistivity but more serrate for the gamma-ray curve because their vertical resolutions differ. Fluid saturations were held constant because of the low porosities measured (Table 2). Later modeling using proprietary rock/fluid substitution capability shows that a wide range of cases of rock porosity/fluid saturations can be built.

Figure 3—Photograph of the Skoorsteenberg location showing prominent sandstones of fans separated by interfan "shales" for Fans 3, 4, and 5. This measured section is the one modeled with pseudologs shown in Figure 4.

To test the value of this approach, a model of Fans 3, 4, and 5 was built at Skoorsteenberg. Total outcrop thickness measured was 205 m on a section transect following the face exposed in the right half of Figure 3. Detailed discussion of each fan and the Skoor-steenberg section appear in Wickens and Bouma (this volume).

Basson (1992) has compared outcrop data in the Tanqua Karoo to seismic data by creating synthetic geophysical logs using actual geophysical logs from the offshore Bredasdorp Basin, South Africa. Rather than building "pseudologs" from the actual outcrop mineralogy and thickness as is done here, he looked at regional distributions of fan and interfan lithologies. Therefore, the scales of resolution related to lithologic changes are different between the two approaches. Both in Basson's work and this model, estimation of tuning thickness is very important to evaluation of seismic character for the individual fans.

Laboratory Measurement of Velocity, Fan 2

Compressional and shear wave velocities were measured on plugs cut horizontal, vertical, and at 45 degree orientation to stratification. Samples from Fan 2 "shales" were collected at Losberg (Rozman, 1998). For the limited number of samples, compressional and shear velocity anisotropies are low (Table 3). Further testing is needed to help resolve the relationship between anisotropy and the mineralogy. If the shales were more clay-rich (e.g., >70% clay) and laminated, greater Vp/Vs anisotropy can be expected.

BUILDING OUTCROP MODELS

Rock-driven seismic calibration models can be built for a variety of different depositional packages provided that outcrops are sufficiently thick (hundreds of meters) and described in detail as has been done for the Permian Ecca Group Skoorsteenberg sequence. The steps 1–7 outlined were followed to build the "pseudologs," which led to subsequent seismic modeling (steps 8–11):

1. Describe outcrops, measure vertical thickness, subdivide into major lithologies.
2. Assign integer values to each lithology subdivision (e.g., shale, siltstone, sandstone).
3. Plot gamma-ray, resistivity, density, and acoustic pseudolog curves using unique ranges of values for each lithology, either obtained from the indurated rocks or, better, from an analogous productive field.
4. Digitize outcrop sequence with unique integers assigned to each lithology, plotted as a "log curve."
5. Edit log suite, depth shifting, and then match to lithologic changes and plot as synthetic well including a lithology curve.

Table 1. Mineralogy of selected samples from Tanqua Karoo fans. Fourier Transform Infrared Spectrometrys analyses (FTIR) on outcrop samples are semiquantitative. Samples from Marot (and Rozman) are qualitative data from visual estimates of thin sections. The very fine grain size of constituents from Rozman made visual identification extremely difficult. Specific collecting localities for Marot's work are unreported.

Outcrop	Quartz (%)	Calcite (%)	Dolomite (%)	Illite (%)	Mxdlayer (%)	Chlorite (%)	Feldspars (%)	Others (%)	Lithic or Matrix	Fan #, Lithology
Losberg	27	0	0	9	0	0	54	10		2, sandst.
	27	2	7	18	9	3	28	6		2, sandst.
	29	3	5	14	2	1	38	8		2, sandst.
Hamerkrans	32	7	12	12	4	1	23	9		2, sandst.
	25	2	4	29	10	8	15	7		4, shale
Kanaalkop	25	0	1	22	11	2	29	10		2, shale
Hamerkrans	15	0	0	47	13	5	14	6		4, shale
	17	0	4	42	10	7	14	6		4, shale
Kleine Riet-fontein	14	1	5	45	15	3	9	8		2, shale
Marot Spls.	50	3	0	trace	not reported	5	10	32 (include lithics)	10	4, sandst.
	40	1	2	2	not reported	1	17	37 (include lithics)	17	3, sandst.
	50	1	trace	trace	not reported	2	14	33 (include lithics)	14	3, sandst.
Losberg (Rozman Spls.)	65	not reported	not reported	not reported	not reported	not reported	3	15	22 (clay/silt size undefined)	2, vfg sandstone
	65	not reported	not reported	not reported	not reported	not reported	1	10	24 (clays/silt size undefined)	2, coarse, argill. stilstone

Table 2. Pseudolog responses assigned for Skoorsteenberg outcrop (lithology-based) model. Each lithology has an assumed mineralogy that can be identified in Tanqua Karoo outcrops. Lithology interfaces are abrupt to gradational and are reflected in curve shape. Values were assigned from data commonly available in logging service company literature.

Lithology	Gamma (cts)	Rock, Integer	Resistivity (o-m)	Density (g/c^3)	Acoustic (ms/ft)	Remarks
Shale	>70	1-2.0	<20	2.72	85	clay rich w/ carbonate
Siltstone	>40–70	>2.0–3.0	20–50	2.68	70	qtz + feld + carb. + clay
Sandstone, Bedded	>10–40	>3.0–4.0	50–90	2.66	55	laminated, turbated, vfg to silty
Sandstone, Massive	<=10	>4.0–6.0	50–90	2.65	58	th. bedded, vfg to fg, sli. porous., qtz.+feld.

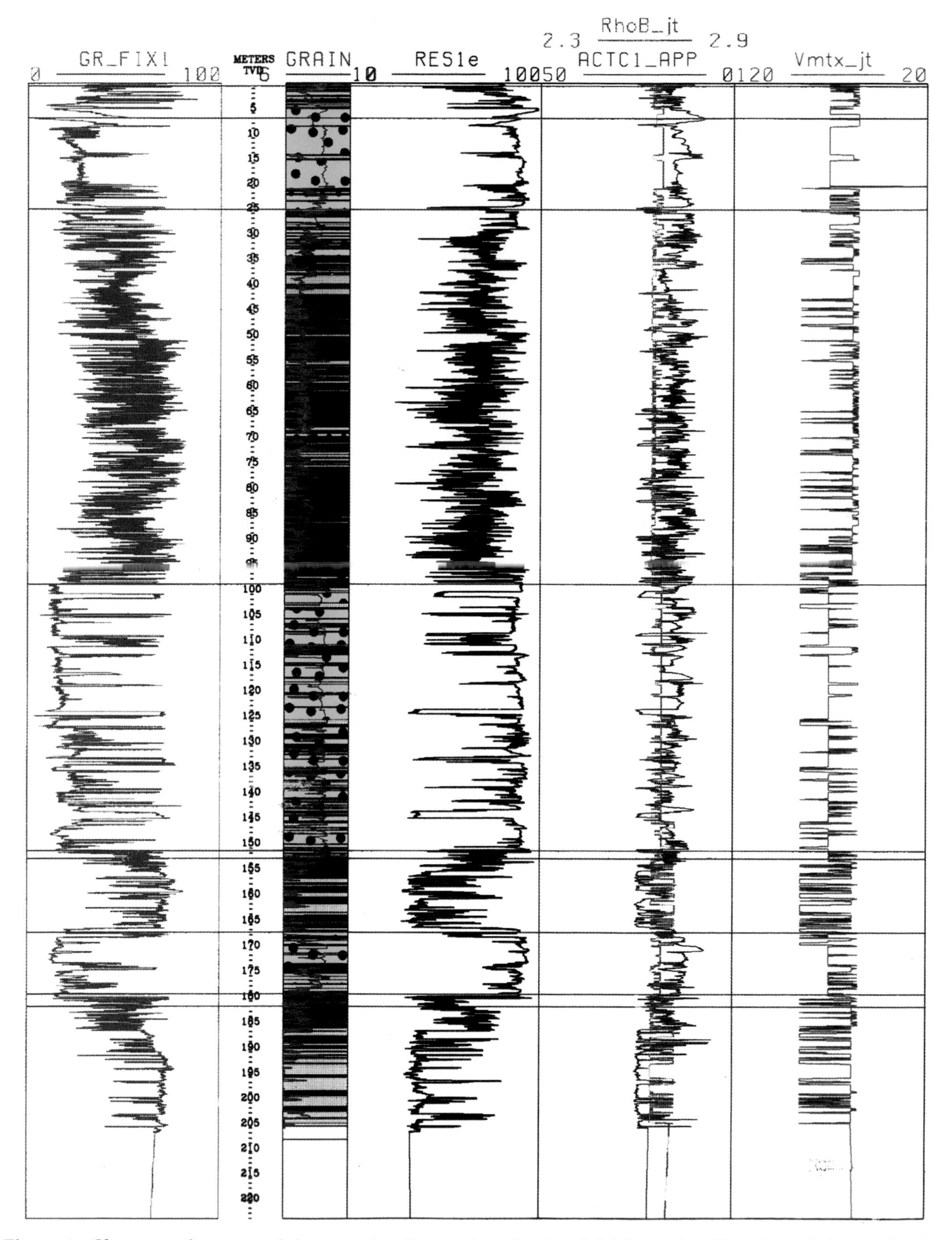

Figure 4—Skoorsteenberg pseudolog covering Fans 3, 4, and 5. Total thickness is 205 m. From left to right, the pseudocurves presented are: gamma ray (track 1), lithology (track 2), resistivity (track 3), density and acoustic velocity (track 4), and matrix velocity (track 5).

Table 3. Laboratory measurements for selected samples provided from Fan 2 at Losberg by Rozman (1998). (A, B) very fine-grained bedded sandstones; (C) laminated, clayey sandstone.

Sample, Orientation	Grain Density (g/c³)	Porosity % (pycnometer)	Vp, saturated (ft/s)	Vp, dry (ft/s @ 500 psi)	Vp, dry (ft/s @ 750 psi)	Vp, dry (ft/s @ 1000 psi)
A, Vertical	2.672	4.63	14604	13169	13373	13460
A, Horizontal	2.667	4.75	15076	14242	14489	14624
A, 45 Degrees	2.668	4.85	14190	13064	13159	13276
B, Vertical	2.672	1.05	15671	13923	13959	13995
B, Horizontal	2.674	1.25	16317	15948	16052	16115
B, 45 Degrees	2.671	2.42	15520	14538	14611	14684
C, Vertical	2.670	2.61	17399	16380	16428	16542
C, Horizontal	2.706	2.40	17498	17307	16677	16911
C, 45 Degrees	2.692	1.18	17140	16453	16596	16778

6. Calibrate "pseudogamma-ray curve" to raw scintillometer count data on multiple outcrops.
7. Calculate and plot acoustic impedance and reflectivity curves using the "pseudodensity and pseudovelocity curves." (Calculated reflectivity is negative at the top of a sandstone overlain by a shale.)
8. Convert depth to time in the workstation environment using a zero phase wavelet of known wavelength, a desired bandpass filter, and a constant time sample interval.
9. Display synthetic zero phase seismogram on the same vertical scale as pseudologs and compare that with the seismic dataset as needed.
10. Edit bandpass filter and wavelength to produce desired seismogram for modeling.
11. Calculate tuning thickness (one-quarter wavelength) and compare to vertical lithologic section. (Note: If well logs from nearby well are available, compare synthetic seismograms.)

AVO or impedance modeling can then be conducted by varying porosity, bulk density, and fluid content. The result is a series of interpretations tied to lithologic changes that are very well documented.

SUMMARY

At the Skoorsteenberg locale in the Tanqua Karoo subbasin, Fans 3, 4, and 5 are well exposed in vertical succession. Changes in bedding thicknesses, lithotypes, and outcrop gamma-ray measurements have been carefully recorded. The gamma-ray outcrop measurements are easily calibrated to the lithotypes, making a model of greater than 205 m thickness possible for seismic modeling purposes.

The general procedure outlined enables a geoscientist to improve the calibration of his or her seismic data beginning with the sequence of deepwater fans of the Skoorsteenberg Formation in the Tanqua Karoo. The critical linkage is generation of "pseudologs" reflecting changes in thickness and mineralogy of an outcrop of "seismic-scale" thickness. Although outcrop gamma-ray measurements confirm that pseudolog creation is valid, it is necessary to analyze compositions of the major lithology types and confirm mineral control on log response. In the depth domain, the match of synthetic seismograms using acoustic impedance and reflectivity is reliable. Conversion to the seismic time domain requires selection of representative wavelength, frequency plus bandpass filters to produce suitable data.

ACKNOWLEDGMENTS

The short cycle time of 8 months to develop and test this modeling approach was built on the industry consortium efforts buided by Geo-Marine Consultants, the research of Drs. Arnold H. Bouma and H. DeVille Wickens and the many students of Stellenbosch, Port Elizabeth, and Louisiana State universities who collected detailed measurements of the fans and interfan exposures. Integrating the model was helped greatly by A. L. Brown, D. J. Rosman, C. F. Bauerschlag, G. Clemenceau, M. L. Suda, C. Rai, and C. Sondergeld. My thanks to all of these colleagues.

REFERENCES CITED

Basson, W. A., 1992, Synthetic seismogram section of the Skoorsteenberg Formation, *in* A. H. Bouma, ed., Tanqua Karoo fluvially dominated, shale-rich submarine fan complex: a report to a consortium of oil companies: Geo-Marine Consultants, Inc., Baton Rouge, LA, v. I, Chapter 12, p. 6–7.

Bouma, A. H., and H. DeV. Wickens, 1991, Permian passive margin submarine fan complex, Karoo Basin, South Africa: possible model of Gulf of Mexico: Gulf Coast Association of Geological Societies Transactions, v. 41, p. 30–42.

Bouma, A. H., and H. DeV. Wickens, 1994, Tanqua Karoo, ancient analog for fine-grained submarine fans, *in* P. Weimer, A.H. Bouma, and B.F. Perkins, eds., Submarine fans and turbidite systems: Sequence stratigraphy, reservoir architecture, and production characteristics: Gulf Coast Section SEPM Foundation 15th Research Conference Proceedings, p. 23–34.

Marot, J. E. B., 1992, Petrography of selected Tanqua Karoo sandstone, *in* A. H. Bouma, ed., Tanqua Karoo fluvially dominated, shale-rich submarine fan complex: a report to a consortium of oil companies: Geo-Marine Consultants, Inc., Baton Rouge, LA, v. 1, Chapter 11, p. 2–6.

McBride, E. F., 1963, A classification of common sandstones: Journal of Sedimentary Petrology, v. 33, p. 664–669.

Rozman, D. J., 1998, Characterization of a fine-grained outer submarine fan deposit, Tanqua–Karoo Basin, South Africa: Unpublished M.S. thesis, Louisiana State University, Baton Rouge, 147 p.

Scott, E. D., 1997, Tectonics and sedimentation: evolution, tectonic influences and correlation of the Tanqua and Laingsburg subbasins, southwest Karoo Basin, South Africa: Unpublished Ph.D. dissertation, Louisiana State University, Baton Rouge, 234 p.

Wickens, H. DeV., 1994, Basin floor fan building turbidites of the southwestern Karoo Basin, Permian Ecca Group, South Africa: Unpublished Ph.D. dissertation, University of Port Elizabeth, Port Elizabeth, South Africa, 233 p.

Hansen, S. M., T. Fett, 2000, Identification and evaluation of turbidite and other deepwater sands using open hole logs and borehole images, *in* A. H. Bouma and C. G. Stone, eds., Fine-grained turbidite systems, AAPG Memoir 72/SEPM Special Publication No. 68, p. 317–338.

Chapter 28

Identification and Evaluation of Turbidite and Other Deepwater Sands Using Open Hole Logs and Borehole Images

Steven M. Hansen
Schlumberger Wireline and Testing
Houston, Texas, U.S.A.

Tom Fett
Schlumberger Wireline and Testing (retired)
Houston, Texas, U.S.A.

ABSTRACT

Deepwater sandstone deposits have become important oil and gas reservoirs. The evaluation of these deposits using standard open hole logs can be difficult because many of these thinly bedded deposits are below the resolution of traditional open hole logs. Borehole images have become an established tool to identify and delineate these deposits. Micro-resistivity images, such as Formation MicroImager* (FMI), can resolve bed thickness down to 1 cm, and can suggest grain-size variations, flow characteristics, sand continuity, and permeability ranges. These images can help to determine structure, faults, unconformities, depositional environment, and sand-body orientations. This high-definition information about reservoir sands, combined with the structural and stratigraphic information, allows the best possible understanding of the reservoir in terms of depositional environment, reservoir quality, and probable productive units.

INTRODUCTION

Turbidites and other deepwater sandstones are major targets for oil and gas exploration in many areas, including the Gulf of Mexico. The evaluation and identification of these deposits with open hole logs is often not straightforward. Even modern "high-resolution" logs have problems with the very thinly (<5–15 cm) bedded sandstones that are commonplace in these environments. Almost all traditional open hole logs require at least 30 cm (1 ft) of thickness to properly evaluate a zone. Borehole images, such as the Formation MicroImager (FMI) tool, allow evaluation of layers down to 1 cm thick and have been widely used in the evaluation of deepwater sandstone deposits. Although the primary

*Represents a patented instrument.

focus of borehole images has been to identify these sands, they can also be used quantitatively to obtain net sand count values for these reservoirs. To fully realize the value of these images, an understanding of the measurement characteristics of these high vertical resolution, three-dimensional measurements and their unique log presentation is important.

Other imaging applications include structural and stratigraphic evaluations. The structural applications include determination of the structural dips and trends, in addition to detecting breaks in these structures that may identify faults and unconformities or sequence boundaries. Using images, faults can be positively identified and fully delineated. Stratigraphic applications include identifying various sedimentary features that aid in determining depositional environments and sand trends. This can often help understand the flow units and flow characteristics of an entire reservoir.

The images can be used not only to determine the net thickness of the individual sandstones, but also whether they have the ability to produce. Even very thin interlaminated sands, when parallel bedded, can have excellent lateral continuity and represent excellent reservoirs when enough gross section is encountered. Disrupted or churned-up sandstone beds, such as bioturbated zones and slumped, contorted beds, usually are not able to drain sufficient areas to be commercial. A comprehension of this type of information from wells in known deepwater settings, along with core and production validation, will allow for a much better understanding of their flow characteristics and evaluation of deepwater sandstone reservoirs. All the examples shown in this paper, while not identified by well, are from known deepwater settings that are considerably below storm wave action at the time of deposition. Numerous references and imaging examples are available. An excellent bibliography listing many of these is available on the Web: one for images at (http://pangea.Stanford.EDU/IDIS/Bib_Im_AB.html) and one for fractures at (http://pangea.Stanford.EDU/IDIS/Bib_Frac_A.html).

Part of the power of image analysis is that, because images are essentially "pictures," anyone with a minimal knowledge of tool characteristics and the presentation formats can appreciate them. The images presented here are resistivity images and compare favorably with photographic (optical) images, either normal or UV light photographs, with each having advantages in particular situations. Images can be readily used by knowing a few basics, including an understanding that:

1. Microresistivity images have very high resolution and are very repeatable.
2. The borehole surface is presented in an oriented format starting with north and proceeding clockwise through east, south, west, and back to north (or in terms of azimuths from 0 to 360 degrees).
3. The tool and borehole deviation and orientation in conjunction with the borehole geometry are measured and accounted for in all calculations.
4. The color scheme for static images becomes darker (from white to beige, to orange, to brown, to black) as micro-resistivity decreases.

AUTHORS' NOTE

In all the examples, selected features will be identified on the images; other features are left for the reader to evaluate. Interpretations may be put forth in an effort to stimulate consideration of the various geological implications. As with all things geological, other interpretations are certainly possible. These "speculations" will be put in quotation marks to reduce controversy.

OPEN HOLE LOGS

Most open hole logs have a vertical resolution of 30 cm (1 ft) or more. The SP, gamma-ray, and porosity logs, including the density, neutron, and sonic types, generally require at least 30–60 cm (1–2 ft) of consistent porosity and lithology to read correctly. Most resistivity logs require even thicker zones to derive the true resistivity. Resolution varies with tool type from 1.2 m (>4 ft) for older tools, to as low as 0.3 m (1 ft) with the latest high-resolution resistivity devices. These measurements can be improved with computer enhancement, but typically they cannot go below a few inches. Older style (nonimaging) dipmeter resistivity curves can be used to identify very thin zones down to 2.5 cm (1 in) or so, but their lack of borehole coverage does not allow continuity determination of these sands around the wellbore. Sidewall cores that are off depth by even a few inches can be very misleading in thinly laminated sequences. Unless cores are acquired before the images are obtained and their actual position is verified on the images, small depth discrepancies can be misleading and condemn a whole zone. Whole core, while invaluable for calibrating the images, is usually not an acceptable alternative to logs due to their prohibitive cost to acquire (including rig time) in most wells.

LOW-RESISTIVITY PAY SANDS

The problem of interpreting low-resistivity pay sands has long been recognized (Moore, 1993) and has spurred new tool developments that try to solve it. Images have provided the best solution. "Low-resistivity pay" can result from at least three sources. They occur when:

1. Very low-resistivity clays are dispersed in the sands.
2. Very low "wet resistivities" result from high salinity waters and high porosity sands. For instance, a 30% porosity sand filled with saturated brine will

have a wet resistivity of 0.1 ohm. This is, in fact, quite common in the Gulf of Mexico and means that a "three to one" pay zone with a water saturation of 58% can have a measured resistivity as low as 0.3 ohm.

3. Thinly bedded pay zones of a few inches or less fall below the resolution of most open hole logs and are not even recognized as reservoir rocks on the log as shown in Figure 1.

IMAGING DEVICES

The very high resolution of modern imaging devices offers a solution to this thin bed interpretation problem. The two basic choices are acoustic and microresistivity images: acoustic images fundamentally detect changes in rugosity (by way of reduced amplitudes); microresistivity images detect changes in resistivity. Some success has been made in delineating permeable zones with acoustical images. In these cases, mud cake reduces the amplitude of the acoustical signal. These zones are presented as darker images and can be used to get "net permeable sand." In general, acoustical images do not have as good a vertical resolution to resolve the sands when compared with electrical images. They are hindered by heavy muds where the acoustical signal is distorted by the high percentage of solids. The major exception to this is when the borehole is filled with oil-base mud and the microresistivity measurements may not be very useful.

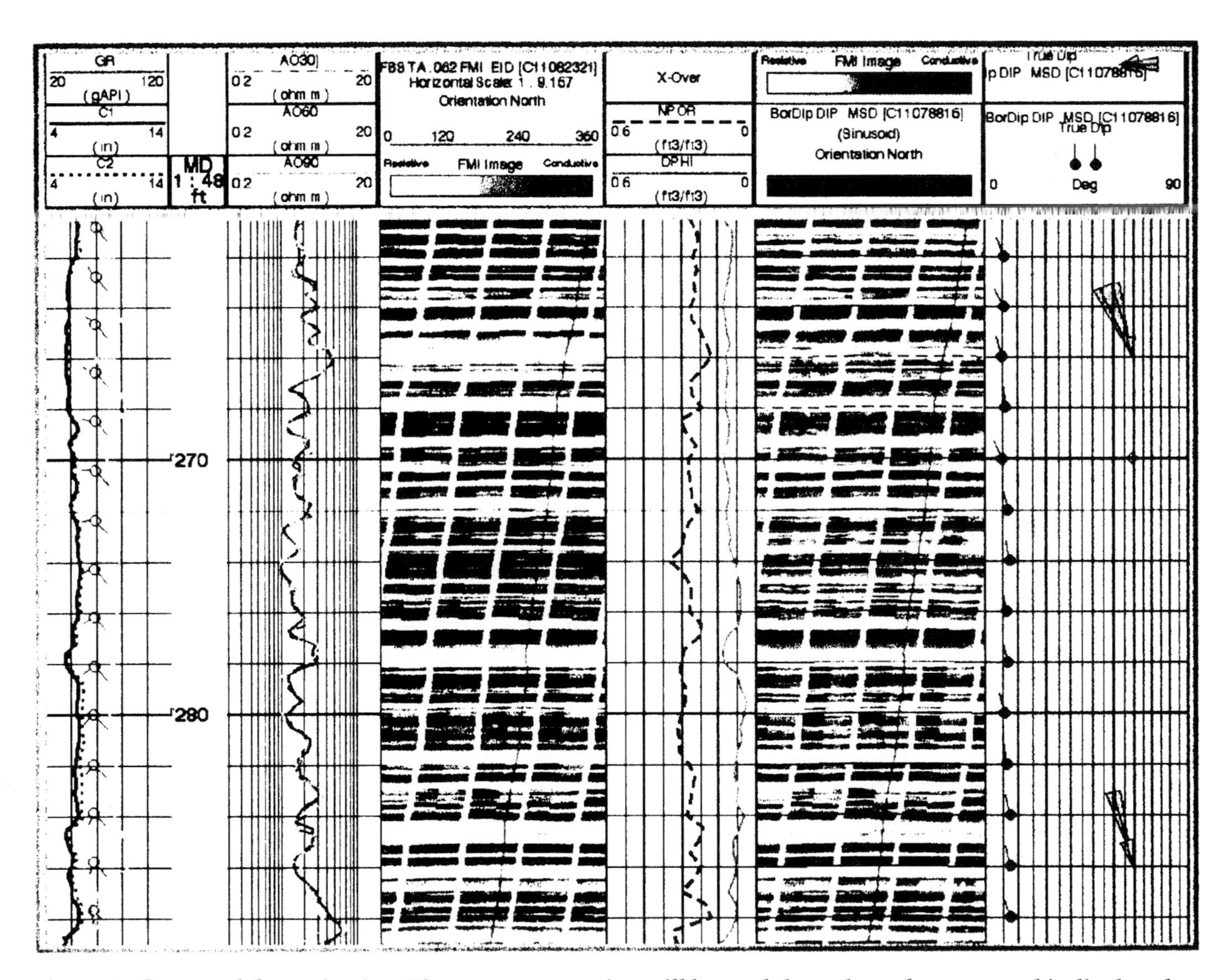

Figure 1—Structural determination. The same presentation will be used throughout the paper and is displayed as follows: track 1—gamma-ray, caliper, and hole orientation; track 2—depth; track 3—resistivity (0.2–20 ohm-m); track 4—static images; track 5—porosity curves (60–0%), density-fed, neutron-blue; track 6—dynamic images; and track 7—dips from images.

Microresistivity tools basically respond to the amount of conductive material in the area in front of the sensors. This is usually either water-saturated sands and silts or conductive clay. For clean sands with consistent porosity, this means that microresistivity images can often be used to identify changes in lithology as described later. Typical microresistivity images are derived from many (≤192) high-resolution measurements spread around the circumference of the borehole. Configurations from four to eight pads are available. All of this information is precisely oriented in three-dimensional space. These tools can contend with any hole orientation or deviation, including horizontal boreholes.

Caliper, borehole, and tool orientation data (three axis accelerometers and magnetometers) have to be acquired simultaneously to correctly orient the data points in space. Gamma-ray or SP logs are usually acquired at the same time for correlation. Other open hole logs are often acquired and used in conjunction with the images for interpretation. The different tool descriptions are readily available from the various service companies that offer these microresistivity imaging services. With all electrical imaging tools, normal dip channels can be extracted and used for typical dip computations as in Figure 2.

Schumberger's FMI tool has four arms with a pad and an associated flap on each arm. Each pad and flap has two rows of 12 sensors. These 192 sensors have a large dynamic range and can detect very small changes in microresistivity. The solid-state orientation data and the four-arm caliper information allow those measurements to be precisely oriented in three-dimensional space. The excellent repeatability of the microresistivity images soon convinces the user that the images computed from these resistivities are real and truly represent the nature of the formation. Image validations from outcrop (Figure 3) or core are very useful. Figure 3 is a comparison of an FMI in a turbidite outcrop. The FMI was actually run behind the outcrop and can be correlated back to the outcrop face. In these environments, some features or lithologies can be correlated over large distances allowing the same features viewed in the outcrop to be seen on the images. Note the two shales at x24 ft and x29 ft in Figure 3. It is important to note that images are acquired after the core has been taken so the small features seen on a core will not be seen on the images that were acquired after coring, and vise versa. It is important to note that in many cases cross-bedding in a clean, uniform grain-size sand will not be distinguishable on normal light photographs, although they can often be observed on the resistivity images. The idea behind comparing images with the hard data (core and outcrop) is to allow the user to feel comfortable in making the correlation between what is detected on the electrical images and what the actual rock structures would look like.

An understanding of the presentation format of this data (Figures 1, 2) is required to properly use the information. For most images, the standard presentation is the same as any log with depth (at various scales) as the Y-axis, but with the images on an azimuthally oriented grid on the X-axis (Figure 2). The gamma-ray, caliper, and other open hole logs are presented in the traditional format and scaled as appropriate for the area (Figure 1). The images are presented on an azimuthal scale starting at an azimuth of zero (north) and continuing to the right through 90 degrees (east), then 180 degrees (south) to 270 degrees (west) and on to 360 degrees (north). The result of presenting the data this way is that planar surfaces (such as bedding planes and faults)

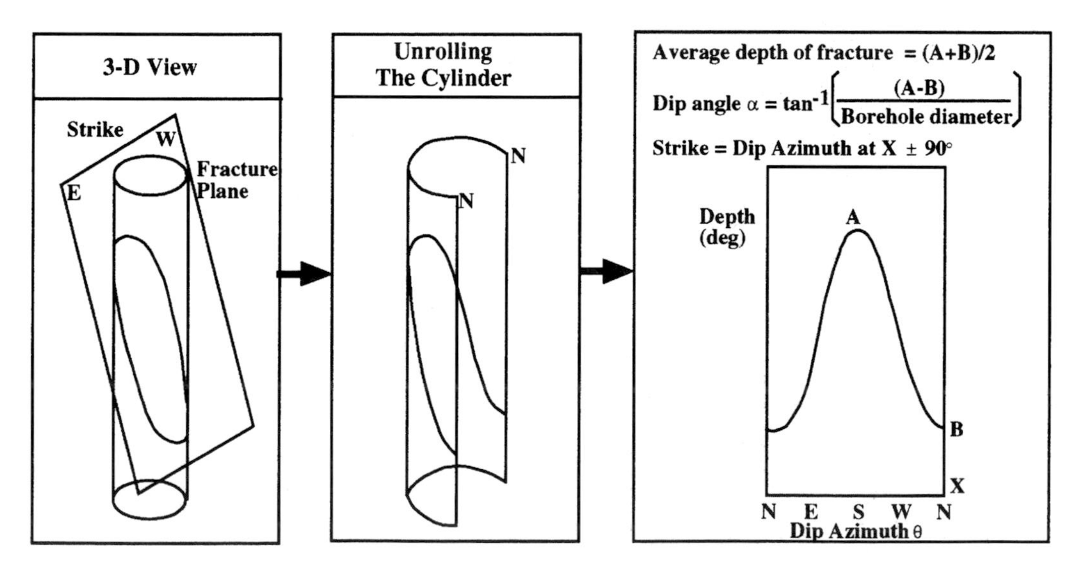

Figure 2—A plane intersecting the borehole when cut and unrolled is seen as a sine wave. In a vertical well the steeper the dip, the larger the sine-wave amplitude will be.

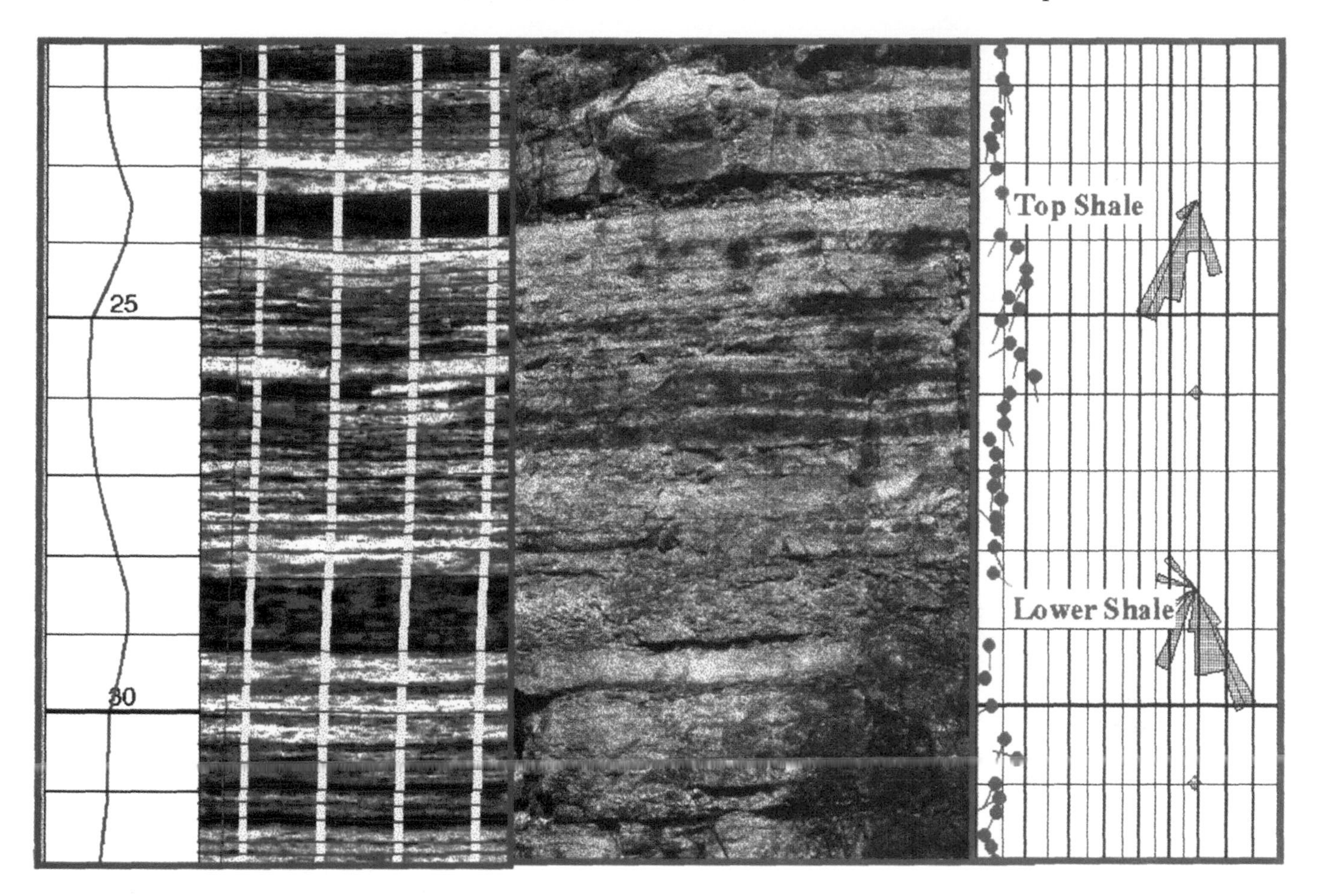

Figure 3—FMI image and outcrop comparison, units are in feet (25 ft = 17.5 m). From left to right the tracks contain: gamma-ray and depth, FMI (Formation MicroImager) images, outcrop photo, and dips computed from the images (photo E. Head).

intersecting the borehole are portrayed as sinusoids (Figure 2). In vertical wells the amplitude (height) of the sinusoid is proportional to the dip magnitude of the bed. Two different normalizations of the image data are often presented, static and dynamic normalization (Figure 4). Static normalization is scaled over the entire interval so resistivities or colors can be compared from zone to zone. Dynamic normalization is rescaled at the processor's discretion, usually every 30–60 cm (1–2 ft), to maximize details. The standard color scheme for static images is for the images to become darker as the resistivity decreases; thus, the sands appear bright and the shales appear dark. When earth-tone colors are used and the resistivity varies from lower to higher, the colors will vary from black to brown to orange to beige to white. Calibrated images are either initially acquired or the static images are calibrated to other shallow investigating resistivity devices. These images are called scaled or calibrated images. On scaled images, each color has a corresponding resistivity.

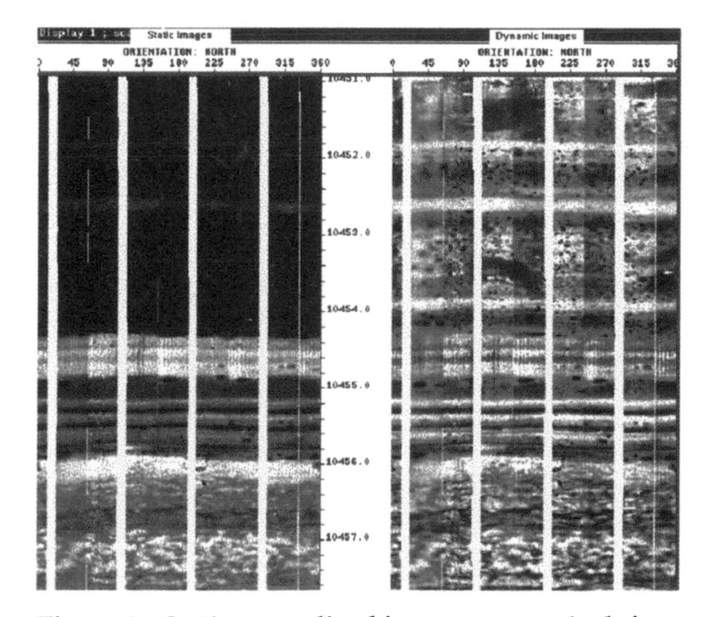

Figure 4—Static normalized images are on the left and dynamic normalized images on the right.

The images presented show the apparent dip as seen by the tool, in the same way a whole core is viewed. The computer makes a correction for hole deviation and orientation and the resultant dip tadpoles presented are the "true dip" (i.e., they are corrected for hole deviation, direction, and to true north) (Figure 5). These dip tadpoles represent a dip magnitude and azimuth. The position of the body of the tadpole is on a scale of zero (horizontal) to 90 degrees (vertical). The tail of the tadpole points in the downdip direction. The tadpoles can be color coded as in Figure 6. (Incidentally, imaging stereonets are almost always presented in the upper hemisphere.) Azimuthal frequency plots over selected intervals are often presented. The plot intervals are delineated by diamonds in the dip track as in Figure 7 at x51 ft and x66 ft.

APPLICATIONS

The many applications of imaging information can be divided into three categories as follows:

1. Structural applications
2. Stratigraphic applications
3. Reservoir descriptions

Structural Applications

Structural applications start with the determination of structural dip. These dips can be computed by the traditional dipmeter methods with batch processing or by interactive workstation dip picking, which often refines the answers (Figures 5–7). Workstations also allow dips to be handpicked (interactively) when

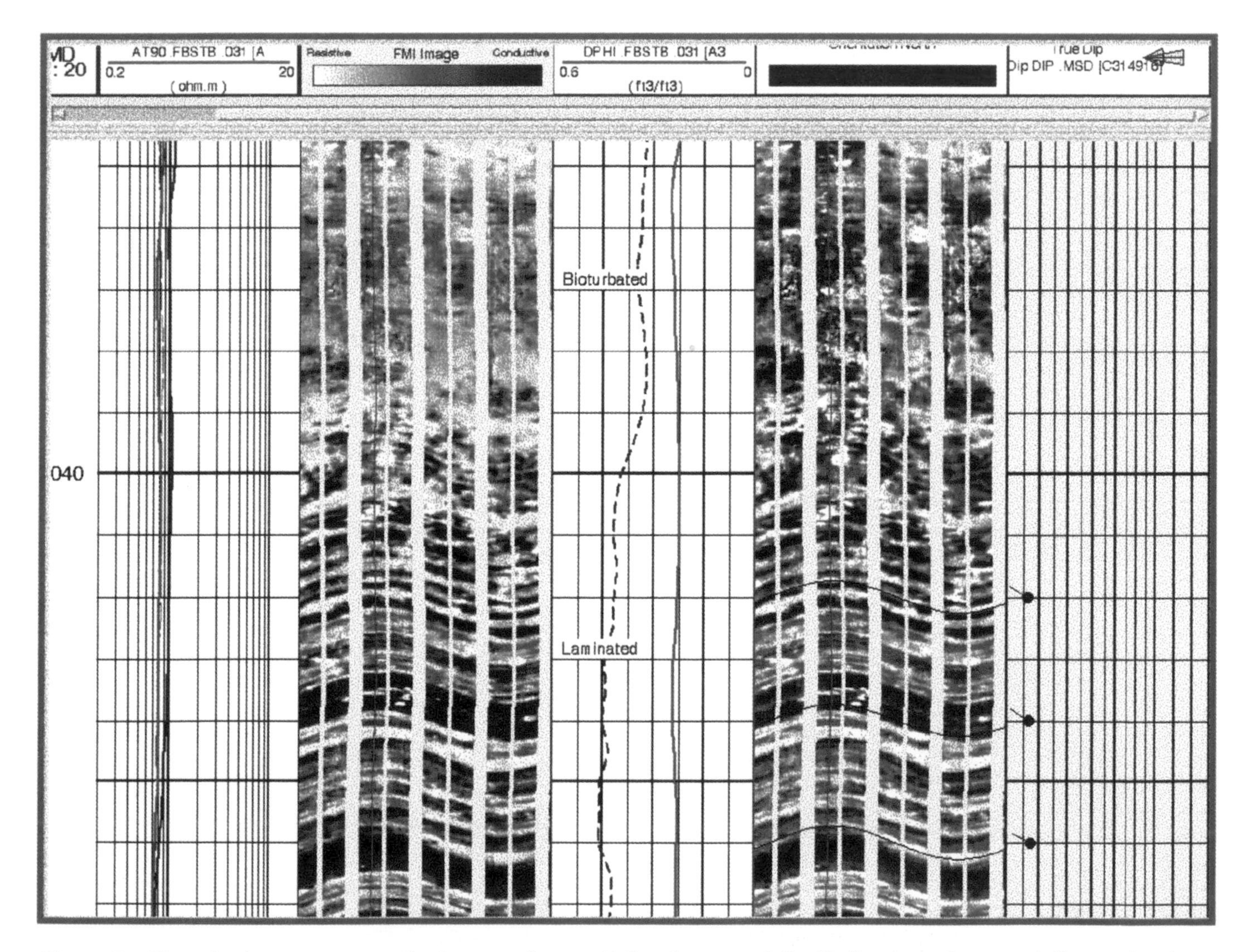

Figure 5—Bioturbation, as seen on the images above x40 ft, takes an originally laminated, potentially productive sequence and turns it into a noncommercial zone. The resistivity does not change with the bioturbation but the neutron porosity does. The porosity curves coming together as seen in this example would normally be considered an indication of a clean sand sequence and not of bioturbation.

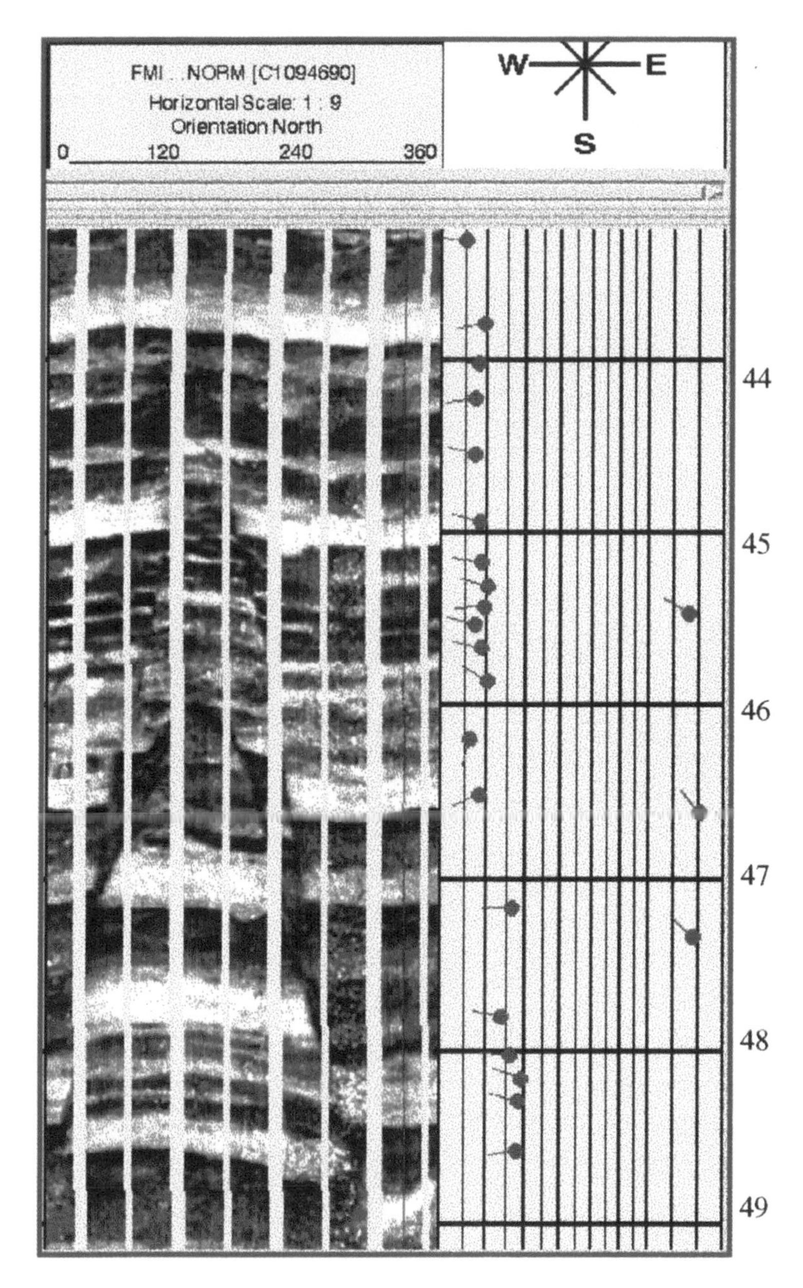

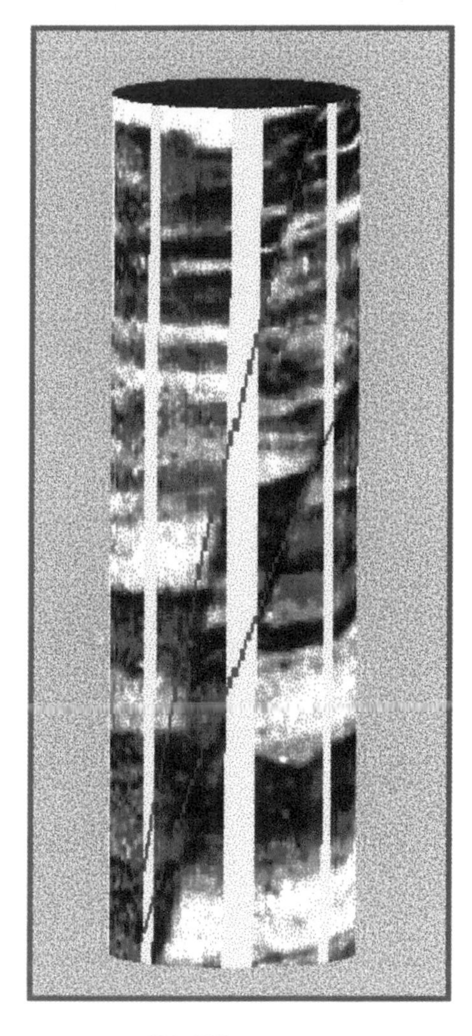

Figure 6—This splinter fault has no distortion or drag above and below the fault. It was only possible to detect and orient the fault on images. The fault plane (red tadpoles) strike is NE–SW, down to the NW.

batch computer correlations are not adequate and did not find correlations that are visible to the user. Workstations can also be used to eliminate poor correlation dips and to classify dips into types, such as stratigraphic, fractures, and faults (Figure 6). These structural dips can be used to delineate the shape and type of the structure and to identify breaks in structure (or missing section) such as faults and unconformities. Structural dips can be rotated out and used to correct stratigraphic dips back to their original orientation at time of deposition.

A lack of computable dips may indicate a change in lithology as shown in Figure 5 at x40 ft. Note the change from mottled or heterogeneous sediments above x40 ft (bioturbated?) to interlaminated sands and shales below x40 ft. Below x40′, it was possible to compute dips indicating a structural dip of 5° down to the WNW (or an azimuth of 290°). Another unusual phenomena in this example are that the neutron-density curves come together in the "bioturbated," nonproductive, section and are separated in the laminated productive zone. This is in direct contrast to what is typically expected to occur, where the neutron-density approach each other in the porous productive zones, occasionally crossing over in gassy intervals.

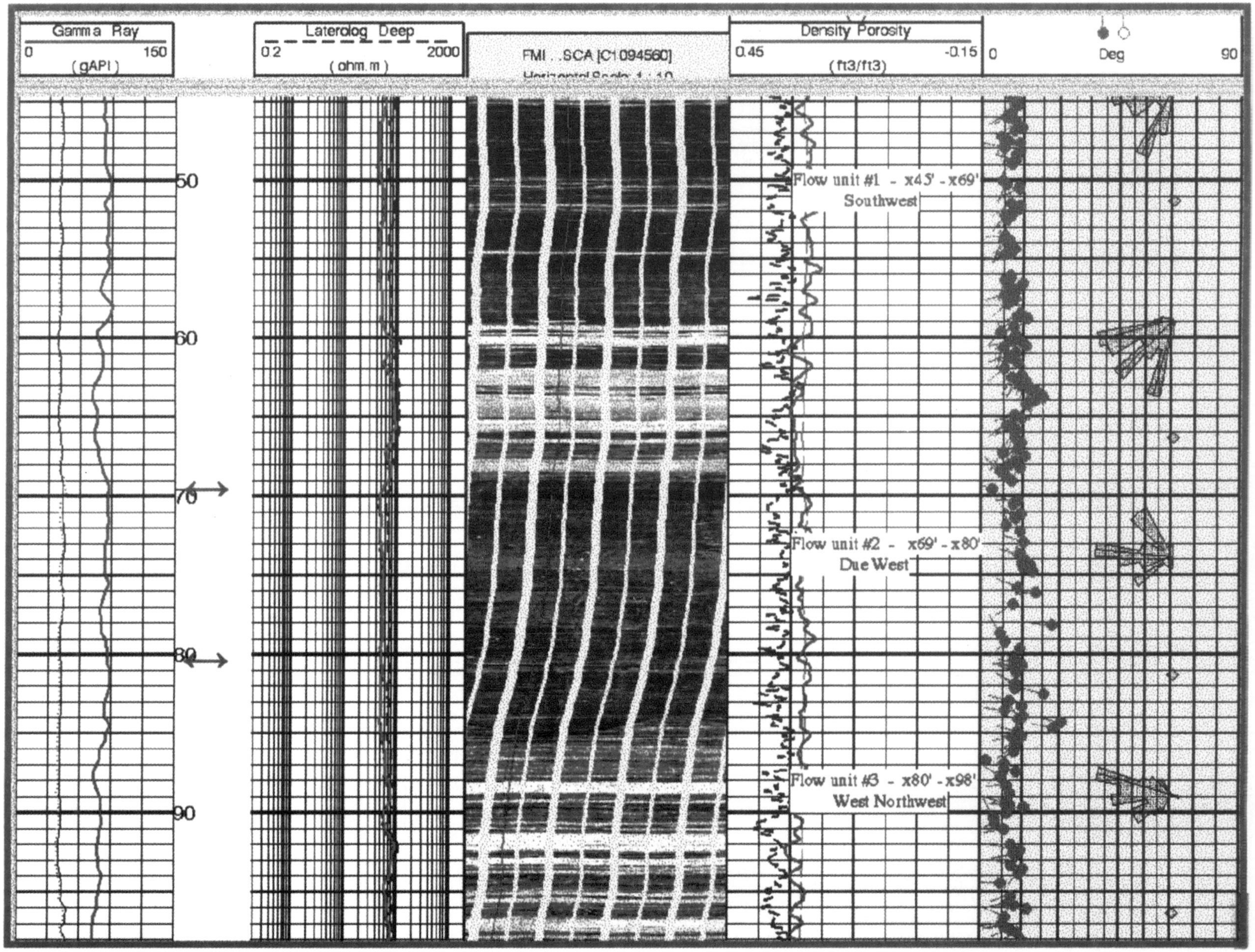

Figure 7—Individual fan sequences that cannot be identified from the open hole logs can be distinguished by the change in dip direction calculated from the images. The upper fan unit from x45 ft to x69 ft has a SW progradation. The middle fan unit from x69 ft to x80 ft has a due west progradation, and the lower fan from x80 ft to x98 ft has a W–NW progradation.

Breaks in structure (x47 ft in Figure 6) suggest possible missing sections. This change is interpreted as a fault because of the planar nature of the surfaces. (The more planar the surface, the better the fit to a sinusoid on the images.) These surfaces are also relatively steep, that is, they have a high dip magnitude (>45°), further indicating a fault. Unconformities, on the other hand, are usually relatively flat lying and not necessarily planar.

A further look at the fault at x47 ft in Figure 6 illustrates the wealth of information available from the images. The workstation "3-D reconstruction" of the image on the right should make it easier to visualize what the images on the left represent. On a workstation this 3-D image can be rotated and viewed from any orientation. The thin-bedded sandstones are the lightest colored zones (>x44 ft). Further note the "fining-upward" nature of the 3-in.- (7.5-m-) thick sand above x44 ft.

The fault can be seen to actually be two or possibly three distinct planes. The red tadpole at x47.3 ft in Figure 7 corresponds to the lower-most fault plane encountered in the borehole. It indicates a surface dipping at 78° from the horizontal and down to the NW. This defines a fault with the following characteristics:

1. Fault depth at x47.3 ft (faults with multiple splays are usually interpreted on the lower-most plane)
2. Fault angle is 78° (from horizontal).
3. Fault is downthrown to NW.
4. Fault strikes NE–SW, perpendicular to the downdip direction.

The very thin, dark area around the fault plane suggests the fault may not be a very good "seal." Fault types, whether rotated or not (normal or growth), can be determined by noting whether the fault drag pattern is in the opposite or the same direction as the fault plane itself. In this example the fault type was not determined due to the lack of drag pattern associated with the fault. A white (high-resistivity) fault plane with a white "halo" above or below the top or bottom of its sinusoid is often a good indicator of a sealing fault. A dark-colored fault plane is not conclusive one way or the other because it could be an open fault filled with drilling fluid, which is nonsealing, or it may contain a low-resistive fault gouge or shale smear, which could be sealing. Unfortunately the amount of displacement of faults cannot be determined unless it is very small and thus can be seen on the images over a few inches. The juxtaposition of the beds across the plane indicates some displacement, making this high angle planar surface a fault. Fracture analysis using images such as this is beyond the scope of this chapter. (See fracture Web site for more references on fracturing.)

The change in texture at x40 ft in Figure 5 could be an unconformity or "sequence boundary." Other log characteristics of unconformities include changes in log characteristics (e.g., gamma-ray hot spots) and the presence of nodules. Changes in structural dip or even a sudden lack of dip, as in this case, are often found. Unconformities will almost always be subparallel to structure and usually do not have planar surfaces associated with them. An understanding of structural dip is needed to do proper stratigraphic interpretations. If structural dip is greater than a few degrees it should be rotated out (dip subtracted). Changes in structural dip can indicate missing sections or stratigraphic changes. For instance, note the changes in dip at x40 ft in Figure 5 and at x68 ft in Figure 7, which are best seen from the azimuth frequency plots.

Stratigraphic Applications

Stratigraphic applications are very numerous, so only a few examples will be discussed. One saving factor here is that because the images are easily visualized, the user can bring his visual experience to bear, whether his background is geology or engineering.

Stratigraphic applications include:

1. Identifying sedimentary features
2. Depositional environment
3. Sand trends

An excellent way to recognize all types of sedimentary features is to review a well-illustrated book on the subject (preferably one with many pictures), such as this volume or that of Scholle and Spearing, (1982).

Stratigraphic interpretations are usually the most speculative and, therefore, controversial interpretations of all. The following *speculations* are offered to stimulate thought. Figure 7 shows three individual "flow units" that are not evident on the open hole logs, even though they can be distinguished by the change in dip direction calculated from the images. The upper unit from x45 ft to x69 ft had a SW progradation, the middle unit from x69 ft to x80 ft had a due west progradation, and the lower unit from x80 ft to x98 ft had a WNW progradation. Figure 8 shows an erosional "channel base" at x138 ft and the subsequent channel-fill deposit, which will allow the orientation of the channel to be determined. Figure 9 looks at another channel base in more detail. The basal erosional surface (yellow tadpole and sinewave) is overlain by the basal channel lag. The largest clast in this lag deposit is approximately 5 cm (2 in.) in diameter. The sharp erosional base, the fining-upward sequence, and the channel lag deposits are distinguishing characteristics of a channel deposit. Figure 10 exhibits a "matrix-supported debris flow or high-density turbidite" from x5496 ft to x5501 ft. The sizes of these clasts are much larger 15–23 cm (6–9 in.) in diameter than those in Figure 11, and they still exhibit some of the original bedding, although distorted, inside the clasts. Orientation of this unit is all but impossible, and would have to be done using the surrounding beds. Figure 11 shows "soft sediment deformation (slumping)" from x51 ft to x52 ft. Notice how the beds above x50 ft and below x52 ft are not disturbed. This type of soft-sediment deformation can indicate the paleoslope direction, which would strike NE–SW and be downslope to the SE. Figure 12 exhibits the lower extreme of detectable sand-body sizes. These individual sand pods are less than

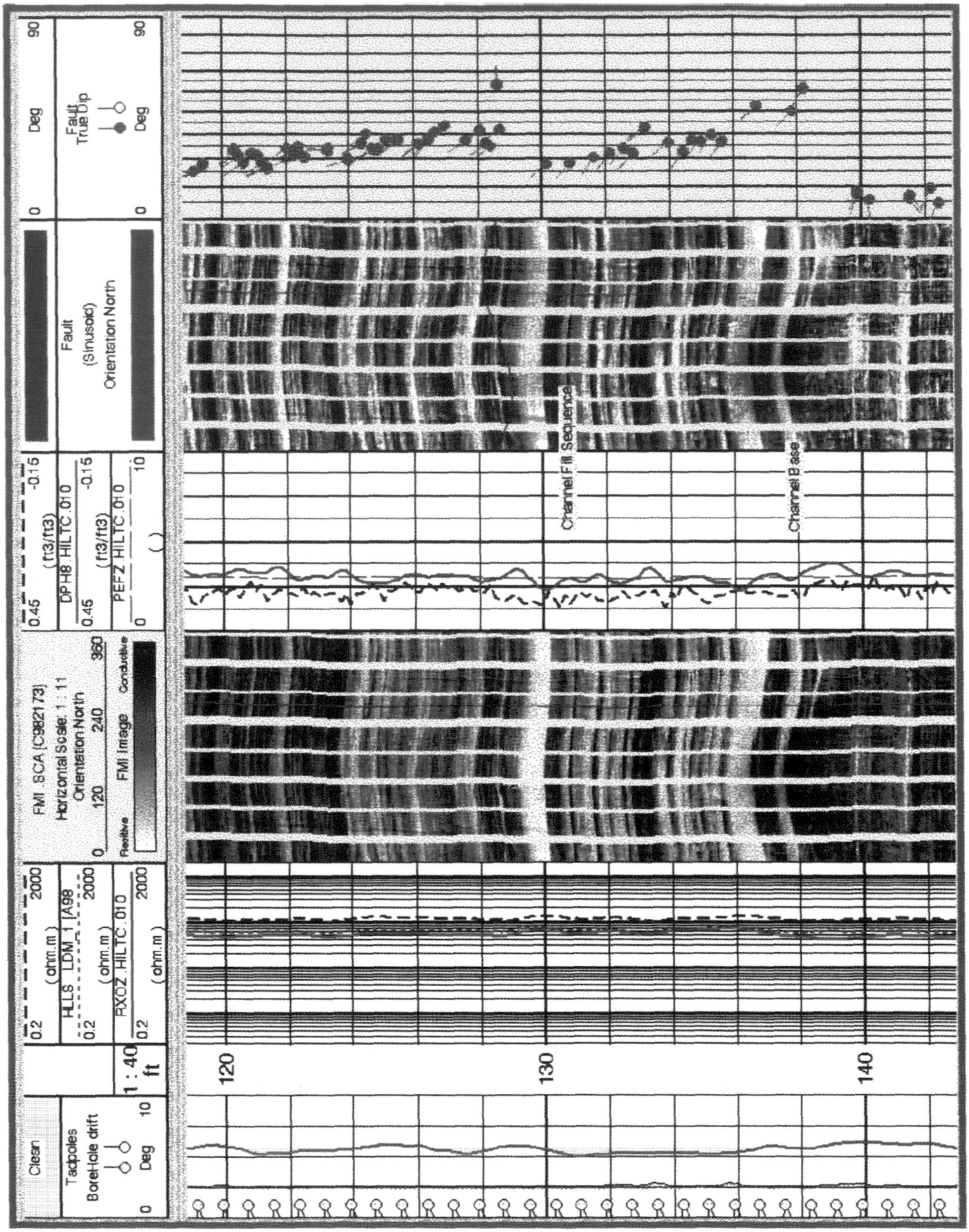

Figure 8—Channel-fill sequence where the channel erodes into the underlying shales at x138 ft. The channel-fill dips allow the orientation of the channel to be calculated.

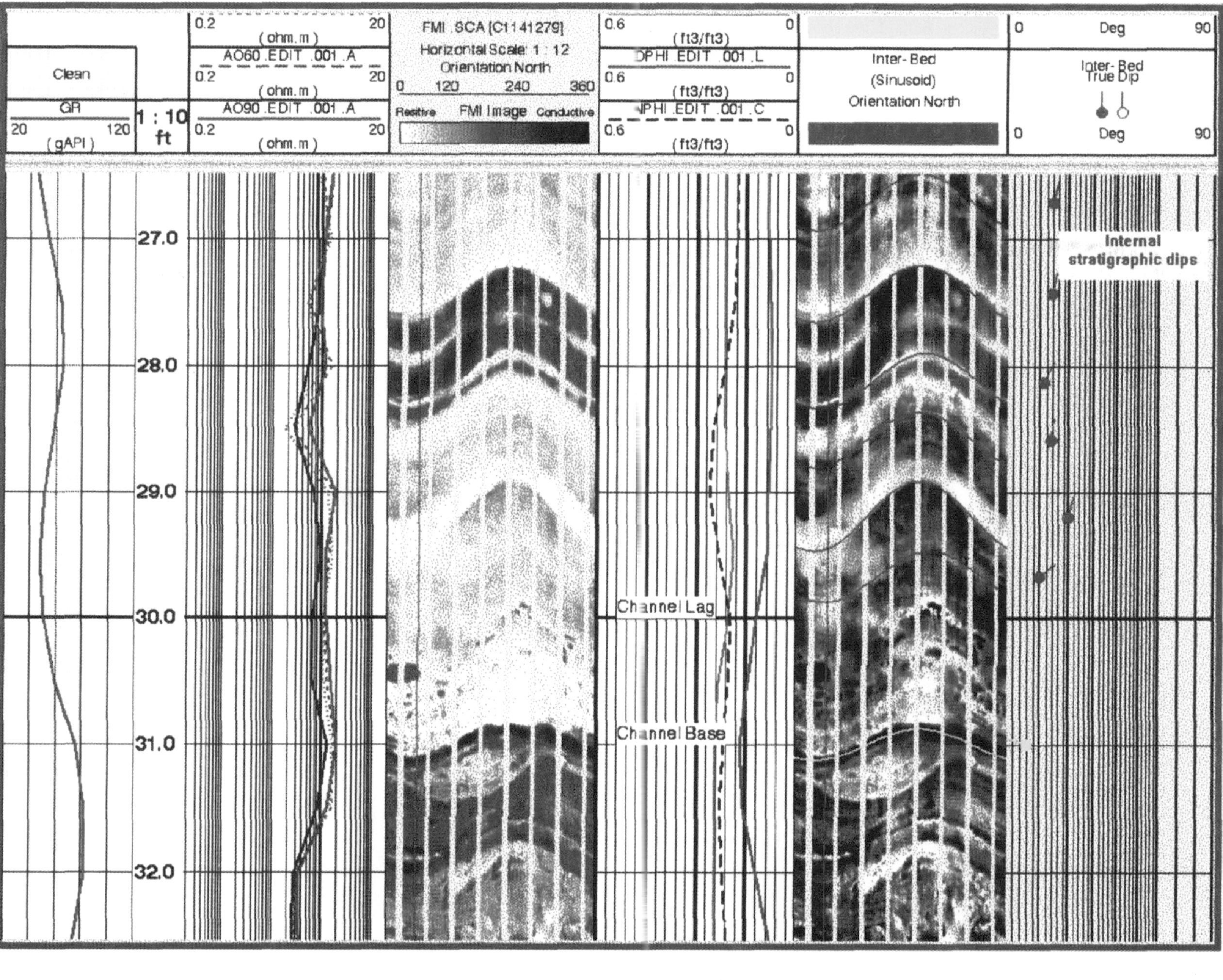

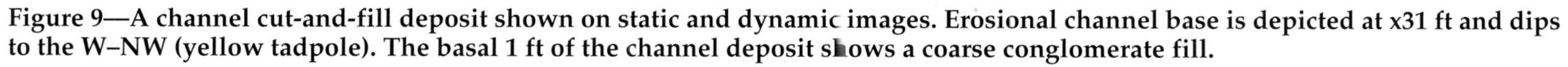
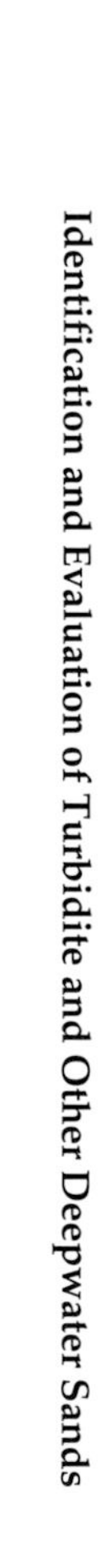

Figure 9—A channel cut-and-fill deposit shown on static and dynamic images. Erosional channel base is depicted at x31 ft and dips to the W–NW (yellow tadpole). The basal 1 ft of the channel deposit shows a coarse conglomerate fill.

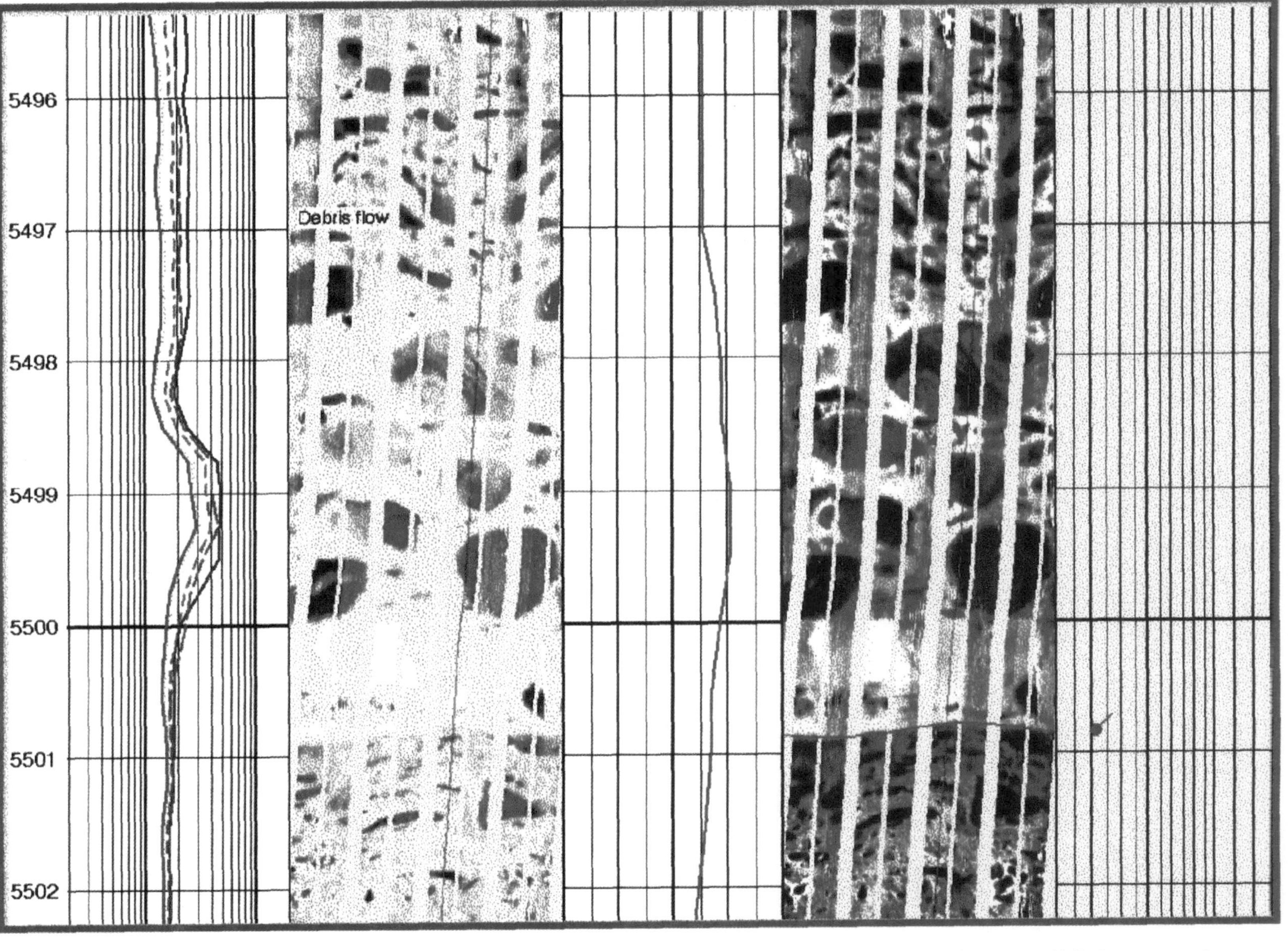

Figure 10—Matrix-supported debris flow. The clast at x5498 ft shows distorted internal bedding and an overall fining-upward sequence of this unit.

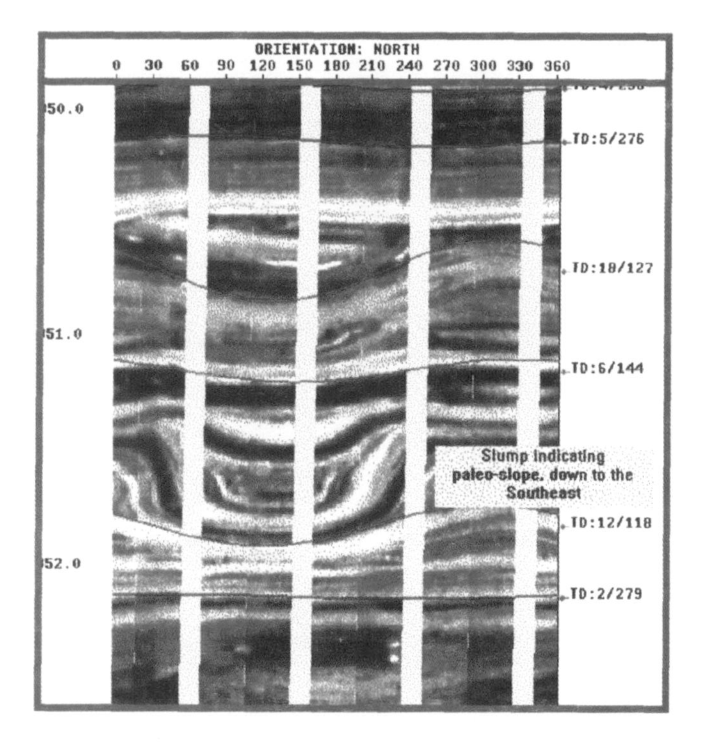

Figure 11—Images show slump to the SE. Slump features on images indicate a paleostrike of NE–SW, as indicated by the SE dip.

30 cm (1 ft) in thickness and do not even extend across the wellbore. It is easy to see how conventional log analysis could mistake an interval like this for a productive thinly bedded reservoir depending on the tool orientation, especially with a unidirectional tool, such as a density or microlog. Figure 13 shows to what extent "soft-sediment deformation" can proceed. These highly contorted sands no longer have any vertical or lateral continuity away from the wellbore. Both Figures 12 and 13 can cause misinterpretation of the reservoir. In each of these examples there is good clean sand present, which in many cases will have hundreds of millidarcys of permeability and good hydrocarbon shows in sidewall cores. In both of the cases the reservoirs are extremely limited due to the lack of vertical and horizontal continuity. The sands in Figure 14 have a very high sand-shale ratio and at first glance appear to be planar across the wellbore. On closer inspection and when viewed in 3-D, the sands are not planar but are actually numerous pods stacked randomly on top of each other. The lack of continuity across the wellbore suggests the same thing will occur away from the wellbore, making this "sand pile" nonproductive.

Core or a good understanding of the depositional environment of the field best validates these features.

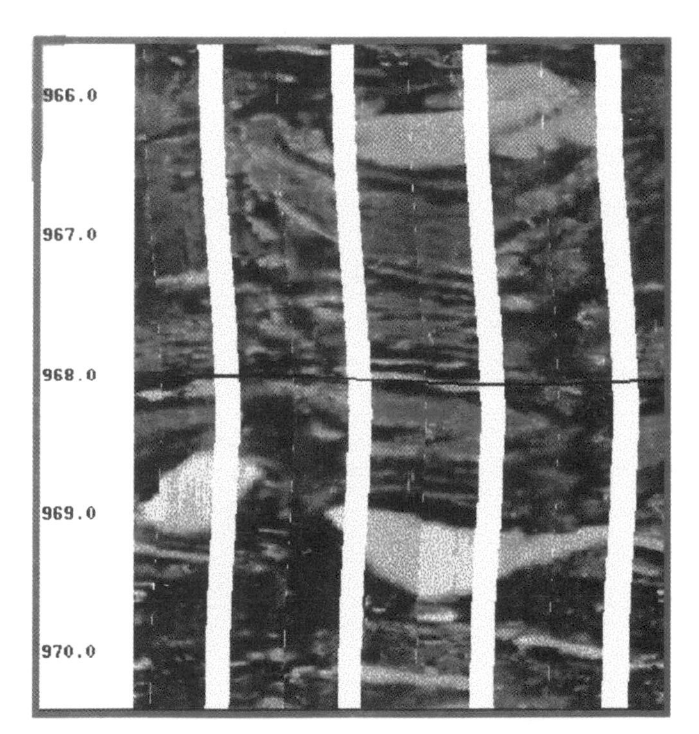

Figure 12—Sand pods seen in this example have excellent porosity, permeability, and oil-saturation. The lack of vertical and lateral continuity make commercial production unlikely.

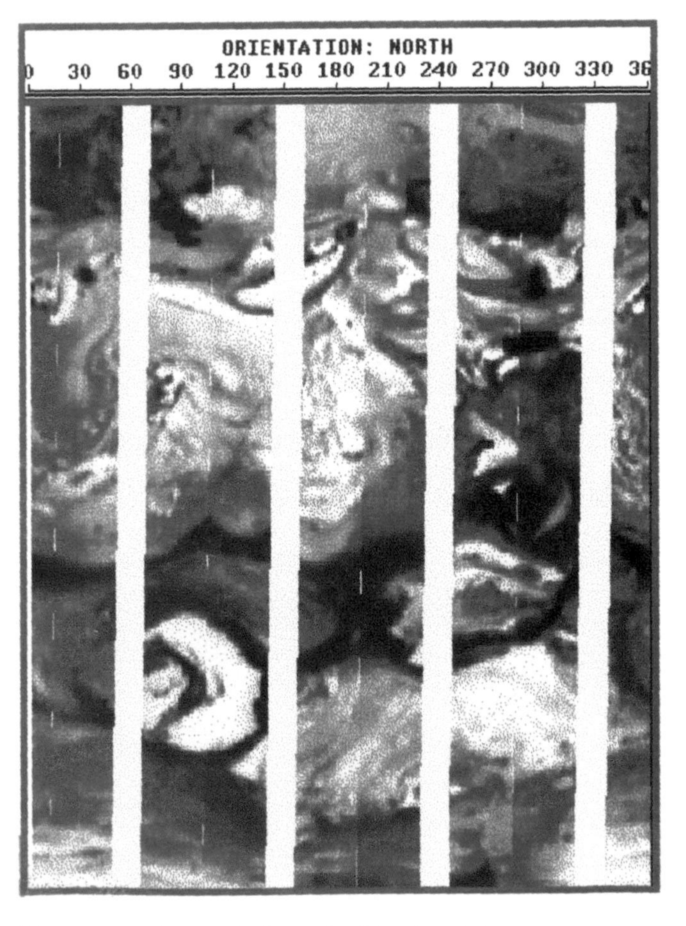

Figure 13—Contorted sands (Pac Man) have no reservoir potential due to the lack of sand continuity.

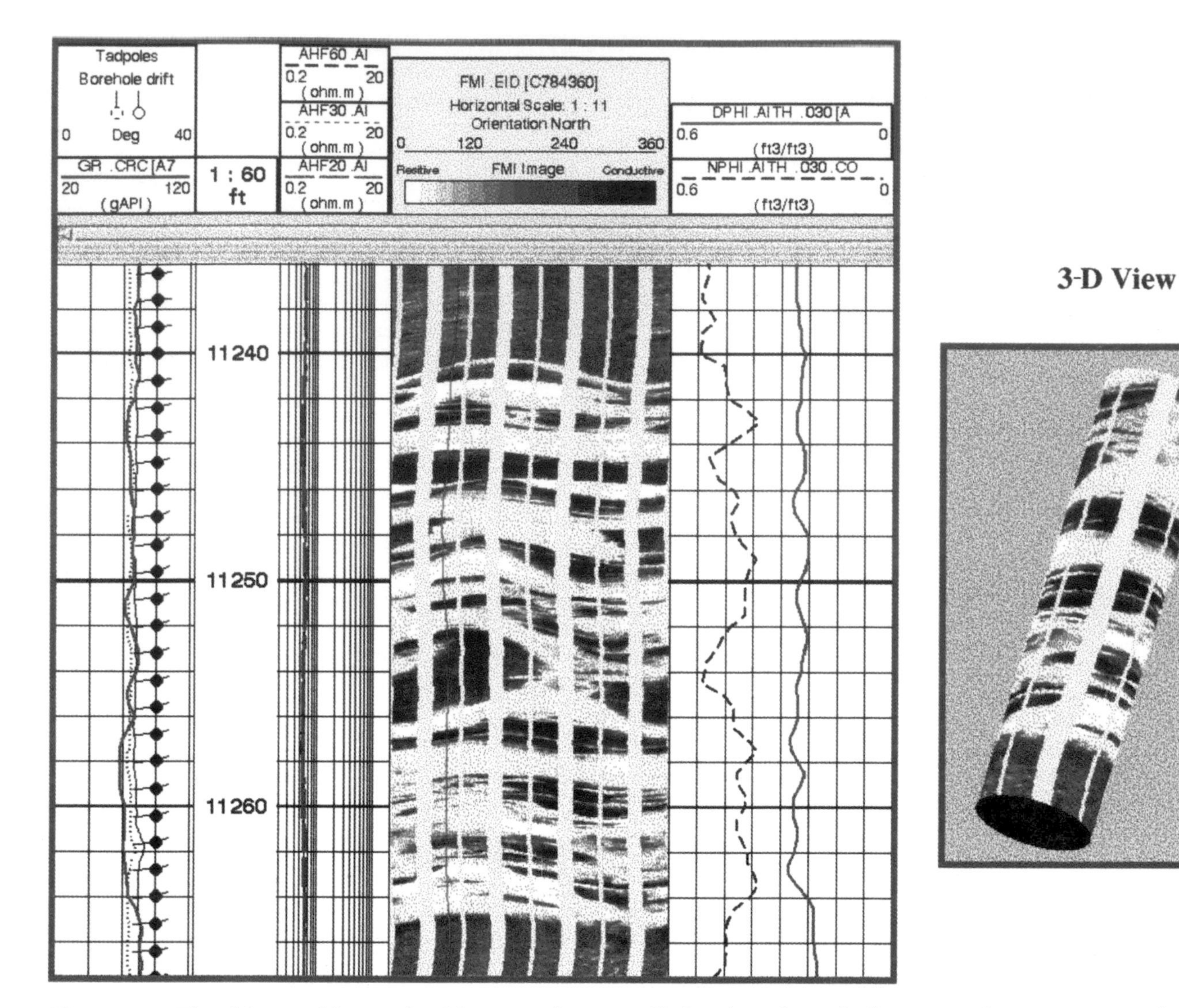

Figure 14—The thin sand lenses in this example are well-developed sands; however, they are not continuous across the wellbore. Sands that have no continuity across the wellbore likely have no continuity away from the wellbore.

When validated, these images lead to insights that can be used to further improve the depositional model. When several consistent directional dip patterns are obtained (as in Figures 7–9 or even by strong individual tadpoles, as in Figures 10, 11), they can be used to predict sand trends. It is not unusual or uncommon to be surprised by the unexpected flow and sand trend directions calculated in these environments. Even locally directed surfaces that are "upregional" are not uncommon in these environments.

Reservoir Description

The primary application of images in deepwater sandstones is the identification and evaluation of potential reservoir-quality rock. For thinly bedded sandstones, images are often the only practical method of determining net pay other than whole coring, which is often cost prohibitive.

Thick sands, greater than 1.20 m (4 ft), are easily evaluated with conventional log evaluation techniques unless they have disposed clay problems or their ability to drain the reservoir is somehow impaired. The churning of the sediments by bioturbation as in Figure 5, or from the contorting of the beds as in Figures 12 and 13, will restrict the lateral extent of the sand and will usually result in noncommercial completions.

Interlaminated or heterolithic zones, such as those found in "turbidite" environments, often look nonproductive due to their thinly bedded nature. Thin beds, such as seen in Figures 1, 2, 5, 6, and 7, can easily be missed altogether even when using modern high-resolution logs. The presence of even a few wiggles on the log curves can often be the only clue to the presence of these thin beds from conventional logs. Any activity on the logs thus becomes of interest and can cause concern that laterally extensive thin pay sands are being missed when encountered on logs. This is, unfortunately, often not the case as in Figures 11–14, where the net-to-gross sand ratio at the wellbore is high enough to be productive, but the sands do not connect either laterally or vertically and cannot properly drain the reservoir. This means the reservoir is measured in square inches instead of in acres. Sidewall cores obviously could also be misleading if they penetrate the sandstone lenses.

Prospective thin-bedded pay sands can easily be identified on images if they exhibit the following: the sands are parallel bedded (i.e., their sinusoids all have the same shape and amplitude), they have good porosity and permeability, and they are hydrocarbon bearing (these type sands often are). With these characteristics they can be evaluated using microresistivity images. The main task then is to determine the net thickness of sand and silt across the sequence. Even zones with individual bed thicknesses no greater than 4–6 cm (2–3 in.) can net a significant amount of pay, especially when they cover hundreds of feet of gross section. If porosity, lithology, and water saturation remain the same, which is often the case, the images can be used to derive excellent net sand and net silt calculations using microresistivity cutoffs.

As mentioned, turbidites are often very clean and well-sorted sands and silts with consistently high porosities. When all of the sands have a similar water saturation (which is also commonly the case), an analysis of the histogram of the microresistivity of the entire intervals usually identifies as many as three or even five "lithology" populations (Figure 15). By carefully integrating the images vs. core data results, either sidewall or whole, these "lithology" populations of such deposits as sands, silty sands, sandy silts, silts, clayey silts, and clays can be identified. In most cases the histogram is divided into only three lithologies: sand, silt, and clay. Wet sands on the histogram will either show up in the shale population or as a separate peak, depending on the formation water resistivity. Computing the histogram only above the hydrocarbon/water contact minimizes this problem. Ranges of permeability can be assigned to these populations based on the microresistivity cutoffs, and corresponding perms can be assumed. In Figure 16 the core permeability (connected dots) is overlain with the microresistivity in order to obtain the proper core to microresistivity relationship

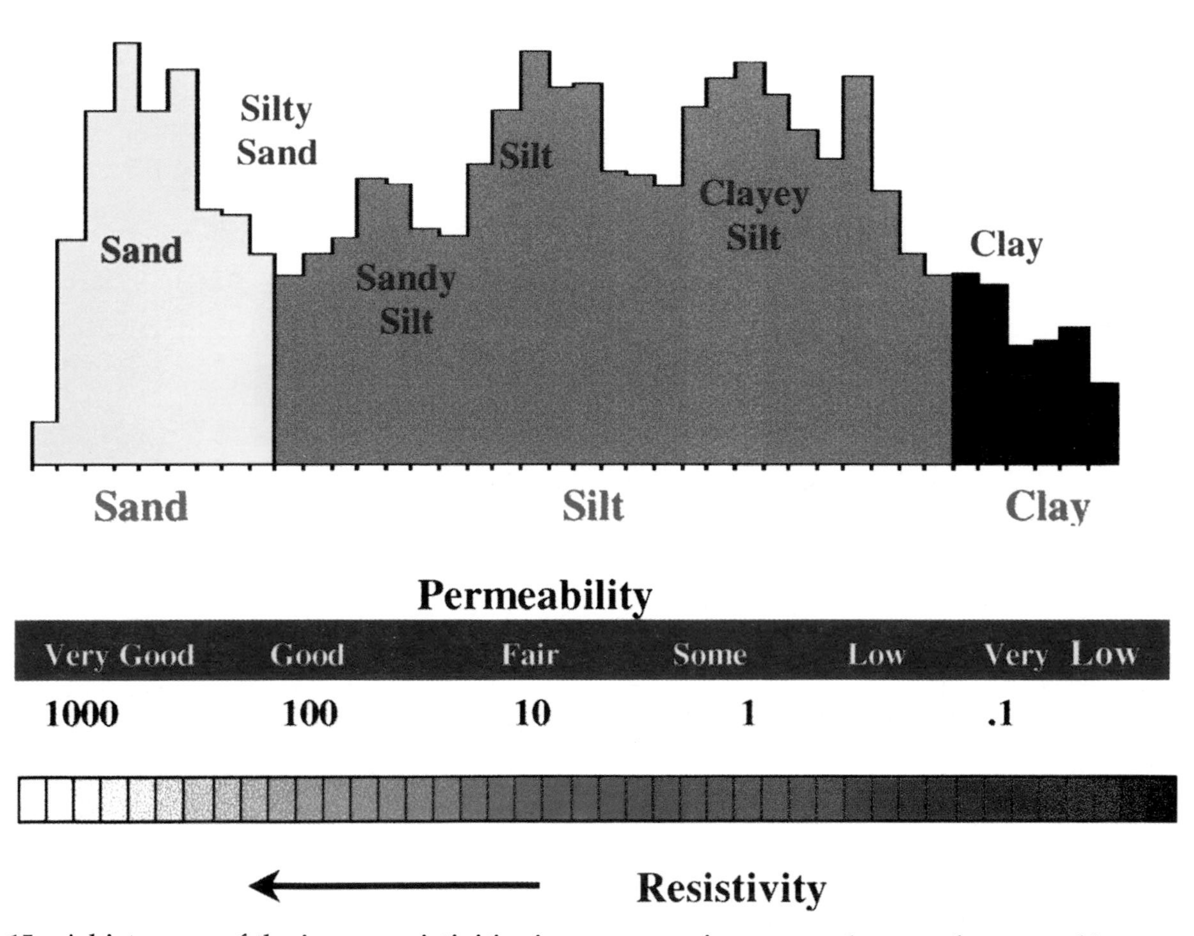

Figure 15—A histogram of the image resistivities in a pay zone is computed across the zone of interest. A correlation between between image resistivity and core permeability allows the data to be subdivided into various lithologies. Normal divisions for evaluation are sand, silt, and clay (shale).

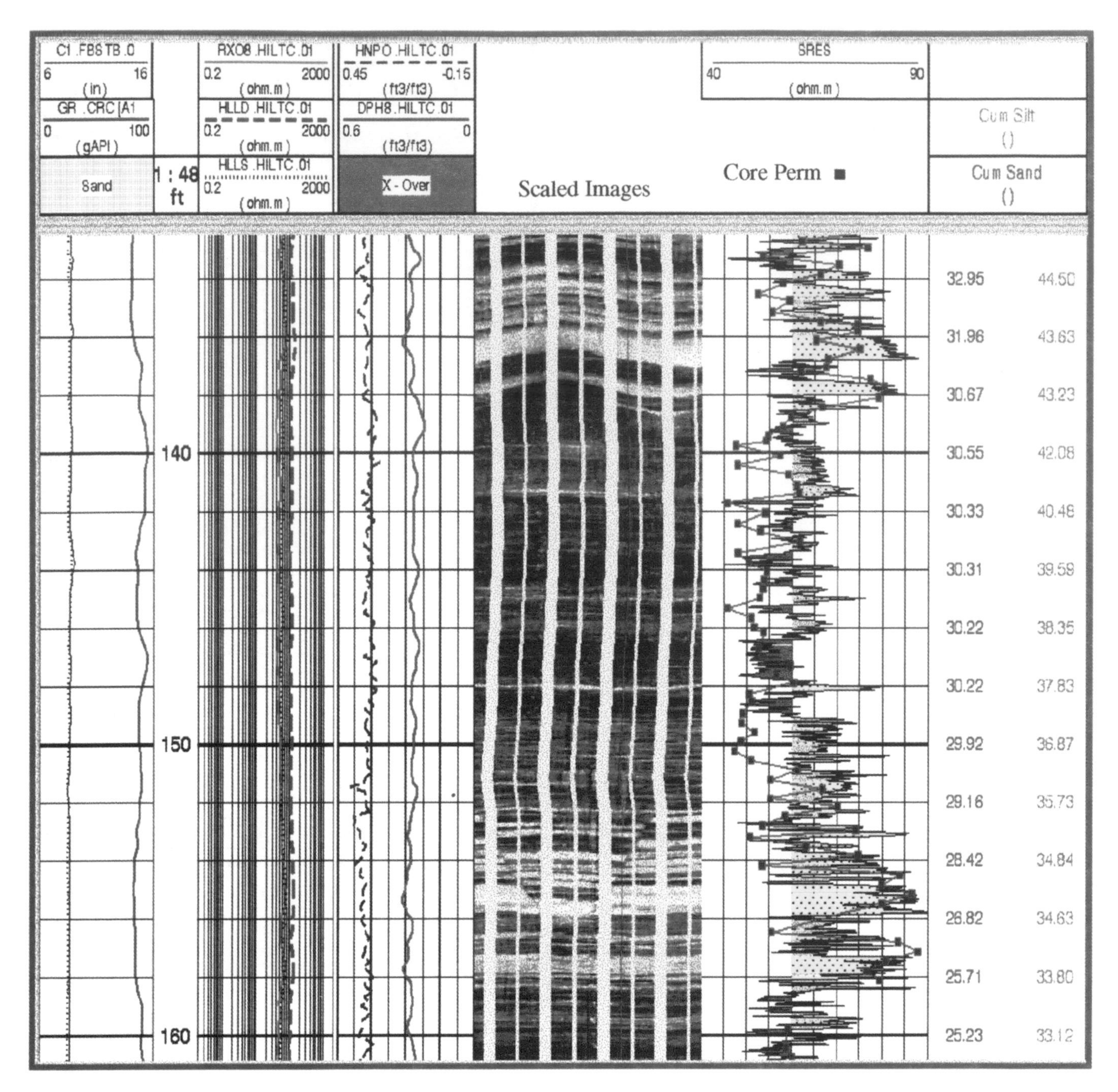

Figure 16—Sand count log displaying the calibrated resistivity (sres), with coded cutoffs in track 6, overlain with core permeability measurements. Cutoffs are determined from this relationship, and the resultants are integrated and displayed in track 7.

and then determine the appropriate cutoffs. In the Gulf of Mexico, a common lithology to permeability relationship established is:

Lithology	Permeability
Sand	>100 md
Silt	10–100 md
Clay (or shale)	<10 md

Net sand and silt counts can then be determined by integrating the microresistivity curve using the calculated cutoffs. In Figure 16 the cumulative sand count in is red, and the cumulative silt count is in gold in the far right-hand track.

Figures 17–21 are examples from the Gulf of Mexico where this method is routinely used. The open hole logs on Figure 17 show what could be a thinly bedded zone from x400 ft to x450 ft, based on the slightly spiky nature to the resistivity curves. A

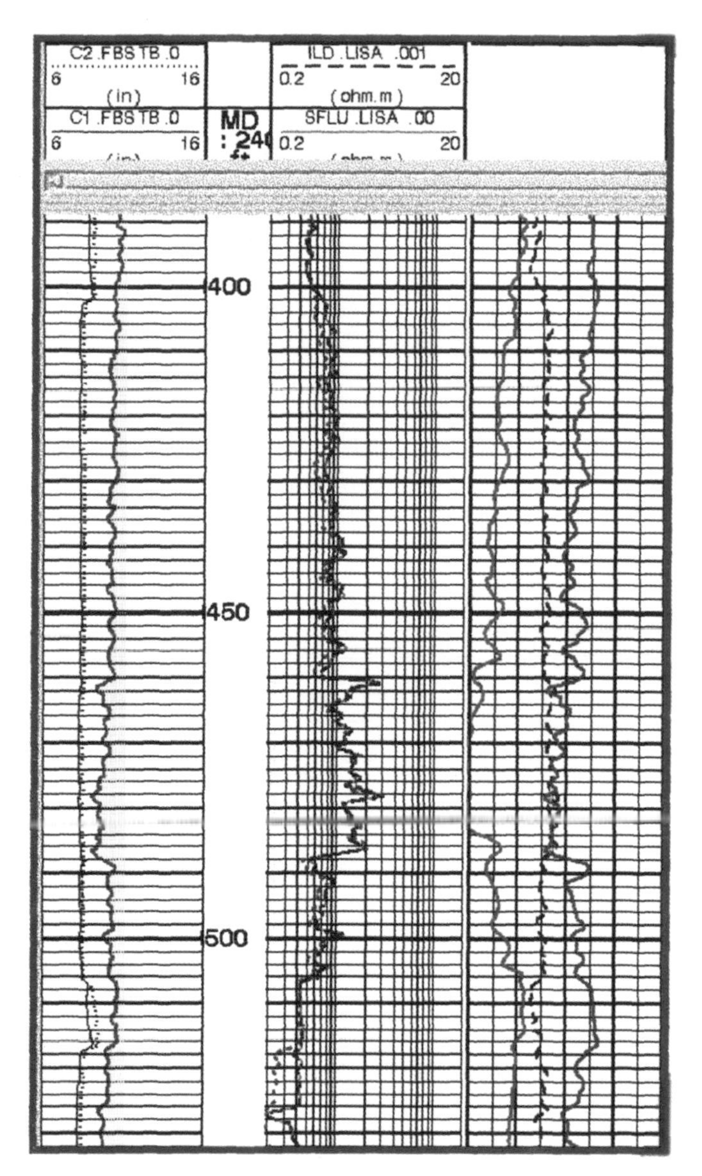

Figure 17—Open hole logs of example #1. Track 1—gammaray and calipers; track 2—depth; track 3—resistivity (0.2–20 ohm-m); and track 4—porosity (60–0%); red-density, blue-neutron, green-sonic.

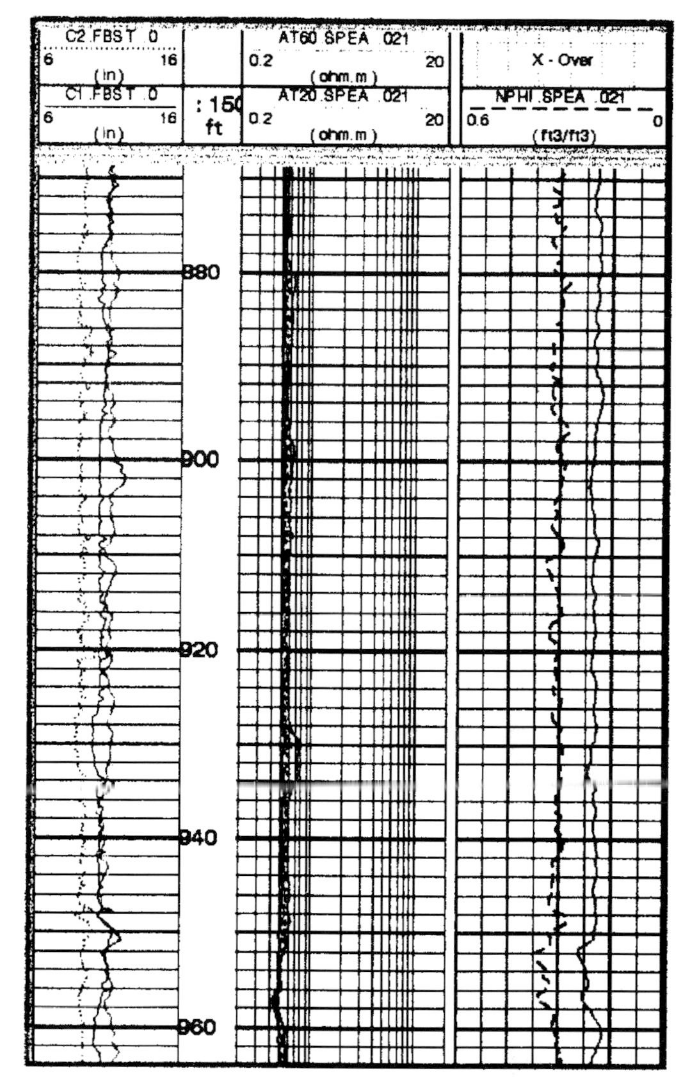

Figure 18—Open hole logs of example #2. Track 1—gammaray and calipers; track 2—depth; track 3—resistivity (0.2–20 ohm-m); and track 4—porosity (60–0%); red-Δdensity, blue-Δneutron.

thicker bedded zone is found from x460 ft to x488 ft. The scaled resistivity is color coded in Figures 19 and 20 to differentiate the three lithologies: clays (brown-black on the images in track 5 = brown coding on scaled resistivity in track 6). The silts are shown as yellow-beige on the images in track 5 = red coding on scaled resistivity in track 6. The sands are shown as white-light yellow on images = yellow coding on scaled resistivity. The sidewall core data are also listed on Figures 19 and 20 with core permeability (in millidarcys) in red and core porosity (in percentages) in blue. Cumulative sand and silt counts were computed over the entire interval. The upper interval (Figure 19) was tested from x402 ft to x442 ft. This zone shows very little shale present (brown coding) and an almost equal portion of sand (yellow coding) and silt (red coding). The net sand or silt for an interval in question is calculated by subtracting the cumulative number at the bottom of the zone of interest from the cumulative number at the top of the zone. This zone tested at 9500 MCF (thousand cubic feet) and 1300 BCPD. The lower interval (x460 ft–x488 ft) (Figure 20) tested has almost no shale present and is predominantly sand (yellow coding) with only about 10% silt (red coding). This zone tested at 13000 MCF and 2500 BCPD. Note that although the lower zone would probably be detected with the conventional logs, the upper one probably would not have been and thus would have been left unproduced. Remember that in these very well-sorted, high-porosity zones, even siltstone can contribute to the overall production, especially in gas reservoirs.

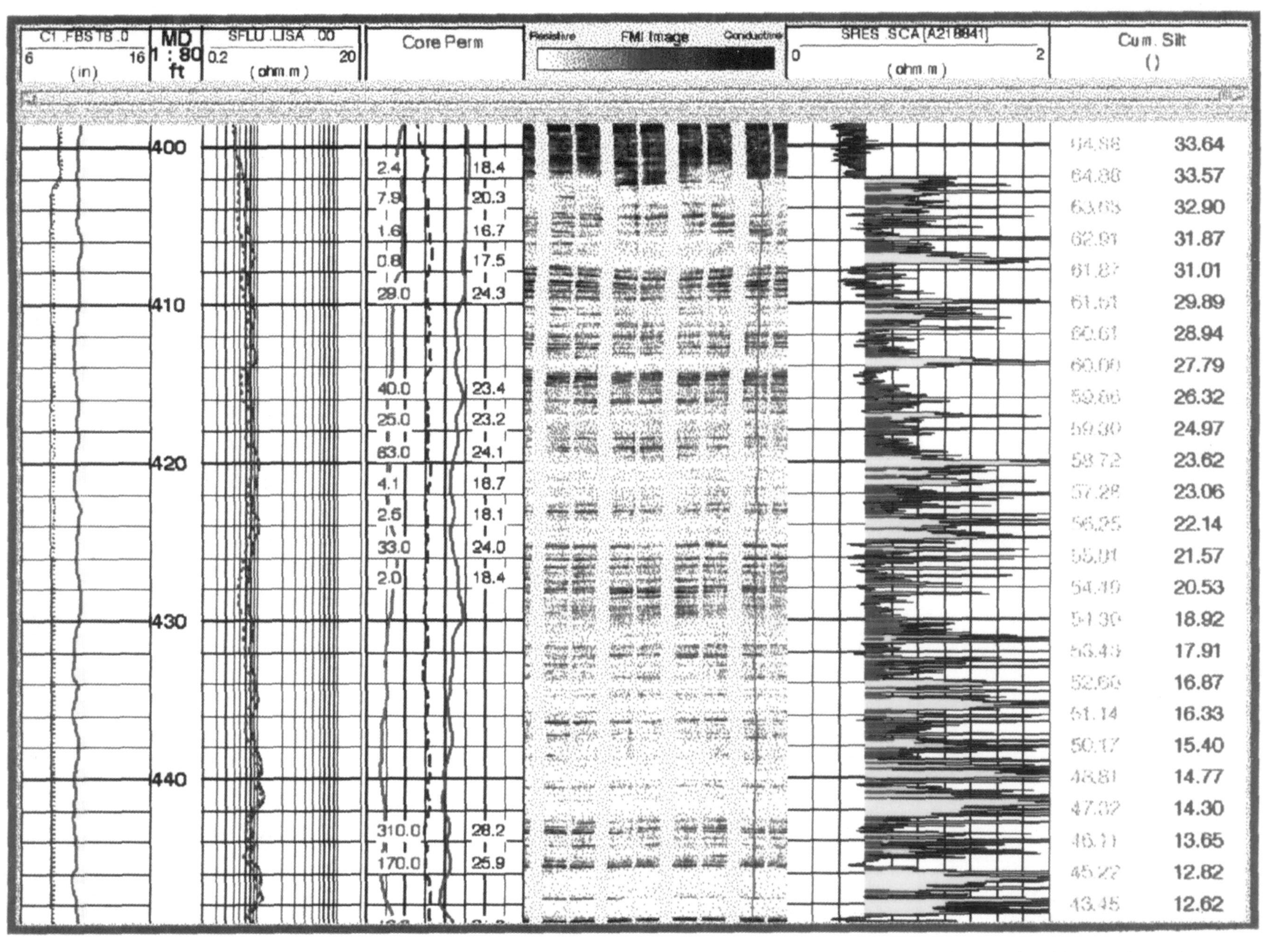

Figure 19—Sand count on lower interval of example #1 in the channel facies. The tested interval was x402 ft–x442 ft. This zone has a larger percentage of silt vs. sand but virtually no shale. The zone tested 9500 thousand cubic feet and 1300 BCPD.

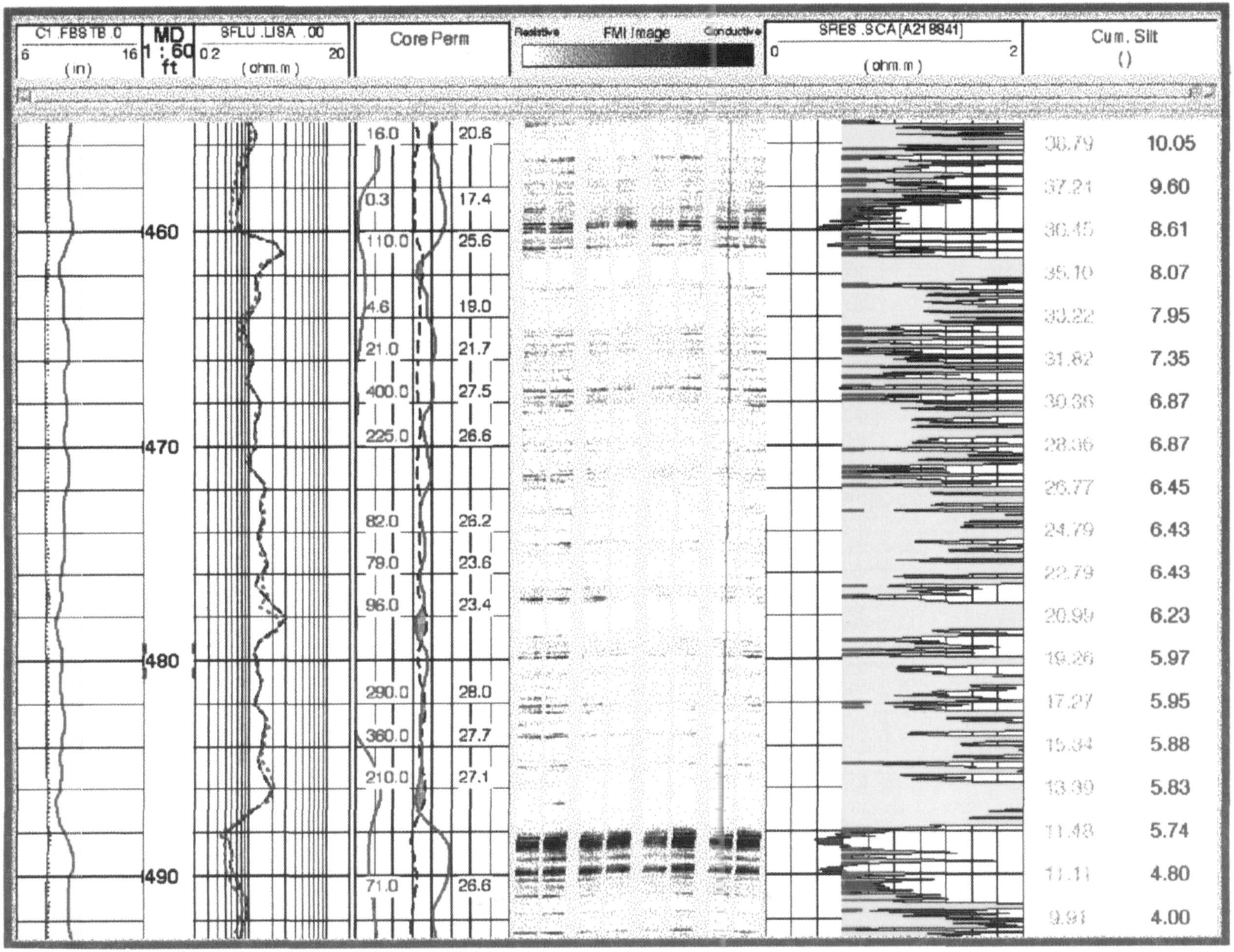

Figure 20—Sand count log on lower interval of example #1 in the channel facies. The tested interval was x460 ft–x488 ft. This zone can be analyzed with the normal open hole logs and is predominantly sand with very little silt. The zone tested 13,000 thousand cubic feet and 2500 BCPD.

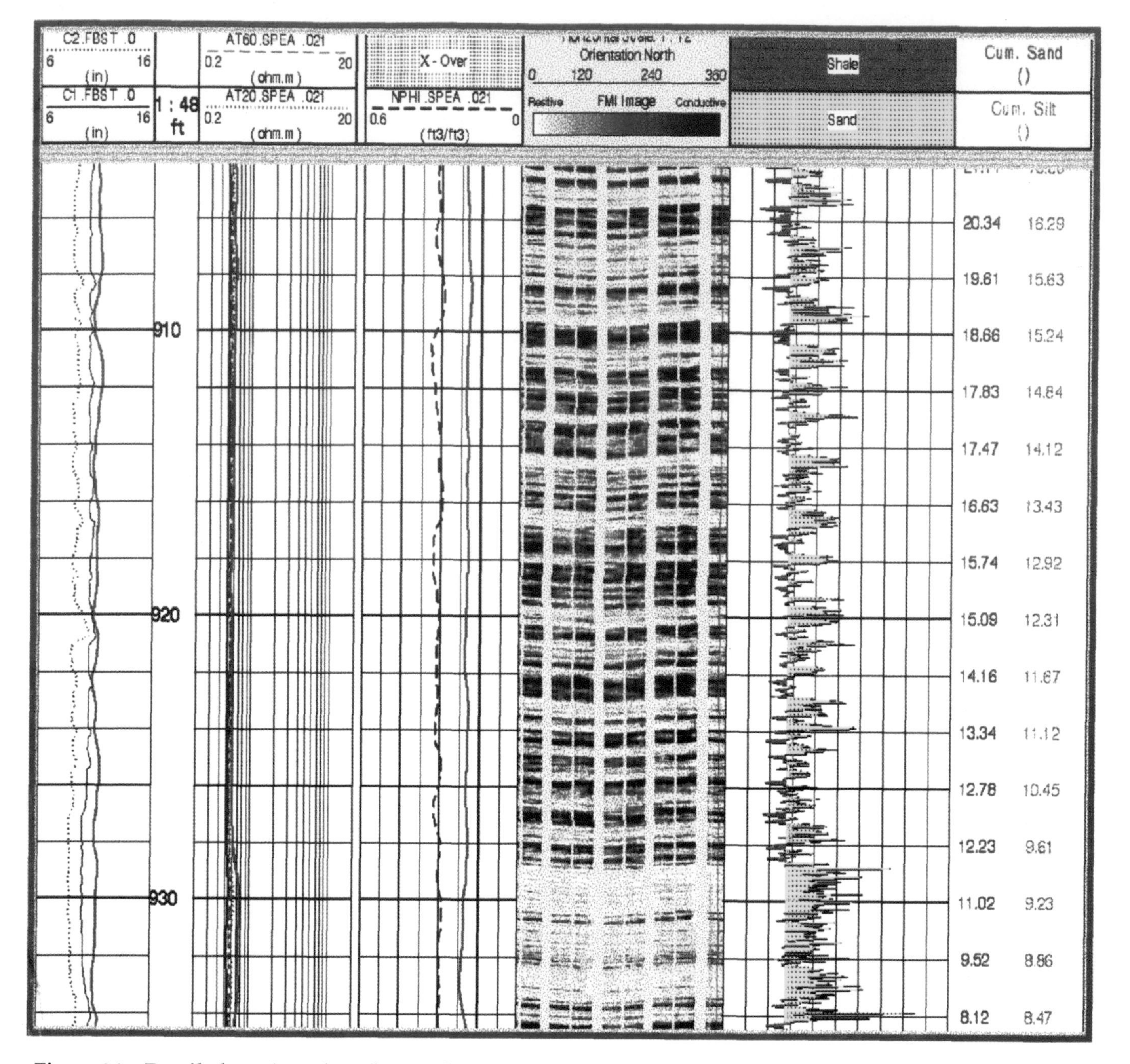

Figure 21—Detailed portion of sand count log on example #2. The tested interval was x875 ft–x960 ft. The images show the thin nature of this laminated sand/shale interval. In this environment, neither the resistivity nor the porosity logs indicate a productive facies. This zone tested 3700 barrels of oil per day.

Figure 18 illustrates another low-resistivity deepwater pay zone that could easily be missed altogether, even with modern high-resolution logs. The resistivity and porosity measurements in this zone give almost no indication of a thin-bedded sand sequence. A sample of the images is shown in Figure 21. This illustrated section has some of the thickest sands present in this interval. The individual sands have an average thickness of 5–7 cm (2–3 in.) and therefore are not resolvable with the conventional open hole logs. The zone was tested from x875 ft to x960 ft and produced 3700 barrels of oil per day, confirming the image based interpretation in this thin-bedded sequence.

CONCLUSIONS

Borehole images can be used to analyze deepwater sediments for structural and stratigraphic application, but currently the primary use for these images is to identify and evaluate the thin-bedded sands that are often overlooked or underestimated by conventional wireline logging suites. These image-based evaluations are best when validated with core or from production information in the area. The authors have been privileged to see many such validations. Many of these are from the deepwater Gulf of Mexico but are unfortunately confidential, so the well identifications cannot be released. The techniques and interpretations described here have nevertheless proven very useful in this area and other similar depositional environments.

ACKNOWLEDGMENTS

The authors wish to thank the various oil and gas companies that have granted permission to show these examples albeit without reference to the well names or fields. We thank Steve Prensky for permission to share his excellent bibliographies on imaging and fracturing. Thanks are extended to Elton Head, Tom Pickens, and Cindy Hansen for reviewing this manuscript. We thank Schlumberger for the opportunity to work with these logs and images over the past 32 years and 21 years, respectively.

REFERENCES CITED

Moore, D. C., 1993, Productive low resistivity well logs of the offshore Gulf of Mexico: New Orleans Geological Society and Houston Geological Society joint publication, 26 p.

Scholle, P. A., and D. R. Spearing, eds.,1982, Sandstone depositional environments: AAPG Memoir 31, 410 p.

Slatt, R. M., D. W. Jordan, and R. J. Davis, 1994, Interpreting formation microscanner log images of Gulf of Mexico Pliocene turbidites by comparison with Pennsylvanian turbidite outcrops, Arkansas, *in* P. Weimer, A. H. Bouma, and B. F. Perkins, eds., Submarine fans and turbidite systems: sequence stratigraphy, reservoir architecture and production characteristics: Gulf of Mexico and international: Gulf Coast Section SEPM Foundation 15th Annual Research Conference, p. 335–348.

Index